TODAY'S TECHNIC[AL]

SHOP MANUAL FOR AUTOMOTIVE ENGINE REPAIR AND REBUILDING

FOURTH EDITION

TODAY'S TECHNICIAN ™

SHOP MANUAL FOR AUTOMOTIVE ENGINE REPAIR AND REBULDING

FOURTH EDITION

Christopher Hadfield
Hennepin Technical College
Brooklyn Park, MN

DELMAR
CENGAGE Learning™

Australia • Canada • Mexico • Singapore • Spain • United Kingdom • United States

DELMAR
CENGAGE Learning™

Today's Technician™: Automotive Engine Repair and Rebuilding, 4th Edition
Christopher Hadfield

Vice President, Career and Professional Editorial: Dave Garza

Director of Learning Solutions: Sandy Clark

Executive Editor: David Boelio

Managing Editor: Larry Main

Senior Product Manager: Matthew Thouin

Editorial Assistant: Jillian Borden

Vice President, Career and Professional Marketing: Jennifer McAvey

Executive Marketing Manager: Deborah S. Yarnell

Marketing Manager: Jimmy Stephens

Marketing Coordinator: Mark Pierro

Production Director: Wendy Troeger

Production Manager: Mark Bernard

Content Project Manager: Cheri Plasse

Art Director: Benj Gleeksman

For product information and technology assistance, contact us at
Cengage Learning Customer & Sales Support, 1-800-354-9706

For permission to use material from this text or product,
submit all requests online at **cengage.com/permissions**.
Further permissions questions can be e-mailed to
permissionrequest@cengage.com.

Library of Congress Control Number: 2009902929

ISBN-13: 978-1435428270

ISBN-10: 1435428277

Delmar
5 Maxwell Drive
Clifton Park, NY 12065-2919
USA

Cengage Learning is a leading provider of customized learning solutions with office locations around the globe, including Singapore, the United Kingdom, Australia, Mexico, Brazil and Japan. Locate your local office at:
international.cengage.com/region

Cengage Learning products are represented in Canada by Nelson Education, Ltd.

For your lifelong learning solutions, visit **delmar.cengage.com**

Visit our corporate website at **cengage.com.**

Notice to the Reader

Publisher does not warrant or guarantee any of the products described herein or perform any independent analysis in connection with any of the product information contained herein. Publisher does not assume, and expressly disclaims, any obligation to obtain and include information other than that provided to it by the manufacturer. The reader is expressly warned to consider and adopt all safety precautions that might be indicated by the activities described herein and to avoid all potential hazards. By following the instructions contained herein, the reader willingly assumes all risks in connection with such instructions. The publisher makes no representations or warranties of any kind, including but not limited to, the warranties of fitness for particular purpose or merchantability, nor are any such representations implied with respect to the material set forth herein, and the publisher takes no responsibility with respect to such material. The publisher shall not be liable for any special, consequential, or exemplary damages resulting, in whole or part, from the readers' use of, or reliance upon, this material.

Printed in Canada
1 2 3 4 5 X X 11 10 09

CONTENTS

Photo Sequence Contents viii

Job Sheets Contents ix

Preface xii

CHAPTER 1 *Safety and Shop Practices* 1
• Personal Safety 2 • Lifting and Carrying 4 • Hand Tool Safety 6 • Power Tool Safety 6
• Compressed Air Equipment Safety 8 • Lift Safety 8 • Jack and Jack Stand Safety 11
• Engine Lift Safety 12 • Cleaning Equipment Safety 12 • Hazardous Waste 16
• Vehicle Operation 17 • Work Area Safety 18 • Case Study 19 • Terms to Know 18
• ASE-Style Review Questions 20 • ASE Challenge Questions 21 • Job Sheets 23

CHAPTER 2 *Basic Testing, Initial Inspection, and Service Writing* 31
• Identifying the Area of Concern 31 • Test-Driving for Engine Concerns 33
• Writing a Service Repair Order 35 • Case Study 36 • Terms to Know 36
• ASE-Style Review Questions 37 • ASE Challenge Questions 38 • Job Sheets 39

CHAPTER 3 *Engine Rebuilding Tools and Skills* 45
• Units of Measure 46 • Engine Diagnostic Tools 48 • Exhaust Gas Analyzers 58
• Engine Analyzer 58 • Engine Measuring Tools 60 • Special Hand Tools 74
• Special Reconditioning Tools and Equipment 77 • Working as a Professional
Technician 90 • Service Information 98 • Case Study 101 • Terms to Know 101
• ASE-Style Review Questions 104 • Job Sheets 105

CHAPTER 4 *Diagnosing and Servicing Engine Operating Systems* . . . 121
• Battery Testing 122 • Battery Testing Series 124 • Starting System Tests and Service 127
• Lubrication System Testing and Service 134 • Cooling System Testing and Service 141
• Gallery and Casting Plugs 151 • Leak Diagnosis 153 • Warning Systems Diagnosis 161
• Case Study 163 • Terms to Know 163 • ASE-Style Review Questions 165
• ASE Challenge Questions 166 • Job Sheets 167

CHAPTER 5 *Repair and Replacement of Engine Fasteners,
Gaskets, and Seals* 196
• Thread Repair 196 • Gasket Installation 204 • Cylinder Head Gasket Installation 204
• Oil Pan Gasket Installation 205 • Valve Cover Gasket Replacement 205
• Seal Installation 205 • Using Sealants 207 • Case Study 208 • Terms to Know 208
• ASE-Style Review Questions 209 • ASE Challenge Questions 210 • Job Sheets 211

CHAPTER 6 *Intake and Exhaust System Diagnosis and Service* 218
• Air Cleaner Service 218 • Vacuum System Diagnosis and Troubleshooting 221
• Vacuum Tests 222 • Intake Manifold Service 224 • Exhaust System Service 229
• Turbocharger Diagnosis 235 • Supercharger Diagnosis and Service 242 • Case Study 245
• Terms to Know 245 • ASE-Style Review Questions 245 • ASE Challenge Questions 246
• Job Sheets 247

CONTENTS

CHAPTER 7 Diagnosing Engine Performance Concerns **259**

• Introduction 259 • Spark Plug Evaluation and Power Balance Testing 259
• Spark Plugs 260 • Power Balance Testing 263 • Engine Smoking 266
• Improper Combustion 272 • Compression Testing 273 • Wet Compression Test 276
• Running Compression Test 277 • Cylinder Leakage Testing 279 • Engine Noise
Diagnosis 282 • Case Study 285 • Terms to Know 285 • ASE-Style Review Questions 286
• ASE Challenge Questions 287 • Job Sheets 289

CHAPTER 8 Engine Removal, Engine Swap, and Engine Installation **303**

• Preparing the Engine for Removal 303 • Removing the Engine 308 • Mounting the Engine
on a Stand (Photo Sequence 16) 312 • Installing a Remanufactured Engine 315 • Installing
the Engine 322 • Engine Start-Up and Break-In 325 • Case Study 329 • Terms to Know 329
• ASE-Style Review Questions 329 • ASE Challenge Questions 330 • Job Sheets 331

CHAPTER 9 Cylinder Head Disassembly, Inspection, and Service **337**

• Disassembling the Cylinder Head 342 • Valve Inspection 349 • Cylinder Head
Inspection 353 • Machine Shop Services 358 • Straightening Aluminum Cylinder
Heads 359 • Valve Guide Repair 360 • Replacing Valve Seats 363 • Refinishing Valves and
Valve Seats 365 • Case Study 372 • Terms to Know 372 • ASE-Style Review Questions 375
• ASE Challenge Questions 376 • Job Sheets 377

CHAPTER 10 Valvetrain Service **397**

• Inspecting and Servicing the Valvetrain 397 • Inspecting and Servicing Valvetrain
Components, OHV Engine 403 • Replacing Rocker Arms Studs 406 • Cylinder Head
Reassembly 413 • Valve Adjustment 418 • Adjusting Mechanical Followers with Shims 421
• Adjusting Valves Using Adjustable Rocker Arms 422 • On-Car Service 424 • Valve Spring
Replacement 429 • Case Study 430 • Terms to Know 430 • ASE-Style Review Questions 430
• ASE Challenge Questions 431 • Job Sheets 433

CHAPTER 11 Timing Mechanism Service **451**

• Introduction 451 • Symptoms of a Worn Timing Mechanism 452 • Symptoms of a Jumped
or Broken Timing Mechanism 452 • Timing Chain Inspection 454 • Timing Belt Inspection
455 • Timing Gear Inspection 456 • Sprocket Inspection 456 • Tensioner Inspection 457
• Timing Chain Guide Inspection 457 • Reuse Versus Replacement Decisions 458
• Timing Mechanism Replacement 458 • Timing Chain Replacement on OHC Engines 458
• Timing Belt Replacement 462 • Timing Chain or Gear Replacement on Camshaft-in-
the-Block Engines 464 • Timing Chain Replacement on Engines with OHC and Balance
Shafts 467 • Timing Chain Replacement on Engines with VVT Systems 469
• Case Study 471 • Terms to Know 471• ASE-Style Review Questions 471 • ASE Challenge
Questions 472 • Job Sheets 473

CONTENTS

CHAPTER 12 *Inspecting, Disassembling, and Servicing the Cylinder Block Assembly* **481**

• Engine Disassembly 481 • Engine Preparation 481 • Cylinder Head Removal 482
• Timing Mechanism Disassembly 484 • Engine Block Disassembly 487
• Inspecting the Cylinder Block 492 • Inspecting the Crankshaft 498
• Inspecting the Harmonic Balancer and Flywheel 501 • Servicing the Cylinder Block Assembly 501 • Camshaft and Balance Shaft Bearing Service 510 • Reconditioning Crankshafts 513 • Installing Oil Gallery Plugs and Core Plugs 515 • Case Study 516
• Terms to Know 516 • ASE-Style Review Questions 517 • ASE Challenge Questions 518
• Job Sheets 519

CHAPTER 13 *Short Block Component Service and Engine Assembly.* **533**

• Inspecting Crankshaft Bearings 533 • Inspecting the Pistons and Connecting Rods 537
• Piston Assembly Service 549 • Engine Block Preparation 557 • Installing the Main Bearings and Crankshaft 561 • Installing the Pistons and Connecting Rods 567
• Completing the Block Assembly 568 • Valve Timing and Installing Valvetrain Components 572 • Checking and Adjusting Valve Clearance 574 • Completing the Engine Assembly 574 • Case Study 580 • Terms to Know 580 • ASE-Style Review Questions 580
• ASE Challenge Questions 582 • Job Sheets 583

CHAPTER 14 *Alternative Fuel and Advanced Technology Vehicle Service* **600**

• Introduction 600 • Servicing Propane Fueled Engines 600 • Servicing CNG Vehicles 601
• Servicing Hybrid Electric Vehicles 603 • Servicing The Toyota Prius 605
• Servicing The Honda Civic Hybrid 606 • HEV Warranties 608 • Case Study 608
• Terms to Know 608 • ASE-Style Review Questions 608 • ASE Challenge Questions 609
• Job Sheets 611

APPENDIX A *ASE Practice Exam* **619**

APPENDIX B *Special Tools Suppliers* **624**

APPENDIX C *Manufacturer's Service Information Centers* **625**

APPENDIX D *Metric Conversion Charts* **626**

APPENDIX E *Engine Specifications Chart* **627**

Glossary **630**

Index **639**

PHOTO SEQUENCES

1. Typical Procedure for Lifting a Vehicle on a Hoist **10**

2. Typical Procedure for Reading a Dial Caliper **65**

3. Typical Procedure for Reading a Micrometer **68**

4. Starter R + R **135**

5. Typical Procedure for Disassembly and Inspection of Rotor-Type Oil Pump **137**

6. Cooling System Testing **158**

7. Typical Procedure for Using a Scan Tool to Retreive DTCs **164**

8. Removing a Broken Bolt and Using a Tap **202**

9. Replacing a Valve Cover Gasket **206**

10. Typical Procedure for Removing and Replacing an Intake Manifold Gasket **227**

11. Typical Procedure for Inspecting Turbochargers and Testing Boost Pressure **239**

12. Typical Procedure for Performing a Cranking Compression Test **275**

13. Typical Procedure for Performing a Cylinder Leakage Test **280**

14. Typical Procedure for FWD Engine Removal **311**

15. Typical Procedure for RWD Engine Removal **313**

16. Typical Procedure for Mounting an Engine on a Stand **314**

17. Typical Procedure for Disassembling an OHC Cylinder Head **347**

18. Typical Procedure for Grinding Valve Seats **373**

19. Typical Procedure for Adjusting Valves **426**

20. Replacing Valve Seals on the Vehicle **428**

21. Typical Procedure for Replacing a Timing Belt on an OHC Engine **465**

22. Typical Procedure for Replacing Timing Chains on a DOHC Engine
    with Balance Shafts **468**

23. Typical Procedure for Checking Main Bore Alignment **497**

24. Typical Procedure for Cylinder Honing **507**

25. Typical Procedure for Checking Main Bearing Oil Clearances **559**

26. Typical Procedure for Installing the Main Bearings and Crankshaft **570**

JOB SHEETS

1. Shop Safety Survey 23

2. Proper Handling of Hazardous Materials 25

3. Lifting a Full Frame Vehicle 27

4. Lifting a Body on Frame Vehicle 29

5. Engine Component Review 39

6. Test-Drive and Basic Inspection 41

7. Writing a Service Repair Order 43

8. Identifying Engine Service Tools 105

9. Using Feeler Gauges 107

10. Reading Dial Calipers 109

11. Reading a Dial Indicator 111

12. Using an Outside Micrometer 113

13. Measuring Cylinder Wall Wear 115

14. Finding Service Information 117

15. Locating and Understanding Technical Service Bulletins 119

16. Performing a Battery Test 167

17. Testing a Starter 169

18. Performing an Oil and Filter Change 171

19. Oil Pump Inspection and Measurement 173

20. Oil Pressure Testing 175

21. Oil Leak Diagnosis 177

22. Testing an Oil Pressure Switch and Sensor 179

23. Test, Drain, and Refill Coolant 181

24. Radiator Replacement 183

25. Diagnosing Coolant Leaks 185

26. Replacing a Water Pump 187

27. Diagnosing and Replacing a Thermostat 189

28. Engine Belt and System Inspection 191

29. Cooling System Fan Inspection 193

30. Using a Scan Tool to Retrieve Engine Codes 195

31. Identifying Engine Fasteners and Torque Specifications 211

32. Loosening a Rusted Fastener 213

JOB SHEETS

33. Replacing an Oil Pan Gasket **215**

34. Identifying Engine Materials **217**

35. Inspect and Service Air Cleaner **247**

36. Diagnosing Vacuum leaks **249**

37. Exhaust System Inspection and Test **251**

38. Turbocharger System Inspection **253**

39. Testing Boost Pressure **257**

40. Diagnosing Engine Noises and Vibrations **289**

41. Analyzing Spark Plugs and Performing a Power Balance Test **291**

42. Oil and Coolant Consumption Diagnosis **295**

43. Performing Compression Tests **297**

44. Performing a Running Compression Test **299**

45. Performing a Cylinder Leakage Test **301**

46. FWD Engine Removal **331**

47. RWD Engine Removal **333**

48. Engine Installation **335**

49. Removing the Harmonic Balancer and Pulley **377**

50. Cylinder Head Removal **379**

51. Cylinder Head Disassembly **381**

52. Cylinder Head Crack Detection **383**

53. Measuring Cylinder Head Warpage **385**

54. Measuring Valve Stem-to-Guide Clearance **387**

55. Insert Valve Seat Replacement **391**

56. Valve Seat Grinding **393**

57. Reconditioning Valves **395**

58. Valve Stem Seal Replacement on an Assembled Engine **433**

59. Valve Spring Inspection **435**

60. Inspecting the Camshaft **437**

61. Inspecting the Lifters (Lash Adjusters) **441**

62. Inspecting the Pushrods **443**

63. Checking and Adjusting Valve Clearance **445**

64. Reassembling the Cylinder Head **447**

JOB SHEETS

65. Inspecting the Timing Chain or Belt 473

66. OHV Valvetrain Timing 475

67. SOHC Valvetrain Timing 477

68. Variable Valve Timing 479

69. Disassembling an Engine Block 519

70. Engine Block Crack Detection 523

71. Measuring Deck Warpage 525

72. Cylinder Block Thread Cleaning 527

73. Inspecting and Measuring the Cylinder Walls 529

74. Inspecting and Measuring Main Bearing Bore Alignment 531

75. Inspecting the Pistons and Connecting Rods 583

76. Inspecting the Bearing Wear Patterns 587

77. Installing the Piston Rings 589

78. Installing the Camshaft 591

79. Installing the Main Bearings and the Crankshaft 593

80. Installing the Pistons and Connecting Rods 597

81. Servicing an LPG or CNG vehicle 611

82. Researching an Advanced Technology Vehicle 613

83. Disabling the High Voltage Battery Pack on a Hybrid Electric Vehicle 615

84. Performing a Compression Test on a Hybrid Electric Vehicle 617

PREFACE

Thanks to the support the Today's Technician™ series has received from those who teach automotive technology. Delmar, a division of Cengage Learning and the leader in automotive related textbooks, is able to live up to its promise to provide new editions regularly. We have listened and responded to our critics and our fans and present this new updated and revised fourth edition. By revising our series regularly, we can and will respond to changes in the industry, changes in technology, changes in the certification process, and to the ever-changing needs of those who teach automotive technology.

The Today's Technician™ series features textbooks that cover all mechanical and electrical systems of automobiles and light trucks (while the heavy-duty trucks portion of the series does the same for heavy-duty vehicles). Principally, the individual titles correspond to the main areas of ASE (National Institute for Automotive Service Excellence) certification. Additional titles include remedial skills and theories common to all of the certification areas and advanced or specific subject areas that reflect the latest technological trends. Each text is divided into two volumes: a Classroom Manual and a Shop Manual.

Unlike yesterday's mechanic, the technician of today and for the future must know the underlying theory of all automotive systems and be able to service and maintain those systems. Dividing the material into two volumes provides the reader with the information needed to begin a successful career as an automotive technician without interrupting the learning process by mixing cognitive and performance learning objectives into one volume.

The design of Delmar's Today's Technician™ series was based on features that are known to promote improved student learning. The design was further enhanced by a careful study of survey results, in which the respondents were asked to value particular features. Some of these features can be found in other textbooks, while others are unique to this series.

Each Classroom Manual contains the principles of operation for each system and subsystem. The Classroom Manual also contains discussions on design variations of key components used by the different vehicle manufacturers. This volume is organized to build upon basic facts and theories. The primary objective of this volume is to allow the reader to gain an understanding of how each system and subsystem operates. This understanding is necessary to diagnose the complex automobiles of today and tomorrow. Although the basics contained in the Classroom Manual provide the knowledge needed for diagnostics, diagnostic procedures appear only in the Shop Manual. An understanding of the basics is also a requirement for competence in the skill areas covered in the Shop Manual.

A spiral bound Shop Manual covers the "how-tos." This volume includes step-by-step instructions for diagnostic and repair procedures. Photo Sequences are used to illustrate some of the common service procedures. Other common procedures are listed and are accompanied with fine line drawings and photos that allow the reader to visualize and conceptualize the finest details of the procedure. This volume also contains the reasons for performing the procedures, as well as when that particular service is appropriate.

The two volumes are designed to be used together and are arranged in corresponding chapters. Not only the chapters in the volumes are linked together, but also the contents of the chapters are linked. This linking of content is evidenced by marginal callouts that refer the reader to the chapter and page that the same topic is addressed in the other volume. This feature is valuable to instructors. Without this feature, users of other two-volume textbooks must search the index or table of contents to locate supporting information in the other volume. This is not only cumbersome, but also creates additional work for an instructor when planning the presentation of material and when making reading assignments. It is also valuable to the students; with the page references, they also know exactly where to look for supportive information.

Both volumes contain clear and thoughtfully selected illustrations. Many of which are original drawings or photos specially prepared for inclusion in this series. This means that the art is a vital part of each textbook and not merely inserted to increase the numbers of illustrations.

The page layout, used in the series, is designed to include information that would otherwise break up the flow of information presented to the reader. The main body of the text includes all of the "need-to-know" information and illustrations. In the wide side margins of each page are many of the special features of the series. Items that are truly "nice-to-know" information include simple examples of concepts just introduced in the text, explanations or definitions of terms that will not be defined in the glossary, examples of common trade jargon used to describe a part or operation, and exceptions to the norm explained in the text. This type of information is placed in the margin, out of the normal flow of information. Many textbooks attempt to include this type of information and insert it in the main body of text; this tends to interrupt the thought process and cannot be pedagogically justified. By placing this information off to the side of the main text, the reader can select when to refer to it.

Jack Erjavec
Series Advisor

HIGHLIGHTS OF THIS EDITION—CLASSROOM MANUAL

The Classroom Manual for this edition of *Today's Technician: Automotive Engine Repair and Rebuilding* has been updated to include a new chapter on the engine repair and rebuilding industry, all new content on hybrid electric vehicles and alternative fuel-related issues, an updated list of FFVs, and new material on VVT systems. Increased coverage on the basics of engine operation is also provided, as well as a new Engine Specification Chart included in Appendix E. The Classroom Manual also features a thoroughly updated art program, now in full color, and will appeal greatly to visual learners.

HIGHLIGHTS OF THIS EDITION—SHOP MANUAL

The fourth edition of Shop Manual now provides all new content on basic testing, initial inspection, and service writing. Information on scan tool usage has been expanded, along with content on diagnostic strategies from real-world scenarios, as experienced by working technicians. The Shop Manual covers 100 percent of the 2008 NATEF Standards, includes nearly 30 new Job Sheets, and features new full-color photo sequences that detail step-by-step diagnostic and repair procedures. ASE questions for the chapters have been updated and are designed to assess students' chapter comprehension and prepare them for future certification testing. As with the Classroom Manual, all photos and illustrations are now full color and offer students outstanding visual support.

CLASSROOM MANUAL

Features of this manual include the following:

Chapter 1
OVERVIEW OF ENGINE PERFORMANCE

UPON COMPLETION AND REVIEW OF THIS CHAPTER, YOU SHOULD BE ABLE TO:

- Describe the relationship of basic engine construction to engine performance.
- Describe the basic function of the cooling, fuel, and air induction systems.
- Explain similarities between distributor ignition systems and distributorless ignition systems.
- Explain the relationship of intake and exhaust systems to engine performance.
- Explain the basic function of the Exhaust Gas Recirculation (EGR) valve.

- Explain the basic function of the Positive Crankcase Ventilation (PCV) valve.
- Describe the function of the catalytic converter.
- Describe the basic function of the evaporative emissions system.
- Examine basic differences between OBD I and OBD II.

COGNITIVE OBJECTIVES

These objectives define the contents of the chapter and define what the student should have learned upon completion of the chapter.

Each topic is divided into small units to promote easier understanding and learning.

INTRODUCTION

Congratulations! Since you are taking a course in *Engine Performance*, you are about to combine most of the knowledge you have gained to date and use it to start diagnosis of "drivability" problems. The Engine Performance Specialist is usually one of the most highly talented people in the shop. It takes a good foundation in electricity and electronics. It requires knowledge of the engine and how it works along with basic transmission operation. You will be surprised at how many transmission complaints are actually related to engine performance. The engine performance person also does most of the electrical diagnosis and scan tool work, so you will also need some knowledge of basic engine testing, computer operation, networking, body control modules, and so forth, which is included to help you in your career. I hope you find this text helpful in your pursuit of the most challenging and rewarding facets of automotive service.

... internal combustion
... the engine. Burning
... sure which powers
... temperatures and

... usually cast from iron
... cylinders which are

WARNING: Do not let the used coolant into the shop drains; collect it in a drain pan. The used coolant should be recycled or removed as a hazardous waste.

Many shops now use a coolant flushing machine that automatically expels the used coolant and refills the system with fresh coolant (Figure 4-28). Many of these pieces of equipment also recycle the coolant to eliminate costly removal from the shop. Follow the instructions provided by the equipment manufacturer to flush the cooling system. Do not use recycled coolant if the manufacturer recommends against it; that could nullify the powertrain warranty.

While the system is drained, it is a good time to replace the thermostat and hoses. Following is a typical procedure for thermostat replacement (Figure 4-29).

Drain the coolant to a level below the thermostat housing. Doing this will reduce the potential to spill coolant onto the floor. Remove the radiator hose from the housing. Some hoses use a worm-gear-type hose clamp. These can be removed by using a flat blade screwdriver to turn the worm gear. Some hoses will have a spring wire clamp; use special clamp pliers or large pliers to remove these clamps. Some engines will have a bypass hose connected to the thermostat housing; if this is the case on your engine, it should be removed at this point.

Loosen, then remove, the fasteners that attach the thermostat housing to the engine. It may be necessary to strike the housing with a rubber mallet to break the seal between the engine and the housing. Do not drive a chisel or flat-blade screwdriver between the mating surfaces; doing so may damage the mating surfaces and prevent the system from sealing.

Remove the thermostat and gasket or O-ring seal. Pay attention to the direction the thermostat was facing. Confirm the proper direction in the service manual. If the thermostat was installed backwards, that could be the cause of overheating.

Being careful not to gouge the metal, use the proper abrasive disc to clean both mating surfaces. Apply the correct sealant, if one is recommended, to the mating surface of the housing. Carefully install the thermostat in the engine block, being sure the pellet is facing

FIGURE 4-28 A coolant flushing machine.

Upper radiator hose   Thermostat housing

FIGURE 4-29 The thermostat sits on the top front of this engine.

144

CAUTIONS AND WARNINGS

Throughout the text, warnings are given to alert the reader to potentially hazardous materials or unsafe conditions. Cautions are given to advise the student of things that can go wrong if instructions are not followed or if a nonacceptable part or tool is used.

CAUTION:
Thermostats have a high failure rate and are a very common cause of cooling system concerns. They are typically replaced during cooling system maintenance, repair, and engine overhaul. Do not install a thermostat with a lower operating temperature rating than specified or without a thermostat. Allowing the engine to run cooler than designed accelerates engine wear and increases emissions and fuel consumption.

fuel is directed. When the injector's solenoid is energized, a normally closed ball valve is lifted (Figure 9-24). Fuel under pressure is then injected at the walls of the throttle body bore just above the throttle plate.

A fuel injector has a moveable armature in the center of the injector, and a pintle with a tapered tip is positioned at the lower end of the armature. A spring pushes the armature and pintle downward so the pintle tip seats in the discharge orifice. The injector coil surrounds the armature, and the two ends of the winding are connected to the terminals on the side of the injector. An integral filter is located inside the top of the injector. When the ignition switch is turned on, voltage is supplied to one injector terminal and the other terminal is connected through the computer. Each time the control unit completes the circuit from the injector winding to ground, current flows through the injector coil and the coil magnetism moves the plunger and pintle upward. Under this condition, the pintle tip is unseated from the injector orifice and fuel sprays out this orifice.

TBI Advantages and Disadvantages. Throttle body systems provide improved fuel metering when compared to carburetors. They are also less expensive and simpler to service. TBI units also have some advantages over port injection. They are less expensive to manufacture, simpler to diagnose and service, and do not have injector balance problems to the extent that port injection systems do when the injectors begin to clog.

However, throttle body units are not as efficient as port systems. The disadvantages are primarily manifold related. Like a carburetor system, fuel is still not distributed equally to all cylinders, and a cold manifold may cause fuel to condense and puddle in the manifold. Like a carburetor, throttle-body injection systems must be mounted above the combustion chamber level, which eliminates the possibility of tuning the manifold design for more efficient operation.

Port Fuel Injection

PFI systems use one injector at each cylinder. They are mounted in the intake manifold near the cylinder head where they can inject a fine, atomized fuel mist as close as possible to the intake valve (Figure 9-25). Fuel lines that run to each cylinder from a fuel manifold are usually referred to as a **fuel rail**. The fuel rail assembly on a PFI system of V6 and V8 engines usually consists of a left- and right-hand rail assembly. The two rails can be connected either by crossover and return fuel tubes or by a mechanical bracket arrangement. A typical fuel rail arrangement is shown in Figure 9-26. Fuel tubes crisscross between the two rails. Figure 9-27 shows a fuel rail on a transverse four-cylinder engine. Since each cylinder has its own injector, fuel distribution is exactly equal. With little or

CROSS-REFERENCES TO THE SHOP MANUAL

Reference to the appropriate page in the Shop Manual is given whenever necessary. Although the chapters of the two manuals are synchronized, material covered in other chapters of the Shop Manual may be fundamental to the topic discussed in the Classroom Manual.

Shop Manual
Chapter 9, page 393

The **fuel rail** delivers fuel to the fuel injectors.

MARGINAL NOTES

These notes add "nice-to-know" information to the discussion. They may include examples or exceptions, or may give the common trade jargon for a component.

FIGURE 1-21 Valve timing failures can be catastrophic.

FIGURE 1-20 The lower crankshaft gear turns at twice the speed of the larger camshaft gear.

A BIT OF HISTORY

This feature gives the student a sense of the evolution of the automobile. This feature not only contains nice-to-know information, but also should spark some interest in the subject matter.

Main **journals** are round machined portions on the centerline of the crankshaft where the crankshaft is held in the main bore.

A Bit of History

Jean Joseph Lenoir created the first workable internal combustion engine in 1860. Nicholas Otto was credited with creating the first four-stroke internal combustion engine back in 1876. All previous internal combustion engines did not compress the air-fuel mixture. They attempted to draw the air-fuel mixture in during a downward

Timing Mechanism

As we discussed, proper valve timing is necessary to keep the valves opening and closing at the correct times. This is achieved by the timing mechanism. A sprocket or gear on the crankshaft is attached by a gear, belt, or chain to a sprocket or gear on the camshaft (Figure 1-20). Timing belts require periodic maintenance. When that is neglected, the timing belt can snap, and serious engine damage can result. In some cases, the valves knock the top of the pistons and bend, and a new set of valves is required (Figure 1-21). In the worst cases, those valves break the pistons, and the whole engine needs to be overhauled. Timing chains also wear out over time, and replacing them requires skill and attention to detail. Variable valve timing systems increase performance and minimize emissions. More and more vehicles employ these systems to vary the times when the valves can be opened. This makes timing mechanism theory and service a new and more complex challenge.

Engine Block

The **engine block** is the main supporting structure of the engine (Figure 1-22). It holds the pistons, connecting rods, crankshaft, and sometimes the camshaft. The cylinder head bolts to the top of the block. The block is bored (drilled) to create cylinders. An eight-cylinder engine has eight-cylinder bores (Figure 1-23). The pistons are pushed up and down in the cylinders by the crankshaft. On the power stroke, the piston is actually pushed down by the force of combustion to turn the crankshaft. Rings on the pistons seal the small clearance between the cylinder walls. This keeps the hot expanding gases from combustion within the combustion chamber, so that all the force acts to push the piston down. The rings also scrape excess oil from the cylinder walls, so the oil doesn't burn inside the combustion chamber. When oil is allowed to burn during combustion, it forms blue smoke that exits the tailpipe. The block also holds the crankshaft in its main bore. The soft main bearings are placed within the main bore to provide a place for the finely machined crankshaft **journals** to ride. The journals are the machined round areas of the crankshaft that allow the bearings and bearing caps to bolt around them and hold the crankshaft in the engine block. The clearance between the journals and the bearings must be just right: enough to allow adequate oil, but not so much that precious oil pressure leaks through excessive clearances.

10

26061_01_ch01_p001-013.indd   10   5/30/09   3:48:55 PM

ing out against the lower p...
Vacuum could be measured in pounds per square inch (p... "U" (in. Hg) are most commonly used for this measurement. Let us assume that a plastic "U" tube is partially filled with mercury, and atmospheric pressure is allowed to enter one end of the tube. If a vacuum is supplied to the other end of the "U" tube, the mercury is forced downward by the atmospheric pressure. When this movement occurs, the mercury also moves upward on the side where the vacuum is supplied. If the mercury moves downward

AUTHOR'S NOTES

This feature includes simple explanations, stories, or examples of complex topics. These are included to help students understand difficult concepts.

Author's Note: As altitude increases, atmospheric pressure decreases. As pressure decreases, the boiling point of a liquid also decreases. For example, water boils at 212°F at sea level but will boil at an incrementally lower temperature as altitude increases. So if you are on a camping trip in the Rocky Mountains (say, 6,000 feet above sea level) and the water boils at a lower temperature (less than 212°F), how long would you have to boil a three-minute egg before it has thoroughly cooked?

FIGURE 2-20 The ideal combustion process.

SUMMARIES

Each chapter concludes with a summary of key points from the chapter. These are designed to help the reader review the chapter contents.

SUMMARY

Terms to Know

Acid
Atmospheric pressure
Atom
Bar
Base
Coefficient of drag (Cd)
Compound
Compressible

- An element contains only one type of atom.
- An atom is the smallest particle of an element.
- Compounds contain two or more types of atoms.
- A molecule is the smallest particle in a compound.
- Protons are positively charged particles at the center, or nucleus, of each atom.
- Neutrons have no electric charge, and they are located in the nucleus of most atoms.
- Electrons are negatively charged particles found in various orbits around the nucleus of an atom.
- Work is the result of applying a force.
- Force is measured in pounds and distance.
- Energy is the ability to do work, and there are six basic types of energy.

Direct drive
Electron
Element
Energy
Friction
Gear reduction
Inertia
Kilopascal
Mass
Molecule
Momentum
Negative pressure
Neutron
Noncompressible
Overdrive
Positive pressure
Power
Pressure
Proton
Torque
Vacuum
Valence ring
Venturi
Volume
Weight
Work

TERMS TO KNOW LIST

A list of new terms appears next to the Summary.

REVIEW QUESTIONS

Short answer essay, fill-in-the-blank, and multiple-choice questions are found at the end of each chapter. These questions are designed to accurately assess the student's competence in the stated objectives at the beginning of the chapter.

REVIEW QUESTIONS

Short Answer Essays

1. Define an element and a compound.
2. Name the four factors that are necessary for complete combustion.
3. Define a molecule.
4. Describe three particles found in an atom, including the electrical charge and location of each particle.
5. Explain why a venturi decreases pressure.
6. What is atmospheric pressure?
7. What happens to air pressure as the temperature of the air changes?
8. Describe Newton's three laws of motion.
9. Describe six different forms of energy.
10. Describe four different types of energy conversion.

Fill-in-the-Blanks

1. At sea level, atmospheric pressure is _____ psi.
2. In nature, a _____ pressure always moves toward a _____ pressure.
3. The nucleus of an atom is comprised of a(n) _____ and a(n) _____.
4. Work is calculated by multiplying _____ times _____.
5. Energy may be defined as the ability to do _____.
6. When one object is moved over another object, the resistance to motion is called _____.
7. Weight is the measurement of the earth's _____ _____ on an object.

SHOP MANUAL

To stress the importance of safe work habits, the Shop Manual also dedicates one full chapter to safety. Other important features of this manual include:

PERFORMANCE-BASED OBJECTIVES

These objectives define the contents of the chapter and define what the student should have learned upon completion of the chapter. These objectives also correspond with the list of required tasks for NATEF certification.

Although this textbook is not designed to simply prepare someone for the certification exams, it is organized around the NATEF task list. These tasks are defined generically when the procedure is commonly followed and specifically when the procedure is unique for specific vehicle models. Imported and domestic model automobiles and light trucks are included in the procedures.

SERVICE TIPS

Whenever a short-cut or special procedure is appropriate, it is described in the text. These tips are generally those things commonly done by experienced technicians.

SPECIAL TOOLS LISTS

Whenever a special tool is required to complete a task, it is listed in the margin next to the procedure.

BASIC TOOLS LISTS

Each chapter begins with a list of the basic tools needed to perform the tasks included in the chapter.

MARGINAL NOTES

These notes add "nice-to-know" information to the discussion. They may include examples or exceptions, or may give the common trade jargon for a component.

PHOTO SEQUENCES

Many procedures are illustrated in detailed Photo Sequences. These detailed photographs show the students what to expect when they perform particular procedures. They also can provide the student a familiarity with a system or type of equipment, which the school may not have.

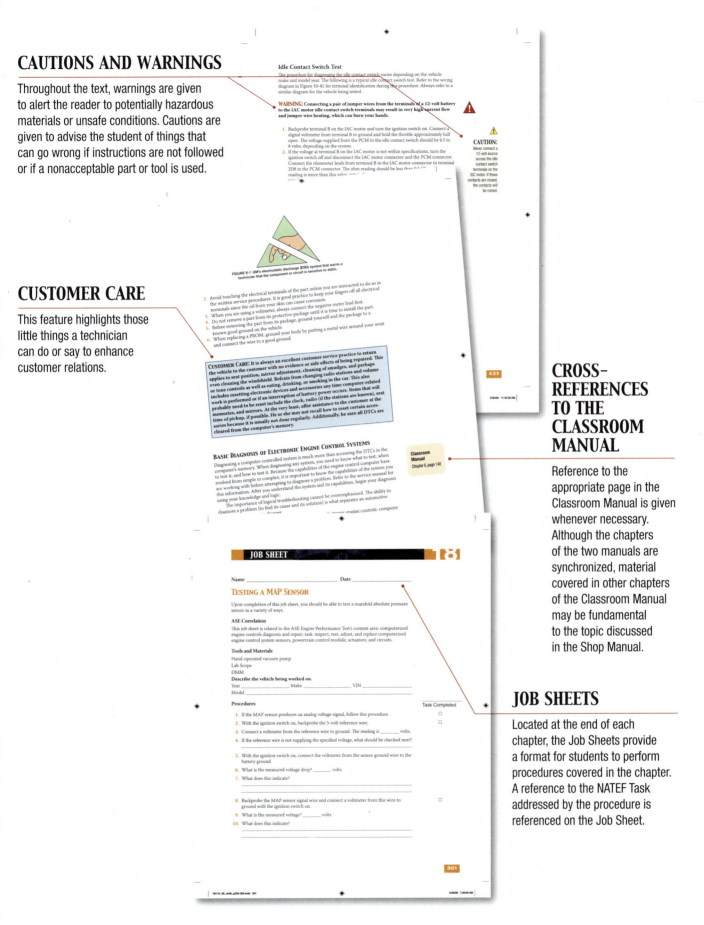

CAUTIONS AND WARNINGS

Throughout the text, warnings are given to alert the reader to potentially hazardous materials or unsafe conditions. Cautions are given to advise the student of things that can go wrong if instructions are not followed or if a nonacceptable part or tool is used.

CUSTOMER CARE

This feature highlights those little things a technician can do or say to enhance customer relations.

CROSS-REFERENCES TO THE CLASSROOM MANUAL

Reference to the appropriate page in the Classroom Manual is given whenever necessary. Although the chapters of the two manuals are synchronized, material covered in other chapters of the Classroom Manual may be fundamental to the topic discussed in the Shop Manual.

JOB SHEETS

Located at the end of each chapter, the Job Sheets provide a format for students to perform procedures covered in the chapter. A reference to the NATEF Task addressed by the procedure is referenced on the Job Sheet.

CASE STUDIES

Case Studies concentrate on the ability to properly diagnose the systems. Beginning with Chapter 3, each chapter ends with a case study in which a vehicle has a problem, and the logic used by a technician to solve the problem is explained.

TERMS TO KNOW LIST

Terms in this list can be found in the Glossary at the end of the manual.

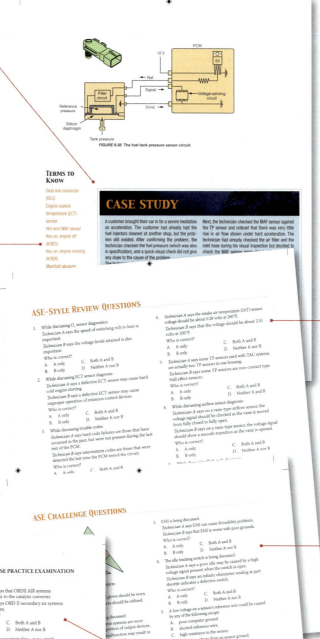

FIGURE 6-85 The fuel tank pressure sensor circuit.

TERMS TO KNOW

Data link connector (DLC)
Engine coolant temperature (ECT) sensor
Hot wire MAF sensor
Key on, engine off (KOEO)
Key on, engine running (KOER)
Manifold absolute

CASE STUDY

A customer brought their car in for a severe hesitation on acceleration. The customer had already had the fuel injectors cleaned at another shop, but the problem still existed. After confirming the problem, the technician checked the fuel pressure (which was also in specification), and a quick visual check did not give any clues to the cause of the problem.

Next, the technician checked the MAF sensor against the TP sensor and noticed that there was very little rise in air flow shown under hard acceleration. The technician had already checked the air filter and the inlet hose during his visual inspection but decided to check the MAF sensor.

ASE-STYLE REVIEW QUESTIONS

1. While discussing O₂ sensor diagnostics:
   Technician A says the speed of switching rich to lean is important.
   Technician B says the voltage levels attained is also important.
   Who is correct?
   A. A only   C. Both A and B
   B. B only   D. Neither A nor B

2. While discussing ECT sensor diagnosis:
   Technician A says a defective ECT sensor may cause hard cold engine starting.
   Technician B says a defective ECT sensor may cause improper operation of emission control devices.
   Who is correct?
   A. A only   C. Both A and B
   B. B only   D. Neither A nor B

3. While discussing trouble codes:
   Technician A says hard code failures are those that have occurred in the past, but were not present during the last test of the PCM.
   Technician B says intermittent codes are those that were detected the last time the PCM tested the circuit.
   Who is correct?
   A. A only   C. Both A and B

6. Technician A says the intake air temperature (IAT) sensor voltage should be about 0.28 volts at 260°F.
   Technician B says that the voltage should be about 2.51 volts at 100°F.
   Who is correct?
   A. A only   C. Both A and B
   B. B only   D. Neither A nor B

7. Technician A says some TP sensors used with TAC systems are actually two TP sensors in one housing.
   Technician B says some TP sensors are non-contact type Hall effect sensors.
   Who is correct?
   A. A only   C. Both A and B
   B. B only   D. Neither A and B

8. While discussing airflow sensor diagnosis:
   Technician A says on a vane-type airflow sensor, the voltage signal should be checked as the vane is moved from fully closed to fully open.
   Technician B says on a vane-type sensor, the voltage signal should show a smooth transition as the vane is opened.
   Who is correct?
   A. A only   C. Both A and B
   B. B only   D. Neither A nor B

ASE-STYLE REVIEW QUESTIONS

Each chapter contains ASE-style review questions that reflect the performance-based objectives listed at the beginning of the chapter. These questions can be used to review the chapter as well as to prepare for the ASE certification exam.

ASE CHALLENGE QUESTIONS

3. EMI is being discussed.
   Technician A says EMI can cause driveability problems.
   Technician B says that EMI is worse with poor grounds.
   Who is correct?
   A. A only   C. Both A and B
   B. B only   D. Neither A nor B

4. The idle tracking switch is being discussed.
   Technician A says a poor idle may be caused by a high voltage signal present when the switch is open.
   Technician B says an infinity ohmmeter reading at part throttle indicates a defective switch.
   Who is correct?
   A. A only   C. Both A and B
   B. B only   D. Neither A nor B

5. A low voltage on a sensor's reference wire could be caused by any of the following except:
   A. poor computer ground.
   B. shorted reference wire.
   C. high resistance in the sensor.
   D. excessive voltage drop on sensor ground.

ASE CHALLENGE QUESTIONS

Each technical chapter ends with five ASE challenge challenge questions. These are not more review questions, rather they test the students' ability to apply general knowledge to the contents of the chapter.

APPENDIX A   ASE PRACTICE EXAMINATION

1. Technician A says a hydrometer reading of 1.200 at 80°F means the battery must be recharged before performing a capacity test.
   Technician B says a capacity test can be correctly performed with the battery cables connected.
   Who is correct?
   A. A only   C. Both A and B
   B. B only   D. Neither A nor B

2. Technician A says engine detonation may be caused by low octane fuel.
   Technician B says that detonation can be caused by engine overheating.
   Who is correct?
   A. A only   C. Both A and B
   B. B only   D. Neither A nor B

3. The PCV system is being discussed.
   Technician A says oil in the crankcase breather filter will confirm that the PCV valve is clogged.
   Technician B says the PCV valve must be disconnected to be checked with the engine operating.
   Who is correct?
   A. A only   C. Both A and B
   B. B only   D. Neither A nor B

4. An EI-equipped vehicle will not start.
   Technician A says a good first step is to make certain that the MIL is illuminated with the key on.
   Technician B says to check for spark at more than one plug wire during diagnosis.
   Who is correct?
   A. A only   C. Both A and B
   B. B only   D. Neither A nor B

5. Electronic fuel injection systems are being discussed.
   Technician A says a hard to start engine may have an open electrical fuel pump relay bypass circuit.
   Technician B says a hesitation when accelerating from idle may be caused by dirty injectors.
   Who is correct?
   A. A only   C. Both A and B
   B. B only   D. Neither A nor B

6. Technician A says that OBDII AIR systems do not supply air to the catalytic converter.
   Technician B says OBD II secondary air systems are all belt driven.
   Who is correct?
   A. A only   C. Both A and B
   B. B only   D. Neither A nor B

7. A vehicle has a consistent slow- or no-crank condition. This may be caused by:
   A. A slipping AC generator drive belt
   B. A defective AC generator
   C. A low regulator voltage
   D. Any of the above

8. A fuel-injected vehicle comes in with a miss on acceleration.
   Technician A says that there could be an ignition problem.
   Technician B says that the intake to MAF duct could be opening up at a tear on acceleration.
   Who is correct?
   A. A only   C. Both A and B
   B. B only   D. Neither A nor B

9. Technician A says static electricity generated by clothing in contact with vehicle upholstery must be discharged before beginning work on electronic devices.
   Technician B says the best control for static electricity is the wearing of a waist or wrist grounding strap.
   Who is correct?
   A. A only   C. Both A and B
   B. B only   D. Neither A nor B

10. Technician A says a camshaft installed one tooth retarded can be compensated by adjusting the ignition's base timing.
    Technician B says to use the starter to crank the engine with the timing belt removed to determine if the engine is interference type or free-wheeling.
    Who is correct?
    A. A only   C. Both A and B
    B. B only   D. Neither A nor B

ASE PRACTICE EXAMINATION

A 50 question ASE practice exam, located in the appendix, is included to test students on the contents of the Shop Manual.

SUPPLEMENTS

INSTRUCTOR RESOURCES

The Instructor Resources DVD is a robust ancillary that contains all preparation tools to meet any instructor's classroom needs. It includes PowerPoint slides with images, video clips, and animations that coincide with each chapter's content coverage, a computerized test bank with hundreds of test questions, a searchable image library with all pictures from the text, theory-based worksheets in Word that provide homework or in-class assignments, the Job Sheets from the Shop Manual in Word, a NATEF correlation chart, and an Instructor's Guide in electronic format.

WEBTUTOR ADVANTAGE

Newly available for this title and to the Today's Technician™ series is the *WebTutor Advantage,* for Blackboard and Angel online course management systems. The *WebTutor* for *Today's Technician: Automotive Engine Repair and Rebuilding, 4e* will include PowerPoint presentations with video clips and animations, end-of-chapter review questions, pre-tests and post-tests, worksheets, discussion springboard topics, Job Sheets, and more. The *WebTutor* is designed to enhance the classroom and shop experience, engage students, and help them prepare for ASE certification exams.

REVIEWERS

The author and publisher would like to extend a special thanks to the instructors who reviewed this text and offered invaluable feedback:

Paul Bean
Utah Valley University
Orem, UT

Brad Lyne
North Central Kansas Technical College
Beloit, KS

Wade Esplin
Southwest Applied Technical College
Cedar City, UT

John Fowler
Iowa Central Community College
Fort Dodge, IA

Gary Ham
South Plains College
Levelland, TX

Tim Reaves
Cosumnes River College
Sacramento, CA

Chapter 1

SAFETY AND SHOP PRACTICES

UPON COMPLETION AND REVIEW OF THIS CHAPTER, YOU SHOULD BE ABLE TO:

- Understand the importance of safety and accident prevention in an automotive shop.
- Explain the basic principles of personal safety, including protective eyewear, clothing, gloves, shoes, and hearing protection.
- Demonstrate proper lifting procedures and precautions.
- Explain the procedures and precautions for safely using tools and equipment.
- Follow safety precautions regarding the use of power tools.
- Demonstrate proper safety precautions during the use of compressed air equipment.
- Demonstrate proper vehicle lift operating and safety procedures.
- Observe all safety precautions when hydraulic tools are used in the automotive shop.

- Follow the recommended procedure while operating hydraulic tools such as presses, floor jacks, and vehicle lifts to perform automotive service tasks.
- Follow safety precautions while using cleaning equipment in the automotive shop.
- Observe all shop rules when working in the shop.
- Operate vehicles in the shop according to shop driving rules.
- Explain what should be done to maintain a safe work area, including handling vehicles in the shop and venting carbon monoxide gases.
- Understand the various government regulations and safety laws.
- Describe how a chemical manufacturer provides safety warnings.

Safety and accident prevention must be a top priority in all automotive shops. There is great potential for serious accidents simply because of the nature of the business and the equipment used. In fact, the automotive repair industry is rated as one of the most dangerous occupations in the country.

Vehicles, equipment, and many parts are very heavy, and parts often fit tightly together. Many components become hot during operation, and high-fluid pressures can build up inside the cooling system, fuel system, or battery. Batteries contain highly corrosive and potentially explosive acids. Fuels and cleaning solvents are flammable. Exhaust fumes are poisonous. During some repairs, technicians can be exposed to harmful dust particles and vapors.

Good safety practices eliminate these potential dangers. A careless attitude and poor work habits invite disaster. Shop accidents can cause serious injury, temporary or permanent disability, and death. Safety is a very serious matter. Both the employer and the employees must work together to protect the health and welfare of all who work in the shop.

This chapter contains many safety guidelines concerning personal safety, tools and equipment, and work area, as well as some of the responsibilities you have as an employee and responsibilities your employer has to you. In addition to these rules, special warnings have been used throughout this book to alert you to situations where carelessness could result in personal injury. Finally, when working on cars, always follow safety guidelines given in service information and other technical literature. They are there for your protection.

PERSONAL SAFETY

Personal safety simply involves those precautions you take to protect yourself from injury. These include wearing protective gear, dressing for safety, and correctly handling tools and equipment.

Eye Protection

Your eyes can become infected or permanently damaged by many things in a shop. Some repair procedures, such as grinding, result in tiny particles of metal and dust that are thrown off at very high speeds. These metal and dirt particles can easily get into your eyes, causing scratches or cuts on your eyeball. Pressurized gases and liquids escaping a ruptured hose or fuel line fitting can spray a great distance. If these chemicals get into your eye, they can cause blindness. Dirt and sharp bits of corroded metal can easily fall down into your eyes while you are working under a vehicle.

Eye protection should be worn whenever you are exposed to these risks. To be safe, you should wear **safety glasses** whenever you are working in the shop. There are many types of eye protection available (Figure 1-1). To provide adequate eye protection, safety glasses have lenses made of safety glass. They also offer side protection. Regular prescription glasses do not offer sufficient protection and therefore should not be worn as a substitute for safety glasses.

You should wear safety glasses at all times. To help develop this habit, wear safety glasses that fit well and feel comfortable. Most safety glasses that provide good protection are rated by the American National Standards Institute (ANSI). The ANSI has established a standard for producing a lens that is durable and safe for most situations. This standard is the **Z-87.1** standard, but most safety glasses that are ANSI approved have a simple Z-87.1 stamping on them.

If chemicals such as battery acid, fuel, or solvents get into your eyes, flush them continuously with clean water. Have someone call a doctor and get medical help immediately.

Safety glasses have lenses that resist breaking and protect the eyes from airborne objects or liquids. If you wear prescription glasses, wear safety goggles that will securely fit over your glasses.

Z-87.1 is the ANSI-established standard for safety glasses.

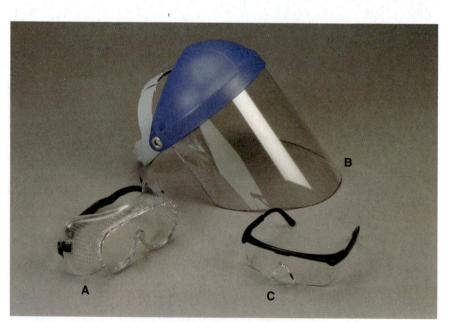

FIGURE 1-1 (A) safety goggles, (B) face shield, (C) safety glasses.

Clothing

Your clothing should be well-fitted and comfortable but made with strong material. Loose, baggy clothing can easily get caught by moving parts and machinery. Some technicians prefer to wear coveralls or shop coats to protect their personal clothing. Cut-offs and short pants are not appropriate for shop work. Many shops will provide you with uniforms.

Automotive work involves the handling of many heavy objects that can be accidentally dropped on your feet or toes. Always wear shoes made of leather or a similar material, or boots with non-slip soles. Steel-tipped safety shoes can give added protection for your feet. Sports sneakers, street shoes, and sandals are not appropriate in the shop.

Good hand protection is often overlooked. A scrape, cut, or burn can limit your effectiveness at work for many days. A well-fitted pair of heavy work gloves should be worn during operations such as grinding and welding or when handling high-temperature components. Always wear approved rubber gloves when handling caustic chemicals. Caustic chemicals are strong and dangerous chemicals. They can easily burn your skin. Be very careful when handling this type of chemical.

Many technicians now choose to wear close-fitting surgical gloves that offer some protection from toxic chemicals such as engine oil, gasoline, brake fluid, and cleaning chemicals (Figure 1-2). It is important to remember that your skin is an organ and it will absorb some of the fluids that are placed on it. These gloves are very inexpensive, disposable, and come in different version, such as latex, powdered, and nitrile. These gloves also save you the time of scrubbing filthy hands every time you want to get into the vehicle or take a road test. One bit of grease in a customer's vehicle can destroy your reputation with the customer, no matter how excellent the rest of your work is. These disposable and inexpensive gloves are available in drug stores and from the major tool manufacturers.

Many mechanics also choose to wear thicker, padded gloves. These gloves are a little bit thicker than the surgical gloves, but offer better protection to sharp or hot objects. Most of the higher quality gloves of this type offer padded protection in high-wear areas, such as the

Caustic chemicals can burn through cloth and skin.

FIGURE 1-2 These tight-fitting latex gloves keep your skin safe from dangerous fluids.

palm and fingers. They are also very comfortable when you have to work on a very hot engine component and you cannot wait for it to cool down. It is a good practice to have many different types of gloves in your toolbox so that you can switch to the one that fits the job.

Ear Protection

Exposure to very loud noise levels for extended periods of time can lead to a loss of hearing. Air wrenches, engines running under a load, and vehicles running in enclosed areas can all generate annoying and harmful levels of noise. Simple ear plugs or earphone-type protectors should be worn in constantly noisy environments.

Hair and Jewelry

Long hair and loose, hanging jewelry can create the same type of hazard as loose-fitting clothing. They can become caught in moving engine parts and machinery. If you have long hair, tie it back or cover it with a cap.

Never wear rings, watches, bracelets, and neck chains. These can easily get caught in moving parts and cause serious injury.

Other Personal Safety Warnings

- Never smoke while working on a vehicle or while working with any machine in the shop.
- Playing around or horseplay is not appropriate for professional technicians. Such things as air nozzle fights, creeper races, or practical jokes can hurt people and have no place in the shop.
- To prevent serious burns, keep your skin away from hot metal parts such as the radiator, exhaust manifold, tailpipe, catalytic converter, and muffler.
- When working with a hydraulic press, make sure that the pressure is applied in a safe manner. It is generally wise to stand to the side when operating the press. Always wear safety glasses.
- Properly store all parts and tools by putting them away in a place where people will not trip over them. This practice not only cuts down on injuries, it also reduces time wasted looking for a misplaced part or tool.

AUTHOR'S NOTE: In my experiences as an automotive technician, I can consider myself lucky. I have come to realize that compared to other technicians I have not been severely injured or harmed while on the job. I have known many technicians who did not wear the appropriate safety items and have been injured as a result; these injuries could have been prevented. Some of these technicians can no longer perform their job, and as a result have been forced to assume a different job at their place of employment or completely change careers. Your success as an automotive technician will depend on you being able to use your hands, eyes, back, and every other part of your body. Be sure to take good care of it.

LIFTING AND CARRYING

Knowing the proper way to lift heavy objects is important. You should also use back protection devices when you are lifting a heavy object. Always lift and work within your ability, and ask others to help when you are not sure that you can handle the size or weight of an object. Even small, compact parts can be surprisingly heavy or unbalanced. Think about how you are going to lift something before beginning. When lifting any object, follow these steps:

1. If the object is to be carried, be sure your path is free from loose parts or tools.
2. Place your feet close to the object. Position your feet so that you will be able to maintain a good balance.

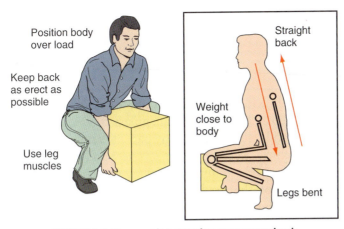

FIGURE 1-3 Use your leg muscles, never your back, to lift heavy objects.

3. Keep your back and elbows as straight as possible. Bend your knees until your hands reach the best place to get a strong grip on the object (Figure 1-3).

4. If the part is in a cardboard box, make sure the box is in good condition. Old, damp, or poorly sealed boxes will tear and the part will fall out.

5. Firmly grasp the object or container. Never try to change your grip as you move the load.

6. Keep the object close to your body and lift it up by straightening your legs. Use your leg muscles, not your back muscles.

7. If you must change your direction of travel, never twist your body. Turn your whole body, including your feet.

8. When placing the object on a shelf or counter, do not bend forward. Place the edge of the load on the shelf and slide it forward. Be careful not to pinch your fingers.

9. When setting down a load, bend your knees and keep your back straight. Never bend forward—this strains the back muscles.

10. Set the object onto blocks of wood to protect your fingers when lowering something heavy onto the floor.

11. If the object is too heavy to lift, try using a wheel dolly to move it (Figure 1-4).

FIGURE 1-4 Use a two-wheel dolly if the load is very heavy.

HAND TOOL SAFETY

Many shop accidents are caused by improper use and care of hand tools. These hand tool safety steps must be followed:

1. Keep tools clean and in good condition. Worn tools may slip and result in hand injury. If a hammer is used with a loose head, the head may fly off and cause personal injury or vehicle damage. Keep all hand tools clean; greasy tools can slip and cause cuts and bruises. If a tool slips and falls into a moving part, it can fly out and cause serious injury to you or the vehicle. Keeping your tools neat and organized can save you time while you work. If you spend ten minutes looking for a particular socket that you didn't put back in its proper spot, that could be ten minutes less that you are paid for.

2. Use the proper tool for the job. Make sure the tool is of professional quality. Using poorly made tools or the wrong tools can damage parts, the tool itself, or cause injury. Do not use broken or damaged tools.

3. Be careful when using sharp or pointed tools. Do not place sharp tools, or other sharp objects, in your pockets. They can stab or cut your skin, ruin automotive upholstery, or scratch a painted surface.

4. Tool tips that are intended to be sharp should be kept in a sharp condition. Sharp tools, such as chisels, will do the job faster with less effort. Dull tools can be more dangerous than sharp tools.

POWER TOOL SAFETY

Power tools are operated by an outside source of power, such as electricity, compressed air, or hydraulic pressure. Safety around power tools is very important. Serious injury can result from carelessness. Always wear safety glasses when using power tools.

If the tool is electrically powered, make sure it is properly grounded. Check the wiring for cracks in the insulation and bare wires before using it. Also, when using electrical power tools, never stand on a wet or damp floor. Disconnect the power source before performing any service on the machine or tool. Before plugging in any electric tool, make sure the switch is off to prevent serious injury. When you are through using the tool, turn it off, and unplug it. Never leave a running power tool unattended.

When using power equipment on a small part, never hold the part in your hand. Always mount the part in a bench vise or use vise grip pliers. Never try to use a machine or tool beyond its stated capacity or for operations requiring more than the rated power of the tool.

When working with larger power tools, like bench or floor equipment, check the machines for signs of damage before using them. Place all safety guards into position (Figure 1-5). A safety guard is a protective cover over a moving part. It is designed to

FIGURE 1-5 Safety guards on a bench grinder.

FIGURE 1-6 A hydraulic press.

prevent injury. Wear safety glasses or a face shield. Make sure there are no people or parts around the machine before starting it up. Keep your hands and clothing away from the moving parts. Maintain a balanced stance while using the machine.

Hydraulic Press

WARNING: When operating a hydraulic press, always be sure that the components being pressed are supported properly on the press bed with steel supports. When hydraulic pressure is supplied to improperly supported components, they may drop off the press, resulting in leg or foot injuries.

WARNING: When using a hydraulic press, never operate the pump handle when the pressure gauge exceeds the maximum pressure rating of the press. If this pressure is exceeded, some part of the press may suddenly break and cause severe personal injury.

When two components have a tight precision fit between them, a **hydraulic press** is used to separate them or press them together. The hydraulic press rests on the shop floor. An adjustable steel beam bed is retained to the lower press frame with heavy steel pins. A hydraulic cylinder and ram is mounted on the top part of the press, with the ram facing downward toward the press bed (Figure 1-6). The component being pressed is placed on the press bed with the appropriate steel supports. A hand-operated hydraulic pump is mounted on the side of the press. When the handle is pumped, hydraulic fluid is forced into the cylinder, and the ram is extended against the component on the press bed to complete the pressing operation. A pressure gauge on the press indicates the pressure applied from the hand pump to the cylinder. The press frame is designed for a certain maximum pressure. This pressure must not be exceeded during operation.

A **hydraulic press** is a tool used to remove and install components with a tight press fit.

COMPRESSED AIR EQUIPMENT SAFETY

Tools that use compressed air are called **pneumatic tools**. Compressed air is used to inflate tires, apply paint, and drive tools. Compressed air can be dangerous when it is not used properly. The shop air supply contains high-pressure air in the shop compressor and air lines. Serious injury or property damage may result from careless operation of compressed air equipment. Follow these guidelines when working with compressed air:

1. Wear safety glasses or a face shield when using pneumatic tools and other compressed air equipment.
2. Wear ear protection when using compressed air equipment.
3. Always maintain air hoses and fittings in good condition. If an end suddenly blows off an air hose, the hose will whip around, and this may cause personal injury or vehicle damage.
4. Do not direct compressed air against the skin. This air may penetrate the skin, especially through small cuts or scratches. If compressed air penetrates the skin and enters the bloodstream, it can be fatal or cause serious health complications. Use only air gun nozzles approved by the Occupational Safety and Health Administration (OSHA).
5. Do not use an air gun to blow off clothing or hair.
6. Do not clean the workbench or floor with compressed air. This action may blow very small parts against your skin or into your eye. Small parts blown by compressed air may cause vehicle damage. For example, if the car in the next stall has the air cleaner removed, a small part may go into the throttle body. When the engine is started, this part will likely be pulled into the cylinder by engine vacuum and could penetrate the top of a piston.
7. Never spin bearings with compressed air, because the bearing will rotate at extremely high speed. Under this condition, the bearing may be damaged or it may disintegrate, causing personal injury or other damage.
8. All pneumatic tools must be operated according to the manufacturer's recommended operating procedure.
9. Follow the equipment manufacturer's recommended maintenance schedule for all compressed air equipment.

LIFT SAFETY

A lift is used to raise a vehicle so a technician can work under the vehicle. The lift arms must be placed under the car at the manufacturer's recommended lift points. Twin posts are used on some lifts, whereas other lifts have a single post (Figure 1-7). Some lifts have an electric motor that drives a hydraulic pump to create fluid pressure and force the lift upward. Other lifts use air pressure from the shop air supply to force the lift upward. If shop air pressure is used for this purpose, the air pressure is applied to fluid in the lift cylinder. A control lever or switch is placed near the lift. The control lever supplies shop air pressure to the lift cylinder, and the switch turns on the lift pump motor. Always be sure that the safety lock is engaged after the lift is raised. When the safety lock is released, a lever can be operated to slowly lower the vehicle.

Be very careful when raising a vehicle on a lift or a hoist. Adapters and hoist plates must be positioned correctly on twin post and rail-type lifts to prevent the vehicle from falling off the lift and causing serious injury to you or severe damage to the vehicle. Improper lift pad placement can also damage the undercarriage of the vehicle. There are specific lift points. These points allow the weight of the vehicle to be evenly supported by the adapters or hoist plates. The correct lift points can be found in the vehicle's service information. Figure 1-8 shows typical locations for frame and unibody cars. These diagrams are for illustration only. Always follow the manufacturer's instructions. Before operating any lift or hoist, carefully read the operating manual and follow the operating instructions.

A lift may be called a hoist.

Some vehicles have electronic suspension systems. In these cases, the air suspension systems must be electrically disabled by a switch (usually located in the trunk) before lifting the vehicle. Not disabling the suspension system may cause damage to it. Refer to *Today's Technician™ Automotive Suspension Systems* for detailed information about air suspension systems.

FIGURE 1-7 A twin post lift.

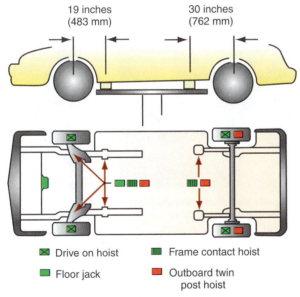

19 inches
(483 mm)

30 inches
(762 mm)

⊠ Drive on hoist   ▥ Frame contact hoist

■ Floor jack   ■ Outboard twin
                            post hoist

FIGURE 1-8 Hoisting and lifting points for a typical
unibody vehicle. Locate your vehicle's lift points in
the service information.

WARNING: Never use a lift or jack to move something heavier than it is designed for. Always check the rating before using a lift or jack. If a jack is rated for two tons, do not attempt to use it for a job requiring five tons. It is dangerous for you and the vehicle.

Before driving a vehicle over a lift, position the arms and supports to provide an unobstructed clearance. Do not hit or run over lift arms, adapters, or axle supports. This could damage the lift, vehicle, or tires.

WARNING: Never disable the lift locking devices or work under a lift where the safety mechanism is broken. The safety handles or locks prevent the lift from falling if the lift fails.

Position the lift supports to contact the vehicle at its lifting points. Raise the lift until the supports contact the vehicle. Then, check the supports to make sure they are in full contact with the vehicle. Jounce the vehicle on the lift to be sure it feels secure. Raise the lift to the desired working height.

Make sure the vehicle's doors, hood, and trunk are closed before raising the vehicle. Never raise a car with someone inside.

 WARNING: **Before working under a car, make sure the lift's locking device is engaged. If the locking device is not engaged, the lift may drop suddenly, resulting in severe personal injury and vehicle damage.**

After lifting a vehicle to the desired height, always lower it onto its mechanical safeties. On some vehicles, the removal (or installation) of components can cause a critical shift of the vehicle's weight, which may cause the vehicle to be unstable on the lift. Refer to the vehicle's service information for the recommended procedures to prevent this from happening.

PHOTO SEQUENCE 1

TYPICAL PROCEDURE FOR LIFTING A VEHICLE ON A HOIST

P1-1 Refer to the service information to determine the proper lift points for the vehicle you are lifting.

P1-2 Center the vehicle over the hoist, considering the vehicle's center of gravity and balance point.

P1-3 Locate the hoist pads under the lift points. Adjust the pads so that the vehicle will be lifted level.

P1-4 Lift the vehicle a couple of inches from the floor. Shake the vehicle while observing it for signs of any movement. If the vehicle is not secure on the hoist or unusual noises are heard while lifting, lower it to the floor and reset the pads.

P1-5 Once the vehicle is at the desired height, lock the hoist. Do not get under the vehicle until the hoist locks have been set.

P1-6 To lower the vehicle, release the locks, and put the control valve into the "lower" position. Once the vehicle is returned to the floor, push the contact pads to a location that is out of the path of the tires.

Make sure tool trays, stands, and other equipment are removed from under the vehicle. Never place oxyacetylene torches under a lift. Release the lift's locking devices according to the instructions before attempting to lower the lift. Refer to Photo Sequence 1.

JACK AND JACK STAND SAFETY

An automobile can be raised off the ground by a hydraulic jack (Figure 1-9). **Jack stands** (Figure 1-10) are supports of different heights that sit on the floor. They are placed under a sturdy chassis member, such as the frame or axle housing, to support the vehicle. Like jacks, jack stands also have a capacity rating. Always use the correct rating of jack stand.

A **floor jack** is a portable unit mounted on wheels. The lifting pad on the jack is placed under the chassis of the vehicle, and the jack handle is operated with a pumping action. This jack handle operation forces fluid into a hydraulic cylinder in the jack, and the cylinder extends to force the jack lift pad upward and lift the vehicle. Always be sure that the lift pad is positioned securely under one of the car manufacturer's recommended lift points. To release the hydraulic pressure and lower the vehicle, the handle or release lever must be turned slowly.

The jack should be removed after the jack stands are set in place. This eliminates a hazard, such as a jack handle sticking out into a walkway. A jack handle that is bumped or kicked can cause a tripping accident or cause the vehicle to fall.

WARNING: **Never use a jack by itself to support a vehicle. Always use a jack stand with a jack and make sure that the jack stands are properly positioned under the vehicle. Jack stands must always be positioned on a concrete floor, not on dirt or in gravel.**

Accidents involving the use of floor jacks and jack stands may be avoided if these safety precautions are followed:

1. Never work under a vehicle unless jack stands are placed securely under the vehicle chassis and the vehicle is resting on these stands.
2. Prior to lifting a vehicle with a floor jack, be sure that the jack lift pad is positioned securely under a recommended lift point on the vehicle. Lifting the front end of a

> **Jack stands** are used to support raised vehicles. Jack stands are also called safety stands.

FIGURE 1-9 Typical hydraulic floor jack.

FIGURE 1-10 Typical jack stands.

vehicle with the jack placed under a radiator support may cause severe damage to the radiator and support.

3. Position the jack stands under a strong chassis member such as the frame or axle housing. The jack stands must contact the vehicle manufacturer's recommended lift points.

4. Since the floor jack is on wheels, the vehicle and jack tend to move as the vehicle is lowered from a floor jack onto jack stands. Always be sure the jack stands remain under the chassis member during this operation, and be sure the jack stands do not tip. All the jack stand legs must remain in contact with the shop floor.

ENGINE LIFT SAFETY

An engine hoist may be called a cherry-picker or engine crane.

If the engine is removed through the hood opening, an engine hoist is used (Figure 1-11). The hoist is attached to the engine by a sling. When attaching the sling, make sure the bolts are strong enough to hold the weight of the engine and are threaded in far enough to prevent the bolts from stripping out. Most service information will provide proper locations for the sling to attach to the engine.

The hoist may have adjustable booms and load legs. Adjust the legs out far enough to prevent the engine and hoist from tipping. When adjusting the boom, remember that the farther out it is extended, the lower its lift capacity. After adjusting the legs and boom, make sure the lock pins are properly installed.

As the engine is being lifted or lowered, stand clear. Never get under an engine supported only by the hoist. Once the engine is out of the vehicle, immediately lower it to the floor, or install it onto an engine stand.

CLEANING EQUIPMENT SAFETY

All technicians are required to clean parts during their normal work routines. Face shields and protective gloves must be worn while operating cleaning equipment. The solution in hot and cold cleaning tanks may be caustic, and contact between this

FIGURE 1-11 An engine hoist.

solution and skin or eyes must be avoided. Parts cleaning often creates a slippery floor, and care must be taken when walking in the parts cleaning area. The floor in this area should be cleaned frequently. When the caustic cleaning solution in hot or cold cleaning tanks is replaced, environmental regulations require that the old solution be handled as hazardous waste. Use caution when placing aluminum or aluminum alloy parts in a cleaning solution. Some cleaning solutions will damage these components. Always follow the cleaning equipment manufacturer's recommendations. Parts cleaning is a necessary step in most repair procedures. Cleaning automotive parts can be divided into four basic categories.

Chemical cleaning relies primarily on some type of chemical action to remove dirt, grease, scale, paint, or rust. A combination of heat, agitation, mechanical scrubbing, or washing may also be used to help remove dirt. Chemical cleaning equipment includes small parts washers, hot/cold tanks, pressure washers, spray washers, and salt baths (Figure 1-12).

Some parts washers provide electromechanical agitation of the parts to provide improved cleaning action. These parts washers may be heated with gas or electricity, and various water-based hot tank cleaning solutions are available, depending on the type of metals being cleaned. For example, Kleer-Flo Greasoff™ Number 1 powdered detergent is available for cleaning iron and steel. Non-heated electromechanical parts washers are also available, and these washers use cold cleaning solutions, such as Kleer-Flo Degreasol™ formulas.

Many cleaning solutions, such as Kleer-Flo Degreasol™ 99R, contain no ingredients listed as hazardous by the Environmental Protection Agency's Resource Conservation and Recovery Act (RCRA). This cleaning solution is a blend of sulphur-free hydrocarbons,

FIGURE 1-12 This hot tank works like a dishwasher
to scrub engine blocks, cylinder heads, and
other parts clean.

FIGURE 1-13 This parts washer sends solvent through the nozzle where you can direct it across the dirty component. Wear gloves and use a brush to thoroughly clean the part.

wetting agents, and detergents. Degreasol™ 99R does not contain aromatic or chlorinated solvents, and it conforms to California's Rule 66 for clean air. Always use the cleaning solution recommended by the equipment manufacturer.

Some parts washers have an agitated immersion chamber under the shelves, which provides thorough parts cleaning. Folding work shelves provide a large upper cleaning area with a constant flow of solution from the dispensing hose. This cold parts washer operates on Degreasol™ 99R or similar cleaning solution (Figure 1-13).

An aqueous parts cleaning tank uses a water-based, environmentally friendly cleaning solution such as Greasoff™ 2, rather than traditional solvents. The immersion tank is heated and agitated for effective parts cleaning. A sparger bar pumps a constant flow of cleaning solution across the surface to push floating oils away, and an integral skimmer removes these oils. This action prevents floating surface oils from redepositing on cleaned parts. Many shops have contracts with chemical companies, which regularly send personnel to service the machine and refresh the solvent. Many of these cleaning chemicals are hazardous waste and must be handled properly by qualified people.

Thermal cleaning relies on heat, which bakes off or oxidizes the dirt. Thermal cleaning leaves an ash residue on the surface that must be removed by an additional cleaning process, such as airless shot blasting or spray washing.

Abrasive cleaning relies on physical abrasion to clean the surface. This includes everything from a wire brush to glass bead blasting, airless steel shot blasting, abrasive tumbling, and vibratory cleaning. Chemical in-tank solution sonic cleaning might also be included here because it relies on the scrubbing action of ultrasonic sound waves to loosen surface contaminants.

CUSTOMER CARE: Prior to disconnecting the battery on some late-model vehicles, it may be necessary to connect an auxiliary power supply to the cigarette lighter socket to maintain power to the onboard computer memories. If an auxiliary power supply is not available, record the radio presets and set the radio and the clock before returning the vehicle to the customer. If battery power has been disconnected from the engine computer, the powertrain control module (PCM), some of the learned memory will be lost. Until the vehicle has been driven several miles, it may not perform exactly the same as it did prior to disconnecting the battery. Be sure to explain this to the customer; he or she should just drive the vehicle normally for a day or two and then the vehicle will perform as usual.

The **Occupational Safety and Health Administration (OSHA)** is a part of the United States government tasked with protecting the health and safety of workers.

WARNING: When removing and replacing an engine, always disconnect the negative battery cable before starting to work on the vehicle. If the positive cable is removed first, the wrench may slip and make contact between the positive battery terminal and ground. This action may overheat the wrench and burn the technician's hand. When installing the engine, do not reconnect the negative battery cable until the installation is complete and you are ready to start the engine.

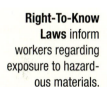

A wide variety of chemical solvents and aerosol sprays are used to clean parts and free up rusted components. Most of these are hazardous to your health and safety. You should understand the risks involved with these products before using them. The **Occupational Safety and Health Administration (OSHA)** established that every employee in a shop be protected by **Right-To-Know Laws** concerning hazardous materials and wastes.

Right-To-Know Laws inform workers regarding exposure to hazardous materials.

The general intent of these laws is for employers to provide a safe workplace, as it relates to hazardous materials. All employees must be trained about their rights under the legislation, including the nature of the hazardous chemicals in their workplace, the labeling of chemicals, and the information about each chemical listed and described on **Material Safety Data Sheets (MSDS)**. These sheets are available from the manufacturers and suppliers of the chemicals. They detail the chemical composition and precautionary information for all products that can present health or safety hazards. An up-to-date notebook containing the MSDS for every hazardous material should be available in each automotive repair shop.

Material Safety Data Sheets (MSDS) provide extensive information about hazardous materials.

Employees must be familiar with the intended purposes of the material, the recommended protective equipment, accident and spill procedures, and any other information regarding the safe handling of hazardous materials. This training must be given to all employees who will work with these materials. The Canadian equivalents to the MSDS are called **Workplace Hazardous Materials Information Systems (WHMIS)**.

Read through the MSDS on the various chemicals used in the shop to learn about the protective equipment required for handling, explosion and fire hazards, incompatible materials, health hazards and first aid procedures, safe handling precautions, and spill and leak procedures.

Workplace Hazardous Materials Information Systems (WHMIS) list information about hazardous materials.

WARNING: Always wear a face shield or safety glasses to avoid dangerous contact with your eyes. Permanent damage to your eyes can result from exposure to certain chemicals.

In general, follow these guidelines when using toxic chemicals and solvents:
- Always wear appropriate eye protection.
- Handle all chemical solvents with care to avoid spillage.

FIGURE 1-14 Flammable liquids should be stored in safety-approved containers.

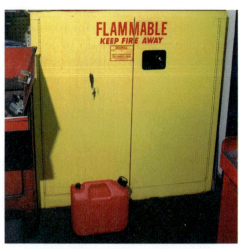

FIGURE 1-15 Store combustible materials in approved safety cabinets.

- Keep all solvent containers closed when not in use.
- Be careful when transferring flammable and toxic materials; only use approved storage containers (Figure 1-14).
- Promptly discard or clean all empty solvent containers; the fumes may be flammable.
- Never use torches or smoke near flammable chemicals and solvents, including battery acid.
- Store solvents, chemicals, and other flammable materials in approved safety cabinets (Figure 1-15).

HAZARDOUS WASTE

An automotive repair shop generates hazardous and environmentally damaging wastes. These include used engine oil, brake fluid, antifreeze, power steering fluid, and transmission fluids. Used oil filters must also be handled properly to minimize their contamination of the environment. The U.S. and Canadian **Environmental Protection Agencies (EPA)** are national agencies that offer guidelines and set regulations for the handling of hazardous waste. Many states and provinces have regulations in addition to those that are instituted by the EPA; be sure to check all applicable laws before disposing of waste that could be hazardous.

Be careful when draining fluids from a vehicle. Do not let fluid stream to the ground and get into the shop drains. Use an appropriate absorbent to promptly clean up spills. Fluids left on the floor create a slipping hazard. Some absorbents may become hazardous waste, depending on the fluid they have absorbed. Check all applicable regulations before disposing of this material. Store used fluids in approved containers. They should be clearly labeled and sealed after each use. Do not mix fluids into one container. Used antifreeze should be recycled whenever possible; if not, it must be disposed of as a hazardous waste. Some shops have in-house antifreeze recovery, recycling, and refilling equipment for cooling system service. Used engine oil is also a hazardous waste that must be recycled or disposed of properly. Some shops use heating systems that are fueled by used engine oil. Be sure that the storage container is in good condition; check regularly for leaks. Use an above-ground storage container whenever possible. Used oil filters should be crushed or drained for twelve hours to remove as much of the toxic oil as possible. Punch a small hole in the domed end of the filter to assist the draining. Recycle used oil filters whenever possible.

VEHICLE OPERATION

When the customer brings a vehicle in for service, certain shop safety guidelines should be followed. For example, when moving a car into the shop, check the brakes before beginning. Then buckle the safety belt. Drive carefully in and around the shop. Make sure no one is near, that the way is clear, and that there are no tools or parts under the car before you start the engine.

When road testing the car, obey all traffic laws. Drive only as far as is necessary to check the automobile. Never make excessively quick starts, turn corners too quickly, or drive faster than conditions allow.

If the engine must be running while working on the car, block the wheels to prevent the car from moving. Place the transmission into park for automatic transmissions or in neutral for manual transmissions. Set the emergency brake. Never stand directly in front of or behind a running vehicle.

Run the engine only in a well-ventilated area, to avoid the danger of poisonous **carbon monoxide (CO)** in the engine exhaust. If the shop is equipped with an exhaust ventilation system (Figure 1-16), use it. If not, use a hose and direct the exhaust out of the building.

WARNING: **To prevent personal injury, never run the engine in a vehicle inside the shop without an exhaust hose connected to the tailpipe.**

Vehicle exhaust contains small amounts of carbon monoxide, which is a poisonous gas. Strong concentrations of carbon monoxide may be fatal for human beings. All shop personnel are responsible for air quality in the shop. Shop management is responsible for providing an adequate exhaust system to remove exhaust fumes from the maximum number of vehicles that may be running in the shop at the same time. Technicians should never run a vehicle in the shop unless a shop exhaust hose is installed on the tailpipe of the vehicle. The exhaust fan must be switched on to remove exhaust fumes.

If shop heaters or furnaces have restricted chimneys, they release carbon monoxide emissions into the shop air. Therefore, chimneys should be checked periodically for restriction and proper ventilation.

Monitors are available to measure the level of carbon monoxide in the shop. Some of these monitors read the amount of carbon monoxide present in the shop air, and other monitors provide an audible alarm if the concentration of carbon monoxide exceeds the danger level.

> **Carbon monoxide (CO)** is an odorless, poisonous gas. When it is breathed in, it can cause headaches, nausea, ringing in your ears, tiredness, and heart flutter. Heavy amounts of CO can kill you.

FIGURE 1-16 Exhaust vent hose to a tailpipe.

TERMS TO KNOW

Abrasive cleaning

Carbon monoxide (CO)

Chemical cleaning

Environmental Protection Agency (EPA)

Floor jack

Hydraulic press

Jack stands

Material Safety Data Sheets (MSDS)

Occupational Safety and Health Administration (OSHA)

Diesel exhaust contains some carbon monoxide, but particulates are also present in the exhaust from these engines. **Particulates** are small carbon particles, which can be harmful to the lungs.

The **sulfuric acid** solution in car batteries is a very corrosive, poisonous liquid. If a battery is charged with a fast charger at a high rate for a period of time, the battery becomes hot, and the sulfuric acid solution begins to boil. Under this condition, the battery may emit a strong sulfuric acid smell, and these fumes may be harmful to the lungs. If this condition occurs in the shop, the battery charger should be turned off or the charger rate should be reduced considerably.

When an automotive battery is charged, hydrogen gas and oxygen gas escape from the battery. If these gases are combined, they form water, but hydrogen gas by itself is very explosive. While a battery is charged, sparks, flames, and other sources of ignition must not be allowed near the battery.

WORK AREA SAFETY

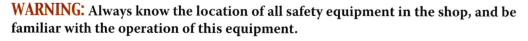

WARNING: Always know the location of all safety equipment in the shop, and be familiar with the operation of this equipment.

Your work area should be kept clean and safe. The floor and bench tops should be kept clean, dry, and orderly. Any oil, coolant, or grease on the floor can make it slippery. Slips can result in serious injuries. To clean up oil, use a commercial oil absorbent. Keep all water off the floor. Water is slippery on smooth floors, and electricity flows well through water. Aisles and walkways should be kept clean and wide enough to move through easily. Make sure the work areas around machines are large enough to operate the machine safely.

Proper ventilation of space heaters, used in some shops, is necessary to reduce the CO levels in the shop. Also, proper ventilation is very important in areas where volatile solvents and chemicals are used. (A volatile liquid is one that vaporizes very quickly.)

Keep an up-to-date list of emergency telephone numbers clearly posted next to the telephone. These numbers should include a doctor, hospital, and fire and police departments. Also, the work area should have a first-aid kit for treating minor injuries. There should also be eye flushing kits readily available.

Gasoline is a highly flammable volatile liquid. Always keep gasoline or diesel fuel in an approved safety can, and never use it to clean your hands or tools. Oily rags should also be stored in an approved metal container. When these oily, greasy, or paint-soaked rags are left lying about or are not stored properly, they can cause spontaneous combustion. Spontaneous combustion results in a fire that starts by itself, without a match.

Make sure that all drain covers are snugly in place. Open drains or covers that are not flush to the floor can cause toe, ankle, and leg injuries.

Know where the fire extinguishers are and what types of fires they put out (Figure 1-17). A multipurpose dry chemical fire extinguisher will put out ordinary combustibles, flammable liquids, and electrical fires. Never put water on a gasoline fire. The water will just spread the fire. Use a fire extinguisher to smother the flames. Remember, during a fire, never open doors or windows unless it is absolutely necessary; the extra draft will only make the fire worse. A good rule is to call the fire department first, and then attempt to extinguish the fire.

To extinguish a fire, stand six to ten feet from the fire. Hold the extinguisher firmly in an upright position. Aim the nozzle at the base and use a side-to-side motion, sweeping the entire width of the fire. Stay low to avoid inhaling the smoke. If it gets too hot or too smoky, get out. Remember, never go back into a burning building for anything.

Shops that employ ASE-certified technicians display an official ASE blue seal of excellence. This blue seal increases the customer's awareness of the shop's commitment to quality service and the competency of certified technicians.

	Class of Fire	Typical Fuel Involved	Type of Extinguisher
Class A Fires (green)	**For Ordinary Combustibles** Put out a class A fire by lowering its temperature or by coating the burning combustibles.	Wood Paper Cloth Rubber Plastics Rubbish Upholstery	Water*[1] Foam* Multipurpose dry chemical[4]
Class B Fires (red)	**For Flammable Liquids** Put out a class B fire by smothering it. Use an extinguisher that gives a blanketing, flame-interrupting effect; cover whole flaming liquid surface.	Gasoline Oil Grease Paint Lighter fluid	Foam* Carbon dioxide[5] Halogenated agent[6] Standard dry chemical[2] Purple K dry chemical[3] Multipurpose dry chemical[4]
Class C Fires (blue)	**For Electrical Equipment** Put out a class C fire by shutting off power as quickly as possible and by always using a nonconducting extinguishing agent to prevent electric shock.	Motors Appliances Wiring Fuse boxes Switchboards	Carbon dioxide[5] Halogenated agent[6] Standard dry chemical[2] Purple K dry chemical[3] Multipurpose dry chemical[4]
Class D Fires (yellow)	**For Combustible Metals** Put out a class D fire of metal chips, turnings, or shaving by smothering or coating with a specially designed extinguishing agent.	Aluminum Magnesium Potassium Sodium Titanium Zirconium	Dry powder extinguishers and agents only

*Cartridge-operated water, foam, and soda-acid types of extinguishers are no longer manufactured. These extinguishers should be removed from service when they become due for their next hydrostatic pressure test.

Notes:

(1) Freeze in low temperatures unless treated with antifreeze solution, usually weighs over 20 pounds, and is heavier than any other extinguisher mentioned.

(2) Also called ordinary or regular dry chemical (sodium bicarbonate).

(3) Has the greatest initial fire-stopping power of the extinguishers mentioned for class B fires. Be sure to clean residue immediately after using the extinguisher so sprayed surfaces will not be damaged (potassium bicarbonate).

(4) The only extinguishers that fight A, B, and C classes of fires. However, they should not be used on fires in liquefied fat or oil of appreciable depth. Be sure to clean residue immediately after using the extinguisher so sprayed surfaces will not be damaged (ammonium phosphates).

(5) Use with caution in unventilated, confined spaces.

(6) May cause injury to the operator if the extinguishing agent (a gas) or the gases produced when the agent is applied to a fire is inhaled.

FIGURE 1-17 Guide to fire extinguisher selection.

CASE STUDY

A technician installed a rebuilt engine in a vehicle. Before completing the installation, the technician reconnected the negative battery cable. While installing the drive belts, the technician leaned on top of the alternator with one arm. The protective boot had been pushed aside on the alternator battery terminal, and the technician's metal wristwatch strap completed the circuit from the alternator battery terminal to ground. This action caused a very high-current flow through the wristwatch strap, and severely burned the technician's arm. From this experience, the technician learned:

1. Do not wear a wristwatch while working in the shop.
2. When installing an engine, do not reconnect the battery until the installation is complete and you are ready to start the engine.

TERMS TO KNOW
(continued)

Particulates
Pneumatic tools
Power tools
Right-To-Know Laws
Safety glasses
Sulfuric acid
Thermal cleaning
Workplace Hazardous
Materials Information
Systems (WHMIS)
Z-87.1

ASE-STYLE REVIEW QUESTIONS

1. *Technician A* says it is recommended that you wear shoes with non-slip soles in the shop.

    Technician B says steel-toed shoes offer the best foot protection.

    Who is correct?

    A.   A only   C.   Both A and B
    B.   B only   D.   Neither A nor B

2. *Technician A* says that some machines can be routinely used beyond their stated capacity.

    Technician B says that a power tool can be left running unattended if the technician puts up a power "on" sign.

    Who is correct?

    A.   A only   C.   Both A and B
    B.   B only   D.   Neither A nor B

3. *Technician A* ties his long hair behind his head while working in the shop.

    Technician B covers her long hair with a brimless cap.

    Who is correct?

    A.   A only   C.   Both A and B
    B.   B only   D.   Neither A nor B

4. *Technician A* uses compressed air to blow dirt from his clothes and hair.

    Technician B says this should only be done outside.

    Who is correct?

    A.   A only   C.   Both A and B
    B.   B only   D.   Neither A nor B

5. While discussing the proper way to lift a heavy object:

    Technician A says that you should bend at your knees if you are going to pick up a heavy object.

    Technician B says that if the object is too heavy then a wheel dolly should be used.

    Who is correct?

    A.   A only   C.   Both A and B
    B.   B only   D.   Neither A nor B

6. While discussing shop cleaning equipment safety:

    Technician A says some hot tanks contain caustic solutions.

    Technician B says some metals such as aluminum may dissolve in hot tanks.

    Who is correct?

    A.   A only   C.   Both A and B
    B.   B only   D.   Neither A nor B

7. While discussing shop rules:

    Technician A says breathing carbon monoxide may cause arthritis.

    Technician B says breathing carbon monoxide may cause headaches.

    Who is correct?

    A.   A only   C.   Both A and B
    B.   B only   D.   Neither A nor B

8. While discussing air quality:

    Technician A says a restricted chimney on a shop furnace may cause carbon monoxide gas in the shop.

    Technician B says monitors are available to measure the level of carbon monoxide in the shop air.

    Who is correct?

    A.   A only   C.   Both A and B
    B.   B only   D.   Neither A nor B

9. While discussing air quality:

    Technician A says a battery gives off hydrogen gas during the charging process.

    Technician B says a battery gives off oxygen gas during the charging process.

    Who is correct?

    A.   A only   C.   Both A and B
    B.   B only   D.   Neither A nor B

10. While discussing hazardous materials:

    Technician A says that used engine oil is considered a hazardous waste.

    Technician B says that spill and leak procedures are covered in the MSDS.

    Who is correct?

    A.   A only   C.   Both A and B
    B.   B only   D.   Neither A nor B

ASE CHALLENGE QUESTIONS

1. *Technician A* says that compressed air can penetrate your skin if you spray yourself with an air nozzle.

   Technician B says that if compressed air gets into your bloodstream it can be fatal.

   Who is correct?

   A. A only   C. Both A and B

   B. B only   D. Neither A nor B

2. *Technician A* says that you should wear Z-87.1 safety glasses while working on or under vehicles.

   Technician B says that if you are hammering on a component on the bench, you should take your safety glasses off so you can see it better.

   Who is correct?

   A. A only   C. Both A and B

   B. B only   D. Neither A nor B

3. *Technician A* says that when a fire starts it is important to open the bay doors for adequate ventilation.

   Technician B says that a Class C fire extinguisher can be used for most ordinary combustibles.

   Who is correct?

   A. A only   C. Both A and B

   B. B only   D. Neither A nor B

4. *Technician A* says to lift vehicles by the front and rear frames.

   Technician B says that some vehicles will permanently twist if lifted improperly.

   Who is correct?

   A. A only   C. Both A and B

   B. B only   D. Neither A nor B

5. *Technician A* says that hazardous waste should be stored underground until pickup.

   Technician B says to check local laws; some hazardous wastes can be poured down the floor drain.

   Who is correct?

   A. A only   C. Both A and B

   B. B only   D. Neither A nor B

Name _____   **Date** _____

SHOP SAFETY SURVEY

As a professional technician, safety should be one of your first concerns. This job sheet should increase your awareness of shop safety rules and safety equipment. As you survey your shop area and answer the following questions, you should learn how to evaluate the safeness of your workplace.

Procedure

Your instructor will review your progress throughout this worksheet and should sign off on the sheet when you complete it.

Task Completed

1. Before you begin to evaluate your work area, evaluate yourself. Are you dressed to work safely?   ☐ Yes   ☐ No

   If no, what is wrong? _____

   _____

2. Are your safety glasses OSHA approved?   ☐ Yes   ☐ No

   Do they have side protection shields?   ☐ Yes   ☐ No

3. Look around the shop, and note any area that poses a potential safety hazard or is an area that you should be aware of.

   _____

   _____

   _____

   Any true hazards should be brought to the attention of the instructor immediately.

4. Are there safety areas marked around grinders and other machinery?
   ☐ Yes   ☐ No

5. What is the line air pressure in the shop? _____ psi
   What should it be? _____ psi

6. Where are the tools stored in the shop? _____

   _____

7. If you could, how would you improve the tool storage area?

   _____

   _____

8. What types of hoists are used in the shop? _____

   _____

9. Ask your instructor to demonstrate the proper use of the hoist.   ☐

10. Where is/are the first-aid kit(s) kept in the work area?

    _____

    _____

11. What is the shop's procedure for dealing with an accident?

_____

_____

_____

12. Where are the fire extinguishers located in your shop?

_____

_____

What class of fires do they extinguish?

_____

13. Where is the electrical emergency shut-off?

_____

_____

14. Is the eyewash station in clear view?   ☐ Yes   ☐ No

☐

Instructor's Response _____

_____

_____

Name _____ **Date** _____

PROPER HANDLING OF HAZARDOUS MATERIALS

Upon completion of this job sheet, you should be familiar with your shop's procedures for handling hazardous materials and hazardous waste. You should also be able to access information from the MSDS book in your shop.

Tools and Materials

Safety glasses
Aerosol solvent or chemical
Used engine oil container
Coolant container or recovery system

Procedure

1. Do you have your safety glasses on?   ☐ Yes   ☐ No

2. Locate the cabinet for flammable chemicals and solvent materials. Note all markings on the cabinet and/or storage area.

   _____

   _____

   _____

3. Remove a chemical spray can such as Brake Kleen from the cabinet. Is it clearly labeled?
   ☐ Yes   ☐ No
   What is the name of the product?

   _____

4. Locate the MSDS book. Is it labeled and stored in an accessible location?
   ☐ Yes   ☐ No
   Where is it located?

   _____

5. Look up the MSDS for the chemical spray you have out. Is the MSDS available?
   ☐ Yes   ☐ No
   Briefly describe the health hazards and safety precautions for this product:

   _____

   _____

   _____

6. Where is the used oil storage receptacle located?

   _____

   Is it clearly labeled?   ☐ Yes   ☐ No
   Is it properly sealed?   ☐ Yes   ☐ No

7. How does your shop deal with used oil filters?

_____

_____

_____

8. How does your shop deal with used coolant?

_____

_____

_____

Instructor's Response _____

_____

_____

JOB SHEET

Name _____ **Date** _____

LIFTING A FULL FRAME VEHICLE

Upon completion of this job sheet, you should be able to properly demonstrate how to lift a full frame vehicle.

Tools and Materials
Full Frame Vehicle
Shop lift

Describe the vehicle being worked on.
Year _____ Make _____ Model _____
VIN _____ Engine type and size _____

Procedure
Your instructor will provide you with a specific vehicle. Write down the information, then perform the following tasks:

1. Locate the instruction guide for the lift you are using, and read through the instructions.
   Explain the operation of the lift and its safety mechanism to your instructor. Is the student explanation satisfactory?   ☐ Yes   ☐ No

2. Locate the lift points for the vehicle you are working on and describe them:

   _____
   _____
   _____
   _____

3. Place the lift pads under the designated spots, and raise the lift until the pads are just contacting the vehicle. Are they properly positioned?   ☐ Yes   ☐ No
   If not, lower the vehicle and reposition the pads.

4. Raise the vehicle one foot off the floor, and jounce the vehicle. Listen for any unusual noises, and watch for improper movement.
   Describe your results: _____
   _____

   Have your instructor check your work. Is the vehicle secure on the lift?
   ☐ Yes   ☐ No

5. Raise the vehicle overhead, and engage any manual locks. Does your lift have manual locks?   ☐ Yes   ☐ No
   Describe the procedure for engaging the lift locks:

   _____
   _____

6. Identify as many of the components under the vehicle as possible:

_____

_____

7. Make sure everyone and everything is clear of the vehicle. Describe the procedure to disengage the lift locks:

_____

_____

_____

8. Lower the vehicle to the floor, and remove the lift arms from under the vehicle.

Instructor's Response _____

_____

_____

Name _____ Date _____

LIFTING A BODY ON FRAME VEHICLE

Upon completion of this job sheet, you should be able to properly demonstrate how to lift a body on frame vehicle.

Tools and Materials
Body on frame vehicle
Shop lift

Describe the vehicle being worked on:
Year _____ Make _____ Model _____
VIN _____ Engine type and size _____

Procedure
Your instructor will provide you with a specific vehicle. Write down the information, then perform the following tasks:

1. Locate the instruction guide for the lift you are using, and read through the instructions. Explain the operation of the lift and its safety mechanism to your instructor. Is the student explanation satisfactory?　　☐ Yes　☐ No

2. Locate the lift points for the vehicle you are working on and describe them:

   _____
   _____
   _____
   _____

3. Place the lift pads under the designated spots, and raise the lift until the pads are just contacting the vehicle. Are they properly positioned?　　☐ Yes　☐ No
   If not, lower the vehicle and reposition the pads.

4. Raise the vehicle one foot off the floor, and jounce the vehicle. Listen for any unusual noises, and watch for improper movement.
   Describe your results: _____

   _____

   Have your instructor check your work. Is the vehicle secure on the lift?
   ☐ Yes　☐ No

5. Raise the vehicle overhead, and engage any manual locks. Does your lift have manual locks?　　☐ Yes　☐ No

6. Describe the procedure for engaging the lift locks:

   _____
   _____

7. With the hoist locks engaged, slowly lower the vehicle so that it is resting on the locks. Do all of the locks engage and hold the vehicle evenly?   ☐ Yes   ☐ No

8. Identify as many of the components under the vehicle as possible:

   _____

   _____

9. Make sure everyone and everything is clear of the vehicle. Describe the procedure to disengage the lift locks:

   _____

   _____

   _____

10. Lower the vehicle to the floor, and remove the lift arms from under the vehicle.

Instructor's Response _____

_____

_____

Chapter 2

BASIC TESTING, INITIAL INSPECTION, AND SERVICE WRITING

UPON COMPLETION OF THIS CHAPTER, YOU SHOULD BE ABLE TO:

- Describe the differences between basic testing and advanced testing.
- Perform a basic (initial) engine and vehicle inspection.
- Complete a service repair order.

- Understand how parts and labor are calculated.
- Explain the importance of the customer concern on a repair order.
- Understand how to use basic hand tools for engine repair.

In the engine repair and rebuilding industry, you will be faced with many different challenges. Each of which will require a different level and type of skill. Before you can read further into this book and complete most of the job sheets at the end of this chapter, you must be able to properly identify the engine components and understand the variety of the designs.

In some cases, the technician may have direct contact with the customer and in some cases they may not. Either way, it is important to understand that the customer's complaint should always be fixed when the job is completed. There are many reported cases where the customer's complaint, as written on the repair order, does not match the real complaint. Many times, this is because the customer and technician do not know how to communicate together—and the wrong thing gets written down. This then is passed on to the technician who may repair the incorrect item.

This is especially true with engine repair. A customer may come into the repair facility with an engine-related noise. The technician may suggest a lengthy and expensive repair, and when the repair is completed the customer may mention that the noise is still there. Although the expensive repair may have been needed, it was not why the customer came in. This is an example of a lack of communication and understanding of many processes that will eventually lead to the customer going somewhere else for their business and having distrust in the industry. Although this chapter is not intended to prepare you fully for a position as a service writer, it is very important to understand the role and communication of one. Remember, when you are a technician you not only represent the repair facility you work for, you represent the entire industry.

IDENTIFYING THE AREA OF CONCERN

There are many parts on the vehicle that wear out, break, fall off, make noise, get dirty, and need repair or replacement. It is your job as a technician to understand almost all of them and determine which area, and component, needs repair.

To do this, you must be able to focus in on one area of the vehicle. In engine repair and rebuilding, you must first be able to focus under the hood. When asking the customer what is wrong with their car, be sure that you understand them clearly. If language is an issue, get an assistant to help with communications.

Being able to communicate with a customer in technical terms proves to be the most challenging of all tasks. When trying to explain something or communicate with a customer, you should avoid using very technical terms and use everyday terms. Avoid using very general terms, like "that thing," as they may make the customer feel very inferior or substandard. Some customers will understand the word "piston" and be able to relate as an engine component, but they may not know what one looks like or what it does. You will have to be the judge of your actions when you are trying to understand what it is the customer needs.

When focusing in on one area, try to ask as many questions as possible. Make sure to get a phone number where the customer can be reached quickly. Having a good working relationship helps here because you can feel comfortable talking with the customer. If you are still trying to establish the area to focus in on with the customer, try test-driving the vehicle with the customer. Ask them to drive the vehicle and attempt to repeat the problem. Sometimes this may be difficult, but remember to be patient and calm (Figure 2-1).

Once you have determined the area to focus in on, make an attempt to zone out other areas that are not related. This is tricky, especially because some other areas may be related to the problem. An example of trying to zone out other nonrelated area is when the customer has an engine noise, but it only makes noise when you accelerate hard from a dead stop. If the customer has a lot of papers and things laying on the dash or center console, those papers may make a lot of annoying noises when accelerating hard. You definitely don't want to clean it up or move it for the customer, but a phone call or simply asking the customer to move them so that you can hear the engine better is polite.

FIGURE 2-1 Sometimes, a test-drive with the customer is necessary.

FIGURE 2-2 A chassis ear is a good tool to help focus in on one area.

Sometimes using a **chassis ear** machine (Figure 2-2) works well when trying to figure out where the noises are coming from. You can attach the ends of the chassis ear to components on the engine and then bring the wires into the passenger compartment and listen to them while driving. The chassis ear is designed for listening to suspension, steering, and chassis problems but can be used for engines as well. Be careful when using this tool because you don't want to hear suspension or steering problems and assume they are engine problems. Make sure to fully read the directions with the chassis ear.

Once you have focused in on the problem area of the engine, ask the customer about their past service history. If they have had other major problems like a head gasket wear, overheating, or oil pressure problems before then ask what repairs they have completed. They may have the old receipts and repair orders from other shops. These are all clues to the puzzle. If things are not adding up, try asking the owner if the vehicle has a salvage or flood title. Although this seems strange, many flood and salvage vehicles can have preexisting damage from a previous accident that was never fully repaired. The body and exterior may be in great shape, but the engine may have suffered damage and never received the proper repairs (Figure 2-3).

A **chassis ear** machine uses several small microphones that can be placed under the hood and then routed to the passenger compartment where the driver can use a set of headphone.

TEST-DRIVING FOR ENGINE CONCERNS

It is often difficult to test-drive a vehicle that has an engine problem. The engine may not start at all, it may not be able to get up to speed, or it may make so much noise that it is simply not safe at all to drive it on the road. Each of these situations may not require a test-drive to determine the problem. But if the engine noise or problem is smaller than that, you may need a test-drive.

During the test-drive, a technician must be aware of everything. Your senses should be heightened during this procedure. Turn off the radio and pay attention. First try looking at the big picture. Looking at the overall condition of the vehicle may give you clues to the age of the engine. Watching the tailpipe for smoke may also give you clues to what problems you may be experiencing (Figure 2-4).

FIGURE 2-3 This car has a flood title. It may look normal from the outside, but the engine has a lot of water damage.

FIGURE 2-4 A lot of thick blue smoke may indicate major engine repairs.

During the test-drive, you may experience a problem that is not engine mechanic related. Maybe you believe the problem is related to the engine computer or fuel system. In these cases, you may want to bring a technician who specializes in engine performance diagnostics with you. You may also want to bring a scan tool with you to watch and record data (Figure 2-5). Also use common sense when test-driving a vehicle and never watch the scan tool while trying to drive.

FIGURE 2-5 It may be helpful to use a scan tool during the test-drive. Always pay attention to the road and use the recording function of the tool.

Most scan tools have a flight recording option so you can look at the data when you return to the shop. The test-drive should last as long as you need to complete a thorough evaluation. However, it should not be longer than it needs to. Often with engine problems, the test-drive is not short because the problem doesn't always occur right away. Only test-drive the vehicle as long as you need to. Too short of a test-drive may mean that the real engine problem didn't occur.

WRITING A SERVICE REPAIR ORDER

Writing a service **repair order** is one of the most important steps of repairing any vehicle. It is always the first step, and it involves good communication with the customer and allows the technician to clearly analyze and focus on repairing what the vehicle actually came in for.

A service repair order is a recorded document that is used for legal, tax, and general recording purposes. Other names for a service repair order are a service record, work order, or repair order (RO). There are usually a few copies of the repair order, one of which is given to the technician. It is the service writer's job to properly communicate with the customer and write down as much related information as possible on it. The service writer may also be the responsible person for providing an estimate and/or completing the money transaction of the sale. The service writer may also need to translate the customer's perception of the problem and ask key questions. For example, the customer may state that the engine makes a loud ticking noise. If this is what is actually written on the repair order for the technician to read, then they will be spending a lot of time looking for the problem. And when they find something wrong, it may not be what the customer is actually complaining about.

The service writer needs to be able to translate and identify this from the customer and ask key questions: such as when does the noise occur, how loud is it, do you hear it when you are accelerating or any other type of driving, does the noise go away or get louder as you drive, and other questions. If the repair order does not clearly state the concern, a misdiagnosis may occur. This could cost the repair facility money, customer trust, and their reputation.

A **repair order** is a written legal document that describes the concern (customer complaint), the cause (what needs repair or what is wrong), and the correction that has been made.

The repair order also contains vital information such as the labor rate, parts cost, estimate of total cost, when the vehicle will be ready, the VIN, year, make, model, license plate number, and the mileage. It is important to check these items just as you bring the vehicle into the shop for their accuracy. The last thing that anyone needs is to get things mixed up and repair the wrong vehicle.

The most important part of the repair order is that it communicates the "3C's": concern, cause, and correction. The concern is what the customer says is wrong with the vehicle. The technician diagnoses the vehicle and completes the "cause" section. This section should be detailed, clean, and well written. The correction is the list of parts and the labor performed that describes how the vehicle was repaired.

The technician may in some cases have to intervene with the service writer and communicate with the customer to find the source of the problem. The technician, the service writer, and the service manager may have to work together using their experiences, the service manual, and online resources to find the cause of the problem. The repair order is a legal document. Make sure that everything is written clearly and correctly.

AUTHOR'S NOTE: Early in my career as a technician, I was preparing to perform a timing belt service on a car. I grabbed the repair order and found the keys hanging up on the key board with all the other vehicle keys. The key wasn't labeled, so I guessed at which one to grab. I went out to the parking lot, found the car, and drove it into the shop. I put it up on the hoist and continued with my repair. Once I was finished with the repair, I test-drove it and then wrote down the "mileage out" on the repair order (some repair orders track the mileage that the repair facility put on the car). I noticed that there must have been a mistake made by the service writer typing in the mileage since it looked like I put on close to 40,000 miles on my test-drive.

When I stepped out of the vehicle, I checked the license plate and the plate number was different than the one on the repair order. I went to the service writer and he mentioned that there are two cars in the lot today and they are the same car, same color, etc., and the only thing that was different was the mileage, license plate, and the work they were in for. After realizing that I made the mistake of grabbing the wrong keys and not verifying the correct vehicle, I pulled the wrong vehicle back into the shop and started to remove the timing belt. The customer came by and I explained the mistake I made. She stopped me right there and said, "my car needed the timing belt replaced soon anyways, keep it on."

CASE STUDY

A close relative of mine had recently brought her van into a service center because the service engine soon light was on and the engine was running rough. The service center called her back a few hours later and told her that the repair would cost around $1,400.00 and that most of the computer parts, ignition parts, and fuel system parts had went bad, all at once. The work was authorized and the parts were installed. After all of the parts were installed, the engine still ran rough and the light was still illuminated. After complaining to the service manager, a technician found that a simple diagnostic step was skipped. After performing this easy step, the technician found a broken valve spring. The customer was very disappointed, as well as the service manager. Often technicians get wrapped up in looking at the most complicated systems and components as the problem and forget to perform the basic tests. This situation could have gone smoother and cost the customer much less if there was better communication and the correct steps were taken in the first place.

ASE-STYLE REVIEW QUESTIONS

1. Some of the vital information on the repair order is:
   A. The VIN
   B. The mileage
   C. The year, make, model
   D. All of the above.

2. The repair order may also be called:
   A. A work order
   B. A service record
   C. Both A and B
   D. Neither A nor B

3. The test-drive:
   A. Allows a technician to always try driving different vehicles
   B. Can be useful when diagnosing engine problems
   C. Should always be short
   D. Is not important

4. The service writer needs to:
   A. Communicate with the technician and the customer.
   B. Have a clear understanding of what the customer's concern is.
   C. Explain the work that is to be performed to the customer.
   D. All of the above.

5. *Technician A* says that writing the repair order is usually the first step in repairing a vehicle.
   Technician B says that the repair order is a legal document.
   Who is correct?
   A. A only
   B. B only
   C. Both A and B
   D. Neither A nor B

6. *Technician A* says that the technician will complete the "cause" section of the repair order.
   Technician B says that the technician may have to communicate with the customer to find the problem.
   Who is correct?
   A. A only
   B. B only
   C. Both A and B
   D. Neither A nor B

7. Who might the technician need to work with to find the cause of a problem?
   A. The service writer
   B. The customer
   C. The service manager
   D. All of the above

8. The "cause" section of the repair order contains:
   A. The parts and labor cost for repairing the vehicle
   B. The amount of sales tax the customer should be charged
   C. The reason the vehicle came into the repair facility
   D. The technician's diagnosis

9. *Technician A* says that basic testing can include the use of a scan tool.
   Technician B says that basic testing can include a chassis ear machine.
   Who is correct?
   A. A only
   B. B only
   C. Both A and B
   D. Neither A nor B

10. A misdiagnosis may occur if the:
    A. Repair order is not clearly written
    B. Year, make, or model are incorrect
    C. Technician is not able to focus in on the area of concern
    D. All of the above

ASE CHALLENGE QUESTIONS

1. The service writer's duties are being discussed.
   Technician A says that the service writer may be the one who provides the repair estimate for the customer.
   Technician B says that the service writer usually completes the repair order.
   Who is correct?
   A. A only
   B. B only
   C. Both A and B
   D. Neither A nor B

2. *Technician A* says that the service writer initiates the repair order.
   Technician B says that the technician initiates the repair order.
   Who is correct?
   A. A only
   B. B only
   C. Both A and B
   D. Neither A nor B

3. When diagnosing an engine concern it is usually best to:
   A. Focus in on one area
   B. Communicate the concern on the repair order with the service writer after the problem was found
   C. Both A and B
   D. Neither A nor B

4. *Technician A* says that the repair order is a legal document.
   Technician B says that the repair order should be used as a communication tool with the customer.
   Who is correct?
   A. A only
   B. B only
   C. Both A and B
   D. Neither A nor B

5. In order, the "3C's" of a repair order are:
   A. Cause, compile, correct
   B. Cause, concern, correction
   C. Concern, cause, correction
   D. Correct, cause, compile

Name _____ **Date** _____

Engine Component Review

Upon completion of this job sheet, you should be able to properly identify several engine components and their basic function.

Tools and Materials

Engine components

Describe the vehicle being worked on:

Year _____ Make _____ Model _____

VIN _____ Engine type and size _____

Procedure

Your instructor will provide you with a variety of engine components. Write down the name of each one and describe one function of each.

1. Name of component _____

   Function _____

   _____

2. Name of component _____

   Function _____

   _____

3. Name of component _____

   Function _____

   _____

4. Name of component _____

   Function _____

   _____

5. Name of component _____

   Function _____

   _____

6. Name of component _____

   Function _____

   _____

7. Name of component _____

   Function _____

   _____

   Instructor's Response _____

   _____

   _____

Name _____ Date _____

TEST-DRIVE AND BASIC INSPECTION

Upon completion of this job sheet, you should be able to properly perform a basic engine and vehicle inspection and understand how to perform a test-drive for engine concerns.

ASE Correlation

This job sheet is related to ASE Engine Repair Test content area: General Engine Diagnosis; Removal and Reinstallation (R&R): Identify and interpret engine concern; determine necessary action.

Tools and Materials

Vehicle

Describe the vehicle being worked on:

Year _____ Make _____ Model _____

VIN _____ Engine type and size _____

Procedure

Complete the job sheet using the vehicle chosen. Consult your instructor before you are ready to perform a test-drive.

Task Completed

1. If there is a customer complaint, list it here.

   _____

2. Open the hood and start the engine. Describe how the engine sounds (good, idles rough, makes noises, etc.).

   _____

   _____

3. Is there any excessive smoke coming from the exhaust tailpipe? _____

4. Turn the engine off and check the major fluid levels, such as the oil and coolant. Describe your findings.

   _____

   _____

5. Perform a quick inspection of the vehicle to make sure that it is road worthy for a test-drive. ☐

6. During your test-drive, you should simulate the problem and make note of what happens during the test-drive. ☐

⚠️ **CAUTION:**
Do not attempt to write when driving. Always follow all traffic laws and drive safe.

7. After your test-drive has been completed, write down your findings. Were you able to repeat the customer's concern? If yes, describe what problems you found during the test-drive.

_____

_____

_____

Instructor's Response _____

_____

_____

Name _____ **Date** _____

WRITING A SERVICE REPAIR ORDER

Upon completion of this job sheet, you should be able to properly write and complete a service order.

ASE Correlation

This job sheet is related to ASE Engine Repair Test content area: General Engine Diagnosis; Removal and Reinstallation (R&R): Complete work order to include customer information, vehicle identifying information, customer concern, related service history, cause, and correction.

Tools and Materials

Vehicle

Service repair work order or specialized computer software program

Describe the vehicle being worked on:

Year _____ Make _____ Model _____

VIN _____ Engine type and size _____

Procedure

Task Completed

Your instructor will assign a vehicle to you. Complete the job sheet using that vehicle by completing a repair order. If one is not available, use the generic form in this job sheet.

1. Complete the customer information section. This should include the name, address, contact numbers, and the best time available to call. ☐

2. Complete the vehicle information. This should include the year, make, model, VIN, mileage, engine type and size, and any other related information. ☐

3. Briefly inspect the vehicle for previously existing body or engine damage. Mark any findings on repair order in the notes section. ☐

4. Ask the customer (or the instructor) what the concern for bringing the vehicle in is. ☐

5. Ask the customer (or the instructor) if there is any previous work or history related to this concern. ☐

6. Ask your instructor to add diagnostic time, parts, and labor to the repair order. Complete the repair order by adding tax. ☐

Customer Information

Name:		Phone:		
Address:		Best time to contact:		
Date/time received:			Date/time promised:	

Vehicle Information

Year:		Model:	
Make:		VIN:	
License:		Engine:	
Mileage:		Notes:	

Customer Concern

Previous Repairs Completed

Repairs Performed

Part #	Amount		Labor Item	Amount

Total Parts	Total Tax	Sales Tax	Total Due

Instructor's Response _____

_____

_____

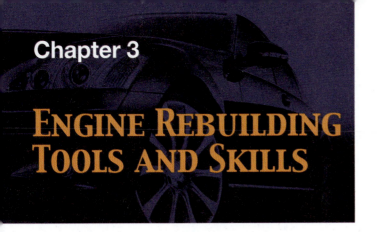

Chapter 3

ENGINE REBUILDING TOOLS AND SKILLS

BASIC TOOLS
Basic mechanic's tool set
Service information

UPON COMPLETION AND REVIEW OF THIS CHAPTER, YOU SHOULD BE ABLE TO:

- Explain the purpose and use of precision measuring instruments.

- Describe the purpose of special measuring instruments, such as the telescoping gauge, small-hole gauge, valve guide bore gauge, valve seat run-out gauge, and cylinder bore dial gauge.

- Describe the types of tools used for cylinder head reconditioning.

- Describe the tools and equipment used to recondition the engine block.

- Explain the purpose and use of connecting rod and piston reconditioning equipment and tools.

- Describe the purpose of special hand tools, including torque wrenches, torque angle gauges, ring expanders, ring-groove cleaners, and ring compressors.

- Describe the purpose and use of a scan tool for basic engine diagnostics.

- Understand the importance of how math is used in engine rebuilding and measuring.

- Research applicable vehicle and service information.

- Describe the importance of professionalism and responsibility in repairing and rebuilding engines.

- Understand typical agreements in the employer/ employee relationship.

- Recognize the challenge and benefits of ASE certification.

During the process of rebuilding or repairing an automotive engine, today's technician will be required to use some specialized tools. Most engine components require accurate measuring to determine if they are within factory specifications. These include the cylinder bores, crankshaft, crankshaft bores, camshaft bores, connecting rod journals, valve guides, and oil pump gears, to name a few. To perform this task, today's technician must be able to use a variety of measuring tools properly.

If wear is outside of specifications, some components may be reconditioned. Special equipment is available to recondition crankshafts, camshafts, connecting rods, valve faces, valve seats, and several other engine components. These operations require the use of line bores, valve machines, grinders, and other specialty tools.

Before you can determine if engine components are within specifications, you must be able to locate these details in the service information. The service information is an invaluable tool.

Repairing and rebuilding today's highly technical engines is challenging and rewarding work. In order to succeed as an engine repair technician you will need to develop numerous personal skills and excellent work habits. Communication skills, professionalism, critical thinking, and attention to detail are just a few of the critical skills required for success in the field. Expertise comes from a combination of well-honed technical and personal talents.

This chapter introduces you to many of the special measuring tools, hand tools, and equipment used to repair or recondition an engine. Many of these tools are essential to properly rebuild the engine. Some are designed for a specific function and may not be required, but make the job faster and easier. A study into the use of service information will assist you in locating the information you need. Finally, we'll discuss how you can become a successful, professional technician in today's repair facilities.

UNITS OF MEASURE

Two systems of weights and measures are commonly used in the United States. One system of weights and measures is the **United States Customary (USC)** system. Well-known measurements for length in the USC system are the inch, foot, yard, and mile. In this system, the quart and gallon are common measurements for volume, and ounce, pound, and ton are measurements for weight. A second system of weights and measures is referred to as the metric system.

In the USC system, the basic linear measurement is the yard, whereas the corresponding linear measurement in the metric system is the meter (Figure 3-1). Each unit of measurement in the metric system is related to the other metric units by a factor of ten. Thus, every metric unit can be multiplied or divided by ten to obtain larger units (multiples) or smaller units (submultiples); for example, the meter may be divided by ten to obtain centimeters (1/100 meter) or millimeters (1/1,000 meter).

The U.S. government passed the Metric Conversion Act in 1975 in an attempt to move American industry and the general public to accept and adopt the metric system. The automotive industry has adopted the metric system, and in recent years, most bolts, nuts, and fittings on vehicles have been changed to metric. Some vehicles still have a mix of USC and metric bolts. Import vehicles have used the metric system for many years. Although the automotive industry has changed to the metric system, the general public in the United States has been slow to convert from the USC system to the metric system. One of the factors involved in this change is cost. What would it cost to change every highway distance and speed sign in the United States to be in kilometers? The answer to that question is probably hundreds of millions or billions of dollars.

Service technicians must be able to work with both the USC and the metric system. One **meter** (m) in the metric system is equal to 39.37 inches (in.) in the USC system. Some common equivalents between the metric and USC systems are these:

1 meter (m) = 39.37 inches

1 centimeter (cm) = 0.3937 inch

1 millimeter (mm) = 0.03937 inch

1 inch = 2.54 cm

1 inch = 25.4 mm

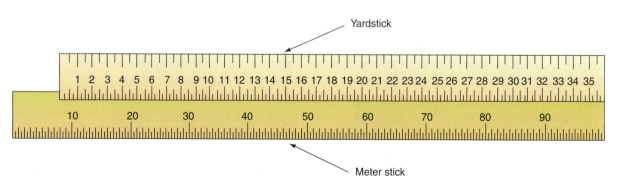

FIGURE 3-1 A meter is slightly longer than a yard.

1 ft. lb. = 1.35 newton meters (Nm)

1 in. lb. = 0.112 Nm

1 in. hg. = 3.38 kilopascals (kPa)

1 psi = 6.89 kPa

1 mile = 1.6 kilometer (km)

1 horsepower (hp) = 0.746 kilowatt (kW)

degrees F − 32 divided by 1.8 = degrees Celsius (C); example, 212 °F − 32 = 180 divided by 1.8 = 100 °C

Appendix D provides a more comprehensive conversion chart.

In the USC system, phrases such as 1/8 of an inch are used for measurements. The metric system uses a set of prefixes; for example, in the word kilometer, the prefix kilo indicates one thousand, and this prefix indicates there are one thousand meters in a kilometer. Common prefixes in the metric system follow:

Name	Symbol	Meaning
mega	M	one million
kilo	k	one thousand
hecto	h	one hundred
deca	da	ten
deci	d	one tenth of
centi	c	one hundredth of
milli	m	one thousandth of
micro	μ	one millionth of

Measurement of Mass

In the metric system, mass is measured in grams, kilograms, or tonnes. One thousand grams (g) = 1 kilogram (kg). In the USC system, mass is measured in ounces, pounds, or tons. When converting pounds to kilograms, 1 pound = 0.453 kilogram.

Measurement of Length

In the metric system, length is measured in millimeters, centimeters, meters, or kilometers. Ten millimeters (mm) = 1 centimeter (cm). In the USC system, length is measured in inches, feet, yards, or miles. When distance conversions are made between the two systems, some of the conversion factors are these:

1 inch = 25.4 millimeters

1 foot = 30.48 centimeters

1 yard = 0.91 meter

1 mile = 1.60 kilometers

Measurement of Volume

In the metric system, volume is measured in milliliters, cubic centimeters, and liters. One cubic centimeter = 1 milliliter. If a cube has a length, depth, and height of 10 centimeters (cm), the volume of the cube is 10 cm × 10 cm × 10 cm = 1,000 cm³ = 1 liter. When volume conversions are made between the two systems, 1 cubic inch = 16.38 cubic centimeters. If an engine has a displacement of 350 cubic inches, 350 × 16.38 = 5,733 cubic centimeters, and 5,733/1,000 = 5.7 liters.

ENGINE DIAGNOSTIC TOOLS

As the trend toward the integration of ignition, fuel, and emission systems progresses, diagnostic test equipment must also keep up with these changes. New tools and techniques are constantly being developed to diagnose electronic engine control systems.

Today's technician must not only keep up with changes in automotive technology, but must also keep up with the new testing procedures and specialized diagnostic equipment. To be successful, a shop must continuously invest in this equipment and the training necessary to troubleshoot today's electronic engine systems.

Not all engine performance problems are related to electronic control systems. Therefore, technicians still need to understand basic engine tests. These tests are an important part of modern engine diagnosis.

Scan Tools

A **scan tool** connects to the vehicle's **diagnostic link connector (DLC)**, (Figure 3-2). It communicates with the **powertrain control module (PCM)**. While a scan tool cannot diagnose internal engine problems, it can give you information about engine performance issues. A scan tool can be very useful in diagnosing fuel and ignition system faults. Scan tools can be used on virtually any vehicle with a computer control system. Cars and light trucks that are of model year 1996 and newer are designated as **OBD II (on board diagnostics)** vehicles. These vehicles meet an EPA requirement for emissions and also have a standardized computer system and **DLC**. The DLC on an OBD II vehicle is found under the dash on the driver's side (Figure 3-3). It is a 16-pin connector (Figure 3-4). There are a select few vehicles that were produced in 1996 and 1997 that were not fully OBD II compliant.

Cars that are computer controlled and produced earlier than 1996 are labeled as OBD I. These cars are not standardized with their computer system and the location of the DLC. Some early OBD I DLCs may be located in uncommon places. There are many types of scan

A **scan tool** communicates with the powertrain control module to display diagnostic trouble codes and engine data.

The **diagnostic link connector (DLC)** is a connector, usually located under the dash on an OBD II vehicle, which provides access to the PCM.

The **powertrain control module (PCM)** is the vehicle's computer that controls engine performance and related functions such as fuel and spark control.

An **OBD II** vehicle meets an EPA requirement for emissions and also has a standardized computer system and DLC location.

FIGURE 3-2 A scan tool communicates with the PCM and displays DTCs and engine data.

FIGURE 3-3 The DLC on an OBD II vehicle is located under the drivers side dash

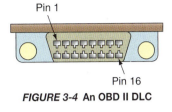

FIGURE 3-4 An OBD II DLC

tools that can be purchased. A global OBD II scan tool will be able to connect and communicate with OBD II-compliant vehicles. Global OBD II scan tools can provide the technician with enough information to repair the most common emissions-related faults. Some manufacturers sell an enhanced OBD II scan tool which can read more information from the vehicle's computer. There is also a manufacture's scan tool available for each vehicle. This scan tool will provide more data from the vehicle's computer, which sometimes makes it easier to diagnose a problem. Many manufacturers are now starting to use laptop and PC-based software to connect to the vehicle. The laptop then takes the place of a dedicated scan tool (Figure 3-5).

When the **malfunction indicator light (MIL)** is illuminated on an OBD II vehicle it means that the PCM has stored a **diagnostic trouble code (DTC)**. These alphanumeric codes can point you in the direction of a system fault. A P0302, for example, means that the PCM has detected a **misfire** on cylinder number two. It doesn't tell you the cause of the

The **malfunction indicator light (MIL)** is an amber light on the dash used to alert the driver of an emissions related powertrain problem.

A **diagnostic trouble code (DTC)** is a coded output from the PCM to identify system faults.

A **misfire** is when a combustion event is incomplete or does not occur at all.

FIGURE 3-5 Laptop computers are now often used as a scan tool

misfire; that is still up to the technician to determine. You may have to perform a compression test to be sure that the valves and piston are sealing the cylinder properly. Connect the scan tool to the DLC, and follow the menu prompts to display DTCs or other engine operating data such as engine coolant temperature, rpm, or vehicle speed.

Compression Testers

Internal combustion engines depend on compression of the air-fuel mixture to maximize the power produced by the engine. The upward movement of the piston on the compression stroke compresses the air and fuel mixture within the combustion chamber. The air-fuel mixture gets hotter as it is compressed. The hot mixture is easier to ignite, and when ignited it will generate much more power than the same mixture at a lower temperature.

If the combustion chamber leaks, some of the air-fuel mixture will escape when it is compressed, resulting in a loss of power and a waste of fuel. The leaks can be caused by burned valves, a blown head gasket, worn rings, slipped timing belt or chain, worn valve seats, a cracked head, and more.

An engine with poor compression (lower compression pressure due to leaks in the cylinder) will not run correctly and cannot perform as designed. To see if a drivability problem is caused by poor compression, a compression test is performed.

A **compression gauge** is used to check cylinder compression. The dial face on the typical compression gauge indicates pressure in both pounds per square inch **(psi)** and metric kilopascals **(kPa)**. The range is usually 0 to 300 psi and 0 to 2,100 kPa. Compression gauges can also have a digital readout.

There are two basic types of compression gauges: the push-in gauge (Figure 3-6) and a screw-in gauge.

The push-in type has a short stem that is either straight or bent at a forty-five-degree angle. The stem ends in a tapered rubber tip that fits any size spark plug hole. The rubber tip is placed in the spark plug hole, after the spark plugs have been removed, and held there while the engine is cranked through several compression cycles. Although simple to use, the push-in gauge may give inaccurate readings if it is not held tightly in the hole.

The screw-in gauge has a long, flexible hose that ends in a threaded adapter (Figure 3-7). This type of compression tester is often used because its flexible hose can reach into areas that are difficult to reach with a push-in type tester. The threaded adapters are changeable and come in several thread sizes to fit 10-mm, 12-mm, 14-mm, and 18-mm diameter

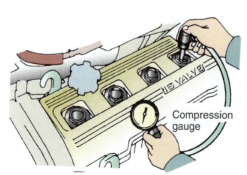

FIGURE 3-6 Push-in compression gauge.

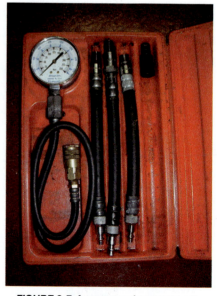

FIGURE 3-7 A compression gauge set.

holes. The adapters screw into the spark plug holes in place of the spark plugs. A compression gauge that reads higher pressures and screws into the injector bore is used on diesel engines. Using a standard compression gauge on a diesel engine will damage the gauge.

Most compression gauges have a vent valve that holds the highest pressure reading on its meter. Opening the valve releases the pressure when the test is complete.

Cylinder Leakage Tester

If a compression test shows that any of the cylinders are leaking, a **cylinder leakage test** can be performed to measure the percentage of compression lost and help locate the source of leakage.

A cylinder leakage tester (Figure 3-8) applies compressed air to a cylinder through the spark plug hole. Before the air is applied to the cylinder, the piston of that cylinder must be at top dead center (TDC) on its compression stroke. A threaded adapter on the end of the air pressure hose screws into the spark plug hole. The source of the compressed air is normally the shop's compressed air system. A pressure regulator in the tester controls the pressure applied to the cylinder. An analog gauge registers the percentage of air pressure lost from the cylinder when the compressed air is applied. The scale on the dial face reads 0 to 100 percent.

A zero reading means that there is no leakage from the cylinder. Readings of 100 percent would indicate that the cylinder will not hold any pressure. The location of the compression leak can be found by listening and feeling for air leaks around various parts of the engine.

Most vehicles, even new cars, experience some leakage around the rings. Up to 20 percent is considered acceptable during the leakage test. When the engine is actually running, the rings will seal much better, and the actual percent of leakage will be lower; however, there should be no leakage around the valves or the head gasket.

Vacuum Gauge

Measuring intake manifold vacuum is another way to diagnose the condition of an engine. Manifold vacuum is tested with a vacuum gauge (Figure 3-9). Vacuum is formed on a piston's intake stroke. As the piston moves down, it lowers the pressure of the air in the cylinder—if the cylinder is sealed. This lower cylinder pressure is called engine vacuum. If there is a leak, atmospheric pressure will force air into the cylinder, and the resultant pressure will not be as low. The reason atmospheric pressure will enter is simply that whenever there is a low and high pressure, the high pressure will always move toward the low pressure.

Vacuum is best defined as a space in which the pressure is significantly lower than an adjacent area of higher pressure. For example, the pressure in the combustion chamber, during the intake stroke, is lower than the pressure of the atmosphere.

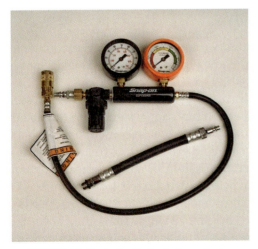

FIGURE 3-8 Typical cylinder leakage tester.

FIGURE 3-9 Vacuum gauge with line adapters.

The vacuum gauge measures the difference in pressure between intake manifold vacuum and atmospheric pressure. If the manifold pressure is lower than atmospheric pressure, a vacuum exists. Vacuum is measured in inches of mercury (in. Hg), kilopascals (kPa), or millimeters of mercury (mm Hg).

To measure vacuum, a flexible hose on the vacuum gauge is connected to a source of manifold vacuum, either on the intake manifold or a point below the throttle plate. Sometimes this requires removing a plug from the manifold and installing a special fitting.

The test is made with the engine cranking and/or running. A good vacuum reading is typically at least 16 in. Hg. However, a reading of 15 to 20 in. Hg (50 to 65 kPa) is normally acceptable. Since the intake stroke of each cylinder occurs at a different time, the production of vacuum occurs in pulses. If the amount of vacuum produced by each cylinder is the same, the vacuum gauge will show a steady reading. If one or more cylinders are producing different amounts of vacuum, the gauge will show a fluctuating reading.

Low or fluctuating readings can indicate many different problems; for example, a low, steady reading might be caused by incorrect valve timing. A sharp vacuum drop at regular intervals might be caused by a burned intake valve.

Vacuum Pumps

There are many vacuum-operated devices and vacuum switches in cars. These devices use engine vacuum to cause a mechanical action or to switch something on or off. The tool used to test vacuum-actuated components is the vacuum pump. There are two types of vacuum pumps: an electrical operated pump and a hand-held pump. The hand-held pump is most often used for diagnostics. A hand-held vacuum pump consists of a hand pump, a vacuum gauge, and a length of rubber hose used to attach the pump to the component being tested. Tests with the vacuum pump can usually be performed without removing the component from the car.

When the handles of the pump are squeezed together, a piston inside the pump body draws air out of the component being tested. The partial vacuum created by the pump is registered on the pump's vacuum gauge. While forming a vacuum in a component, watch the action of the component. The vacuum level needed to actuate a given component should be compared to the specifications given in the factory service information.

The vacuum pump is also commonly used to locate vacuum leaks. This is done by connecting the vacuum pump to a suspect vacuum hose or component and applying vacuum. If the needle on the vacuum gauge begins to drop after the vacuum is applied, a leak exists somewhere in the system.

Leak Detectors

A vacuum or compression leak might be revealed by a compression check, a cylinder leak down test, or a manifold vacuum test; however, finding the location of the leak can often be very difficult.

A simple but time-consuming way to find leaks in a vacuum system is to check each component and vacuum hose with a vacuum pump. Simply apply vacuum to the suspected area, and watch the gauge for any loss of vacuum. A good vacuum component will hold the vacuum that is applied to it. A smoke leak detector is an excellent tool to find vacuum and other leaks (Figure 3-10). The smoke leak detector uses shop air and a nontoxic solution to blow low pressure smoke through the system being tested. To test for vacuum leaks use a plug (provided with the tool) to block off the throttle opening. Then connect the smoke line to a vacuum port, and activate the smoke tester. Smoke will stream out of any leaking area for easy detection. Follow the tool manufacturer's guidelines for proper operation and applications.

FIGURE 3-10 A smoke leak detector.

FIGURE 3-11 An ultrasonic leak detector picks up the sound frequencies of the leak.

WARNING: When using a smoke leak detector to check for leaks on evaporative emissions systems, use nitrogen rather than shop air to pressurize the system to below 1 psi. Be sure to follow the manufacturers' recommended procedures to prevent damage to the system or a hazardous situation for yourself and others.

Another method of leak detection is done by using an ultrasonic leak detector (Figure 3-11). Air rushing through a vacuum leak creates a high frequency sound, higher than the range of human hearing. An ultrasonic leak detector is designed to hear the frequencies of the leak. When the tool is passed over a leak, the detector responds to the high frequency sound by emitting a warning beep. Some detectors also have a series of light-emitting diodes (LEDs) that light up as the frequencies are received. The closer the detector is moved to the leak, the more LEDs light up or the faster the beeping occurs. This allows the technician to zero in on the leak. An ultrasonic leak detector can sense leaks as small as 1/500 inch and accurately locate the leak to within 1/16 inch.

Cooling System Pressure Tester

A cooling system pressure tester contains a hand pump and a pressure gauge (Figure 3-12). A hose is connected from the hand pump to a special adapter that fits on the radiator filler neck. This tester is used to pressurize the cooling system and check for coolant leaks. Additional adapters are available to connect the tester to the radiator cap. With the tester connected to the radiator cap, the pressure relief action of the cap may be checked.

Coolant Hydrometer

A coolant hydrometer is used to check the amount of antifreeze in the coolant. This tester contains a pickup hose, coolant reservoir, and squeeze bulb. The pickup hose is placed in the radiator coolant. When the squeeze bulb is squeezed and released, coolant is drawn into the reservoir. As coolant enters the reservoir, a pivoted float moves upward with the coolant level. A pointer on the float indicates the freezing point of the coolant on a scale located on the reservoir housing (Figure 3-13). The reading must then be adjusted based on the

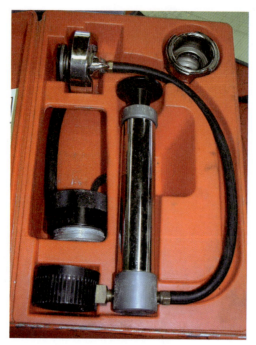

FIGURE 3-12 A cooling system pressure tester.

FIGURE 3-13 Coolant hydrometer.

temperature of the coolant. Some hydrometers have a built-in temperature gauge. There are separate readings (and sometimes separate hydrometers) for the different types of coolant.

Coolant Refractometer

A coolant refractometer is an instrument that is used to determine the refractive index of the coolant sample that is placed on it. A refractometer works by having light pass through a slit. The light is refracted (bent) by the coolant sample. The index of refraction is the ratio of the speed of light in a vacuum divided by the speed of light in the sample. The light is bent and put into two sample tubes, which is viewed in the eyeglass piece (Figure 3-14).

This process will determine the specific gravity of the coolant solution; by understanding the specific gravity, we can understand the strength (percentage of the mixture) of

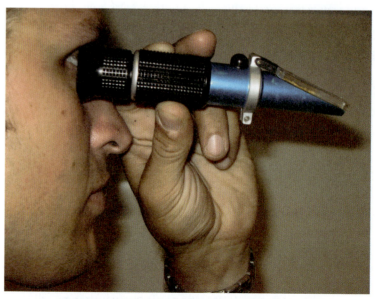

FIGURE 3-14 Hold the refractometer upto your eye.
Almost any light source will do.

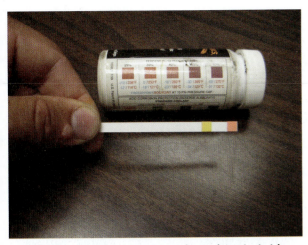

FIGURE 3-15 You can test the coolant using a test strip.

FIGURE 3-16 This oil pressure gauge is threaded into the oil pressure sending unit's bore.

coolant and the freezing/boiling point. A refractometer is more accurate than a hydrometer because there is no temperature correction. A coolant refractometer can also be used with all major types of coolant.

Coolant Test Strips

A coolant test strip is a piece of litmus paper that can test both the percentage of coolant and the acidity of the solution. This type of test is very important because if the antifreeze additives start to break down, corrosion and electrolysis will start to occur in the system. This type of information is good for the customer as well. If you can flush or replace the coolant before it does any major harm, you can save the customer a lot of money. This type of test is also easy to show the customer because it works well as a visual aid. A coolant test strip has separate areas for the different types of coolant as well (Figure 3-15).

Oil Pressure Gauge

The oil pressure gauge may be connected to the engine to check the oil pressure (Figure 3-16). Various fittings are usually supplied with the oil pressure gauge to fit different openings in the lubrication system.

Belt Tension Gauge

A belt tension gauge is used to measure timing or drive belt tension. The belt tension gauge is installed over the belt and indicates the amount of belt tension (Figure 3-17).

A different type of tool called a cricket can be used to check a serpentine, micro-vee, or vee belt's tension. Simply place the tool on an unsupported part of the belt and push down with your finger until you hear the tool crack. The noise the tool makes is similar to the sound a cricket makes. When this happens, you take the tool off and read the gauge (Figure 3-18).

Stethoscope

A stethoscope is used to locate the source of engine and other noises. Sound vibrations travel through the metal rod of the stethoscope, through the tube, and into the earpiece. The sound is then heard in both ears. The stethoscope pickup is placed on the suspected component, and the stethoscope receptacles are placed in the technician's ears (Figure 3-19).

FIGURE 3-17 The belt tension gauge may be used to check or adjust timing belt tension.

FIGURE 3-18 The belt cricket can be used to check belt tension. There are different ones used for either micro-vee or vee belts.

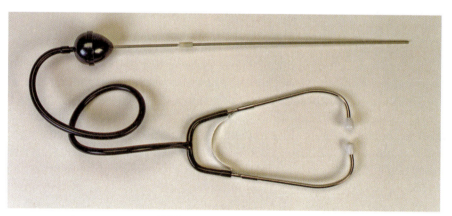

FIGURE 3-19 Stethoscope.

AUTHOR'S NOTE: For a long time as a technician, I used a long skinny screwdriver to listen to abnormal engine noises. The screwdriver does work the same as the stethoscope, but it is not calibrated and designed to do so. After being introduced to using a stethoscope by a fellow technician, I discontinued my major use of the screwdriver for listening to engines. Having both earpieces is like having surround sound and your senses are able to tune in better. It also silences the distracting noises that are around you, like an exhaust leak or a car in the next bay over.

Fuel-Pressure Gauge

A **fuel-pressure gauge** (Figure 3-20) is used to measure the pressure in the fuel system. Fuel injection systems rely on very high fuel pressures, from 35 to 70 psi. A drop in fuel pressure will reduce the amount of fuel delivered to the injectors and result in a lean air-fuel mixture.

FIGURE 3-20 A fuel pressure gauge.

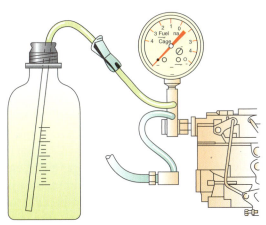

FIGURE 3-21 Checking fuel pump discharge volume.

A fuel pressure gauge is used to check the discharge pressure of fuel pumps, the regulated pressure of fuel-injection systems, and injector pressure drop. This test can identify faulty pumps, regulators, or injectors, and can identify restrictions present in the fuel delivery system. Restrictions are typically caused by a dirty fuel filter, collapsed hoses, or damaged fuel lines.

Some fuel pressure gauges also have a valve and outlet hose for testing fuel pump discharge volume (Figure 3-21). The manufacturer's specification for discharge volume will be given as a number of pints or liters of fuel that should be delivered in a certain number of seconds.

WARNING: While testing fuel pressure, be careful not to spill gasoline. Gasoline spills may cause explosions and fires, resulting in serious personal injury and property damage.

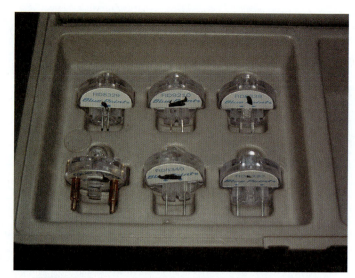

FIGURE 3-22 A set of noid lights to fit a variety of connectors.

Noid Light

A special test light called a **noid light** can be used to determine if a fuel injector is receiving its proper voltage pulse from the computer. The wiring harness connector is disconnected from the injector, and the noid light is plugged into the connector (Figure 3-22). When the engine is turned over by the starter motor, the noid light will flash rapidly if the voltage signal is present. No flash usually indicates an open in the power feed or ground circuit to the injector. Noid lights may also be used in the connectors to many coil on plug ignition systems. Crank the engine over to see if the coil is receiving its voltage pulse from the PCM or ignition module.

Exhaust Gas Analyzers

Exhaust gas analyzers are very valuable diagnostic tools. By looking at the quality of an engine's exhaust, a technician is able to look at the effects of the combustion process. Any defect can cause a change in exhaust quality. The amount and type of change serves as the basis of diagnostic work.

Modern exhaust gas analyzers are either four- or five-gas analyzers (Figure 3-23). A four-gas analyzer measures the level of hydrocarbons (HC) and carbon monoxide (CO) in the exhaust stream. These toxic pollutants are regulated by the EPA, and the level in the exhaust is measured in many states as part of an emissions inspection program. The four-gas analyzer also measures the level of oxygen (O_2) and carbon dioxide (CO_2). Seeing the level of these gases in the exhaust stream, in combination with the pollutants, can give an experienced technician a lot of information about the quality of combustion. If an engine is running too rich, the level of CO will be higher than normal and the oxygen will be lower. A few possible causes of a rich mixture are a faulty fuel-pressure regulator or oxygen sensor, a restricted exhaust, or low compression. An engine with an ignition misfire will emit high levels of unburned fuel in the form of hydrocarbons, and the oxygen level will also be high. Worn spark plugs or spark plug wires or faulty coils are common culprits causing ignition misfire. A five-gas analyzer also measures the level of oxides of nitrogen (NO_x). This pollutant is formed when combustion temperatures are too hot; it contributes to the formation of ground level ozone, smog.

Engine Analyzer

When performing a complete engine performance analysis, an engine analyzer may be used. An engine analyzer houses much of the necessary test equipment. Although the term engine analyzer is often loosely applied to any multipurpose test meter, a complete engine

Exhaust gas analyzers are often called infrared testers. This is because many analyzers use infrared light to analyze the exhaust gases.

FIGURE 3-23 A five-gas analyzer.

analyzer will incorporate most, if not all, of the test instruments mentioned in this chapter. Most engine analyzers are based on a computer that guides a technician through the tests. Most will also do the work of the following tools:

- Compression gauge
- Pressure gauge
- Vacuum gauge
- Vacuum pump
- Tachometer
- Timing light/probe
- Voltmeter
- Ohmmeter
- Ammeter
- Oscilloscope
- Computer scan tool
- Emissions analyzer

With an engine analyzer, you can perform tests on the battery, starting system, charging system, primary and secondary ignition circuits, electronic control systems, fuel system, emissions system, and the engine assembly. The analyzer is connected to these systems by a variety of leads, inductive clamps, probes, and connectors. The data received from these connections is processed by several computers within the analyzer.

The microprocessors in some computerized engine analyzers are programmed with specifications for specific model vehicles. Diagnostic trouble codes have also been loaded into the analyzer's memory. Based on the input from the leads and connectors, the microprocessors can identify worn, misadjusted, or faulty components in all major engine systems. The analyzer will also list the probable causes of specific performance problems and will prompt, or guide, the technician step by step through a troubleshooting procedure designed to verify and correct the problem.

Commands and specifications can be entered into the analyzer on a computer-like keyboard. Specifications, commands, and test results are displayed on the CRT screen. Some analyzers will graphically display test results on their CRT screen. The analyzer's printer can also print out copies of the information that appears on the screen.

Many engine analyzers will perform a complete series of tests and record the results automatically. The analyzer compares all the test results to the vehicle manufacturer's specifications. When the test series is completed, the analyzer prints a report indicating those readings that were not within specifications. Many analyzers also provide diagnostic assistance for the problems indicated by the readings that were not within specifications.

However, the technician may select any test function, or functions, separately. Engine analyzer test capabilities vary depending on the equipment manufacturer, and technicians must familiarize themselves with the engine analyzer in their shop; for example, some analyzers have oscilloscope patterns and digital ignition readings, and others use digital ignition readings exclusively.

Some engine analyzers have vehicle specifications on diskette, and the technician enters the necessary information, such as model year and engine size, for the vehicle being tested. Specifications may be updated simply be obtaining a new diskette from the equipment manufacturer.

A phone modem is contained in some engine analyzers to provide networking capabilities. This phone modem allows the technician to unload all technical reports and pattern reports of a specific problem vehicle to off-location technical support teams.

ENGINE MEASURING TOOLS

Proper operation of the engine requires that most internal components be manufactured with very precise tolerances. To measure these components properly, special micrometers capable of determining 10-thousandths (0.0001) inch or thousandths (0.001) millimeter are used.

WARNING: Measuring tools require special care. They are delicate instruments and cannot withstand abuse. Handle them with care to prevent dropping, striking, or other abusive actions.

Some components of the engine have wider tolerance ranges and measurements are not as critical. In these areas, less delicate tools may be used; however, accuracy is not as precise. Tools used for these areas include feeler gauges, machinist's rule, calipers, micrometers, and dial indicators.

Today, most measurement specifications are stated using the metric system. It is best to have the proper tools to measure in the system stated by the specifications. However, it can get expensive having to purchase duplicate tools. Specifications and measurement readings can be converted between the two systems, allowing the use of one set of tools (Figure 3-24).

Feeler Gauges

A **feeler gauge** is a thin metallic strip of a known thickness.

A **feeler gauge** can be used to measure valve clearance, crankshaft end play, connecting rod side clearance, piston ring side clearance, and other measurements when exacting measurements are not critical (Figure 3-25). The thickness of the gauge is stamped on the blades in thousandths of an inch or tenths of a millimeter.

The most common style of feeler gauge is the blade type. This gauge is constructed of metal strips approximately 12 millimeters wide. Several different types of feeler gauges are available to perform some specific functions.

To Find		Multiply	x	Conversion Factors
millimeters	=	inches	x	25.40
centimeters	=	inches	x	2.540
centimeters	=	feet	x	32.81
meters	=	feet	x	0.3281
kilometer	=	feet	x	0.0003281
kilometer	=	miles	x	1.609
inches	=	millimeters	x	0.03937
inches	=	centimeters	x	0.3937
feet	=	centimeters	x	30.48
feet	=	meters	x	0.3048
feet	=	kilometers	x	3048
yards	=	meters	x	1.094
miles	=	kilometers	x	0.6214

FIGURE 3-24 Common English/metric conversion factors.

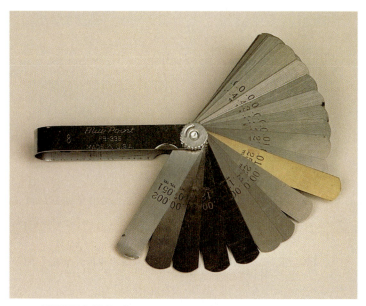

FIGURE 3-25 Feeler gauge set used to measure clearances.

Using Feeler Gauges

Proper use of a feeler gauge requires the technician to develop a feel for the drag on the gauge when it is removed from the gap. To measure gap width, start with a gauge thinner than the gap and continue to increase the gauge size. The measurement of the clearance is the blade that has a slight drag when it is moved in and out. The size can be confirmed by selecting the next size gauge and attempting to slide it into the gap; it should not fit.

Feeler gauges are also used when adjusting clearances, such as valve lash. In this instance, use the gauge the same size as specified in the service information. With the lock nut on the rocker arm backed off, adjust the lash until you feel a slight drag as you move the feeler gauge back and forth. Tighten the lock nut, being careful not to change the gap.

SPECIAL TOOLS
Feeler gauge set

Machinist's Rule

The **machinist's rule** usually contains four different **scales**, two on each side of the rule. Common English scales include 1/8, 1/16, 1/32, and 1/64 inch (Figure 3-26). A decimal machinist's rule is also available with the scales divided into decimal intervals of 0.1, 0.02, and 0.01. Common scales on metric rules are divided into centimeters and millimeters.

Calipers

Calipers are used to take inside, outside, and depth measurements not requiring accuracy to a thousandth of an inch (Figure 3-27). Typical usages include measuring valve springs, piston diameters, and bearing thicknesses. A graduated steel beam contains a fixed jaw with a sliding scale. There are two types of calipers available for use in automotive repair: vernier and dial. Either one may be capable of displaying both English and metric measurements.

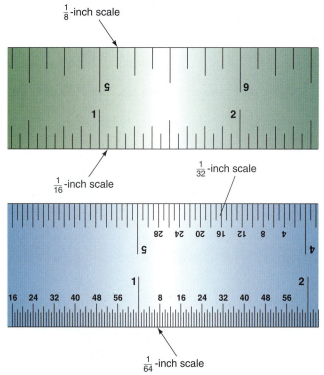

FIGURE 3-26 Typical scales used on a machinist's rule.

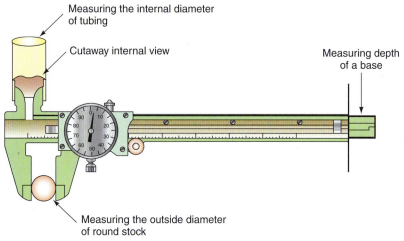

FIGURE 3-27 Dial calipers can be used to measure many different parts.

Vernier Calipers. The **vernier caliper** is a basic measuring instrument. The base (bar) scale on the beam is divided into ten equal parts. Each part is equal to one-tenth (0.1) of an inch. Small numbers from 1 to 9 appear between each inch. These numbers have a value of 100 thousandths (0.100) of an inch. Between each of these numbers are four graduations or marks. Since each number is equal to 0.100 inch, the value of these small graduations is 25 thousandths (0.025) of an inch. The vernier scale is divided into 0.001-inch increments with twenty-five graduations. The scale is marked with the numbers 5, 10, 15, 20, and 25. These numbers represent values of 0.005, 0.010, 0.015, and 0.025 inch, respectively. Metric calipers have a base scale in millimeters, with the vernier scale divided into 0.02 mm. Both calipers work identically. The end jaw is fixed to the scale, while the other slides on it. The piece being measured is located between the jaws. A thumb screw moves the adjustable jaw. The length of the scale determines the size of the calipers. The most common are 6 and 12 inches. Measurements are obtained using a combination of the base scale and the vernier scale.

To read a vernier caliper, the zero on the left edge of the vernier scale shows the base measurement. If the zero aligns with an increment mark of the beam, the last digit of the measurement is zero (Figure 3-28). If the zero is not aligned with any marks, look down the vernier scale to find the number that aligns perfectly with any line on the main scale. Use the vernier scale to determine the value to be added to the base measurement. There will always be one set of lines that will align with each other (Figure 3-29).

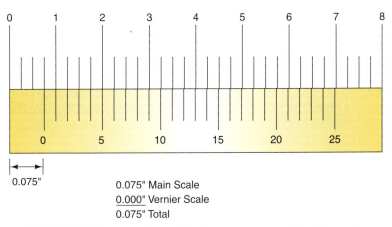

0.075" Main Scale
0.000" Vernier Scale
0.075" Total

FIGURE 3-28 If the zero aligns with a base scale graduation, the vernier scale is not used. In this case, the reading is 0.075 inch.

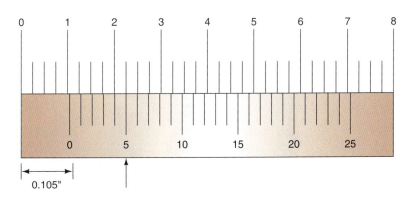

0.100" Main Scale
0.005" Vernier Scale
0.105" Total

FIGURE 3-29 Since the zero does not align with a base scale graduation, the vernier scale is used. In this case, the reading is 0.105 inch since the 5 is aligned.

For example, if the measured component were 0.034 inch, then the caliper scales would be read as follows:

Base measurement—0.025 inch

Vernier scale—0.009 inch

Total measurement—0.034 inch

Dial calipers are similar to vernier calipers, except the dial performs the function of the vernier scale.

Dial Calipers. Some calipers are fitted with a dial to make measurement reading easier. To measure a component, use the thumb wheel to open the jaws, then close the jaws over the component. The dial uses a rack and pinion mechanism to transfer movement of the jaws to the dial needle. Like the vernier caliper, the base measurement is taken from the beam, and the dial reading is added to it. The base scale on an English dial caliper is marked in 0.100-inch increments, and the dial provides the thousandths (0.001) of an inch readings. Most metric calipers have base scale increments of 2 millimeters and dial increments of 0.01 millimeter. Follow the steps in Photo Sequence 2 to accurately measure a component using an English system dial caliper.

The latest innovation in calipers provides a digital readout. An internal microprocessor calculates the position of the jaws and shows the reading in a display window (Figure 3-30).

Micrometers

A **micrometer** may be called a "mike." A micrometer is a precision measuring tool; some can measure to 0.0001 inch and measure the outside or inside diameter or depth of an object.

A **micrometer** is used to make precise measurements (Figure 3-31). Most micrometers will measure to 0.001 inch or 0.01 mm. However, to perform some measurements in engine applications, a micrometer capable of measuring within 0.0001 inch or 0.001 mm may be required. The micrometer has a range it will work within (Figure 3-32). Unlike calipers, different micrometers are used to take outside, inside, and depth measurements.

The principle of the micrometer is to record the advancement of a screw for a number of turns or parts of a turn. The major components of a micrometer include (Figure 3-33):

- Frame
- Anvil
- Spindle
- Lock
- Sleeve
- Sleeve reading line
- Thimble

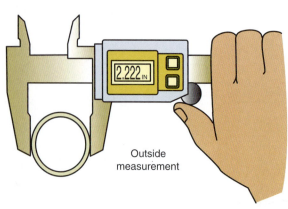

FIGURE 3-30 Electronic digital caliper makes quick work of measuring.

FIGURE 3-31 You will use a 0–1-inch micrometer to measure many engine components.

TYPICAL PROCEDURE FOR READING A DIAL CALIPER

P2-1 Set the dial caliper in your right hand so that the gauge can be easily read and your thumb is on the wheel.

P2-2 Make sure the jaws are clean and close them all the way. Zero the gauge if needed.

P2-3 Using your thumb to slide the jaws of the caliper, move it into the position desired to measure.

P2-4 Make sure that you are using the beveled edge of the caliper jaws as the contact point.

P2-5 Once you have taken your measurement, lock it in place and remove it from the part. This makes reading the measurement easier.

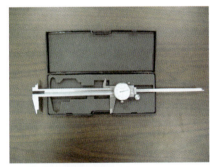

P2-6 The dial caliper being used in this example has a range from 0 to 6 inches. Each division between the inches is 1/10th of an inch (0.1 in.).

P2-7 View and record the measurement on the sliding ruler. In this example, the caliper reads 0.4 in.

P2-8 After reading the inches and first decimal place, you can read the dial. In this example, each mark on the dial measures 1/1000th of an inch (0.001 in.). This will give you the second and third decimal places. In this example, the dial reads 55. That means that 0.055 in. is added to the first number.

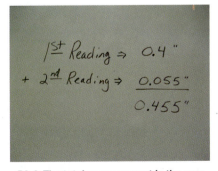

P2-9 The total measurement is the sum of all the recorded numbers.

FIGURE 3-32 A set of micrometers with a range up to four inches.

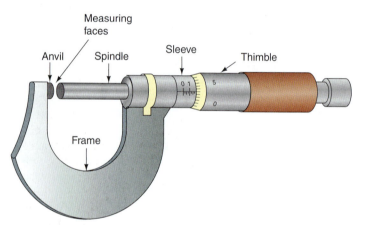

FIGURE 3-33 Parts of an outside micrometer.

FIGURE 3-34 The sleeve is usually marked in 0.025-inch increments. To read the sleeve, count the visible lines by 0.025.

English system micrometers have marks on the sleeve in 0.025-inch increments (Figure 3-34). The marks on the thimble are in 0.001-inch increments. One complete revolution of the thimble equals 0.025-inch movement of the spindle. If the micrometer is capable of measuring to the ten-thousandths inch, a vernier scale is also on the sleeve (Figure 3-35). Metric micrometers mark the sleeve in 0.5-millimeter increments and the thimble scale in 0.01-millimeter increments.

Reading a Micrometer

Every time a micrometer is used, its calibration should be checked and, if needed, adjusted. Calibration is checked by cleaning the faces of the anvil and spindle, then closing them (on 0-to-1-inch micrometers). The scale should indicate a zero reading. Micrometers greater than a 1-inch range will require that a standard test gauge be placed between the anvil and spindle (Figure 3-36). When the faces are closed against the standard, the scale should read zero.

FIGURE 3-35 A micrometer with a 0.0001-inch scale on the upper left of the sleeve.

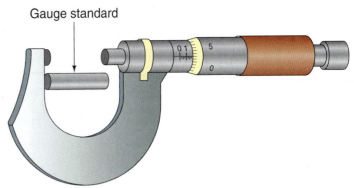

FIGURE 3-36 A standard test gauge is used to calibrate outside micrometers over 1 inch.

If adjustments are necessary, perform the following procedure on a typical micrometer:

1. Clean both faces of the spindle and anvil.
2. Close the faces together or against the appropriate standard gauge.
3. Remove the knurled cap on the end of the thimble and loosen the set screw.
4. Make the zero line on the thimble coincide with the zero line on the barrel.
5. Lightly tighten the set screw.
6. Separate the faces by turning the spindle.
7. Hold the micrometer by the thimble and securely tighten the set screw.
8. Recheck zero calibration again.
9. Replace the thimble cap.

SPECIAL TOOLS
Micrometer set

Reading inside or outside micrometers is done using the same procedures. Follow the steps in Photo Sequence 3 to accurately measure a component using an English system micrometer. Before making any measurements, clean the part and the micrometer. When taking the measurements, the anvil and spindle must be at right angles to the part being measured. To assure proper contact, slightly rock the micrometer as the spindle is turned. Do not overtighten the spindle. The spindle and anvil should just come into contact with the part, so a slight drag is felt when the micrometer is removed. On many micrometers, the end has a slip gauge. If you tighten the micrometer by the end of the thimble, it will stop rotating when a set resistance is reached.

As with the caliper, measurement readings are obtained by adding the markings together. Begin by finding the largest number shown on the sleeve. If no other marks on the sleeve are visible and the zero on the thimble aligns with the reading line, the sleeve shows the size of the component (Figure 3-37). Any marks that are visible after the number on the

Outside micrometers are designed to measure the outside diameter or thickness of a component. Inside micrometers are used to measure the inside diameter of a hole. Depth micrometers are used to measure the depth of a bore.

FIGURE 3-37 Since the zero on the thimble scale aligns on the reading line and the last sleeve graduation is 5, the reading is 0.500 inch.

TYPICAL PROCEDURE FOR READING A MICROMETER

P3-1 Tools required to perform this task are a micrometer set and a clean shop towel.

P3-2 Select the correct micrometer. If the component measurement is less than 1 inch, use the 0- to -1-inch micrometer.

P3-3 Check the calibration of the micrometer.

P3-4 Locate the component between the anvil and spindle of the micrometer, and rotate the thimble to slowly close the micrometer around the component. Tighten the thimble until a slight drag is felt when passing the component in and out of the micrometer. If the micrometer is equipped with a ratchet, it can be used to assist in maintaining proper tension.

P3-5 Lock the spindle to prevent the reading from changing.

P3-6 Remove the micrometer from the component.

P3-7 Each number on the sleeve is 0.100 inch, and each graduation represents 0.025 inch. To read a measurement, count the visible lines on the sleeve.

P3-8 The graduations on the thimble define the area between the lines on the sleeve in 0.001 increments. To read this measurement, use the graduation mark that aligns with the horizontal line on the sleeve.

P3-9 Add the reading obtained from the thimble to the reading on the sleeve to get the total measurement.

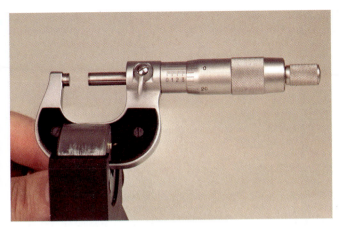

FIGURE 3-38 If the graduation marks are visible after a number on the sleeve scale, add the graduation value to the number. In this case, the micrometer reads 0.375 inch.

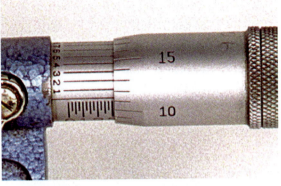

FIGURE 3-39 Use the thimble scale to determine the thousandths to be added to the sleeve scale. This micrometer is reading 0.3112 inch.

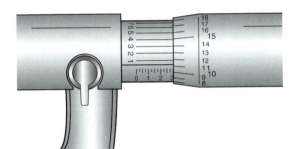

FIGURE 3-40 Some micrometers have vernier scales that allow readings to a ten thousandth of an inch. The 2 on the vernier scale is the only one that aligns with a thimble graduation mark. The reading is 0.3112 inch.

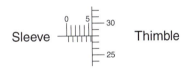

Reading 5.78 mm

FIGURE 3-41 Reading a metric micrometer is much like reading a standard micrometer. This scale indicates 5.78 mm.

SPECIAL TOOLS
Metric micrometer set

Telescoping gauges are used to measure the inside diameter of a hole. They are sometimes called snap gauges.

CAUTION:
Do not back off the lock screw more than necessary. Continued rotation of the lock screw will disassemble the gauge period.

sleeve are added to the number (Figure 3-38). If the zero on the thimble scale does not align with the reading line, add the value to the thimble reading (Figure 3-39). Some micrometers are capable of reading ten-thousandths (0.0001) inch. The vernier scale is used if no lines on the thimble align with the reading line. To read the vernier scale, find the line on the vernier scale that aligns with a line on the thimble scale. Add the number from the vernier scale to the sleeve and thimble readings (Figure 3-40).

Setting up the metric micrometer is the same. Each number on the sleeve represents 5 millimeters, with graduations of 0.5 mm between them. The thimble is divided into 50 equal divisions, each being 0.01 millimeter. The reading obtained on the thimble is added to the reading on the sleeve (Figure 3-41). In this example, the sleeve reading is 5.5 millimeters, and the thimble reading is 0.28 millimeter. The total reading is obtained by adding the sleeve reading to the thimble reading; in this case, the total is 5.78 millimeters.

Telescoping Gauge

Telescoping gauges are used to measure the inside diameters of bores (Figure 3-42). The plungers are retracted and locked with the lock screw. The gauge is then inserted into the bore. When the lock screw is loosened, the plungers extend under spring pressure. Lock the lock screw, then rock the gauge to assure it is at the largest diameter. If any corrections must be made, loosen the lock screw and repeat. With the gauge removed from the bore, the plunger distance is measured with an outside micrometer (Figure 3-43).

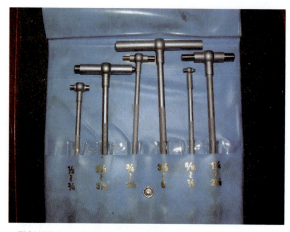

FIGURE 3-42 Telescoping gauges are used to measure inside bore diameters.

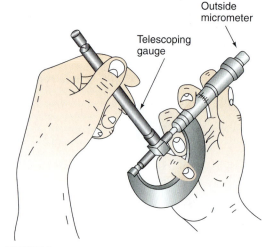

FIGURE 3-43 To determine the diameter of the bore, use an outside micrometer to read the width of the gauge.

Small-Hole Gauge

A **small-hole gauge** consists of a round head, wedge, and a lock handle (Figure 3-44). The round head is inserted into the bore, and the handle is rotated to expand the head outward. When the split ball contacts the bore walls, the gauge is removed. An outside micrometer is used to measure the diameter of the head.

Dial Indicators

Dial indicators can be used to measure valve lift, shaft out-of-round, and end play (Figure 3-45). The dial indicator consists of a dial face with a needle. The dial is usually calibrated in 0.001-inch increments. Most metric system dial indicators measure in 0.01-millimeter increments. A spring-loaded plunger or toggle lever transfers movement to the dial needle. To achieve accurate readings, the plunger must be preloaded.

Using a Dial Indicator

To use the dial indicator, attach it to its stand. The stand is attached to the indicator by a lug on the back of the housing. Stands can be clamp type (using a small C-clamp) or permanent magnet type. It is important to assure a rigid mounting of the dial indicator stand to achieve accurate measurements.

A **small-hole gauge** is used to measure smaller holes or bores than a telescoping gauge can measure.

The **dial indicator** measures the travel of a plunger in contact with a moving component.

SPECIAL TOOLS
Dial indicator

FIGURE 3-44 A small-hole gauge set.

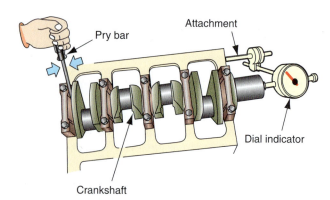

FIGURE 3-45 A dial indicator can be used for several measurements, including crankshaft end play.

FIGURE 3-46 Rotate the face of the dial to align the pointer on zero.

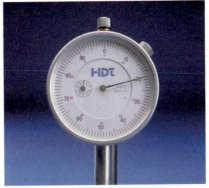

FIGURE 3-47 The pointer will indicate the amount of movement of the plunger.

Adjust the stand until the dial indicator plunger contacts the piece being measured. Try to set the support arms as short as possible to keep the setup rigid. Mount the indicator so the plunger is at a 90-degree angle with the part. If the plunger is mounted at an angle, an accurate reading will not be obtained. In addition, to obtain accurate readings, preload the plunger until about one-half of the plunger is outside of the indicator. If the dial is equipped with a movable face, adjust it so the needle is pointing to zero (Figure 3-46). Movement of the workpiece will be measured on the dial (Figure 3-47). Look at the dial straight on; looking at an angle will give false readings. If the needle travels completely around the dial, add that value to your final reading to get the total measurement. When reading runout measurements, the needle may move in both directions from zero. The total indicated reading (TIR) is the amount of movement on both sides of zero added together; for example, if the needle indicates movement of 0.015 inch (0.40 mm) on one side of zero and 0.025 inch (0.65 mm) on the other side, the TIR is 0.040 inch (1.0 mm).

Valve Guide Bore Gauge

Valve guide bore can be measured using a small-hole gauge; however, **valve guide bore gauges** are available for quick, accurate measurements. The slim probe is inserted into the guide, and the bore diameter is displayed by the dial indicator. Moving the gauge up and down the bore provides taper measurements, while rotating the gauge provides out-of-round measurements.

Valve Seat Runout Gauge

A **valve seat runout gauge** consists of a dial indicator, an arbor, and an indicator bar. The arbor centers the tool into the valve guide, and the indicator bar rests on the valve seat. The tool is rotated while observing the dial indicator.

Cylinder Bore Dial Gauge

Cylinder bore taper and out-of-round can be measured using an inside micrometer or telescoping gauge. For faster measurements of the cylinder, most high volume shops use a **cylinder bore dial gauge** (Figure 3-48). This gauge can quickly provide cylinder bore diameter, taper, and out-of-round. Most rocking type gauges consist of a handle, guide blocks, lock, indicator contact, and dial indicator. To make the gauge universal, extensions can be added to the indicator contacts. The sliding guides center the plunger in the bore. While slightly rocking the gauge up and down in the bore, observe the smallest reading obtained (pointer reverses direction). This is the diameter of the bore.

Another type of bore gauge is the sled type. When it is inserted into the bore, the plunger spring pushes the sled against the cylinder wall. After the dial is set to zero, slide the

The **valve guide bore gauge** provides a quick measurement of the valve guide. It can also be used to measure taper and out-of-round.

The **valve seat runout gauge** provides a quick measurement of the valve seat concentricity.

SPECIAL TOOLS
Cylinder bore dial gauge
Setting fixture

FIGURE 3-48 A common type of cylinder bore dial gauge.

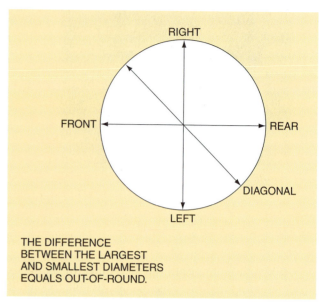

THE DIFFERENCE BETWEEN THE LARGEST AND SMALLEST DIAMETERS EQUALS OUT-OF-ROUND.

FIGURE 3-49 A cylinder bore does not wear round. To measure the amount of out-of-round, measure in three directions toward the top of the cylinder.

gauge up and down and rotate it around the bore. This will provide taper and out-of-round measurements.

As a cylinder wears, it becomes out-of-round and tapered. Out-of-round is caused by the thrusts applied to the piston forcing it into the cylinder wall. Taper occurs as the rings travel against the cylinder wall. If the cylinder is worn excessively, it may require cylinder boring. To measure the amount of cylinder out-of-round, measure the bore in three directions at the same bore depth (Figure 3-49). Subtracting the smallest reading from the largest provides the amount of cylinder out-of-round. Taper is measured by comparing the readings obtained at the top of ring travel with the reading at the bottom of the ring travel. The difference between the two readings is the amount of cylinder taper.

Bore gauges do not read the actual size of the bore, instead they must be set up for a specified size, and the dial will indicate the amount of deviation from this size. To use a cylinder bore dial gauge, first obtain the bore size specifications in the service information. Adjust the gauge setting fixture to the specified size by placing the correct size spacer bar for the inch increments. Use the micrometer to set the fractions of an inch.

Once the setting fixture is set up, the bore gauge is ready to be set. Begin by placing the gauge in the setting fixture and attaching the correct size extension. The extension must push the plungers on the gauge back into the base. Set the dial to zero by turning the extension in and out. Lock the dial at zero by tightening the lock nut on the extension. Fine tune the zero adjustment by rotating the dial faceplate using the serrated rim. This sets the gauge to read zero at the specified size of the bore. Insert the gauge into the bore, and measure the diameter.

Out-of-Round Gauge

Out-of-round gauges are used to measure the concentricity of connecting rod and main bores.

Engine bearings will conform to the shape of the main bore. For this reason, the bore must be perfectly round. This measurement can be made using an inside micrometer; however, a special **out-of-round gauge** provides quicker measurements. The out-of-round gauge consists of a base with a dial indicator (Figure 3-50). An adjustable slide is located on the bottom of the base. To use, set the locking slides to the diameter of the bore, and set the dial

FIGURE 3-50 An out-of-round gauge checks connecting rod bores quickly.

indicator to zero. Next, rotate the gauge in the bore. The dial indicator will provide a plus or minus reading indicating any variance in the bore diameter.

Aligning Bar

Alignment of the crankshaft bores must be checked to determine warpage. A straightedge and feeler gauge set can be used to measure for warpage. Another method is to use an **aligning bar**. With all main bearings removed from the crankshaft bores, the correct size aligning bar is placed onto the bearing saddles. Next, the bearing caps are installed and torqued to specifications. If the bar can be rotated with the use of a 12-inch lever, the saddle bores are aligned. If there is a variation in saddle placement, the bar will not rotate.

Plastigage

Plastigage (Figure 3-51) is commonly used to measure the oil clearance between the main bearings and the crankshaft, and between the crankshaft and the connecting rod bearings. The plastigage is placed on the crankshaft prior to installing the bearing half and cap (Figure 3-52). The bearing half and cap are installed, and the fasteners are torqued to the

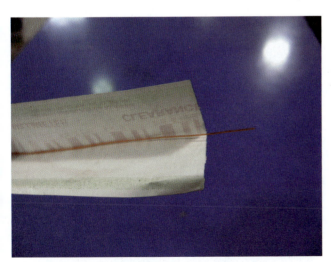

FIGURE 3-51 Plastigage is the small plastic strip inside the envelope. The outside of the envelope has the measuring scale on it.

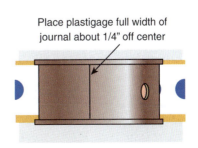

Place plastigage full width of journal about 1/4" off center

Installing plastigage

FIGURE 3-52 A piece of plastigage is laid across the crankshaft journal.

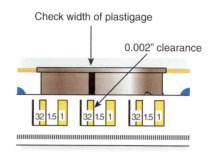

Check width of plastigage

0.002" clearance

Measuring plastigage

FIGURE 3-53 Use the scale on the wrapper to compare the width of the plastigage.

proper specifications. The clamping force causes the plastigage to flatten. The amount of flattening depends on the oil clearance. The bearing half and cap are then removed and the thickness of the plastigage is measured using the scale provided on the package (Figure 3-53). A metric scale is provided on one side of the package, while an English scale is on the reverse side. Plastigage is available in different diameters for measuring a variety of clearances.

SPECIAL HAND TOOLS

There are a few special hand tools designed to make the operation of engine rebuilding or repairing easier. Although most of these are optional, there are some that are essential to perform an engine repair correctly. These include torque wrenches and torque angle gauges.

SPECIAL TOOLS

Torque wrench

Torque Wrenches

Most of the fasteners used to attach components to the engine block require proper torquing; for example, the cylinder head must be properly torqued to set the bolts and apply the correct amount of pressure to seal the combustion chamber and water passages. In addition, proper torquing.sets the stress into the block. If the bolts were not torqued properly, the cylinder head or block could warp.

One of the most important tools used during an engine service, reconditioning, or rebuild is the **torque wrench**. Never guess at a torque value. The torque wrench will give you a precise indication of the torque being applied either by a click, a light, a beam, or a dial (Figure 3-54).

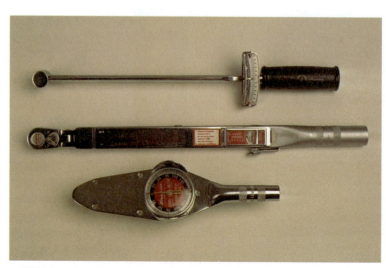

FIGURE 3-54 Different styles of torque wrenches: (l–r) beam type, click type, and dial indicator type.

Using a Torque Wrench

There are many different types of torque wrenches available. Since each type is used differently, always refer to the manufacturer's instructions when using a torque wrench. There will be times when an adapter is required to gain access to a fastener. Some adapters will affect the reading obtained by the torque wrench.

If an adapter that increases the length of the torque wrench between the fastener and handle is required, the reading must be corrected. This is required since a longer bar changes the leverage applied to the fastener. Torque wrenches are calibrated for the length they are designed. Use the following formula to correct the reading if an adapter is required:

$$\text{Wrench reading} = \frac{\text{Torque at fastener} \times \text{Wrench length}}{\text{Wrench length} + \text{Adapter length}}$$

For example, if the torque specification is 50 foot-pounds, and a 6-inch extension is installed onto a 16-inch wrench, the calculated reading on the wrench would be 36 foot-pounds. At 36 foot-pounds, the actual applied torque to the fastener is 50 foot-pounds.

Torque Angle Gauge

A torque angle gauge is not like a torque wrench; it does not measure torque. It is a gauge that measures the amount of circular movement (in degrees) as you tighten down a fastener with a ratchet or breaker bar (Figure 3-55). Most torque angle gauges have a square fitting for a 1/2 in. or 3/8 in. ratchet and a similar-sized socket fitting on the other end. A torque angle gauge is used when you are tightening torque-to-yield bolts. Cylinder head bolts are one example of where torque-to-yield bolts might be used.

Torque-to-Yield

Torque-to-yield bolts must be tightened using the procedure specified by the vehicle manufacturer. A typical method is to tighten the bolts in two steps:

1. Tighten to the specified torque value.
2. Tighten the fastener an additional amount as stated in the specifications.

Extensions that increase the *height* of the wrench from the fastener will not affect the torque reading, provided they do not twist.

SPECIAL TOOLS
Torque angle gauge

A bolt that has been torqued to its yield point can be rotated an additional amount without any increase in clamping force. When a set of **torque-to-yield** fasteners is used, the torque is actually set to a point above the yield point of the bolt. This assures that the set of fasteners will have an even clamping force.

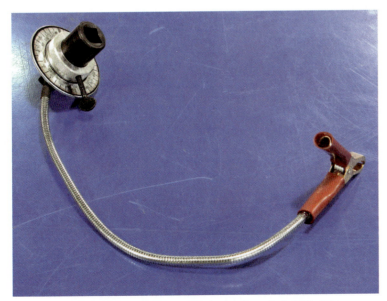

FIGURE 3-55 The torque angle has a clamp to hold it still while turning the ratchet.

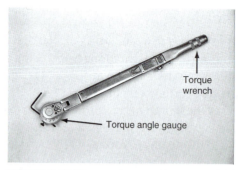

FIGURE 3-56 Many engines use torque-to-yield fasteners that require a torque angle gauge to tighten properly.

Torquing to the specifications may require more than one step. The first step may be about one-half the total specification, with the final torque being obtained in additional steps. The fastener is then stretched by turning it an additional amount. You should not guess, or estimate by using your eye, how much angle your ratchet has turned. To accurately measure the amount of additional angle being applied, a torque angle gauge should be used (Figure 3-56). Attach the gauge to the wrench and secure the rod against a surface. Turning the wrench will result in the needle indicating the amount of rotation.

WARNING: In general, do not reuse torque-to-yield bolts; they have been stretched during installation. There are some manufacturers that allow their torque-to-yield bolts to be reused two times. As always, check the manufacturer's service information to perform the procedure correctly.

Ring Expanders

To prevent damage to the rings during disassembly and assembly, a special ring expander can be used (Figure 3-57). The ring expander will safely expand the rings wide enough to allow them to be slipped out of the groove and over the piston dome.

FIGURE 3-57 Ring expanders help in removal and installation of the rings.

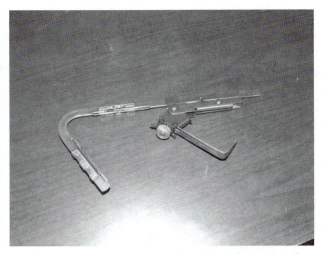

FIGURE 3-58 The safest way to clean piston ring grooves is with a special-groove cleaner.

FIGURE 3-59 A common ring compressor.

Ring-Groove Cleaners

Before installing rings onto the piston, it is important to remove any carbon or other contamination from the ring groove. A ring groove cleaner provides a fast method of thoroughly cleaning the grooves (Figure 3-58). If a ring-groove cleaner is not available, break an old ring and use the sharp edge to clean inside the groove. Do not use enough force to scratch the piston surfaces.

Ring Compressors

To prevent ring or piston damage during installation of the piston assembly into the block, the rings must be compressed to a size smaller than the cylinder bore. There are several different types of ring compressors available to perform this task (Figure 3-59).

SPECIAL RECONDITIONING TOOLS AND EQUIPMENT

There are various special tools and equipment used in rebuilding and reconditioning automotive engines. Some equipment is used by specialty machine shops, while other equipment is found in most automotive repair facilities. This portion of the chapter discusses the basics of most of these types of equipment. Even though you may not operate the equipment, it is important to understand what is being performed and what can be done to correct a defective component.

Cylinder Head Reconditioning Equipment

Cylinder heads contain areas that are subject to wear and damage due to the immense pressures and temperatures to which they are exposed. There are many types of tools and equipment available to recondition the cylinder head, including:

- Valve spring compressor
- Valve spring tension tester
- Valve guide renewing equipment
- Seat grinder
- Seat cutter
- Seat inserters
- Valve grinding equipment
- Head resurfacing equipment

Valve spring tension testers are used to measure the open and closed valve spring pressures. Proper closed pressure is required for a tight seal. Proper open pressure overcomes the effects of inertia and assures proper valve closing.

Installed height is sometimes referred to as closed height.

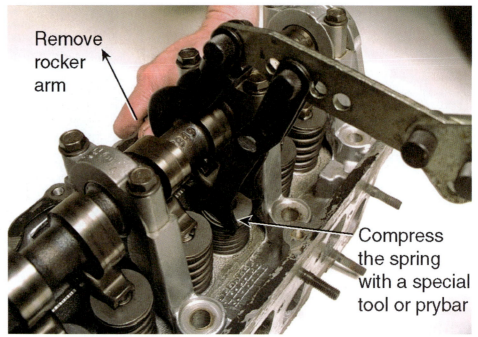

Remove rocker arm

Compress the spring with a special tool or prybar

FIGURE 3-60 Compressing valve springs using a pry bar–type compressor.

Valve Spring Compressor. To remove the valve springs, they must first be compressed so the keepers can be removed. There are many types of spring compressors available. Some are designed to allow valve spring removal without first removing the cylinder head from the block. Other designs are only used with the cylinder head removed from the engine. In addition, there are special spring compressors used for OHC engines.

A pry–type spring compressor is one design that provides valve spring removal with the cylinder head attached to the block (Figure 3-60). With the piston at TDC, the cylinder is filled with compressed air to hold the valve in place and prevent it from falling into the cylinder. The hole in the pry bar fits over the rocker arm stud and provides a pivot point to compress the spring.

OHC engines may require a special type of spring compressor (Figure 3-61). This style can usually be used with the head installed or removed from the engine block. This type bolts to the cylinder head and uses adapters attached to a threaded extension. Turning the T-handle forces the adapter down onto the spring and compresses it. Another version uses a pry bar–type handle to apply the pressure.

A C-clamp–type compressor is used to remove the valve springs with the cylinder head removed (Figure 3-62).

WARNING: Always wear eye protection whenever using a valve spring compressor. It is possible for something to slip or come loose, resulting in airborne parts.

Valve Spring Tension Testers. Before valve springs are used, they must be tested to determine if they are within the manufacturer's specifications for tension. Specifications are usually supplied for open and closed spring tensions. Two types of tension testers are used to make these measurements. The first uses a needle indicating torque wrench (Figure 3-63). The adjusting wheel is used to set spring height, and pressure is applied to the torque wrench until a click is heard. At this point, the value on the torque wrench is read and the reading is multiplied by two. This measurement should be within 10 percent of the valve closed specification.

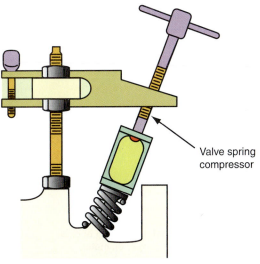

FIGURE 3-61 Some overhead camshaft engines require a special type of valve spring compressor. This type can be used with the cylinder head installed or removed from the block.

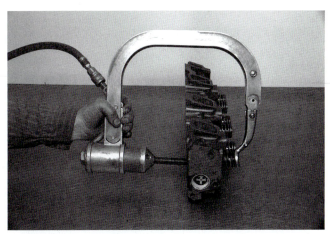

FIGURE 3-62 A C-clamp type spring compressor is used when the cylinder head is removed from the engine. This compressor is air operated, others may be hand operated.

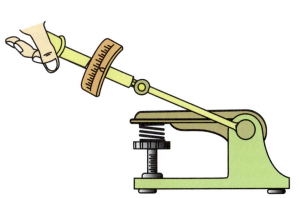

FIGURE 3-63 This style of valve spring tension tester uses a torque wrench to determine tension at the specified height.

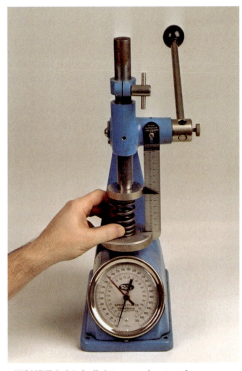

FIGURE 3-64 A dial type spring tension gauge.

CAUTION:
In order to obtain proper spring tension measurements, the spring tension must be checked at the installed spring height. If a shim is used, it must be inserted under the spring during this check.

The second type of tester uses a scale with a dial reading (Figure 3-64). Pressure is applied against the spring until it is compressed to the spring installation height specification. A scale on the right of the tester is used to indicate when this height is achieved. At this time, the dial is read to obtain tension measurements. Next, the pressure is increased until the valve open height is obtained, and the tension gauge is read again.

Valve Guide Renewing Equipment. Whenever the cylinder head and valves are reconditioned, always check the valve guide for wear. If the guide is worn or damaged, there are three common methods of renewing the guide: knurling, inserts, and replacement.

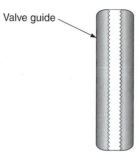

FIGURE 3-65 The knurling bit raises areas of the valve guide to decrease the bore diameter.

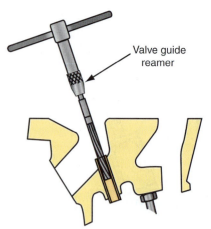

FIGURE 3-66 After knurling, the bore is reamed to the proper diameter.

Knurling is a two-step process of returning the valve guide to its proper bore. First, a knurl bit is driven into the guide using a drive reduction drill motor. The knurl bit forces the metal to rise and decrease the bore diameter (Figure 3-65). The peaks of the rises create a diameter smaller than the diameter of the valve stem. The next step is to ream the bore to specifications (Figure 3-66).

If the guide is excessively worn, it cannot be resized by knurling. Also, knurled guides wear faster than other methods of guide repair. The preferred method is to install valve guide inserts or replacement guides. These can be installed using hand tools; however, high volume shops may have cylinder head reconditioning machines that perform most operations at one setup.

The insert can be a bronze or cast-iron spiral coil, sleeve, or thin wall tube type. If the cylinder head is equipped with inserts, they can be driven out and the new ones installed. The new insert must be reamed to the proper size.

To install the bronze spiral coil insert, a special tap is used to cut an oversized thread into the guide. Next, the proper length of spiral coil is threaded into the guide bore. The coil is forced into the guide by use of a swaging tool. A reamer is then used to restore the correct bore.

The bronze sleeve and thin wall tube-type inserts are installed by reaming the guide to the proper diameter required for the insert. The sleeve is then driven or pressed into the guide. A swaging tool is used to force the sleeve into contact with the guide. The last step is to ream the insert to the proper diameter.

On most aluminum cylinder heads, the old guides can be driven out using a guide punch and a hammer. Some manufacturers specify that the head be heated before guide removal; always follow the manufacturer's service information. The new guides are already properly sized and are driven or pressed in.

Seat Grinder. Worn or damaged valve seats may be reconditioned. A common method of repair is to grind the seat to provide a new surface. The most common type of valve seat grinder is the concentric style (Figure 3-67). It provides a full contact on the seat, allowing the entire circumference of the seat to be ground in one motion.

The beveled grinding wheel is attached to a special holding fixture with a hollow center. The hollow center is placed over a pilot shaft inserted into the valve guide. The pilot shaft centers the stone onto the valve seat (Figure 3-68). The grinding stone is driven by a special high speed motor that turns about 1,800 rpm.

Grinding stones are available with different angles. Use the angle specified by the engine manufacturer. To properly resurface the valve seat, the grinding stone must be dressed before use. It may be necessary to dress the stone again during the resurfacing operation.

Most valve guide reamers come in sizes 0.001 inch apart.

CAUTION:
Before grinding the valve seat, check the valve guides for wear. It is important that the pilot shaft be properly centered. A worn guide will result in off-center grinding of the valve seat.

FIGURE 3-67 Typical valve seat grinder.

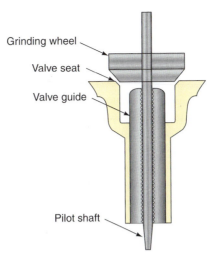

FIGURE 3-68 The seat grinding stone is centered onto the seat by a pilot shaft inserted into the valve guide.

WARNING: Always wear eye protection whenever using a seat grinder. Particles discharged from the seat grinding stone may cause eye injury. Also, do not use any machinery or equipment without first receiving proper instruction.

Seat Cutter. Another method of resurfacing the valve seats is to use a seat cutter. The valve cutter uses carbide cutters to cut a new surface (Figure 3-69). The cutter can be hand operated or driven by a low speed motor turning about 60 rpm. Three angle blades are available to cut the valve seat in one operation.

WARNING: Always wear eye protection whenever using a seat cutter. Small metal particles may fly upward toward your face.

Seat Inserters. If there is not enough material to resurface the valve seat, the valve seat is excessively worn or cracked, or if the valve stem tip installed height has increased more

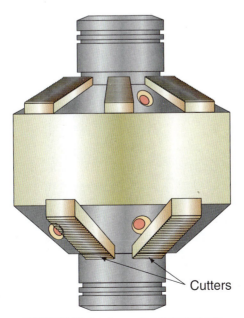

FIGURE 3-69 A valve seat cutter uses carbide bits to cut a new seat surface.

FIGURE 3-70 A special counterbore bit is used to prepare the cylinder head for valve seat inserts.

than 0.060 inch, it may be possible to install a seat insert. If the cylinder head is equipped with insert seats, they are knocked out first. The replacement insert is slightly larger than the original. Before installing the new insert, the counterbore must be enlarged using a special bit (Figure 3-70). This bit is also used to counterbore cylinder heads with integral seats.

Once the counterboring is completed, a driver is used to install the insert. A guide is installed to assure proper centering of the insert; then a hammer is used to impact the driver and seat the insert. Most inserts are interference fit to prevent movement, usually 0.006 in. (0.150 mm) to 0.008 in. (0.200 mm) for inserts installed into aluminum heads and 0.004 in. (0.100 mm) to 0.005 in. (0.125 mm) for inserts installed into cast-iron heads. This means that the diameter of the seat insert is actually larger than the diameter of the bore it will be pressed into.

 WARNING: Always wear eye protection whenever performing any grinding, cutting, or driving.

Valve Grinding Equipment. The valve face and stem tip may be reconditioned using special valve grinding equipment (Figure 3-71). A chuck accepts and holds the valve stem. The chuck is mounted on a sliding surface that is moved by a lever. The angle of contact between

FIGURE 3-71 Typical valve grinder used to recondition the valve face and tip.

the valve face and grinding stone is adjustable. Use the engine manufacturer's specifications to determine the correct face angle.

The valve rotates with the chuck when the motor is turned on. The operator uses the lever to move the valve face across the grinding stone. Some valve grinding equipment has a second grinding wheel used to refinish the valve stem tip.

WARNING: **Always wear eye protection whenever using any valve grinding equipment.**

Head Resurfacing Equipment. The mounting surface of the cylinder head to the block must be flat. Over a period of time, or if the engine overheats, this surface may become warped. Other reasons to resurface the cylinder head are to raise the compression ratio and to square the deck to the bores.

Some shops resurface the cylinder head with a **belt surfacer**. This method uses a belt that operates similar to a conveyor belt. The head is placed on the belt and held in position by a restraint rail. The operator applies hold-down pressure while slightly moving the cylinder head on the belt to achieve the desired surface finish (Figure 3-72).

Using the belt surfacer has disadvantages. This method does provide uniform finishing, but it does not square the head (Figure 3-73). For these reasons, the preferred method is to resurface the cylinder head using cutting bits. The milling machine uses large rotating cutters to remove thin layers of material (Figure 3-74). The cylinder head is mounted to a table. Depending on the type of milling machine, either the table is moved below the cutters, or the cutters move across the head. Usually about 0.005 in. (0.127 mm) is removed with each pass. The desired finish cut is achieved by changing the speed and depth of cut.

If the cutters are mounted below the table, the machine is called a **broach**. Like the milling machine, the cutters move in a horizontal plane. The cylinder head is mounted to the table, which moves over the cutters.

Another common style of resurfacer is the **surface grinder**. The surface grinder can be used to resurface cylinder heads or engine blocks (Figure 3-75). The head is mounted to an

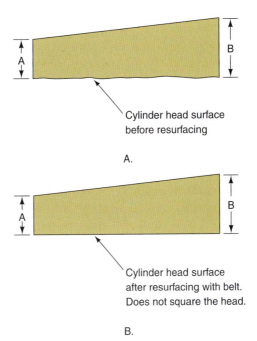

FIGURE 3-73 A belt resurfacer will not square the cylinder head.

FIGURE 3-72 Using a belt type sander.

FIGURE 3-74 A milling machine uses cutters to remove layers of material from a cylinder head or block.

FIGURE 3-75 Some surface grinders can be used to resurface the engine block.

adjustable table, and coolant is pumped onto the work piece as the table moves below the grinder wheel. About 0.002 in. (0.050 mm) is removed with each pass. When the grinder completes a pass resulting in a smooth surface, additional passes are made without any further metal being removed. This provides a finish on the surface and assures the surface is straight.

WARNING: Always wear eye protection whenever resurfacing a cylinder head or block.

Profilometers can be used to measure the surface roughness after the machining operation is completed. A stylus is moved across the surface, and the profilometer measures the distance between the peaks and valleys of the cuts. For proper head gasket sealing, a finish between 60 and 120 **microinch** is desirable for most applications. Some manufacturers specify a surface finish as smooth as 20 microinch, so always check the service information for the application you are working on.

Engine Block Reconditioning Equipment

During engine rebuilding, special tools and equipment are used to properly disassemble, assemble, and recondition the engine block. Some of these tools include:

- Camshaft bearing remover and installer
- Ridge reamer
- Cylinder boring equipment
- Cylinder hone
- Align boring equipment
- Block resurfacing equipment

Camshaft Bearing Remover and Installer. The type of camshaft bearing remover and installer used depends on the engine design. Most OHV engines, with the camshaft in the engine block, will use a bushing driver and hammer (Figure 3-76). The correct size mandrel is selected to fit the bearing. Turning the handle tightens the mandrel against the camshaft bearing. Then the bearing is driven out by hammer blows. The same tool is used to replace the bearings. Some OHC engines require a special puller/installer (Figure 3-77).

Ridge Reamer. As the piston moves within the cylinder, the rings cause wear on the cylinder walls. The greatest amount of wear is near the top of the travel due to the high pressures, high temperatures, and little lubrication in this area (Figure 3-78). The resultant ridge must be removed before attempting to remove the pistons from the block. This ridge

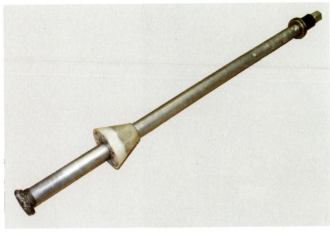

FIGURE 3-76 Typical camshaft bearing tool used for camshaft bearings in the block.

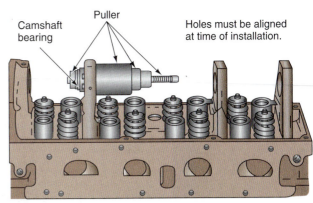

FIGURE 3-77 A special camshaft bearing puller/installer is used on OHC engines.

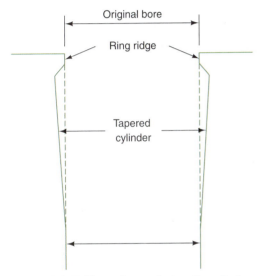

FIGURE 3-78 Piston ring contact on the cylinder wall causes a ridge to be formed.

FIGURE 3-79 Use a ridge reamer before removing the pistons to prevent damage to the pistons.

A **ridge reamer** is a cutting tool used to remove the ridge at the top of the cylinder.

CAUTION:
Attempting to remove the pistons before removing the ridge may result in damage to the rings and piston lands. Also, damage to new rings may result if they are installed without the ridge removed.

is removed using a **ridge reamer** (Figure 3-79). When using a ridge reamer, do not remove any more material than necessary to remove the pistons.

WARNING: Always wear eye protection whenever using a ridge reamer.

Cylinder Boring. If the cylinder bore is excessively worn, tapered, or out-of-round, it may be bored to the next piston size. In addition, cylinder boring may be performed to increase the displacement of the engine. The **boring** process causes surface fractures that extend to a depth from 0.0005 to 0.001 in. (0.01 to 0.02 mm). For this reason, the cylinder is bored to a size 0.003 in. (0.05 mm) smaller than required. The additional material is removed during the honing process. The bore machine has a feed mechanism that moves the rotating cutters through the cylinder.

Boring is the process of enlarging a hole.

WARNING: Always wear eye protection whenever using a cylinder bore.

Cylinder Hone. After the cylinder is bored, it must be honed to obtain proper ring and cylinder seating. The rigid cylinder hone uses two to four abrasive stones to remove cylinder wall material (Figure 3-80). When the adjustment nut is turned, it forces the stones outward

FIGURE 3-80 A cylinder bore hone.

toward the cylinder walls. The nut is adjusted until the stones contact the walls, then the hone is driven by a heavy duty drill motor. As the stones are rotated in the cylinders, the hone is also moved up and down.

The first passes may be done with coarse stones, while finishing cuts are performed with finer stones. The final finish should have diamond-shaped grooves that cross at 30–40-degree angles (Figure 3-81).

Another style of hone is the brush hone (Figure 3-82). The round stones at the end of the bristles cut away the material much like the rigid hone. This hone is simpler to use and provides good results; however, this type of hone cannot remove taper or out-of-round

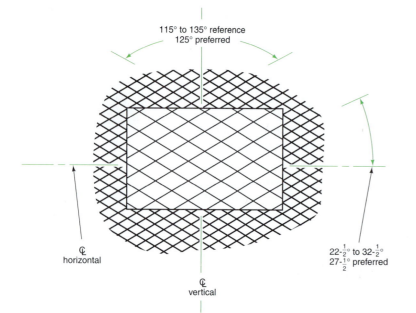

FIGURE 3-81 Desired finish pattern after honing the cylinder.

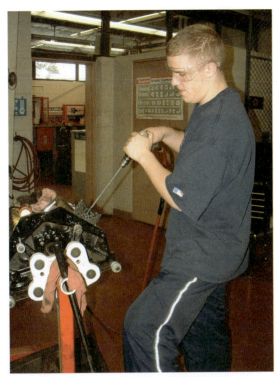

FIGURE 3-82 Use a brush hone to de-glaze the cylinder wall.

from the cylinder bore. The brush hone is used to break the glaze in the cylinder and provide a new surface finish. The smoothness of the surface finish depends on what type of rings will be used. Some manufacturers do not recommend the use of a brush hone. Refer to service information from the manufacturer and the ring supplier to achieve the proper cylinder finish.

Torque Plates. During the boring and honing process, the cylinder head is removed. When the head is reinstalled and torqued, it sets up stress that results in dimensional changes. To prevent any alignment problems, many engine manufacturers require a **torque plate** be attached to the engine block prior to boring and honing of the cylinders (Figure 3-83). The torque plate is installed with the same torque required to bolt the cylinder head to the block. This stresses the block in the same manner as the head. The main bearing caps should also be installed and torqued. Any boring and honing operations will now be true.

Torque plates are metal blocks about 2 inches thick. They are bolted to the cylinder block at the cylinder head mating surface to prevent twisting during honing and boring operations.

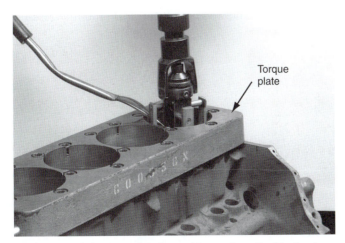

Torque plate

FIGURE 3-83 A torque plate is used to prestress the block prior to boring or honing the cylinders.

Align Boring Equipment. Over a period of time, the main bearing saddles may become warped and out of alignment. To correct this, a boring bar or align boring machine is used. A boring bar is centered in the two main bearing saddles at either end of the block. Before installing the main bearing caps, about 0.002 in. (0.05 mm) is removed from the base of the cap. The boring bar uses cutters to remove material from the main bearing saddles and caps to realign the crankshaft bearing bores.

If the crankshaft bores are not excessively misaligned, it may be possible to use an **align hone** instead of a boring bar. The hone has a long arbor bar with spring-loaded honing stones. First, remove 0.002 to 0.003 in. (0.05 to 0.07 mm) of material from the parting surface of the main bearing caps. With the caps installed to the engine block, insert the align hone. A special motor rotates the hone while the arbor is moved back and forth in the bores.

The **align hone** is able to hone all bearing bores at the same time to realign them.

WARNING: Always wear eye protection whenever using an align bore bar or machine.

Block Resurfacing Equipment. As with the cylinder head, the engine block deck may require resurfacing at the mating surfaces of the head gasket. Generally, the same equipment used to mill or grind the cylinder head is used to resurface the block.

Crankshaft Reconditioning Equipment

To correctly determine if the crankshaft is straight, a dial indicator is used in conjunction with V-blocks. With the dial indicator on the center main bearing journal, the crankshaft is rotated through one complete revolution. Total movement of the indicator needle is the bow of the crankshaft. Always check the results with the manufacturer's specifications. While the crankshaft is set into the V-blocks, check the flywheel flange and vibration dampener mounting surface for runout.

If the crankshaft is bent, it is usually replaced. In some cases when the crankshaft is antique or custom-made it can be straightened using a specialized press.

Crankshaft Grinder. If the main and connecting rod journals are worn or damaged, the surface can be reconditioned by using a crankshaft grinder (Figure 3-84). The grinder gradually removes material from each journal until it is perfectly round and smooth.

Crankshaft Polishers. After the crankshaft journals have been resurfaced, they must be polished. The crankshaft polisher is like a belt sander with fine grit paper and may be attached to the crankshaft grinder (Figure 3-85). The polisher smooths the surface of the journals to provide a low friction surface for the bearings.

Crankshaft Balancers. A crankshaft that is out of balance will result in vibrations. When balancing the crankshaft, there are two major concerns: the reciprocating weight of the pistons and connecting rods, and the rotating weight that turns around the axis of the crankshaft.

CAUTION: Straightening the crankshaft may result in cracks. Carefully inspect the crankshaft after straightening. If there are any doubts, do not use it. The crankshaft may be magnetically tested for cracks.

Classroom Manual Chapter 12

FIGURE 3-84 Crankshaft grinder.

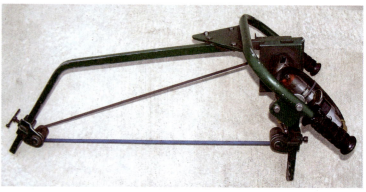
FIGURE 3-85 Crankshaft polisher.

FIGURE 3-86 A special crankshaft balancing machine.

To balance the crankshaft, it is installed to a special balancer (Figure 3-86). If the crankshaft is out of balance due to a heavy counterweight, a hole is drilled in the counterweight to lighten it. If the counterweight is too light, a hole is drilled in the counterweight and a heavy metal plug is inserted.

Connecting Rod and Piston Reconditioning Equipment

There may be instances in which the connecting rod or piston may need to be reconditioned. The bore of the connecting rod may need to be resized due to wear. The most common tools and equipment used to recondition connecting rods and pistons include:

- Pin drift
- Ring end-gap grinders
- Pin hole machine
- Rod aligner
- Rod cap grinder
- Rod heaters

Pin Drift. A piston pin drift is used to remove press fit piston pins (Figure 3-87). Pin drifts are available in assorted sizes to accommodate the different sizes of piston pins. When using the pin drift, it is important to properly support the piston assembly to prevent damage to the piston.

A piston pin may be called a wrist pin.

Ring End-Gap Grinders. If the ring end gap is smaller than specifications, in some cases it can be enlarged by removing some of the ring material. This may be accomplished by use of a file or by a special ring end-gap grinder (Figure 3-88). Many ring end-gap grinders are

FIGURE 3-87 Piston pin drift set.

FIGURE 3-88 This type of ring grinder helps when correcting ring end gap.

FIGURE 3-89 Connecting rod honer.

FIGURE 3-90 A rod aligner can be used to check and straighten some connecting rods.

portable units that are secured into a vise. A crank is turned to rotate a small grinding wheel that removes material from the ring.

WARNING: **Always wear eye protection whenever using a grinder.**

Pin Hole Machine. Worn connecting rod piston pin bosses or big end bores can be corrected using a special **rod honer** (Figure 3-89). Like honing the main bearing bores, this process removes very little material with each pass. An abrasive stone is rotated inside the bore while cooling oil is pumped over the work piece. The operator moves the connecting rod over the full length of the stone. Measurements are taken regularly until the desired diameter is obtained.

A **rod aligner** measures connecting rod twist and bend.

Rod Aligner. A connecting rod that is twisted or bent will result in early failure of the rings, piston, cylinder wall, and rod bearings. A **rod aligner** makes it possible to straighten some bent connecting rods (Figure 3-90).

Rod Cap Grinder. Worn connecting rod saddles and caps may be restored to their original size. To do this, material is removed from the cap and the connecting rod saddle at the parting surfaces. Usually about 0.002 in. (0.05 mm) is removed from each surface. This makes the bore smaller than specifications. The bore is restored to standard size by honing.

Rod Heaters. Pressing interference fit piston pins into the connecting rod bore can damage the piston. A rod heater is used to heat the small end of the connecting rod and provide easier assembly of the rod to the piston without having to press in the pin. When the small end of the rod is heated, the bore expands, allowing the piston pin to be pushed into place. When the rod cools, the pin is locked into place.

WORKING AS A PROFESSIONAL TECHNICIAN

Before you are given the responsibility of rebuilding an engine, you will need to prove yourself as an excellent general service technician. Once you have proven that you consistently perform quality work and pay attention to details, you may be given major engine work. Overhauling an engine is an expensive proposition for the consumer, and no shop owner wants to have major work come back with a defect. In today's service industry, many engines are replaced with remanufactured units rather than rebuilt in house. These come with a written warranty that protects both the consumer and the shop owner if the engine fails. Whether you rebuild or replace an engine, you may generate repair bills in excess of $3,000. Your work is expected to hold up for many thousands of miles. A seemingly small oversight during diagnosis, assembly, or installation could cause catastrophic engine failure. When your physical work is complete, you will need to describe your work effectively to the manufacturer, your service advisor, and/or the customer. Engine repair work is performed by professionals; learn to work and act as a professional technician.

Education

Your training to become a professional automotive technician is critical to your success. You must have a solid technical education and a well-rounded general education. You are required to understand a huge amount of technical information and to regularly interpret new information. As soon as you leave school, you will need to find ways to keep your training up-to-date (Figure 3-91). Specialized skills required to repair engines must be accompanied by a thorough understanding of automotive systems.

Your ability to perform mathematical functions is essential; an error of one thousandth of an inch (0.001 in.) could destroy an engine repair. Accuracy in measurements, interpretation of specifications, and conversion of numbers are all required skills. Additionally, the problem solving and analytical thinking used in math will be useful in almost every repair job. When you approach a vehicle problem, you will need to analyze the situation and filter information to guide you through a logical process of determining the cause. When an engine comes in with a serious internal defect, you must analyze potential causes and be sure that you correct the source of the problem not just the symptom. Overheating, for example, can seriously damage an engine. If you were to install a remanufactured engine without replacing the restricted radiator, you could wind up with a very unhappy customer and boss. Your diagnostic process must be logical, thorough, and effective for you to be successful.

You also need excellent communication skills. When performing a **warranty repair**, it is your responsibility to effectively describe the customer concern, the cause of the problem, and all the necessary corrections. The manufacturer relies on the information you provide on the repair order to determine whether it will reimburse the dealer for the parts and labor required for the job. Failure to properly document your engine work could cost your

A **warranty repair** means that the manufacturer reimburses the shop for the parts and labor to make a repair.

FIGURE 3-91 Reading professional periodicals can help you keep your knowledge up-to-date.

FIGURE 3-92 Explain your repair fully and professionally to gain respect and trust from your customers and co-workers.

employer thousands of dollars. Repair information is also kept on file so that vehicle history is available when other repairs are needed. It is also extremely important that you can communicate effectively with your service advisors, service manager, and customers. You will need to be able to explain technical information in a way that lay people can understand. Your service advisors rely on your description of your work to provide information to the customer. Often, particularly with large jobs such as an engine overhaul or replacement, the customer will want to speak directly with the technician (Figure 3-92). You must be able to professionally and effectively describe your work to customers.

Professionalism

From the moment you begin repairing vehicles, either as a student or a paid professional, the image you portray will help define your role in the shop, your rate of pay, your opportunities for advancement, and your overall success in the automotive industry. Clean work uniforms, proper language and behavior, and excellent communication skills will help develop your image as a professional service technician. Your presentation helps you as an individual and the service industry as a whole. When you dress, behave, and speak like a professional you will be treated and paid accordingly.

There will be times when you are frustrated with a job. Take a short break and walk away before losing your cool. Profanities will not loosen a bolt or help your career. Your employer, a potential employer such as a manufacturer's representative, or a customer could be in earshot at any time. It will damage your image and your potential if they hear a string of foul language coming from you.

Keep yourself and your work area clean; you should always be prepared to present yourself to a customer. Neither your employer nor your customers want a filthy technician in a nice vehicle. Customers' vehicles are typically their second largest investments; they want a professional to provide service. You can perform the best engine repair, but leave a spot of grease on the driver's seat and the customer will be unhappy with the whole job (Figure 3-93).

Your ability to communicate well with your service advisors, your employer, and your customers is essential to your career. If you can't describe your work professionally, you simply won't get that work. The more complex and in-depth the job, the more critical it is that you can explain your diagnosis and repair to management and your customers. When performing engine repairs, you'll need to explain why the problem occurred, how you found it, and what you did to correct it. If you are discussing an engine rebuild, you are

FIGURE 3-93 This customer would not appreciate a smudge of grease on her spotless interior.

trying to "sell" thousands of dollars of work. Your employer relies on your explanations to get authorization for the job. Imagine the different effects the two following explanations might have on a customer deciding to have the work performed at your shop. In the first case, the technician says, "The engine blew; I'll have to totally overhaul it. It will probably cost around two thousand dollars." In the second case, the technician says, "Two cylinders have low compression caused by worn piston rings. This is the cause of your lack of power. Usually when the rings are worn, there are other components in the engine that will need service or replacement. Once we disassemble and evaluate the engine, we can offer you specific repair options. In the worst case, a thorough engine overhaul or replacement may be needed; this could cost upwards of two thousand dollars." Who is the customer more likely to trust? If you were the boss, which technician would you choose to perform the work? Offer your customers and your management respectful and professional communication; they will respond in kind.

Compensation

You have the opportunity to enter a rewarding profession. Your skills are in high demand; rarely will a competent technician have any difficulty securing work. Skilled technicians can make a very comfortable wage. Your pay will increase as your skill level improves.

Most technicians start out working "**straight time**." This means that you will be paid a dollar amount for every hour that you are at work. This gives you a chance to learn the shop practices and develop some speed in repairing vehicles. Some technicians will work for an hourly wage throughout their careers. A technician who performs primarily diagnostic work or electrical work will often be paid straight time. Sometimes those who overhaul engines are paid straight time too. The employer understands that the speed of the overhaul is not nearly as critical as thoroughness and attention to detail.

Many shop owners will change your method of payment to "**flat rate**" pay when they determine that you will be productive on this system. When you are paid flat rate, it means

Straight time pay means that you are paid for every hour you are physically at work, regardless of your productivity.

Flat rate pay means that you are paid for each hour of billable work as determined by the manufacturer, an aftermarket labor-estimating guide, or your shop.

you get paid by the job hour. If, for example, a vehicle needs a headgasket, the service advisor will look up the flat rate time and quote that amount of labor to the customer. When you do the job in six hours rather than the quoted eight hours, you will be paid for eight hours. Your billable hours for the week are added up to calculate your pay. If you produced fifty-six hours worth of work while actually working for only forty-five, you will be paid for fifty-six hours. In a strictly flat rate shop, this can also work to your disadvantage. If a few jobs take you much longer than the posted flat rate time, you may actually be paid for fewer hours than you worked that week. This could happen, for example, if you replace a timing belt on a vehicle one week and the next week it comes in with a snapped timing belt. Many shops will not pay you again to repair a "come back"; if your service failed, you can be held accountable and repeat the job without getting paid hours for it. This is just one of the many reasons why it is critical to perform quality work.

Some shops use a combination of the two forms of payment. They may guarantee you a certain number of base hours per week and pay you extra for increased production. This ensures that the technician is adequately compensated every week, regardless of the type or quantity of work available. It also serves as an incentive to be as productive as possible.

ASE Certification

The **National Institute of Automotive Service Excellence (ASE)** offers a well-recognized national certification program for automotive technicians. The standardized ASE tests are offered in eight basic areas of automotive repair as well as in several specialized and advanced areas. When you pass all eight basic tests, you become an ASE-certified master automotive technician. Engine repair is one of the eight core tests needed to become a master technician.

The ASE tests are written tests offered twice a year at test sites all across the country. The tests check your practical knowledge of automotive systems and of diagnostic and repair techniques. To become certified, you must pass the test and prove that you have two years of work experience. If you complete a two-year college program in automotive technology, it will substitute for one year of experience. Once you are certified, you will receive a certificate, a wallet card, and a shoulder patch documenting your achievement (Figure 3-94). Technicians must retake the tests every five years to keep their certifications current.

The **National Institute of Automotive Service Excellence (ASE)** offers nationally recognized certifications to automobile and heavy duty truck technicians through a combination of biannual testing and proof of professional experience.

FIGURE 3-94 This ASE-certified technician wears his ASE patch proudly; it is a sign of accomplishment.

The engine repair ASE test asks questions about general engine diagnosis, cylinder head service, engine block service, cooling and lubrication system repair, and a few questions on fuel, electrical, ignition, and exhaust systems. A free preparation booklet is available from ASE. It lists the tasks covered in the test and offers some sample questions. Many of the questions are in the format we use in this shop manual:

Technician A says that a compression test can locate a weak cylinder. *Technician B* says that a cylinder leakage test can pinpoint the cause of a low compression reading. Who is correct?

a. Technician A only
b. Technician B only
c. Both Technician A and Technician B
d. Neither Technician A nor Technician B

In this question, both technicians are correct, as you'll learn in upcoming chapters. Successfully answering these questions requires a thorough knowledge of the subject and careful reading. While many technicians find the questions tricky, if you really know the information, the correct answer will usually be clear. The engine repair test is made up of fifty multiple-choice questions in the following areas:

Content Area	Number of Questions	Percentage of Test
General engine diagnosis	15	30%
Cylinder head and valvetrain diagnosis and repair	10	20%
Engine block diagnosis and repair	10	20%
Lubrication and cooling systems diagnosis and repair	8	16%
Fuel, electrical, ignition, and exhaust systems inspection and service	7	14%

AUTHOR'S NOTE: Don't get stumbled by an ASE-style question. There are many of them, both in this book and on the ASE tests, and you should take your time when answering them. Try to use this method when formulating an answer.

First read the sentence(s) at the beginning of the question. Sometimes there is not an introductory sentence, but if there is it should be considered as factual and pertinent to the question. Then read what *Technician A* says and ask yourself if that technician's statement is correct. Do not read any further until you have formulated an answer. Knowing if the first technician is correct or not will allow you to eliminate half of the answers. Then read *Technician B*'s statement and ask the same question, Is he correct? Now you should be able to answer the question easily because you have already eliminated half of the answers. Try this technique and see if it works for you. It may help you not get confused even though you may know the answer.

The ASE tests are written by industry experts: technicians; technical service experts; vehicle, parts, and tool manufacturers; and automotive instructors. Each question must be accepted by each member of the board and then tested on a sample of technicians. The tests do a good job of checking your technical knowledge. Certification is voluntary, but most employers prefer to hire ASE-certified technicians. ASE certification shows your employer that you are knowledgeable and motivated to excel (Figure 3-95).

FIGURE 3-95 This tech and her shop boast the blue seal of excellence.

Skills Required for Engine Repair

Special skills are required to succeed at engine repair. You must thoroughly understand engine operation and design to effectively diagnose problems. For example, you'll need to remember how critical engine breathing is to proper performance when diagnosing a low power condition as a restricted exhaust system. Your study and practice of the information and procedures described in this text will provide you with a solid base of engine operation, diagnosis, and repair knowledge.

You must learn to thoroughly diagnose an engine problem before attempting a repair. Don't attempt to guess, test. We describe many different diagnostic procedures to pinpoint the cause of engine problems. Study and practice these tests now, so you will be competent at them in the shop. The customer should receive a realistic quote for engine work; you should not get halfway through the job and *then* realize that more significant work is needed. Performing the proper tests before disassembly will allow you and your service advisor to estimate the required work. They will also help you locate the root cause of the problem, so that your repair is 100 percent effective.

There are many repair procedures using special tools that you must master in order to become skilled at engine repair. An engine overhaul requires thorough preliminary diagnosis, accurate measurement with special tools, precision machining and refinishing, and careful torque and assembly techniques. You must be competent at all the techniques to complete the overhaul. In many cases, a significant portion of these repairs will be performed at an engine machine shop.

In many cases, your diagnosis will uncover engine problems that may be more easily and cost effectively resolved with a rebuilt engine. These "crate" engines are delivered to your shop fully assembled except for all the accessories such as water pumps, generators, and electronic components. This type of repair is clearly the trend in the industry, due to the high cost of labor and individual parts. The crate engines come with a warranty, so the

FIGURE 3-96 Careful measurement and observation are essential for quality engine work.

shop is not responsible for the cost of repairs if the rebuild were to fail. This makes your job of diagnosis even more critical; you certainly wouldn't want to install a "new" engine if the problem is not actually an internal engine failure. You will carry much of the responsibility for determining whether the engine should be repaired or replaced.

Equally as important as your technical skills is your attitude. An engine repair technician must be patient enough to be extremely thorough. You cannot rush through critical engine measurements and expect accurate results. You need to thoroughly evaluate the problem and complete each step required for repair (Figure 3-96). Your attention to detail is essential to your success. Locating a cracked cylinder head requires close inspection. Forgetting to analyze the engine bearing wear patterns could cause you to miss a problem that would damage your freshly rebuilt engine. Aside from taste, all your senses are useful during engine diagnosis. Listen for abnormal noises before disassembly; smell the engine oil for signs of contamination or overheating; look carefully at each part for clues about the causes of failure; and touch the cylinder walls, crank journals, and valve stems to feel for wear.

Use logical steps to analyze and repair faults. A haphazard approach to engine repair will likely cause you to miss something critical. You will learn all the steps required to fully analyze and overhaul an engine; keep notes on the steps that you can refer to when you perform your first several overhauls. If you forget one step, such as measuring crankshaft end play, the engine could come apart within the first few hundred miles of operation.

Keep yourself well-organized. Label brackets and wires and hoses when you remove the engine or cylinder head. Many technicians put the bolts removed during disassembly loosely back in position or label them in bags. Locating the proper bolts for the timing cover could become a time-consuming and frustrating task if you just dump all the hardware into one bin. Organize your engine measurements into a "blueprint," so that you can look at all the data to determine what repairs need to be made. If you don't write your results down as you go along, you will forget some information and have to repeat the procedure. You will need documentation to justify your repairs for warranty purposes or for the customer. As an engine repair technician, you will need patience to work your way through the big job of diagnosing and repairing an engine's mechanical failure.

Honesty is essential when building your reputation. No one wants to pay for unneeded work. Your employer does not want to be caught in a lie to the customer. Be realistic about your estimation of what is required for the repair. If you minimize the work, you will have to come back to the customer later to get authorization for the additional work. If you overestimate the necessary repairs, the customer could get a more realistic quote elsewhere. That customer will never return. If you make an honest mistake, offer an honest explanation. Don't try to cover up mistakes with unprofessional work. All technicians make mistakes; you will too. What separates the excellent technicians from the shoddy ones is how they deal with the mistake.

Mastery as an engine repair technician requires technical knowledge, hands-on experience, and good work habits. If you enjoy engine work, learn and practice the diagnostic and repair techniques presented here. Then, always remember how essential your professional attitude is to your success. Engine repair is a challenging and rewarding part of automotive service; aim for excellence.

SERVICE INFORMATION

Service information is one of the most important tools for today's technician. It provides information concerning engine identification, service procedures, specifications, and diagnostic information. In addition, service literature provides information concerning wiring harness connections and routing, component location, and fluid capacities. Service information can be supplied by the vehicle manufacturer or through aftermarket suppliers.

Manufacturers supply most of their service information electronically. Service information is provided on CDs or DVDs and the material is updated regularly (Figure 3-97). Most dealerships retain an archive of service information for older vehicles. Manufacturers also have their service information available on the Internet. This information is available to technicians working in the manufacturers' dealerships. It is also available to any individual for a fee. A list of automobile manufacturers' service information websites is available at www.nastf.org. Pricing is based on a one day, one month, or an annual subscription. Manufacturers also provide training materials to their service technicians. Many use Internet-based training for new product and new system updating. Additionally, manufacturers may offer hands-on training seminars in different regions.

Technicians working in aftermarket repair facilities have a more difficult time accessing adequate service information for all vehicles. There are a few comprehensive subscription-based information providers. Shop Key and All Data, for example, are service information systems that provide repair information for foreign and domestic vehicles back to 1982. Service information for different years of vehicles or for particular types

FIGURE 3-97 Computers are replacing paper service information in many shops.

of service is also available. Some shops subscribe to a telephone-based troubleshooting service. When a technician has a difficult diagnosis he can call a tech line specialist who will help troubleshoot the problem.

To obtain the correct information, you must be able to identify the engine you are working on. This may involve using the vehicle's vehicle identification number (VIN). This number has a code for model year and engine. Which numbers are used varies between manufacturers, but the service information will provide instructions for proper VIN usage. The service information may also assist you in engine identification through the interpretation of casting numbers and marks on the block or cylinder head.

With the engine properly identified, the required information can be retrieved. In the past, each manufacturer and manual publisher used its own method of organizing its manuals. Recent guidelines now require manufacturer service information to have a standard organization (Figure 3-98).

Procedural information provides the steps necessary to perform the task. Most service information provides illustrations to guide the technician through the task (Figure 3-99). To get the most out of the service information, you must use the correct information for the specific vehicle and system being worked on, and follow each step in order. Some technicians lead themselves down the wrong trail by making assumptions and skipping steps.

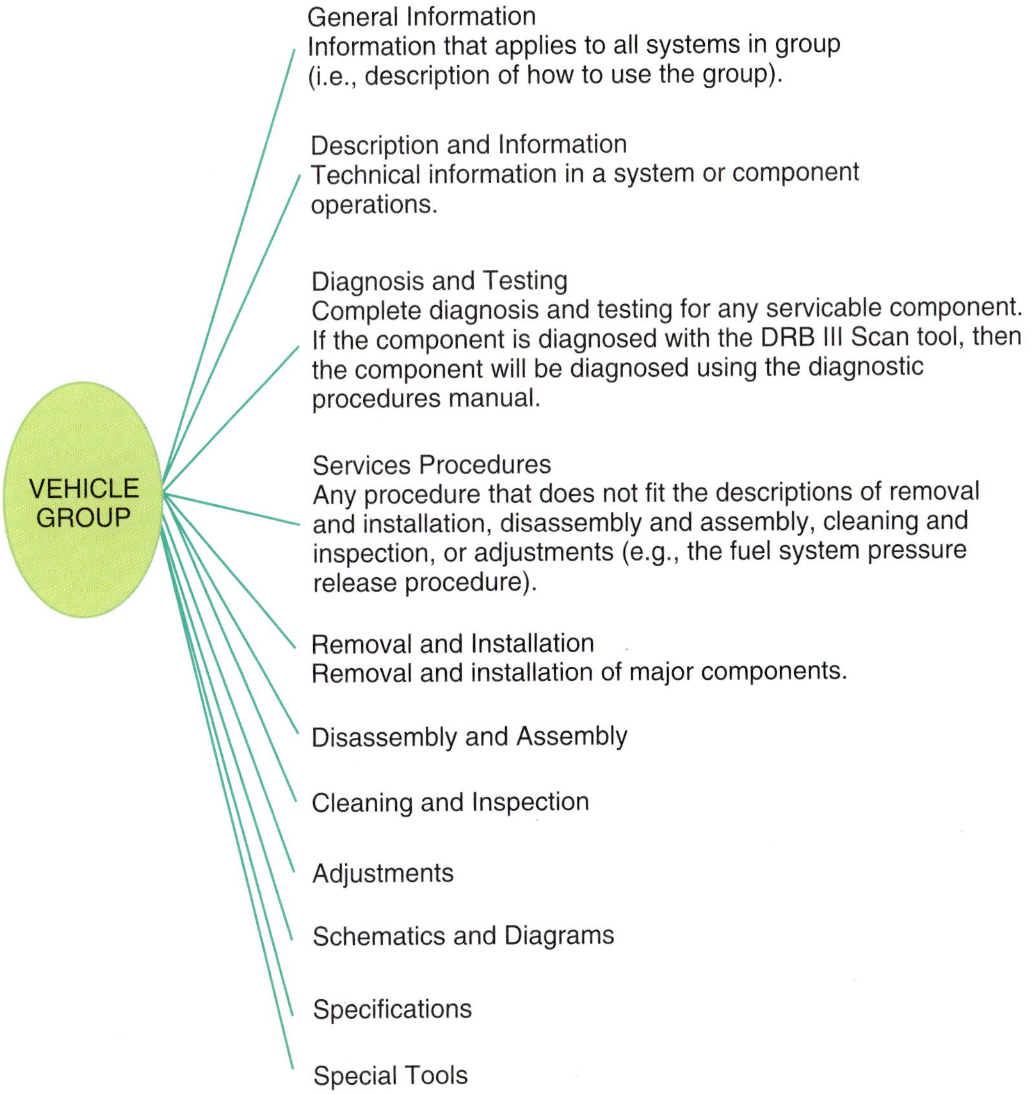

VEHICLE GROUP

General Information
Information that applies to all systems in group (i.e., description of how to use the group).

Description and Information
Technical information in a system or component operations.

Diagnosis and Testing
Complete diagnosis and testing for any servicable component. If the component is diagnosed with the DRB III Scan tool, then the component will be diagnosed using the diagnostic procedures manual.

Services Procedures
Any procedure that does not fit the descriptions of removal and installation, disassembly and assembly, cleaning and inspection, or adjustments (e.g., the fuel system pressure release procedure).

Removal and Installation
Removal and installation of major components.

Disassembly and Assembly

Cleaning and Inspection

Adjustments

Schematics and Diagrams

Specifications

Special Tools

FIGURE 3-98 Uniform service information layout.

1. Measure the diameter of the piston pin.
   Piston pin diameter:
   Standard 20.994 to 21.00 mm
   (0.8265 to 0.8268 in.)

3. Measure the piston-to-pin clearance. If the piston pin clearance is greater than 0.022 mm (0.0009 in.), remeasure using an oversized piston pin.

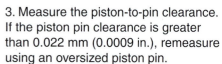

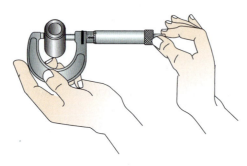

2. Zero the dial indicator to the piston pin diameter.

4. Check the difference between piston pin diameter and connecting rod small end diameter.

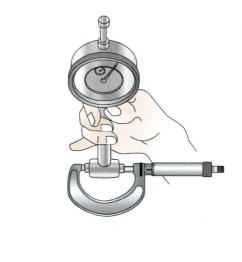

FIGURE 3-99 A step-by-step procedure for inspecting the piston assembly components.

Torque, end play, and clearance specifications may be located within the text of the procedural information. In addition, specifications are provided in a series of tables (Figure 3-100). The heading above the table provides a quick reference to the type of specification information being provided.

Diagnostic procedures are often presented in a chart form or a tree (Figure 3-101). The tree guides you through the process as system tests are performed. The result of a test then directs you to a branch. Keep following the steps until the problem is isolated.

Since the service information is divided into several major component areas, a table of contents is provided for easy access to the information. Each component area of the vehicle is covered under a section in the service information (Figure 3-102). Using the table of contents identifies the section to refer to. Once in the appropriate section, a smaller, more specific table of contents will direct you to the page on which the information is located.

TORQUE SPECIFICATIONS

Fan Assembly Motor-to-Fan.. 3.3 Nm (29 lbs. in.)
Fan-to-Radiator Support Bolt .... 9 Nm (80 lbs. in.)
Hose Clamps
    Heater Hose ....1.7 Nm (15 lbs. in.)
    Radiator Hose...3.4 Nm (30 lbs. in.)
Lower Air Deflector to Impact Bar... 2 Nm (18 lbs. ft.)
Throttle Body Inlet Pipe Bolt 3.1 liters (VIN T) .... 25 Nm (18 lbs. ft.)
Transmission Oil Cooler Fittings at Radiator .... 27 Nm (20 lbs. ft.)
Trans. Oil Cooler Pipe Connections (Alum. Radiator)............. 20 Nm (15 lbs. ft.)
Radiator Outlet Pipe to Block 2.01 (VIN K)............................ 27 Nm (20 lbs. ft.)
Radiator to Radiator Support Bolts.. 10 Nm (90 lbs. in.)
Thermostat Housing Bolts
    2.0 liters (VIN K) & 3.1 liters (VIN T)................................ 27 Nm (20 lbs. ft.)
Coolant Pump-to-Block Bolts
    2.0 liters (VIN K).. 25 Nm (19 lbs. ft.)
    3.1 liters (VIN T).. 24 Nm (18 lbs. ft.)
Coolant Pump-to-Front Cover Bolts
    3.1 liters (VIN T) .... 10 Nm (90 lbs. in.)
Coolant Pump Pulley-to-Pump Bolts
    2.0 liters (VIN K) .... 24 Nm (17 lbs. ft.)
    3.1 liters (VIN T) .... 21 Nm (15 lbs. ft.)
Surge Tank Bolts .... 4 Nm (15 lbs. ft.)
Surge Tank Pipe to Block Bolt 3.1 liters (VIN T)...................... 8 Nm (70 lbs. in.)
Temperature Sending or Gauge Unit .... 27 Nm (20 lbs. ft.)

FIGURE 3-100 Service information provides specification tables.

Service procedures may change in mid-year production, or problem components may be replaced with new designs, and so on. Technical service bulletins (TSBs) provide up-to-date corrections, additions, and information concerning common problems and their fixes. TSBs can save hours of work for the technician. These days many symptoms may be cured by reprogramming the PCM. Trying to resolve the customer concern without the appropriate TSB may not even be possible.

CASE STUDY

A customer hears a knocking noise from the engine when it is cold. The technician believes the cause is piston slap but needs to confirm the diagnosis. With the cylinder heads removed, the technician uses a dial bore gauge to measure bore size and out-of-round. After comparing the measurements with specifications provided in the service information, the technician finds that a cylinder was oversized. A technical service bulletin concerning this condition recommends replacing the piston with a modified unit. After receiving customer approval, the technician performs the needed repair, making sure to properly torque all fasteners to the specified value.

TERMS TO KNOW

Align hone
Aligning bar
ASE
Belt surfacer
Boring
Broach
Calipers
Compression gauge
Cylinder bore dial gauge
Cylinder leakage test
Diagnostic link connector (DLC)
Diagnostic trouble code (DTC)
Dial calipers
Dial indicators
Exhaust gas analyzer
Feeler gauge
Flat rate pay
Fuel-pressure gauge
Knurling
kPa
Machinist's rule
Malfunction indicator light (MIL)
Meter
Microinch
Micrometer
Misfire
Noid light
OBD II
Out-of-round gauge
Plastigage
Powertrain control module (PCM)
Profilometer
psi
Ridge reamer
Rod aligner
Rod honer
Scale
Scan tool

TERMS TO KNOW

(continued)

Small-hole gauge

Straight time pay

Surface grinder

Telescoping gauges

Torque plate

Torque-to-yield

Torque wrenches

United States
Customary (USC)

Valve guide bore
gauge

Valve seat runout
gauge

Valve spring tension
testers

Vernier calipers

Warranty repair

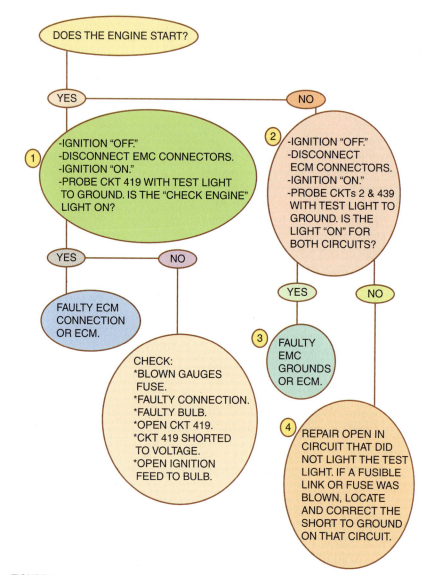

FIGURE 3-101 A diagnostic tree helps the technician pinpoint the cause of the problem.

TABLE OF CONTENTS	SECTION NUMBER
GENERAL INFO. AND LUBE	
General Information	0A
Maintenance and Lubrication	0B
HEATING AND AIR COND.	
Heating and Vent. (non A/C)	1A
Air Conditioning System	1B
V-5 A/C Compressor Overhaul	1D3
BUMPERS AND FRONT BODY PANELS	
Bumpers (See 10-4)	
Fr. End Body Panels (See 10-5)	
STEERING, SUSPENSION, TIRES, AND WHEELS	
Diagnosis	3
Wheel Alignment	3A
Power Steering Gear & Pump	3B1
Front Suspension	3C
Rear Suspension	3D
Tires and Wheels	3E
Steering Col. On-Vehicle Service	3F
Steering Col. - Std. Unit Repair	3F1
Steering Col. - Tilt, Unit Repair	3F2
DRIVE AXLES	
Drive Axles	4D
BRAKES	
General Info. - Diagnosis and On-Car Service	5
Compact Master Cylinder	5A1
Disc Brake Caliper	5B2
Drum Brake - Anchor Plate	5C2
Power Brake Booster Assembly	5D2
ENGINES	
General Information	6
2.0 Liter I-4 Engine	6A1
3.1 Liter V6 Engine	6A3
Cooling System	6B
Fuel System	6C
Engine Electrical - General	6D
Battery	6D1
Cranking System	6D2
Charging System	6D3
Ignition System	6D4
Engine Wiring	6D5
Driveability & Emissions - Gen.	6E
Driveability & Emissions - TBI	6E2
Driveability & Emissions - PFI	6E3
Exhaust System	6F

TABLE OF CONTENTS	SECTION NUMBER
TRANSAXLE	
Auto. Transaxle On-Car Serv.	7A
Auto. Trans. - Hydraulic Diagnosis	3T40-HD
Auto. Trans. - Unit Repair	3T40
Man. Trans. On-Car Service	7B
5-Sp. 5TM40 Man. Trans. Unit Repair	7B1
5-Sp. Isuzu Man. Trans. Unit Repair	7B2
Clutch	7C
CHASSIS ELECTRICAL, INSTRUMENT PANEL & WASHER WIPER	
Electrical Diagnosis	8A
Lighting and Horns	8B
Instrument Panel and Console	8C
Windshield Wiper/Washer	8E5
ACCESSORIES	
Audio System	9A
Cruise Control	9B
Engine Block Heater	9C
BODY SERVICE	
General Body Service	10-1
Stationary Glass	10-2
Underbody	10-3
Bumpers	10-4
Body Front End	10-5
Doors	10-6
Rear Quarters	10-7
Body Rear End	10-8
Roof & Convertible Top	10-9
Seats	10-10
Safety Belts	10-11
Body Wiring	10-12
Unibody Collision Repair	11-1
Welded Panel Replacement	11-2
INDEX	
Alphabetical Index	

FIGURE 3-102 The table of contents directs you to the major component area of the service information.

ASE-STYLE REVIEW QUESTIONS

1. *Technician A* says the machinist's rule offers more precise measurements than a micrometer.

   Technician B says a dial indicator can be used to measure movement.

   Who is correct?

   A. A only   C. Both A and B

   B. B only   D. Neither A nor B

2. The use of torque wrenches is being discussed.

   Technician A says if an adapter is required that increases the length of the torque wrench between the fastener and the handle, the reading must be corrected.

   Technician B says extensions that increase the height of the wrench from the fastener will not affect the torque reading, provided they do not twist.

   Who is correct?

   A. A only   C. Both A and B

   B. B only   D. Neither A nor B

3. *Technician A* says a small-hole gauge can be used to measure the bore of a valve guide.

   Technician B says a valve guide bore gauge can be used to measure the bore of a valve guide.

   Who is correct?

   A. A only   C. Both A and B

   B. B only   D. Neither A nor B

4. The use of a cylinder bore dial gauge is being discussed.

   Technician A says when using the rocking-type gauge, the largest reading obtained is the diameter of the bore.

   Technician B says most cylinder bore dial gauges are not accurate enough to provide good taper and out-of-round measurements.

   Who is correct?

   A. A only   C. Both A and B

   B. B only   D. Neither A nor B

5. Valve spring tension testers are being discussed.

   Technician A says when using the type with a needle indicating torque wrench, spring height must be set first.

   Technician B says to apply pressure on the torque wrench until a click is heard, then multiply the torque wrench reading by two.

   Who is correct?

   A. A only   C. Both A and B

   B. B only   D. Neither A nor B

6. *Technician A* says to measure the amount of out-of-round of a cylinder, measure the bore in three directions at the same bore depth.

   Technician B says when measuring the bore, subtracting the smallest reading from the largest, at the same bore depth, will provide the amount of the out-of-round.

   Who is correct?

   A. A only   C. Both A and B

   B. B only   D. Neither A nor B

7. *Technician A* says that a refractometer can show you the mixture and freezing point of the coolant mixture.

   Technician B says that you can use a coolant test strip to check the acidity level of a coolant.

   Who is correct?

   A. A only   C. Both A and B

   B. B only   D. Neither A nor B

8. *Technician A* says a ridge reamer is used to remove the top ridge in the cylinder.

   Technician B says the top ridge must be removed prior to removing the pistons.

   Who is correct?

   A. A only   C. Both A and B

   B. B only   D. Neither A nor B

9. *Technician A* says torque-to-yield fasteners are used to provide a uniform clamping pressure.

   Technician B says the use of torque wrenches is only for the inexperienced technician as a learning tool.

   Who is correct?

   A. A only   C. Both A and B

   B. B only   D. Neither A nor B

10. *Technician A* says that the VIN can be used to identify the engine.

    Technician B says that a TSB may provide updated service information.

    Who is correct?

    A. A only   C. Both A and B

    B. B only   D. Neither A nor B

Name _____ **Date** _____

IDENTIFYING ENGINE SERVICE TOOLS

Upon completion of this job sheet, you should be able to properly identify several common engine service repair tools.

Tools and Materials

Engine service tools

Describe the vehicle being worked on:

Year _____ Make _____ Model _____

VIN _____ Engine type and size _____

Procedure

Your instructor will provide you with a variety of common engine service repair tools. Write down the name of each one and describe one function of each.

1. Name of component _____
   Function _____
   _____

2. Name of component _____
   Function _____
   _____

3. Name of component _____
   Function _____
   _____

4. Name of component _____
   Function _____
   _____

5. Name of component _____
   Function _____
   _____

6. Name of component _____
   Function _____
   _____

7. Name of component _____

Function _____

_____

Instructor's Response _____

_____

_____

JOB SHEET

Name _____ **Date** _____

USING FEELER GAUGES

Upon completion of this job sheet, you should be able to accurately use a feeler gauge to measure valve clearance.

ASE Correlation

This job sheet is related to ASE Engine Repair Test content area: Cylinder Head and Valvetrain Diagnosis and Repair; Task: Adjust valves on engines with mechanical or hydraulic lifters.

Tools and Materials

Feeler gauge set
Technician's tool set

Describe the vehicle being worked on:

Year _____ Make _____ Model _____

VIN _____ Engine type and size _____

Procedure

Using the engine provided, in the spaces below record the valve clearance.

1. Locate the valve clearance specifications in the service information.
   Intake _____ Exhaust _____

2. What valves can be adjusted with piston number one at TDC compression stroke?
   _____
   _____
   _____

3. What piston position is required to adjust the remaining valves?
   _____

4. Set the engine to number one at TDC compression stroke and measure the valve clearance. Record your results below:
   _____
   _____
   _____

5. Rotate the engine crankshaft until the correct piston listed in step 4 is located at TDC compression stroke. Measure and record the remaining valve clearance below.
   _____
   _____
   _____

6. Are any of the valves' clearances out of specification? ☐ Yes ☐ No
If yes, which ones?

_____

_____

_____

Instructor's Response _____

_____

_____

Name _____ Date _____

READING DIAL CALIPERS

Upon completion of this job sheet, you should be able to accurately measure parts with a dial caliper.

ASE Correlation

This job sheet is related to ASE Engine Repair Test content area: Cylinder Head and Valvetrain Diagnosis and Repair; Task: Inspect and measure camshaft journals and lobes.

Tools and Materials

Dial caliper

Describe the vehicle being worked on:

Year _____ Make _____ Model _____

VIN _____ Engine type and size _____

Procedure

Your instructor will provide you with a variety of parts. Use the dial calipers to measure length, width, inside diameter, outside diameter, and depth for each part as needed. Record the measurements in the space provided.

1. Part number one.

   _____
   _____
   _____

2. Part number two.

   _____
   _____
   _____

3. Part number three.

   _____
   _____
   _____

4. Part number four.

   _____
   _____
   _____

Name _____ Date _____

READING A DIAL INDICATOR

Upon completion of this job sheet, you should be able to accurately measure the camshaft lobe lift with a dial indicator.

ASE Correlation

This job sheet is related to ASE Engine Repair Test content area: Cylinder Head and Valvetrain Diagnosis and Repair: Inspect and/or measure camshaft for runout, journal wear, and lobe wear.

Tools and Materials

Dial indicator
Camshaft
Vee blocks

Describe the vehicle being worked on:

Year _____ Make _____ Model _____

VIN _____ Engine type and size _____

Procedure

Task Completed

Your instructor will provide you with a camshaft and vee blocks. Use the dial indicator and a base stand (magnetic or clamp style) to measure the camshaft lobe lift. Record the measurements in the space provided.

1. Attach the base stand to the dial indicator. ☐

2. Put the camshaft on the vee blocks and position the dial indicator needle on the base of a camshaft lobe and zero the dial. ☐

3. Rotate the camshaft and record the amount of lobe lift. _____

4. Repeat this process on the other camshaft lobes and record your findings below.

_____

_____

Instructor's Response _____

_____

_____

Name _____ Date _____

USING AN OUTSIDE MICROMETER

Upon completion of this job sheet, you should be able to accurately measure parts with an outside micrometer.

Tools and Materials

Outside micrometer set

Describe the vehicle being worked on:

Year _____ Make _____ Model _____

VIN _____ Engine type and size _____

Procedure

Your instructor will provide you with a variety of internal engine components. Use the micrometer to measure outside diameter or thickness for each part as needed. Record the measurements in the space provided.

1. Part number one—Outside Diameter.

   _____

   _____

   _____

2. Part number two—Outside Diameter.

   _____

   _____

   _____

3. Part number three—Thickness.

   _____

   _____

   _____

4. Part number four—Thickness.

   _____

   _____

   _____

Instructor's Response _____

_____

_____

JOB SHEET

Name _____ **Date** _____

MEASURING CYLINDER WALL WEAR

Upon completion of this job sheet, you should be able to accurately measure cylinder wall wear using different methods to measure taper and out-of-round.

ASE Correlation

This job sheet is related to ASE Engine Repair Test content area: Engine Block Diagnosis and Repair; Task: Inspect and measure cylinder walls; remove cylinder wall ridges; hone and clean cylinder walls; determine need for further action.

Tools and Materials

Telescoping gauge
Inside micrometer
Cylinder bore gauge
Outside Micrometer

Describe the vehicle being worked on:

Year _____ Make _____ Model _____
VIN _____ Engine type and size _____

Procedure

Your instructor will provide you with an engine block. Use the method directed to measure a cylinder for taper and out-of-round. Record the measurements in the space provided.

1. Cylinder number one, measure using a telescoping gauge and micrometer.
   a. Measure the bore diameter in three directions at the top of ring travel.

   _____
   _____
   _____

   b. What is the amount of out-of-round? _____
   c. Measure the diameter of the bore at the bottom of ring travel. _____
   d. What is the amount of taper? _____

2. Cylinder number two, measure using a cylinder bore gauge.
   a. Set the bore gauge to the correct bore size. Bore size _____
   b. Set the dial to zero and lock the extension.
   c. Measure the bore diameter in three directions at the top of ring travel.

   _____
   _____
   _____

d. What is the amount of out-of-round? _____

e. Measure the diameter of the bore at the bottom of ring travel. _____

f. What is the amount of taper? _____

Instructor's Response _____

_____

_____

JOB SHEET

Name _____   Date _____

FINDING SERVICE INFORMATION

Upon completion of this job sheet, you should be able to properly locate specific service details on the service information or on a computer system.

Tools and Materials

Service information

Computer with automotive service software (i.e., Mitchell on Demand, Shop Key, or AllData)

Describe the vehicle being worked on:

Year _____ Make _____ Model _____

VIN _____ Engine type and size _____

Procedure

Your instructor will provide you with a specific vehicle and engine package. Write down the information, then locate the following.

1. What are the bore and stroke of the assigned engine?

   _____
   _____
   _____

2. What are the torque specifications for the cylinder head?

   _____
   _____
   _____

3. Are the cylinder head bolts torque-to-yield?   ☐ Yes   ☐ No

4. What is the oil capacity of the engine with a new filter?

   _____

5. What is the oil clearance specification for the main bearings?

   _____

6. Locate a TSB applicable to your vehicle, and list the title below:

   _____

Instructor's Response _____

_____

_____

Name _____ Date _____

LOCATING AND UNDERSTANDING TECHNICAL SERVICE BULLETINS

Upon completion of this job sheet, you should be able to properly locate specific service details on the service information or on a computer system.

ASE Correlation

This job sheet is related to ASE Engine Repair Test content area: General Engine Diagnosis; Removal and Reinstallation (R&R): Research applicable vehicle and service information, such as internal engine operation, vehicle service history, service precautions, and technical service bulletins.

Tools and Materials

Service information

Computer with automotive service software (i.e., Mitchell on Demand, Shop Key, or AllData)

Describe the vehicle being worked on:

Year _____ Make _____ Model _____

VIN _____ Engine type and size _____

Procedure

Task Completed

Your instructor will provide you with a specific vehicle and engine package. Write down the information, then locate the following.

1. Locate an engine-specific technical service bulletin for the vehicle and engine package you are assigned to.   ☐

2. What is the bulletin number and the date it was released?

   _____

2. Are there any updates to this bulletin?

   _____

3. Describe the process needed to find the technical service bulletin.

   _____

   _____

   _____

4. Explain how you might find this information useful when working on engines.

_____

_____

_____

Instructor's Response _____

_____

_____

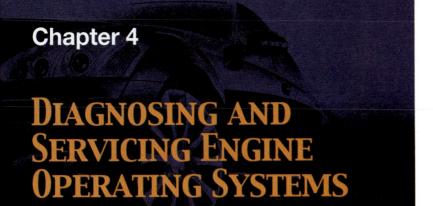

BASIC TOOLS

Basic mechanic's tool set Service information

Classroom Manual
Chapter 3, page 49

UPON COMPLETION AND REVIEW OF THIS CHAPTER, YOU SHOULD BE ABLE TO:

- Diagnose customer complaints associated with engine operation properly.
- Perform an open circuit test on the battery and accurately interpret the results properly.
- Test the capacity of the battery to deliver both current and voltage and to accurately interpret the results.
- Perform a battery conductance test.
- Perform a systematic diagnosis of the starting system.
- Determine what can cause slow crank and no-crank conditions.
- Determine if a slow or no-cranking condition is engine related.
- Perform a quick-check test series to determine the problem areas in the starting system.
- Remove and reinstall a starter motor.
- Inspect engine assembly for fuel, oil, coolant, and other leaks; determine necessary action.
- Perform an oil and filter change.
- Perform oil pressure tests and accurately interpret the results.

- Inspect oil pump gears or rotors and other internal components; repair as needed.
- Inspect auxiliary oil coolers; replace as needed.
- Test coolant; drain, flush, and refill coolant; bleed as required.
- Diagnose customer concerns associated with the cooling system.
- Perform cooling system, cap, and recovery system diagnostic tests.
- Inspect and replace cooling system hoses, thermostat, water pump, and radiator.
- Inspect and test cooling fans, shrouds, and dams.
- Inspect, test, and replace oil temperature and pressure switches and sensors.
- Inspect, replace, and adjust drive belts, tensioners, and pulleys.
- Connect a scan tool to an OBD II vehicle and check for DTCs.
- Interpret some of the data stream from an OBD II vehicle's PCM.

The engine's operating systems provide the means for starting the engine and keeping the engine running, and provide longevity of the engine. These systems require periodic maintenance to assure proper operation. Proper diagnosis of the operating systems can prevent costly failures in the engine.

The battery, starting system, cooling system, lubrication system, and fuel and ignition systems must be operating properly for the engine to perform as designed. Before diagnosing an engine mechanical problem, you must be sure that an engine support system is not causing engine performance symptoms. Proper maintenance of these systems is important to engine longevity and customer satisfaction.

In addition, if the engine is being rebuilt, prior to assembling the engine, it is very important to make sure the lubrication and cooling systems are inspected and any faulty components are replaced or repaired. Scrimping in this area can lead to early failure of the engine.

Many of the inspection and test procedures used during engine rebuilding can be used to diagnose and repair the systems with the engine still in the vehicle. When rebuilding an engine, many of the lubrication and cooling system components are replaced as a common practice. The remaining components must be cleaned and inspected before returning them to service.

BATTERY TESTING

Because of its importance, the battery should be checked whenever the vehicle is brought into the shop for service. A faulty battery can cause engine performance problems or leave a customer stranded. Checking the battery regularly can save diagnostic time and prevent vehicle breakdowns. By performing a battery test series, the state of charge and output capacity of the battery will determine if it is good, in need of recharging, or must be replaced.

There are many different manufacturers of test equipment designed for testing the battery (Figure 4-1). Always follow the procedures given by the manufacturer of the equipment you are using (Figure 4-2).

Battery Inspection

Before performing any electrical tests, the battery should be inspected, along with the cables and terminals. The complete visual inspection of the battery will include the following items:

1. *Battery date code.* This provides information as to the age of the battery.
2. *Condition of battery case.* Check for dirt, grease, and electrolyte condensation. Any of these contaminants can create an electrical path between the terminals and cause the battery to drain. Also check for damaged or missing vent caps and cracks in the case. A cracked or buckled case could be caused by excessive tightening of the holddown fixture, excessive under-hood temperatures, buckled plates from extended overcharged conditions, freezing, or excessive charge rate.

Classroom Manual
Chapter 4

SPECIAL TOOLS
Fender covers
Safety glasses

FIGURE 4-1 This volt amp tester (VAT) can test the capacity of the battery, starting and charging system.

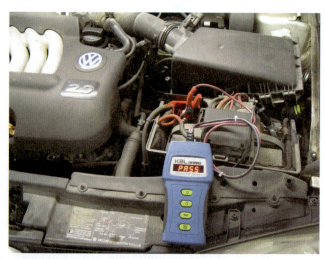

FIGURE 4-2 This type of tester will test the condutance of the battery. It is quick and can usually be done in less than a minute.

3. *Electrolyte level, color, and odor.* If necessary, add distilled water to fill to 1/2 inch above the top of the plates. After adding water, charge the battery before any tests are performed. Discoloration of electrolyte and the presence of a rotten egg odor indicate an excessive charge rate, excessive deep cycling, impurities in the electrolyte solution, or an old battery.

WARNING: Do not attempt to pry the caps off of a maintenance-free battery; this can ruin the case. You cannot check the electrolyte level or condition on maintenance-free batteries.

WARNING: Always wear safety glasses when working around a battery. Electrolyte can cause severe burns and permanent damage to the eye.

4. *Condition of battery cables and terminals.* Check for corrosion, broken clamps, frayed cables, and loose terminals (Figure 4-3).
5. *Battery abuse.* This includes the use of bungee cords and 2 × 4s for holddown fixtures, too small a battery rating for the application, and obvious neglect to periodic maintenance. In addition, inspect the terminals for indications that they have been hit upon by a hammer and for improper cable removal procedures; also check for proper cable length.
6. *Battery tray and holddown fixture.* Check for proper tightness. Also check for signs of acid corrosion of the tray and holddown unit. Replace as needed.
7. *If the battery has a built-in hydrometer, check its color indicator* (Figure 4-4). Green indicates the battery has a sufficient charge, black indicates the battery requires charging before testing, clear indicates low electrolyte level, and yellow indicates the battery should be replaced.

SPECIAL TOOLS
Safety glasses
Voltmeter Terminal
pliers Terminal
puller Terminal and
clamp cleaner
Fender covers

A **conductance test** sends a low current AC signal into the battery. The return signal is then analyzed.

A **capacity (load) test** uses a carbon pile to draw energy out of the battery. The voltage level of the battery should not drop below 9.6 Volts during the test.

When the battery and cables have been completely inspected and any problems corrected, the battery is ready to be tested further. For many of the tests to be accurate, the battery must be fully charged.

FIGURE 4-3 These battery connections could easily cause a no-start condition.

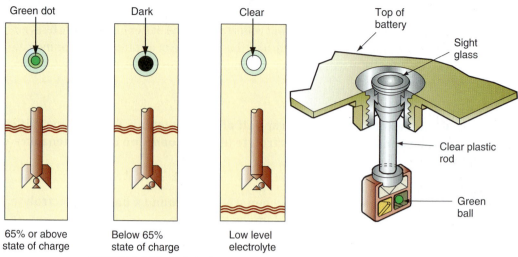

Green dot   Dark   Clear

| 65% or above state of charge | Below 65% state of charge | Low level electrolyte |

Top of battery

Sight glass

Clear plastic rod

Green ball

FIGURE 4-4 A built-in hydrometer indicates the battery state of charge.

BATTERY TESTING SERIES

SERVICE TIP:
Grid growth can cause the battery plate to short out the cell. If there is normal electrolyte level in all cells but one, that cell is probably shorted.

Battery Conductance Test. A newer method of testing a battery is called a **conductance test**. A conductance tester will test the battery's ability to conduct electricity (Figure 4-2). This is a good indicator of a battery's health and whether it will be able to provide adequate service when installed in the vehicle. A conductance test is considered to be more accurate than the previous method of testing batteries. During a conductance test, the tester will send a low current AC signal into the battery. The return signal that is generated is then analyzed. Some conductance battery testers also include a battery charger. These types of testers can recharge a battery if it is not ready to be tested accurately (Figure 4-5). During recharging, the tester can measure key items that will indicate if the battery will be strong enough to hold a good charge when it is done.

Battery Capacity Test. A **capacity test** (also known as a load test) is still a valid battery test, but it has been proven to be less accurate in determining if a battery will provide good service. A carbon pile tester is used to perform this test (Figure 4-6). There are many types and brands of testers available, but all of them have a voltmeter and an ammeter. A load test machine works by connecting the high current clamps of the tester to the battery terminals

FIGURE 4-5 This Midtronics tester is becoming an industry standard.

FIGURE 4-6 This is a VAT (voltage amperage yester) machine. It contains a carbon pile in the backside and can perform a capacity (load) test.

and the inductive pickup lead (amp clamp) to the negative tester cable. The battery must be fully charged before performing the test. For best results, the battery electrolyte should be close to 80°F (26.7°C). The tester then draws 1/2 the battery's CCA rating from the battery for 15 seconds. On some testers this is done automatically and on others there is a control knob for the amperage. If the voltmeter does not drop below 9.6 volts during the 15-second test, then the battery passes the load test.

Battery Terminal Voltage Drop Test. This simple test will establish the condition of the terminal connection. It is a good practice to perform this test any time the battery cables are cleaned or disconnected and reconnected to the terminals. By performing this test, come-backs due to loose or faulty connections can be reduced.

Set your digital multimeter (DMM) to DC volts. Connect the negative test lead to the cable clamp, and connect the positive meter lead to the battery terminal (Figure 4-7). Disable the ignition system to prevent the vehicle from starting. This may be done by removing the ignition coil primary wires (Figure 4-8). (You can also remove the ignition system fuse, PCM fuse, or fuel pump fuse.) On vehicles with distributor-less ignition, disconnect the coil primary wires (these are the small gauge wires connected to the coils).

Crank the engine and observe the voltmeter reading. If the voltmeter shows over 0.5 volt, there is a high resistance at the cable connection. Remove the battery cable using the terminal puller (Figure 4-9). Clean the cable ends and battery terminals or *replace* as needed (Figure 4-10).

Open Circuit Voltage Test. To obtain accurate test results, the battery must be stabilized. If the battery has just been recharged, perform the capacity test, then wait at least 10 minutes to allow battery voltage to stabilize. Connect a voltmeter across the battery terminals, observing polarity (Figure 4-11). Measure the open circuit voltage, taking the reading to the one-tenth volt.

To analyze the open circuit voltage test results, consider that a battery at a temperature of 80°F, in good condition, should show at least 12.45 volts (Figure 4-12). If the state of charge is 75 percent or more, the battery is considered charged (Figure 4-13).

Battery Case Test. A dirty battery can allow voltage to leak across the top of the battery and cause discharging. To test a battery for this condition, place the positive voltmeter lead on a battery post. Place the negative lead at spots across the top of the battery. Repeat the test on the other battery post. Any reading above 0.5 volt indicates that the battery is dirty and allowing discharging. Clean the top of the battery with a mixture of baking soda and water or use a battery cleaning solvent. Retest to verify your repair.

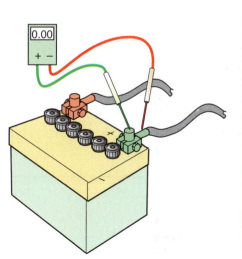

FIGURE 4-7 Test connections of the battery terminal voltage drop test.

FIGURE 4-8 Disconnecting the coil primary connections to prevent the engine from starting and to protect the coil.

FIGURE 4-9 Use battery terminal pullers to remove the clamp from the battery post. Do not pry the clamp.

FIGURE 4-10 The terminal cleaning tool is used to clean the clamp and terminal.

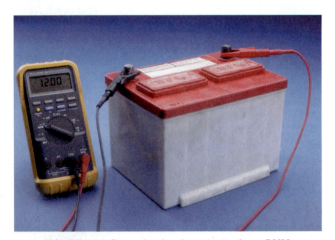

FIGURE 4-11 Open circuit voltage test using a DMM.

Open Circuit Voltage	State of Charge
12.6 or greater	100%
12.4–12.6	70–100%
12.2–12.4	50–70%
12.0–12.2	25–50%
0.0–12.0	0–25%

FIGURE 4-12 The results of the open circuit voltage test indicate the state of charge.

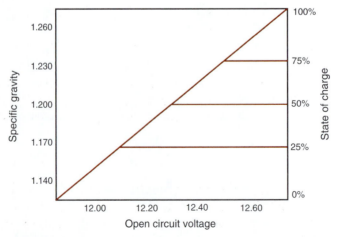

FIGURE 4-13 Open circuit voltage test results relate to the specific gravity of the battery cells.

STARTING SYSTEM TESTS AND SERVICE

The starter motor must be capable of rotating the engine fast enough so it can start and run under its own power. The starting system is a combination of mechanical and electrical parts working together to start the engine. The starting system includes the following components:

1. Battery
2. Cable and wires
3. Ignition switch
4. Starter solenoid or relay
5. Starter motor
6. Starter drive and flywheel ring gear
7. Starting safety switch

Starting System Troubleshooting

Customer complaints concerning the starting system generally fall into four categories: no-crank, slow cranking, starter spins but does not turn engine, and excessive noise. As with any electrical system complaint, a systematic approach to diagnosing the starting system will make the task easier. First, the battery must be in good condition and fully charged. Perform a complete battery test series to confirm the battery's condition. Many starting system complaints are actually attributable to battery problems. If the starting system tests are performed with a weak battery, the results can be misleading, and the conclusions reached may be erroneous and costly.

Before performing any tests on the starting system, begin with a visual inspection of the circuit. Repair or replace any corroded or loose connections, frayed wires, or any other trouble sources. The battery terminals must be clean, and the starter motor must be properly grounded.

The diagnostic chart shows a logical sequence to follow whenever a starting system complaint is made (Figure 4-14). The tests to be performed are determined by whether or not the starter will crank the engine.

If the customer complains of a no-crank situation, attempt to rotate the engine by the crankshaft pulley nut. Rotate the crankshaft in a clockwise direction two full rotations, using a large socket wrench. If the engine does not rotate, it may be seized due to operating with lack of oil, **hydrostatic lock**, or broken engine components.

A slow crank or no-crank complaint can be caused by several potential trouble spots in the circuit (Figure 4-15). Excessive voltage drops in these areas will cause the starter motor to operate slower than required to start the engine. The speed the starter motor rotates the engine is important to engine starting. If the speed is too slow, compression is lost and the air-fuel mixture draw is impeded. Most manufacturers require a speed of approximately 250 rpms during engine cranking.

If the starter spins but the engine does not rotate, the most likely cause is a faulty starter drive. If the starter drive is at fault, the starter motor will have to be removed to install a new starter or drive mechanism. Before faulting the starter drive, also check the starter ring gear teeth for wear or breakage, and for incorrect gear mesh of the ring gear and starter motor pinion gear.

Most noises can be traced to the starter drive mechanism. The starter drive can be replaced as a separate component of the starter.

Testing the Starting System

As with the battery testing series, the tests for the starting system are performed on the volt-ampere tester (VAT). Since the starter performance and battery performance are so closely related, it is important that a full battery test series be done before trying to test the starter system. If the battery fails the load test and is fully charged, it must be replaced before doing any other tests.

The **open circuit voltage test** is used to determine the battery's state of charge.

SPECIAL TOOLS

Battery charger
Voltmeter Safety
glasses Fender
covers

CAUTION:
Some engines turn counterclockwise as viewed from the front crankshaft pulley. These engines are rare, but damage may occur if they are turned the other way. Check the service manual for warnings or ask your instructor.

Hydrostatic lock is the result of attempting to compress a liquid in the cylinder. Since liquid is not compressible, the piston is not able to travel in the cylinder.

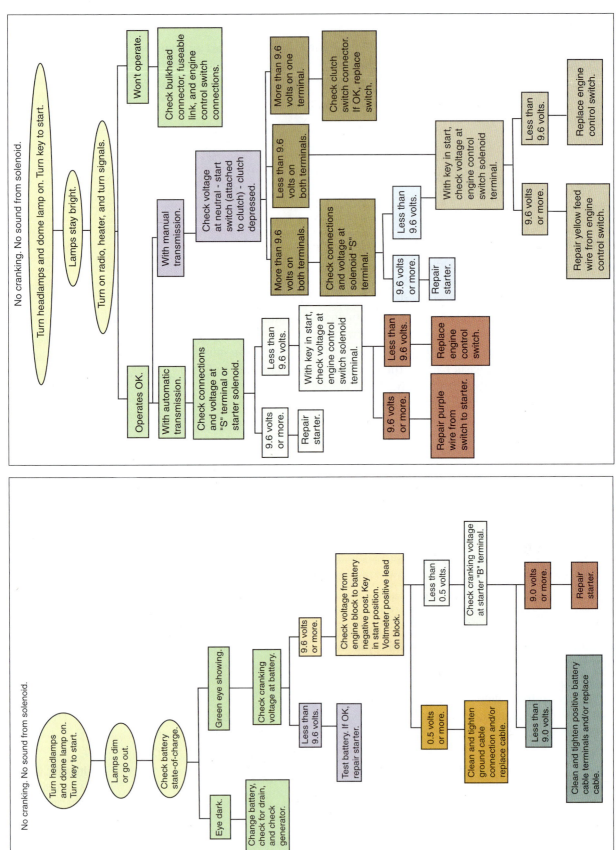

FIGURE 4-14 **Diagnostic chart used to determine starting system problems.**

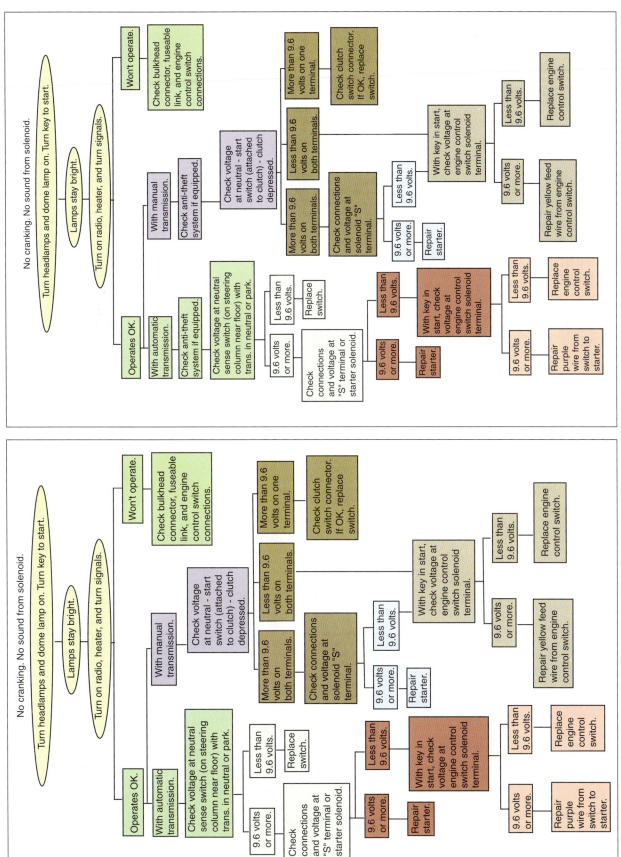

FIGURE 4-14 (Continued)

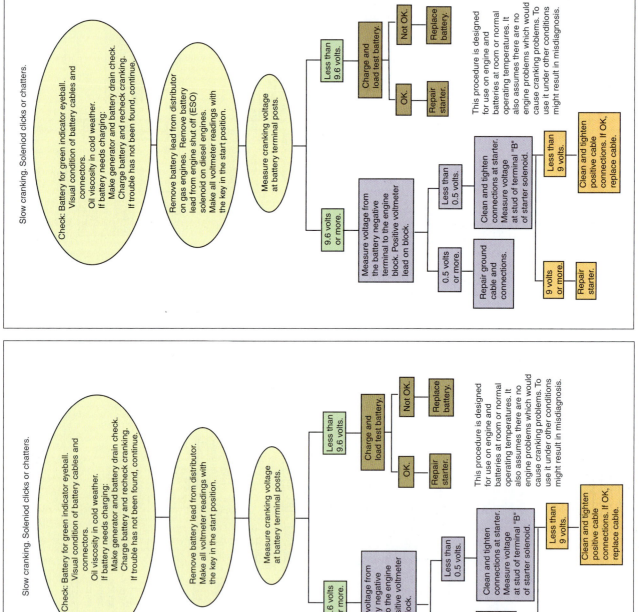

Slow cranking. Soleniod clicks or chatters.

Check: Battery for green indicator eyeball.
Visual condition of battery cables and
connectors.
Oil viscosity in cold weather.
If battery needs charging:
Make generator and battery drain check.
Charge battery and recheck cranking.
If trouble has not been found, continue.

Remove battery lead from distributor
on gas engines. Remove battery
lead from engine shut off (ESO)
solenoid on diesel engines.
Make all voltmeter readings with
the key in the start position.

Measure cranking voltage
at battery terminal posts.

Less than 9.6 volts.

Charge and load test battery.

OK.

Repair starter.

Not OK.

Replace battery.

This procedure is designed
for use on engine and
batteries at room or normal
operating temperatures. It
also assumes there are no
engine problems which would
cause cranking problems. To
use it under other conditions
might result in misdiagnosis.

9.6 volts or more.

Measure voltage from
the battery negative
terminal to the engine
block. Positive voltmeter
lead on block.

Less than 0.5 volts.

0.5 volts or more.

Repair ground cable and connections.

Clean and tighten connections at starter. Measure voltage at stud of terminal "B" of starter solenoid.

Less than 9 volts.

Clean and tighten positive cable connections. If OK, replace cable.

9 volts or more.

Repair starter.

Slow cranking. Soleniod clicks or chatters.

Check: Battery for green indicator eyeball.
Visual condition of battery cables and
connectors.
Oil viscosity in cold weather.
If battery needs charging:
Make generator and battery drain check.
Charge battery and recheck cranking.
If trouble has not been found, continue.

Remove battery lead from distributor.
Make all voltmeter readings with
the key in the start position.

Measure cranking voltage
at battery terminal posts.

Less than 9.6 volts.

Charge and load test battery.

OK.

Repair starter.

Not OK.

Replace battery.

This procedure is designed
for use on engine and
batteries at room or normal
operating temperatures. It
also assumes there are no
engine problems which would
cause cranking problems. To
use it under other conditions
might result in misdiagnosis.

9.6 volts or more.

Measure voltage from
the battery negative
terminal to the engine
block. Positive voltmeter
lead on block.

Less than 0.5 volts.

0.5 volts or more.

Repair ground cable and connections.

Clean and tighten connections at starter. Measure voltage at stud of terminal "B" of starter solenoid.

Less than 9 volts.

Clean and tighten positive cable connections. If OK, replace cable.

9 volts or more.

Repair starter.

FIGURE 4-14 (Continued)

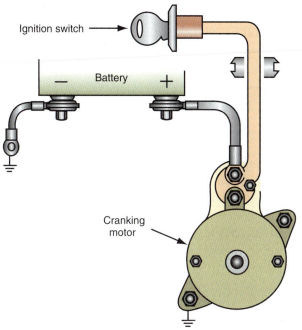

Ignition switch →

Battery − +

Cranking motor

FIGURE 4-15 Excessive wear, loose electrical connections, or excessive voltage drop in any of these areas can cause a slow crank or no crank condition.

Starter Quick Testing. If the starter does not turn the engine at all and the engine is in good mechanical condition, the **starter quick test** can be performed to locate the problem area. To perform this test, make sure the transmission is in neutral, and set the parking brake. Turn on the headlights. Next, turn the ignition switch to the START position while observing the headlights.

There are three things that can happen to the headlights during this test:

1. They will go out.
2. They will dim.
3. They will remain at the same brightness level.

If the lights go out completely, the most likely cause is a poor connection at one of the battery terminals. Check the battery cables for tight and clean connections. It will be necessary to remove the cable from the terminal and clean the cable clamp and battery terminals of all corrosion.

If the headlights dim when the ignition switch is turned to the START position, the battery may be discharged. Check the battery condition. If it is good, then there may be a mechanical condition in the engine preventing it from rotating. If the engine rotates when turning it with a socket wrench on the pulley nut, the starter motor may have internal damage. A bent starter armature, worn bearings, thrown armature windings, loose pole shoe screws, or any other worn component in the starter motor that will allow the armature to drag can cause a high current demand.

If the lights stay brightly lit and the starter makes no sound (listen for a deep clicking noise), there is an open in the circuit. The fault is in either the solenoid or the control circuit. To test the solenoid, bypass the solenoid by bridging the BAT and S terminals on the back of the solenoid. The B1 terminal is a large terminal on the solenoid with a heavy gauge wire coming from the battery. Also on the solenoid, the S terminal has a smaller gauge wire coming to it from the ignition switch or starter relay. Use heavy jumper cables to connect the two terminals. If the starter spins, the ignition switch, starter relay, or control circuit wiring is faulty. If the starter does not spin, the starter motor and/or solenoid are faulty.

SPECIAL TOOLS
Fender covers
Jumper cables

The **starter quick test** will isolate the problem area; i.e., if the starter motor, solenoid, or control circuit is at fault.

If the engine does not crank and the headlights do not come on, check the fusible link.

Check the ignition switch and the safety starting switch for proper operation and adjustment.

SPECIAL TOOLS
Fender covers
VAT-40
Jumper wires

The **current draw test** measures the amount of current the starter draws when actuated. It determines the electrical and mechanical condition of the starting system.

WARNING: The starter will draw up to 400 amperes. The tool used to jump the terminals must be able to carry this high current and it must be have an insulated handle. A jumper cable that is too light may become extremely hot, resulting in burns to the technician's hand.

Current Draw Test. If the starter motor cranks the engine, the technician should perform the **current draw test**. The following procedure is used for volts/amps tester:

1. Connect the large red and black test leads on the battery posts, observing polarity.
2. Zero the ammeter.
3. Connect the ampere inductive probe around the battery ground cable. If more than one ground cable is used, clamp the probe around all of them (Figure 4-16).
4. Make sure all loads are turned off (lights, radio, and so on).
5. Disable the ignition system to prevent the vehicle from starting. This may be done by removing the ignition coil secondary wire from the distributor cap and putting it to ground, by removing the ignition system fuse, fuel pump fuse, PCM fuse, or by disconnecting the primary wires from the coils of an EI system.
6. Crank the engine and note the voltmeter reading.
7. Read the ammeter scale to determine the amount of current draw.

After recording the readings from the current draw test, compare them with the manufacturer's specifications. If specifications are not available, as a rule, correctly functioning systems will crank at 9.6 volts or higher. Amperage draw is dependent upon engine size. Most V8 engines have an ampere draw of about 200 amperes, 6-cylinder engines about 150 amperes, and 4-cylinder engines about 125 amperes.

Higher-than-specified current draw test results indicate an impedance to rotation of the starter motor. This includes worn bushings, a mechanical blockage, internal starter motor damage, and excessively advanced ignition timing. Lower-than-normal current draw test results indicate excessive voltage drop in the circuit, a faulty solenoid, or worn brushes.

If the current test results are high and the engine does not turn, the starter may not need replacement. Hydrostatic lock or engine seizure may have occurred. In order to check for this, you should put a socket and ratchet on the crankshaft pulley bolt and attempt to turn the engine over manually by hand. If the engine turns easily, it is most likely a problem with the starter. If the engine does not turn, it is an internal engine problem.

To determine what type of internal problem, start by removing the spark plugs from the engine and attempt to turn the crankshaft again. If it starts to turn easier, look for coolant, fuel, or oil that might be coming out of the spark plug holes (Figure 4-17). These fluids may

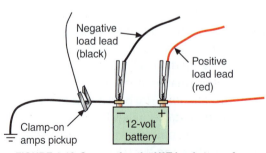

FIGURE 4-16 Connecting the VAT leads to perform the starter current draw test.

FIGURE 4-17 This spark plug is fouled with coolant.

have gotten into the combustion chamber by a blown head gasket, leaking fuel injector, or a cracked cylinder head or block. If the engine does not turn and no fluids come out of the spark plug hole, the engine may be mechanically seized.

Since the readings obtained from the current draw test were taken at the battery, these readings may not be an exact representation of the actual voltage at the starter motor. Voltage losses due to bad cables, connections, relays (or solenoids) may diminish the amount of voltage to the starter. Before removing the starter from the vehicle, these should be tested.

Starter Motor Removal

If the tests indicate the starter motor must be removed, the first step is to disconnect the battery from the system.

WARNING: **Remove the negative battery cable. It is a good practice to wrap the cable clamp with tape or enclose it in a rubber hose to prevent accidental contact with the battery terminal. If the positive cable is removed first, the wrench may slip and make contact between the terminal and ground. This action may overheat the wrench and burn the technician's hand.**

It may be necessary to place the vehicle on a lift to gain access to the starter motor. Before lifting the vehicle, disconnect all wires, fasteners, and so on that can be reached from the top of the engine compartment. Disconnect the wires leading to the solenoid terminals. To prevent confusion, it is a good practice to use pieces of tape to identify the different wires.

WARNING: **Check for proper lift pad–to–frame contact after the vehicle is a few inches above the ground. Shake the vehicle. If there are any unusual noises or movement of the vehicle, lower it and reset the pads. If the lift pads are not properly positioned on the specified vehicle lift points, the vehicle may slip off the lifts, resulting in technician injury and vehicle damage.**

On some vehicles, it may be necessary to disconnect the exhaust system to be able to remove the starter motor. Spray the exhaust system fasteners with a penetrating oil to assist in removal. Loosen the starter mounting bolts and remove all but one. Support the starter motor, remove the remaining bolt, then remove the starter motor.

WARNING: **The starter motor is heavy. Make sure that it is secured before removing the last bolt. If the starting motor is not properly secured, it may drop suddenly, resulting in leg or foot injury.**

To install the starter motor, reverse the procedure. Be sure all electrical connections are tight. If you are installing a new or remanufactured starter, remove any paint that may prevent a good ground connection. Be careful not to drop the starter. Make sure that it is supported properly. Refer to Photo Sequence 4 for the typical starter removal and reinstallation.

Some General Motors' starters use shims between the starter motor and the mounting pad (Figure 4-18). To check this clearance, insert a flat blade screwdriver into the access slot on the side of the drive housing. Pry the drive pinion gear into the engaged position. Use a piece of wire that is 0.020 in. in diameter to check the clearance between the gears (Figure 4-19).

If the clearance between the two gears is excessive, the starter will produce a high pitched whine while the engine is being cranked. If the clearance is too small, the starter will make a high pitched whine after the engine starts and the ignition switch is returned to the RUN position.

CAUTION:
Always refer to the manufacturer's service manual for the correct procedure for disabling the ignition system. Using an improper procedure may damage ignition components.

The specification for current draw is the maximum allowable, and the specification for cranking voltage is the minimum allowable.

SPECIAL TOOLS
Fender covers
Battery cable puller

CAUTION:
Do not operate the starter motor for longer than 15 seconds. Allow the motor to cool between cranking attempts. Operating the starter for more than 15 seconds may overheat and damage starter motor components.

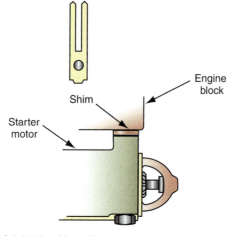

A 0.015-in. shim will increase the clearance approximately 0.005 in. More than one shim may be required.

FIGURE 4-18 Shimming the starter to obtain the correct pinion–to–ring gear clearance.

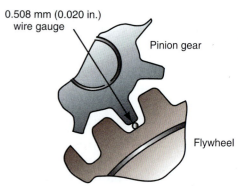

FIGURE 4-19 Checking the clearance between the pinion gear and ring gear.

LUBRICATION SYSTEM TESTING AND SERVICE

Oil Pressure Testing

Proper oil pressure is essential to engine life. Oil pressure is dependent upon oil clearances and proper delivery. If the clearance between a journal and the **bearing** becomes excessive, pressure is lost. Not all low oil pressure conditions are the result of bearing wear though. Other causes include improper oil level, improper oil grade, and oil pump wear. Another common cause of low oil pressure is thinning oil as a result of excessive temperatures or gas dilution.

If low oil pressure is suspected, begin by checking the oil level. Too low a level will cause the oil pump to aerate and lose volume. If the oil level is too high, it may be gasoline is entering the crankcase as a result of a damaged fuel pump, ignition misfire, leaking injector, or engine flooding. If the oil level and condition are satisfactory, check oil pressure using a shop gauge.

To perform an **oil pressure test**, remove the oil pressure sending unit from the engine (Figure 4-20). Using the correct size adapters, connect the oil pressure gauge to the oil passage. Start the engine and observe the gauge as the engine idles. Watch the gauge as the engine warms to note any excessive drops due to temperature. Increase the engine rpm to 2,000 while observing the gauge. Compare the test results with the manufacturer's specifications. Manufacturers provide oil pressure specifications with the engine at normal

FIGURE 4-20 Remove the factory equipped oil pressure sending unit to connect an oil pressure gauge.

STARTER R + R

P4-1 Disconnect the negative battery cable first to prevent electrical arcing.

P4-2 If necessary, raise the vehicle to gain access to the starter.

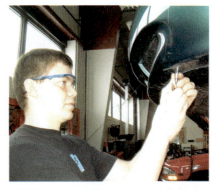

P4-3 Remove under-car protective covers as needed to access the starter wiring and mounting hardware.

P4-4 Disconnect the battery cable and the other wires from the starter.

P4-5 Loosen the starter attaching bolts and remove the starter.

P4-6 Bench test the old starter to confirm your diagnosis. Also test the new or remanufactured unit to be sure it functions properly.

P4-7 Work the new starter back into the proper position.

P4-8 Clean the fasteners and tighten each of them securely. The starter is often grounded through the mounting bolts.

P4-9 Do a professional job of reassembling each of the components you had to remove to reach the starter.

STARTER R + R

P4-10 **Reconnect the battery.**

P4-11 **Verify that the new starter operates properly and without excess noise by cranking the engine over a few times.**

operating temperature; be sure the engine is fully warmed up. After the test is complete, reinstall the oil pressure sending unit, start the engine, and confirm there are no leaks.

No oil pressure indicates the oil pump drive may be broken, the pickup screen is plugged, the gallery plugs are leaking, or there is a hole in the pickup tube. Lower-than-specified oil pressure indicates improper oil viscosity, a worn **oil pump**, a plugged oil pickup tube, a sticking or weak oil pressure relief valve, or worn bearings.

It is possible to have oil pressure that is too high. This can be caused by a stuck-closed pressure relief valve, by a blockage in an oil gallery, or by contaminated oil.

Oil Pump Service

Since the oil pump is so vital to proper engine operation, it is usually replaced along with the pickup tube and screen whenever it fails or when the engine is being rebuilt or replaced. The relatively low cost of an oil pump is considered cheap insurance by most technicians, so replacement is the general rule. However, there may be instances when the oil pump will have to be disassembled, inspected, and repaired. Proper inspection requires the oil pump to be disassembled and cleaned. Photo Sequence 5 is a typical procedure for disassembling and inspecting a rotor-type oil pump that is driven by the crankshaft. Disassembly procedures for gear-type pumps are very similar to that of rotor pumps.

If the pump housing is not damaged, a pump rebuild kit containing new rotors or gears, a relief valve and spring, and seals may be available.

Since the oil is delivered to the pump before it goes through the oil filter, the pump is subject to wear and damage as contaminated oil passes through it. The particles passing through the gears or rotors of the pump wear away the surface area, resulting in a reduction of pump efficiency. If the particles are large enough, they may cause the metal of the rotor or gear surfaces to raise, resulting in pump seizure. In addition, these larger particles can force a wedge in the pump and cause it to lock up.

When removing the oil pressure **relief valve**, note the direction of the valve in the housing. Before removing the gears or rotor, mark adjacent teeth so they can be indexed in the same location when reassembled. Some manufacturers provide index marks.

After the oil pump has been disassembled and cleaned, it can be inspected. This usually requires a thorough visual inspection, measuring clearances, and measuring parts. Carefully inspect the drive shaft for the oil pump. If the corners are rounded or the slot is damaged, replace the shaft.

Gear-type oil pumps also require end clearance checks (Figure 4-21). In addition, each gear's clearance must be checked (Figure 4-22). Check gear clearance at several locations

TYPICAL PROCEDURE FOR DISASSEMBLY AND INSPECTION OF ROTOR-TYPE OIL PUMP

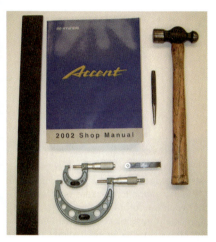

P5-1 The tools required to perform this task include a micrometer, feeler gauge, straightedge, center punch, hammer, and appropriate service manual.

P5-2 Remove the relief valve cap, relief valve, and spring. Note the direction the valve is installed into the oil pump housing for installation.

P5-3 Remove the oil pump cover screws and remove the cover.

P5-4 If the rotors do not have index marks, use a center punch to mark the rotors for installation.

P5-5 Wash all parts in solvent.

P5-6 Visually inspect the housing and rotors for signs of wear or scoring. If the housing is scored, replace the entire pump assembly. The mating surfaces of the rotors should be smooth.

P5-7 Place a straightedge across the pump cover surface and attempt to insert a 0.003 in. (0.076 mm) feeler gauge under the straightedge. The cover should be replaced or machined if it is excessively warped.

P5-8 Use an outside micrometer to measure the diameter of the outer rotor. Compare the measurement with specifications.

P5-9 Use an outside micrometer to measure the thickness of the outer rotor. Compare the results with specifications.

TYPICAL PROCEDURE FOR DISASSEMBLY AND INSPECTION OF ROTOR-TYPE OIL PUMP

P5-10 Place the rotors back into the oil pump housing and measure the side clearance of the outer rotor. If the measurement is greater than specifications and the rotor outer diameter is correct, replace the pump housing.

P5-11 Use a feeler gauge to measure the clearance of the inner rotor by aligning one of its lobes with a lobe of the outer rotor. If the measurement is greater than specifications, replace both rotors.

P5-12 With the rotors installed, place a straightedge across the housing and use a feeler gauge to measure the distance between the rotors and the straightedge. If the clearance is excessive and the rotor thickness is within specifications, replace the pump assembly.

P5-13 Visually inspect the relief valve for indications of scoring. Light scoring may be removed using 400 grit emery cloth.

P5-14 Measure the relief valve spring free height and pressure. Compare the test results with specifications.

SPECIAL TOOLS

Feeler gauge set Micrometer Straightedge

The **relief valve** is used to prevent excessive oil pressure.

CAUTION:

Installing the relief valve backwards will result in no oil pressure.

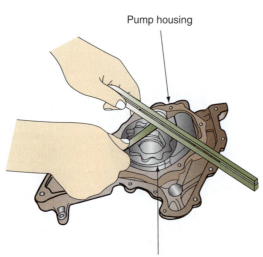

Pump housing

Outer rotor

FIGURE 4-21 Use a straight edge and feeler gauge to measure clearance.

FIGURE 4-22 Gear-type oil pumps must have gear-to-housing clearances checked before reusing them.

around each gear. If the clearance is greater than 0.005 in. (0.13 mm) in any location, replace the pump assembly. Always consult the manufacturer's service information for proper specifications.

When assembling the oil pump, make sure to align the index marks of the gears or rotors. The cover gasket maintains proper gear or rotor clearance. Use the appropriate thickness as supplied by the manufacturer. After the cover is installed, torque the bolts to the specified value. Improper torquing may result in cover warpage and low pump output. The **pickup tube** and screen should be replaced whenever the oil pump is replaced or rebuilt. If the tube attaches to the oil pump, install the new tube using light taps from a hammer. Make sure the pickup tube is properly positioned to prevent interference with the oil pan or crankshaft. Bolt-on pickup tubes may use a rubber O-ring to seal them. Do not use a sealer in place of the O-ring. Lubricate the O-ring with engine oil prior to assembly, then alternately tighten the attaching bolts until the specified torque is obtained.

Prime the oil pump before installing it by submerging it into a pan of clean engine oil and rotating the gears by hand. When the pump discharges a stream of oil, the pump is primed.

If a gasket is installed between the pump and the engine block, check to make sure no holes or passages are blocked by the gasket material. Soak the gasket in oil to soften the material and allow for good compression. When installing the oil pump, make sure the drive shaft is properly seated, and torque the fasteners to the specified value.

> The **pickup tube** is used by the pump to deliver oil from the bottom of the oil pan. The bottom of the tube has a screen to filter larger contaminants.

Oil Jet Valves

Some engines use **jet valves** to spray oil into the underside of the piston to cool the top of the piston. These valves must be inspected and cleaned before they are returned to service. The valve consists of a check ball, spring, and nozzle. Check the nozzle opening by inserting a small drill bit into the hole. Do the same for the oil intake. With the drill bit installed in the intake hole, the check ball should be able to be moved about 0.160 in. (4.0 mm). Finally, operation of the jet valve can be checked using air pressure set at 30 psi. At this pressure, the check ball should lift off its seat. If the nozzle is bent or damaged, replace the jet valve.

> **Jet valves** are used by some manufacturers to direct a stream of oil to the underside of the piston head. Jet valves are common on turbocharged engines to keep the piston cool to prevent detonation and piston damage.

Lubrication System Maintenance

Lubrication system maintenance consists of periodic oil and filter changes. The oil pan is equipped with a drain plug to remove the oil from the crankcase. The engine must be warmed up before draining the oil to be sure the oil has picked up all debris from the engine. The oil will also drain faster when it is warm.

The oil filter usually threads to the engine block or adapter. A band-type wrench or special socket is used to remove the filter from the engine. A rubber seal is located at the top of the oil filter to seal between the filter and block. Sometimes this seal may come off the filter

SPECIAL TOOLS
Oil filter wrench

139

and remain on the block. Be sure to remove the seal from the block if this occurs. Failure to remove the oil seal will result in oil leaks. Before installing the new oil filter, lubricate the seal, and fill the filter with oil. Install the oil filter. Do not overtighten the filter. Turn the filter 3/4 turn after the seal makes contact. Do not use the oil filter wrench to tighten the filter. Only hand pressure is required.

While the vehicle is in the air, check its underside for fluid leaks and make a note on the repair order if you see problems. Also make a quick safety check of the exhaust, suspension, and driveline. This may be the only time a customer brings his vehicle in for routine maintenance; the customer will appreciate your finding problems, so he doesn't end up stranded on the road. Check and adjust all tire pressures, including the spare tire. Many shops will have an oil change safety checklist for you to follow. Communicate any problems you find to the service advisor or customer; the customer may choose to have repairs done immediately.

Some engines require pre-oiling before they are started after an oil change. This is especially true of engines equipped with turbochargers. The turbocharger rotates at a high rate of speed and requires lubrication to prevent damage. After the oil is drained and the oil filter replaced, it may take several moments before oil reaches the turbocharger. Preoiling the engine will assure all components receive oil before the engine is started. Use the following procedure to pre-oil the engine:

1. Make sure the oil filter and crankcase are filled with oil.
2. Disable the ignition system using the manufacturer's recommended procedure.
3. Crank the engine for 30 seconds, then allow the starter to cool, and crank the engine for another 30 seconds.
4. Repeat this procedure until oil pressure is indicated (oil pressure indicator light off or oil pressure gauge indications).
5. Enable the ignition system and start the engine.
6. Observe the oil pressure indicator. If oil pressure is not indicated, shut off the engine.
7. Shut off the engine and recheck the oil level.

After filling the engine with oil and starting the vehicle, shut the engine off and recheck the fluid level and top off as needed. Lift the vehicle and double-check for leaks before letting the vehicle go. Lower the vehicle and check and adjust the other fluid levels. Install an oil sticker on the windshield or door jam instructing the customer when to have the oil changed again (Figure 4-23).

CAUTION:
Overtightening the oil filter may cause seal damage and result in oil leakage.

CAUTION:
Make sure the crankcase has sufficient oil before cranking the engine, because the oil level will go down when oil flows through the engine lubrication system.

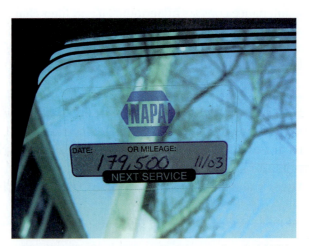

FIGURE 4-23 Always put an oil sticker on the windshield or in the driver's door jam to remind the customer when to have their oil changed again.

COOLING SYSTEM TESTING AND SERVICE

It is the function of the engine's cooling system to remove heat from the internal parts of the engine and dissipate it into the air. Engine overheating is a common cause of internal engine damage, cracked cylinder heads, and blown head gaskets. Proper engine operation and service life depend upon the cooling system functioning as designed.

Since the cooling system requires periodic maintenance, most service procedures can be performed with the engine still in the vehicle. Whenever the engine is rebuilt, it is a good practice to remove the **radiator** and replace it or have it professionally cleaned. In addition, the block and cylinder head passages should be thoroughly cleaned.

Most of the remaining cooling system components are replaced whenever the engine is rebuilt. These components include the

- water pump
- thermostat
- radiator hoses
- heater hoses
- radiator cap
- coolant
- drive belt

Cooling System Maintenance

Cooling system maintenance consists of keeping the coolant clean and at the proper level, inspecting for leakage, and checking the condition of the belts and hoses.

Checking Coolant Level. Checking coolant level on older vehicles without a recovery system should only be done with the engine cool. These vehicles require the radiator cap to be removed so the technician can see if the coolant level is above the radiator tubes. Regardless of the system used, removing the radiator cap on a hot engine will release the pressure in the system, causing the coolant to boil immediately. This can cause coolant to be expelled from the radiator at a high rate, resulting in severe burns.

Today's vehicles are equipped with a coolant recovery system, so the radiator cap usually does not need to be removed to check or add coolant to the system. The coolant reserve system provides quick visual checks of coolant level through the translucent bottle. Simply observe that the level is between the ADD and FULL marks while the engine is idling and warmed to normal operating temperature. If coolant needs to be added, the coolant should be added directly to the coolant recovery tank.

On vehicles that use a remote mounted pressure tank or reservoir, look at the minimum and maximum level markings on the tank and make sure the fluid is between the two marks. Be careful when opening the tank to top off a system; the tank will be under pressure if the system is hot.

Manufacturers will specify the correct coolant for each vehicle. Many coolants are not compatible with each other or a particular cooling system. Use the recommended coolant and do not mix coolants. If the improper coolant is present in the system, drain it thoroughly or flush it and then refill with the correct coolant.

Cooling System Inspection. The antifreeze content in the coolant may be tested with a coolant hydrometer, refractometer (Figure 4-24), or coolant test strips. The vehicle manufacturer's specified antifreeze content must be maintained in the cooling system, because the antifreeze contains a rust inhibitor to protect the cooling system. Inspect the coolant for rust, scale, corrosion, and other contaminants such as engine oil or automatic transmission fluid. Aside from a visual check of coolant condition, a paper test strip can be used to test the coolant pH level. This will give you a visual indication of whether the coolant has become

Classroom Manual

Chapter 4

The **radiator** is a series of tubes and fins that transfers the heat in the coolant to the air.

SERVICE TIP:
It is a good idea to flush the heater core prior to reinstalling the engine into the vehicle.

Classroom Manual

Chapter 4

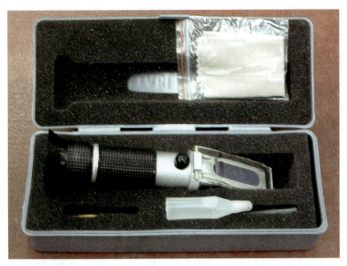

FIGURE 4-24 This refractometer can be used to determine the antifreeze protection level of any type of coolant.

FIGURE 4-25 This coolant reservoir is covered in sludge; the system should be flushed and refilled with clean coolant.

too acidic and requires flushing. If the coolant is contaminated, the cooling system must be drained and flushed (Figure 4-25). If oil is visible floating on top of the coolant, the automatic transmission cooler may be leaking. This condition also contaminates the transmission fluid with coolant. If the vehicle has an external engine oil cooler, it may also be a source of oil contamination in the coolant. These coolers may be pressure tested to determine if they are leaking. Testing or repairing radiators and internal or external oil coolers is usually done by a radiator specialty shop. Inspect the air passages through the radiator core and external engine oil cooler for contamination from debris or bugs. These may be removed from the heater core with an air gun and shop air, or water pressure from a water hose.

Visually inspect the radiator and all cooling system hoses for leaks. Leaking components must be replaced or repaired. Inspect the heater core area for signs of coolant leaks. Some vehicles have the heater core mounted under the dash, while other vehicles have this component under the hood. If the heater core is mounted under the dash, a leaking core may cause coolant to drip on the floor mat. Check all the radiator and heater hoses for soft spots, cracks, bulges, deterioration, and oil contamination. Replace hoses that indicate any of these conditions. Inspect the radiator and heater hoses for contact with other components that could rub a hole in the hose. Reroute other components away from the hoses if necessary. Be sure all the hose clamps are tight and in satisfactory condition. Inspect the engine for coolant leaks at locations such as the thermostat housing, core plugs, and water pump. Coolant leaks at the water pump usually appear at the water pump drain hole or behind the pulley (Figure 4-26). A leaking water pump must be replaced.

Weep
hole

FIGURE 4-26 This water pump had been leaking from the weep hole.

Each type of coolant reacts differently when it comes in contact with oxygen. If the cooling system was not bled properly, oxygen will enter the system. Excess oxygen can cause the coolant to do many things, the worst of which may be gelling. When engine coolant starts to gel up, it becomes thicker. It will start to stick to the walls of the cooling system, causing the heat transferring capability to be diminished, acting like a thermal blanket. This may cause certain engine parts to overheat, but the coolant temperature gauge (or warning light) will not show any severe change in fluid temperature. Always check the manufacturer's service manual for interval changing periods and frequently check technical service bulletins for information regarding specific models.

Inspect the water pump drive belt for cracks, missing chunks, fluid contamination, fraying, and bottoming in the pulley. If the belt indicates any of these conditions, belt replacement is necessary. Belt tension may be tested with the engine stopped and a belt tension gauge installed over the belt at the center of the longest belt span. When the belt tension is not within specifications, adjust the belt as necessary. Many ribbed V-belts have an automatic tensioner and do not require adjusting. Some automatic belt tensioners have a wear indicator that indicates the amount of belt wear.

Draining the Cooling System. Since most manufacturers recommend periodic coolant changes, a drain plug is usually provided at the bottom of the radiator (Figure 4-27). Following is a typical procedure for draining the cooling system:

1. Allow the engine to cool. Never open the radiator cap, open the drain plug, or disconnect a hose while the system is pressurized.
2. Start the engine.
3. Move the temperature selector of the heater control panel to the full heat position.
4. Shut off the engine before it begins to warm.
5. Without removing the radiator cap, loosen the drain plug.
6. The coolant recovery tank should empty first. Then remove the radiator cap to drain the rest of the system.
7. On most engines, it will be necessary to drain the block separate from the radiator. Remove the drain bolt from the side of the engine block to drain the block and heater core.

Starting the engine is required to move the heat control valve on vehicles that are actuated by vacuum.

Rusted fins   Drain port

FIGURE 4-27 The small tube at the bottom right of the radiator serves as a drain tube. This radiator has badly damaged fins and caused overheating, which blew a head gasket.

WARNING: Do not let the used coolant into the shop drains; collect it in a drain pan. The used coolant should be recycled or removed as a hazardous waste.

Many shops now use a coolant flushing machine that automatically expels the used coolant and refills the system with fresh coolant (Figure 4-28). Many of these pieces of equipment also recycle the coolant to eliminate costly removal from the shop. Follow the instructions provided by the equipment manufacturer to flush the cooling system. Do not use recycled coolant if the manufacturer recommends against it; that could nullify the powertrain warranty.

While the system is drained, it is a good time to replace the thermostat and hoses. Following is a typical procedure for thermostat replacement (Figure 4-29).

Drain the coolant to a level below the thermostat housing. Doing this will reduce the potential to spill coolant onto the floor. Remove the radiator hose from the housing. Some hoses use a worm-gear-type hose clamp. These can be removed by using a flat blade screwdriver to turn the worm gear. Some hoses will have a spring wire clamp; use special clamp pliers or large pliers to remove these clamps. Some engines will have a bypass hose connected to the thermostat housing; if this is the case on your engine, it should be removed at this point.

Loosen, then remove, the fasteners that attach the thermostat housing to the engine. It may be necessary to strike the housing with a rubber mallet to break the seal between the engine and the housing. Do not drive a chisel or flat-blade screwdriver between the mating surfaces; doing so may damage the mating surfaces and prevent the system from sealing.

Remove the thermostat and gasket or O-ring seal. Pay attention to the direction the thermostat was facing. Confirm the proper direction in the service manual. If the thermostat was installed backwards, that could be the cause of overheating.

Being careful not to gouge the metal, use the proper abrasive disc to clean both mating surfaces. Apply the correct sealant, if one is recommended, to the mating surface of the housing. Carefully install the thermostat in the engine block, being sure the pellet is facing

FIGURE 4-28 A coolant flushing machine.

Upper radiator hose   Thermostat housing

FIGURE 4-29 The thermostat sits on the top front of this engine.

the correct direction (usually the pellet faces into the block). A small indented ridge is generally provided for the outer portion of the thermostat to seat into. Make sure the thermostat does not slip during the rest of the installation process or a leak may result.

Install the gasket or O-ring seal. Some manufacturers use a locating tab or notch for proper indexing of the O-ring seal. Carefully locate the housing over the thermostat, and align the holes with the block. Install and torque the fasteners.

Install all hoses to the thermostat housing. It is a good practice to use new clamps to assure no leaks will occur. If using the worm-gear-type clamps, do not overtighten them: doing so will cut the hose and cause a weak spot that may begin to leak in a few months. After confirming all connections are tight, refill the cooling system.

Filling the Cooling System.   The cooling system is usually filled with a mix of 50/50 anti-freeze to water. To fill the cooling system, make sure all hoses are installed and clamps are tight. Also, close the drain plug before filling. Look up the cooling capacity in the service manual. Fill the system half of capacity with 100 percent antifreeze through the radiator cap opening. Then complete filling with water. Since many vehicles are designed with the radiator lower than the engine, a bleed valve is used when filling the system (Figure 4-30). Loosen the bleed valve while the radiator is being filled. Close the valve when coolant begins to flow out in a steady stream without bubbles. Leave the radiator cap off, and start the engine and let the engine warm up. Continue to fill the radiator as needed as the engine warms. When the radiator is full, install the radiator cap, and allow the system to pressurize while observing for leaks. It will probably be necessary to add additional mix to the recovery tank. It may take as many as four warm-up cycles before all of the air is removed from the system and the recovery tank equalizes.

Cooling System Flushing.   The effectiveness of antifreeze and the additives mixed with it decreases over time. All manufacturers have a maintenance requirement for the cooling system. The recommended schedule for drain, flush, and refill ranges from once a year to once every five years. At the same time that the refill is performed, the thermostat is generally replaced along with the water pump drive belt.

While many shops are now using automatic flushing machines, flushing of the cooling system can also be accomplished by using pressurized water forced through the cooling system in a reverse direction of normal coolant flow. A special flushing gun mixes low air pressure with water. Reverse flushing causes the deposits to dislodge from the various components. They can then be removed from the system. The engine block and radiator should be flushed separately. To flush the radiator, drain the radiator, then disconnect the upper and lower radiator hoses. A long hose may be attached to the upper hose outlet to deflect the water. Disconnect and plug any heater hoses that are attached to the radiator.

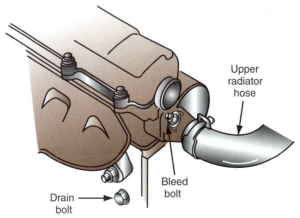

FIGURE 4-30 A bleeder bolt may be located on the thermostat housing or on another high spot in the cooling system.

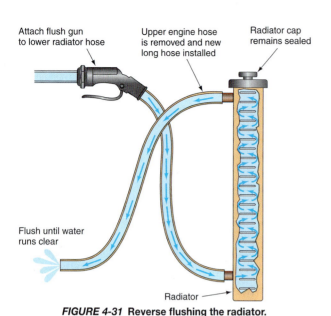

FIGURE 4-31 Reverse flushing the radiator.

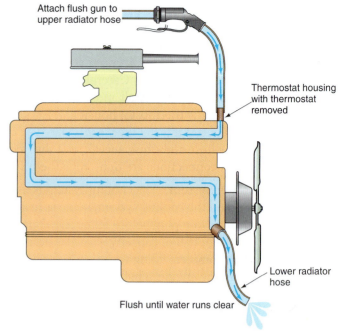

FIGURE 4-32 Reverse flushing the engine block.

SPECIAL TOOLS
Flush gun

CAUTION:
Do not allow internal pressures in the radiator to increase over 20 psi (138 kPa) or damage to the radiator could result.

Aluminum hydroxide deposits result from corrosion products being carried to the radiator and deposited when cooled off. They appear as dark gray when wet and white when dry.

Fit the flushing gun to the lower hose opening. This causes the raditor to be flushed upward (Figure 4-31). Fill the radiator with water and turn on the gun in short bursts. Continue to flush the radiator until the water being expelled from the upper hose outlet is clean.

To flush the engine block, disconnect the radiator upper and lower hoses. Also, remove the thermostat and reinstall the thermostat housing. Install the flushing gun to the thermostat housing hose (Figure 4-32). Turn on the water until the engine is full. Then turn on the air in short bursts. Allow the engine to refill with water between blasts of air. Repeat this process until the water runs clean.

Water is usually sufficient to remove most contaminants from the cooling system. However, **aluminum hydroxide deposits** require a two-part cleaner to remove them. The two parts are an oxalic acid and a neutralizer. The acid is added to the system first. Then the engine is allowed to idle for the specified length of time. The cooling system is then flushed. After the flush is completed, the neutralizer is added to prevent the acid from damaging metal components.

If chemical cleaning fails to remove the deposits, the radiator will need to be removed and "boiled out." Care should be taken to select the proper solutions and solvents to clean the radiator. Clean it in a solvent that is not destructive to the type of materials used to construct the radiator. After removing the radiator from the solvent, flush it until all of the solvent is removed. Flush the transmission cooler too, if the radiator is equipped with one. It may be necessary to use a neutralizer if an acid solvent is used. Check the fins for damage, and straighten any that are bent. After the radiator is cleaned and flushed, pressure test it to assure there are no leaks.

Cooling System Component Testing

Testing the Thermostat. If a customer brings in a vehicle with an overheating problem, it is possible the thermostat is not opening. In addition, an engine that fails to reach normal operating temperature can be caused by a faulty thermostat. A thermostat that is not opening will cause the coolant temperature gauge to rise steadily into the red zone after several minutes of operation. Sometimes the thermostat will stick closed intermittently. The customer may report that the coolant gauge spiked into the red and then came back down to a normal reading one or more times. The thermostat is inexpensive; problems caused by engine overheating are not. It is wise to replace the thermostat if you suspect that it is sticking. Thermostats have a high failure rate and are often replaced simply based on the symptoms described or as part of routine maintenance.

To attempt to verify a sticking thermostat, start the engine while it is relatively cool. Hold the upper radiator hose toward the radiator as the engine warms up. Periodically check the coolant temperature gauge to make sure the engine does not overheat. If the thermostat is working properly, the upper radiator hose should get quite hot and feel pressurized as the temperature gauge rises into the normal zone. If the gauge starts to read hot and you have not felt a distinct temperature increase in the hose temperature, the thermostat is sticking closed. Replace a faulty thermostat. You can also test for a stuck-open thermostat using a similar method. Start the engine after it has cooled. Hold the upper radiator hose toward the radiator end. Let the vehicle run for a few minutes. If the hose starts to get warm within the first few minutes of operation or before the engine reaches a normal temperature, it is sticking open.

Sometimes you may have to remove the thermostat to try to verify that it is faulty and causing the system overheating. Check the rating of the thermostat, and confirm it is the proper one for the engine application. Also confirm it was installed in the right direction. It is possible to test thermostat operation by submerging it in a container of water and heating the water while observing the thermostat. Use a thermometer so the temperature when the thermostat opens can be determined. At the rated temperature of the thermostat, it should begin to open.

WARNING: Do not open a hot radiator. The radiator is under pressure, and opening the cap will cause hot coolant to spray out of the filler tube.

Cooling Fan Inspection. The fan can be driven by a drive belt off of the crankshaft, or it can be operated electrically (Figure 4-33). Regardless of the design used, inspect the fan blades for stress cracks. The fan blades are balanced to prevent damage to the water pump bearings and seals. If any of the blades are damaged, replace the fan.

WARNING: An electric fan may start at any time, even after the engine has stopped. Either let the engine cool down or disconnect the fan electrically before touching the fan.

The drive belts should be inspected for glazing, cracking, and proper tension. A belt that is worn or loose will not turn the fan at a sufficient speed to draw the optimum mass of air over the radiator.

SPECIAL TOOLS
Thermometer

Classroom Manual
Chapter 4

SPECIAL TOOLS
Jumper wires
Scan tool

Cooling fans force airflow through the radiator to help in the transfer of heat from the coolant to the air.

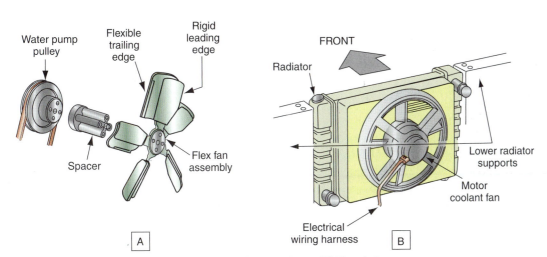

FIGURE 4-33 (A) Belt driven fan. (B) Electric fan.

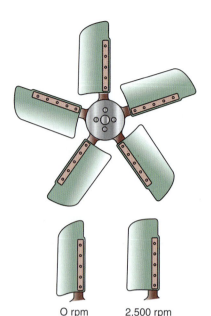

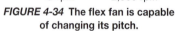
O rpm   2,500 rpm

FIGURE 4-34 The flex fan is capable of changing its pitch.

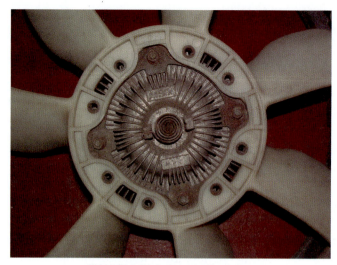

FIGURE 4-35 A viscous clutch fan.

Most modern engines using belt-driven fans use either **flex fans** (Figure 4-34) and/or a viscous fan clutch (Figure 4-35). Some newer engines may use an electrically operated viscous fan clutch (Figure 4-36). Flex fans are inspected for stress cracks as any other fan. The viscous fan clutch should be visually inspected for indication of silicone leakage. Next, observe movement of the thermostatic spring coil and shaft. If the amount of movement is out of specifications, replace the clutch assembly. Also, the shaft should rotate with the coil. To check the thermostatic spring, use a screwdriver to lift one end of the spring out of its retaining slot (Figure 4-37). Rotate the spring counterclockwise until it comes to its stop. Measure the distance from the retainer clip to the end of the spring, and compare to specifications (Figure 4-38). Return the coil spring end to its retainer slot.

Electric fans are also inspected for damage and looseness. If the fan fails to turn on at the proper temperature, the problem could be the temperature sensor, the fan motor, the fan control relay, the circuit wires, or the PCM. To isolate the cause of the malfunction, attempt to operate the fan by bypassing its control. On many modern vehicles, this can be done by using a scan tool to activate the fan. If the fan operates, the problem is probably in the

FIGURE 4-36 This clutch fan is electrically controlled by the PCM.

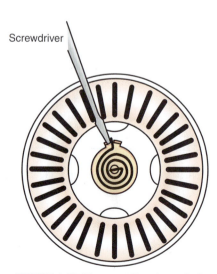
Screwdriver

FIGURE 4-37 Disconnecting the end of the thermostatic spring.

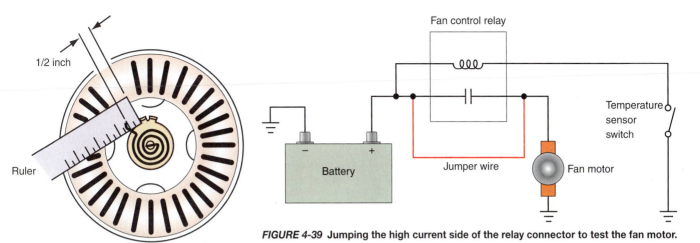

1/2 inch

Ruler

FIGURE 4-38 Measuring the gap between the spring and the clip.

Fan control relay

Temperature sensor switch

Battery

Jumper wire

Fan motor

FIGURE 4-39 Jumping the high current side of the relay connector to test the fan motor.

coolant sensor circuit. It is also possible to check fan function by jumping the fan relay to attempt operating the fan motor (Figure 4-39). You must be careful to only jump the power waiting terminal to the switched power out terminal; these are often labeled number 30 and number 87, respectively, on the relay. If the fan operates, the relay may be the faulty component; however, additional tests will have to be performed on the control circuit of the relay.

To direct airflow more efficiently, many manufacturers use a shroud (Figure 4-40). Proper location of the fan within the shroud is also required for proper operation. Generally, the fan should be at least 50 percent inside the shroud. If the fan is outside the shroud, the engine may experience overheating due to hot under-hood air being drawn by the fan instead of the cooler outside air. If the shroud is broken, it should be repaired or replaced. Do not drive a vehicle without the shroud installed.

Also inspect the idler pulley and belt tensioner, if equipped (Figure 4-41). Many manufacturers use idler pulleys so components such as generators, air pumps, air-conditioning compressors, and so forth will be provided a greater area of belt contact on their pulleys (Figure 4-42). Tensioners are used to properly maintain drive belt tension. Most tensioners are spring- or silicone-filled to provide self-adjustment. Test the pulley for free rotation and inspect for any wear grooves or looseness. If the pulley fails inspection, it must be replaced. There is no service available on these units.

If the belt squeals just after the engine starts or as the engine is quickly accelerated, the belt tension may be less than it needs to be. A squealing belt means that it is slipping. If the belt slips, the accessories are not spinning as fast as they should. This may include critical

SERVICE TIP:

If the shroud is not severely damaged, it can be repaired using hot-air or airless welding. If this welding equipment is not available, it will be necessary to glue the shroud. It is hard to find a glue to hold polyethylene or polypropylene plastic. One method is to apply SuperGlue® to the seams of the break, and while applying pressure to hold the two pieces together, spread baking soda onto the crack. The baking soda creates a chemical action that results in a harder setting of the glue.

(continued)

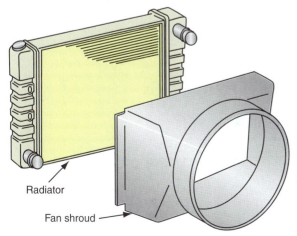

Radiator

Fan shroud

FIGURE 4-40 The fan shroud is used to control airflow direction.

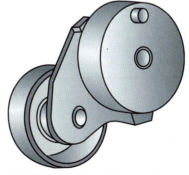

FIGURE 4-41 Typical belt tensioner.

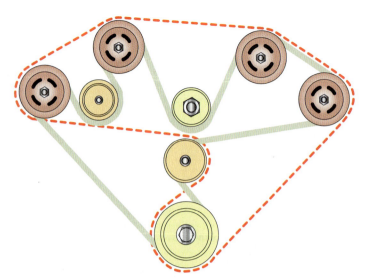

FIGURE 4-42 Idler pulleys provide for increased surface area contact of the drive belt. The dotted line indicates belt routing without idler pulleys. The added surface contact helps prevent belt slippage.

FIGURE 4-43 Depending on the design, you may be able to use a wrench or ratchet to move the belt tensioner.

SERVICE TIP:
(continued)
Apply additional glue on the top and bottom of the crack and add more baking soda as needed. The baking soda will become extremely hard and can be sanded, then painted as needed. If the shroud is constructed of fiberglass, it can be glued with epoxy.

Classroom Manual

Chapter 4

items such as the water pump and alternator. If the belt system uses an automatic tensioner, you can test one by first removing the belt. Then place a ratchet, socket (or in some cases a special wrench) on the tensioner and attempt to move it back and forth. It should move back and forth smoothly (Figure 4-43). It should also not have any movement on its mounting bolt and not have any grease or silicone leaking from it.

Water Pump and Radiator Removal

Water pump removal is usually performed with the engine in the vehicle. This is usually performed due to coolant loss through the water pump seal, or because the water pump is noisy. Radiator removal is usually performed only if the radiator requires special flushing, replacement, or the engine is being removed.

On some vehicles, it is necessary to remove the radiator prior to servicing the water pump, so this procedure will be discussed first. It is impossible to cover all of the procedures for removing the radiator on all vehicles. Always refer to the proper service information for the vehicle you are working on. Following is a general guide for radiator removal for vehicles equipped with electric fans (Figure 4-44):

1. Disconnect the negative battery terminal at the battery. This is necessary on vehicles using electric fans, to prevent them from turning on unexpectedly.
2. Drain the cooling system using the procedure covered earlier.

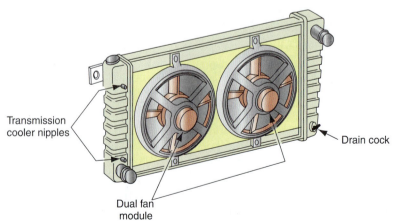

FIGURE 4-44 Radiator with dual fans.

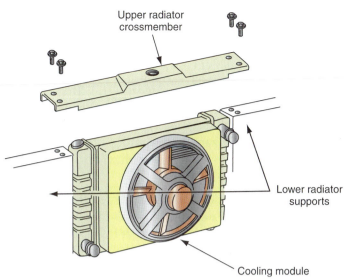

FIGURE 4-45 The radiator is usually supported by an upper mounting that must be removed.

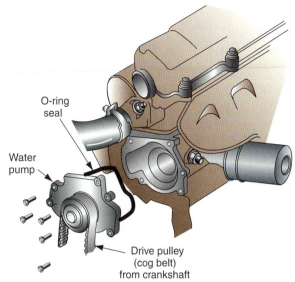

FIGURE 4-46 A gasket or seal is used between the water pump and the engine block. This water pump is driven by the timing belt; a bit more work is involved in this replacement.

3. Loosen or remove the hose clamps, then remove the upper and lower hoses from the radiator.
4. Disconnect the transmission cooler lines and plug them off, if equipped.
5. Disconnect the electric fan motor connector, if equipped.
6. Remove the fasteners attaching the fan to the radiator.
7. On some vehicles equipped with air-conditioning systems, it may be necessary to discharge the system. This is the case if the radiator and condenser cannot be separated in the vehicle.
8. Remove the upper radiator cross member or mounts (Figure 4-45).
9. Disconnect and plug the air-conditioning lines at the condenser, if needed.
10. Remove the radiator and fan as one unit if possible.
11. Separate the fan from the radiator.
12. If required, separate the radiator from the condenser.

It is not always necessary to remove the radiator to service the water pump; however, on many vehicles, the fan shroud must be separated from the radiator to gain access to the fan mounting bolts. Once the fan is separated from the water pump, the shroud and fan assembly can be removed together. Most water pumps can be removed after disconnecting the lower radiator hose at the pump (drain the cooling system first). Remove any other hoses that attach to the pump, such as bypass hoses. When removing the bolts attaching the pump to the block, note the location of each bolt in the pump. There are usually several different lengths of bolts used. Installing the wrong bolt in the wrong location can cause block damage or leaks. Most water pumps have gaskets or seals located at the mating surface to the block (Figure 4-46). Be sure to thoroughly clean the surfaces and to use the approved sealer.

GALLERY AND CASTING PLUGS

The oil gallery and coolant passages are important parts of the lubrication and cooling systems. Plugs are located on the block and cylinder head assemblies to cap the openings used for casting the block or for drilling the passages. Over time these plugs can rust through or work loose. In many cases, plugs can be replaced while the engine is still in the vehicle; however, any time the engine is removed for service, it is common practice to replace all plugs.

⚠️ **CAUTION:**
You must have special tools and be certified to perform operations on the air-conditioning system that require refrigerant handling.

SPECIAL TOOLS
Plug driver
Bore brush

Plugs can be either threaded in or pressed in. There are three basic designs of press-in plugs:

1. Disc-type plugs are installed with the convex side facing out (Figure 4-47). As the plug is driven into its bore, the crown is flattened. This causes the sides to push outward, providing a tight seal.

2. Cup-type plugs are installed with the convex side facing in (Figure 4-48). The sides of the plug are tapered so the widest part is at the outer edge. The taper assists in the installation of the plug and works to seal the plug.

3. Expansion-type plugs have tapered sides with the narrow end at the outer edge (Figure 4-49). The convex side of this plug is installed facing out. As the plug is driven into place, the sides will expand outward and seal the plug.

Pressed-in gallery plugs can be removed using a slide hammer with a self-threading screw. Center drill the plug, then thread in the self-threading screw. With the slide hammer attached to the screw, work the sliding weight until the plug is removed. It is also possible to install a self-threading screw into the plug and to pry the plug out by gripping the screw with a pair of pliers or side cutters.

Core plugs can be removed using the previous method, or by tapping one edge with a punch to turn the plug 90 degrees in the hole (Figure 4-50). Use a pry bar or pliers to force the plug out of the hole. If the plug falls into the water jacket, it must be removed. The plug can usually be gripped with a pair of pliers and pried out of the opening.

Threaded gallery plugs may be difficult to remove. Over the life of the engine, these plugs expand and contract several times. In addition, contamination can accumulate around the threads, and cause them to "seize" to the cylinder head. Whenever the block or cylinder head is serviced, these plugs must be removed for proper cleaning. To remove a seized gallery plug, it may be necessary to heat it, then spray it with penetrating oil after it cools.

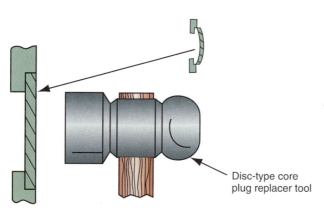

FIGURE 4-47 Installing a disc-type core plug.

Disc-type core plug replacer tool

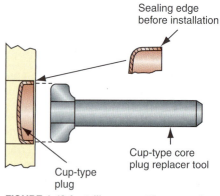

Sealing edge before installation

Cup-type plug

Cup-type core plug replacer tool

FIGURE 4-48 Installing a cup-type core plug.

Sealing edge before installation

Expansion-type plug

Expansion-type core plug replacer tool

FIGURE 4-49 Installing an expansion-type core plug.

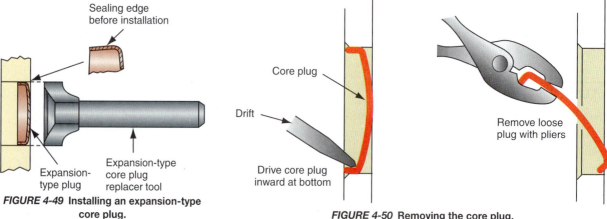

Core plug

Drift

Drive core plug inward at bottom

Remove loose plug with pliers

FIGURE 4-50 Removing the core plug.

Another method is to heat the plug, then lay a piece of paraffin wax against the exposed threads of the plug. The wax will melt and work its way down the threads, lubricating them. Plugs with damaged heads or threads can be removed using the same type of equipment used to remove damaged fasteners. Common methods include using left-handed drill bits, or center drilling the plug and using a screw extractor to remove it. Refer to Chapter 5 of this book for the procedures to extract damaged fasteners.

Oil gallery and core plugs are under pressure when the engine is running, so proper sealing is important. Coat the threads or flanges with a water-resistant sealer prior to installing the plug. Be careful to keep any chemical sealers off the end of the plug, where it can be washed into the engine oil. Start threaded gallery plugs by hand, then tighten them to the proper torque. Most threaded plugs use a taper pipe thread. These plugs will not fully seat until they are tightened.

> **CUSTOMER CARE:** Always protect the customer's vehicle. The customer has worked hard to be able to afford the vehicle. Respect the owner and his possession by using fender covers, seat covers, and floor mats to protect the vehicle's finish and interior. Keep your hands and clothes clean if you need to enter the vehicle. Attempt to return the vehicle to the customer as clean or cleaner than when he left it in your care.

LEAK DIAGNOSIS

Oil and coolant leaks can cause serious engine problems. Running an engine low on oil or coolant can quickly turn the engine into scrap metal. Most external leaks are the result of deteriorating seals or gaskets. Additional causes include defective components and cracks in the cylinder block or head.

Oil Leaks. The most common locations for oil leakage are at the valve cover, oil pan, oil filter, front seal, and rear main seal. Other locations include the camshaft expansion plugs and timing chain cover (Figure 4-51). When inspecting these areas, look for wash where the oil has left a trail. Oil contains cleaning agents, and many times the area around the leak will be washed clean by the oil.

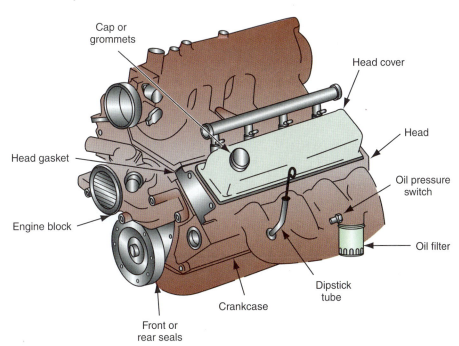

FIGURE 4-51 Common locations for external oil leaks.

WARNING: Follow all safety rules associated with hoist or floor-jack and safety-stand use if the vehicle must be lifted from the floor.

Regardless of the location of the leak(s), the technician must be aware that even the smallest of leaks can result in major oil loss. It is estimated three drops of oil leaking every 100 feet results in a total loss of 3 quarts every 1,000 miles.

Oil leakage occurs under two different conditions: normal and abnormal crankcase pressures. Seals and gaskets can leak when crankcase pressures are normal. Most of this leakage is due to normal wear; however, some leakage resulting from normal crankcase pressures is not the result of normal wear, including damage to seals and gaskets as a result of overheating or lack of oil. This damage is usually identified by excessive warping and discoloring of metallic components.

If the crankcase pressures are excessive, early failure of seals will result. Generally, excessive crankcase pressure is developed when the **positive crankcase ventilation system** inlet becomes plugged. The PCV system is designed to control normal engine oil vapor **blowby** without venting the vapors to the atmosphere (Figure 4-52). If the PCV valve malfunctions, it will not allow ventilation of the crankcase blowby vapors, and the result is excessive pressure.

There are several methods used to locate the cause of an oil leak. The first is a complete visual inspection. Look for oil leaking onto the block, pan, and bell housing (Figure 4-53). It may be helpful to clean the engine, then start it in order to observe where the oil leaks from.

The **positive crankcase ventilation (PCV) system** is an emission control system that routes blowby gases and unburned oil/fuel vapors to the intake manifold to be added to the combustion process.

Blowby is the unburned fuel and products of combustion that leak past the piston rings and enter the crank-case.

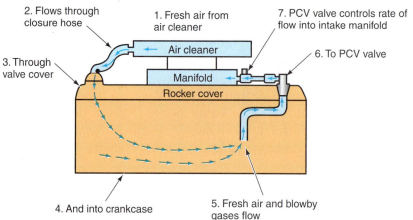

2. Flows through closure hose
1. Fresh air from air cleaner
7. PCV valve controls rate of flow into intake manifold
Air cleaner
3. Through valve cover
6. To PCV valve
Manifold
Rocker cover
4. And into crankcase
5. Fresh air and blowby gases flow

FIGURE 4-52 Normal operation of the PCV system reduces crankcase pressures.

FIGURE 4-53 Fresh oil on the pan indicates the close proximity of an external leak.

WARNING: Be careful to avoid hot exhaust manifolds and other engine components when looking for the cause of an oil leak, since the engine must be at normal operating temperatures for the inspection to be successful. Moving parts, such as fans, pulleys, and belts, are also a danger.

There are two things the technician should keep in mind when performing a visual inspection for an oil leak. First, when the vehicle is stopped, the oil will seek a low point before it drips. Second, when the vehicle is moving, airflow over the engine will spread the oil to other under-car components. In most cases, oil will travel down and somewhat to either side. If the oil finds its way into a rail or ledge, it will continue to follow that path until it gets to the lowest point.

WARNING: Whenever running the engine in the shop, be sure to connect the shop's exhaust system to the vehicle's exhaust pipe. Carbon monoxide is deadly.

If oil is spread throughout the undercar components, visual inspection becomes more difficult. In this case, clean the engine and undercar components with a steam cleaner or solvent. Allow the cleaning agents to dry completely. Start the engine, and observe suspect areas while the engine is brought up to normal operating temperatures. Sometimes a leak is more detectable with a cold engine, and in other instances, the leak may not appear until the engine is warm.

If the visual inspection fails to locate the cause of the leak, a chemical dye can be added to the engine's oil (Figure 4-54). First, clean the outside of the engine with a steam cleaner or solvent. Next, add the recommended amount of dye to the oil. Run the engine for about 15 minutes to allow the oil to circulate throughout the engine. On vehicles with a small leak, it may be necessary to run the vehicle for a longer period or to have the customer return in a day or two. Shut off the engine, and inspect for traces of the dye with a black light. The dye will show fluorescent yellow or orange when the black light is used.

A variation of the dye procedure is the use of a leak tracing powder. The powder will turn brown or black when it comes into contact with oil. Before applying the powder to the outside of the engine, all traces of oil must be removed from the engine. Use a steam cleaner or solvent to thoroughly clean the engine, and allow it to dry completely. Apply the powder to the suspected areas and start the engine. While the engine is running, observe the powder; it will change color as the oil comes into contact with it.

SERVICE TIP:
Aerosol-type powder, such as foot powder, can be used to find the leak. Spray the area, then operate the engine. Inspect the powder for traces of oil.

CAUTION:
Be sure to use regulated air pressure at 5 psi in the crankcase. Using higher pressure can *cause* seal failure.

FIGURE 4-54 Use of a chemical dye and a black light may help in locating the source of a leak.

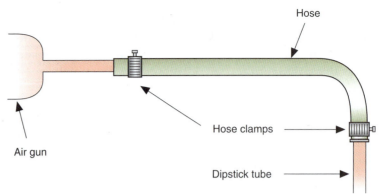

FIGURE 4-55 Use an adapter connected to the dipstick tube to perform a low pressure air test.

Low and high pressure air testing is another method of locating the cause of an oil leak. Low pressure air testing uses a regulated air pressure of 5 psi (34.5 kPa) and soapy water to locate the leak.

A hose and clamp arrangement can be made to adapt the air hose to the dipstick tube (Figure 4-55). Pull and plug the PCV valve and breather tubes. Lock or tape the blow gun into the ON position. With a constant amount of air pressure entering the engine, apply the soap and water solution to likely leak areas. A leak is indicated by bubbles created by the escaping air. If the leak is bad enough, it may be possible to hear the air escaping.

If low pressure testing fails to find the location of the leak, use high pressure air testing. This test is done by first removing the oil pressure sending unit from the engine. Select the correct size adapters to tap into the threaded hole. Set the regulated air pressure between 80 and 100 psi, and install the blow gun to the adapters. Lock or tape the blow gun ON, and apply the soapy water solution around oil galley plugs, oil filter seals, and any other suspected areas.

Once the location of the leak is isolated, the technician should make a determination of the cause. Gaskets and seals can fail due to several factors. One of the most common causes for gasket and seal failure is overheating. Before replacing a faulty gasket or seal, inspect the area for signs of overheating. These include blueing or discoloring of the metal, hardening and cracking of the gasket or seal, and warping of the surface areas.

If overheating is not the cause of the failure, check for excessive pressure in the engine crankcase. If the PCV valve malfunctions, it will not allow ventilation of the crankcase blowby vapors and the result is excessive pressure. Before faulting the PCV valve, check for a plugged filter in the breather line from the valve cover to the air cleaner housing.

Quick-check the PCV valve by removing it from the engine and shaking it. A rattle should be heard; if not, the valve is stuck and needs to be replaced. A valve that does not rattle must be replaced; however, a valve that rattles may still be plugged enough to cause pressure.

Next, remove the PCV valve from the engine, and allow it to hang free. Listen for a hissing sound from the valve. No sound indicates the PCV valve or hose is plugged. A vacuum should be felt if you place your thumb over the valve opening, and the engine speed should decrease about 50 rpm.

If there are no signs of external oil leaks, yet the engine is using oil, remove the spark plugs and inspect them for indications of oil burning. Oil on the spark plugs indicates valve seal leakage, worn valve guides, or worn piston oil control rings.

Coolant Leaks. Cooling system leaks or loss of pressure can occur at many locations (Figure 4-56). When inspecting the engine for coolant leaks, check it both cold and at normal operating temperatures. Sometimes coolant will only leak at a location when the

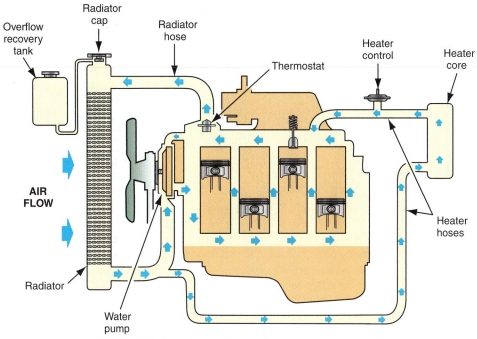

FIGURE 4-56 Areas from which coolant or pressure can leak.

engine is cold. The cold components are contracted, and gaps appear. As the engine warms, the gaps may close and stop the leak. Other leaks may not appear until the engine is warm and pressures are increased in the cooling system.

WARNING: **Electric fans can come on at any time. Keep fingers and hands away from the fans. Also, make sure you have no loose clothing or jewelry that can get tangled in the fans.**

Engine coolant leaks that are not easily identified by a visual inspection may be located using a cooling system pressure tester (Figure 4-57). To pressure test the cooling system, begin by using the correct service manual for the vehicle and engine being tested, to locate the cooling system pressure specifications. With fender covers properly placed to protect the vehicle's finish, perform a visual inspection of the hose connections, and tighten any loose clamps (see Photo Sequence 6).

SERVICE TIP:
The cooling system pressure is often indicated on the radiator cap; just be sure it is original equipment. Most systems use 13- to 17-lb. caps. Do not increase pressure over 16 lbs. unless specifications call for higher test pressures.

FIGURE 4-57 A cooling system pressure test is used to test for internal and external leaks.

COOLING SYSTEM TESTING

P6-1 Visually inspect the coolant and use a coolant analysis strip to test the coolant.

P6-2 Dip the strip in the coolant and compare the colors to the indications in the bottle to determine if the coolant needs to be replaced.

P6-3 With many coolants you can use a traditional anitfreeze protection level tester.

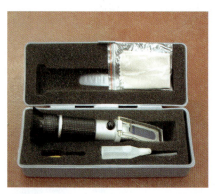

P6-4 On some newer long life coolants, a refractometer provides a more accurate analysis of the coolant.

P6-5 Test antifreeze protection level whenever cooling system maintenance is performed.

P6-6 Use a cooling system pressure tester to test the system for leaks and to test the function of the radiator cap.

P6-7 Pressurize the system to the pressure stated on the cap and make sure the system holds pressure while you look for leaks.

P6-8 Test the radiator cap to be sure it holds pressure. A weak cap can cause boiling over and overheating. The cap should also release pressure above its rating.

P6-9 You can use a chemical block tester to be sure that no combustion gases are leaking into the cooling system from a faulty head gasket.

COOLING SYSTEM TESTING

P6-10 Use a pyrometer to be sure that the engine is running at normal operating temperature.

P6-11 You can check your gauge accuracy by comparing your readings on the engine.

WARNING: Severe burns can result from heated coolant. Wait until the upper tank of the radiator is cool to the touch before opening the pressure cap. Remove the cap slowly to release any pressure in the system.

Remove the radiator cap, and inspect the spring washer for indications of weakness. If the spring is weak, replace the cap. Also, inspect the relief valve gasket on the cap and inner sealing surface of the filler neck for damage or contamination. Clean or replace as needed.

Clean the cap and install it on a pressure tester. Pump the pressure tester slowly to obtain minimum holding pressure. A reading less than that specified as the minimum holding pressure in the service manual indicates a defective cap.

Remove the cap from the pressure tester, and install the tester on the radiator filler neck. Use the pump to increase the pressure in the cooling system to the specified level. Once the specified pressure has been obtained, a correctly sealed cooling system will maintain it for 15 minutes. If the pressure drops, there is a leak. Visually inspect all possible areas for external leaks. If the leak cannot be found with a pressure tester, use a fluorescent dye similar to that used in locating oil leaks. Use a black light to pinpoint any leaks. If there are no visible signs of external leakage, the cause of the pressure drop is internal.

WARNING: Relieve the pressure applied by the tester before removing the test equipment.

Inspect the oil dipstick and oil filler cap for signs of water, milky-colored oil, or rust. Also, inspect for water on the spark plugs and oil deposits on the radiator filler neck. Any of these indicate coolant has entered the crankcase.

WARNING: Be careful of moving parts. Fans, pulleys, and belts can cause severe injury.

To test for cylinder head gasket leakage, leave the cooling system pressure tester connected to the radiator filler neck and apply a low pressure of approximately 5 psi. Run the engine at idle while observing the gauge. Excessive pressure buildup indicates combustion chamber pressures are entering the cooling system.

FIGURE 4-58 A combustion leak detector checks for combustion gases in the coolant.

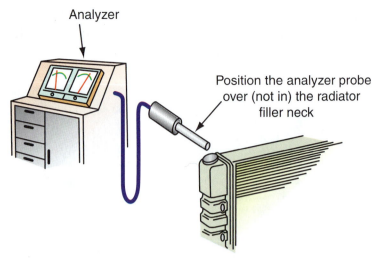

FIGURE 4-59 A four or five gas exhaust analyzer can be used to detect combustion gases in the cooling system.

SPECIAL TOOLS

Infrared exhaust analyzer

A four-gas exhaust analyzer enables the technician to look at the effects of the combustion process. It will measure hydrocarbons, carbon monoxides, carbon dioxide, and oxygen levels in the exhaust.

A damaged cylinder head gasket or cracked block can also be verified by the use of a combustion leak detector (Figure 4-58). Lower the coolant level in the radiator to allow a 1-inch space above the fluid. Start the engine, and add the special test fluid into the tester. Squeeze the bulb on the tester to draw in gases from above the coolant through the test fluid. If there are combustion chamber gases present in the coolant, the test fluid will change color.

It is also possible to use an infrared *exhaust gas analyzer*. Place the probe over the filler neck with the engine running (Figure 4-59). Do not place the analyzer tip in the fluid. The presence of combustion chamber gases will be indicated by a hydrocarbon (HC) and/or CO reading.

A third method is to use a chemical tester. The test kit comes with a blue-colored paper. A strip of the paper is soaked in the coolant. If combustion gases are present, the paper turns yellow.

> **CUSTOMER CARE:** Gallery plugs or soft plugs that are rusted through indicate excessive water in the coolant mixture. If this condition is found, the customer should be advised of the correct coolant mixture to prevent a recurrence of the problem. The cooling system should be flushed when repairs are made.

Occasionally, the crack in the head gasket, head, or block is very small and will cause less-dramatic symptoms. When the engine is cool, the metals in either of the suspect parts contract and enlarge the crack. Coolant may leak into the affected cylinder and foul the spark plug or cause small amounts of white smoke to exit the tailpipe. Symptoms may occur for a short time after startup; then the engine heat can cause enough expansion in the metal to seal the gap. To confirm your diagnosis, remove the spark plugs, allow the engine to cool *thoroughly* and place rated system pressure on the coolant using a pressure tester. Let the vehicle sit for as long as is practically possible but at least one half hour. Disable the fuel system; have an assistant crank the engine over and watch closely for coolant spray out of a cylinder. Sometimes it is helpful to place a paper towel in front of each plug hole, so you are able to check each cylinder. Any coolant from a cylinder indicates a sealing failure. If you are uncertain of the result from a cylinder, you can perform a cylinder leakage test as described in Chapter 7 of the Shop Manual.

WARNING SYSTEMS DIAGNOSIS

Engine warning systems must function properly to avoid costly engine failures. Every time you change the oil and filter, you should check that the oil pressure warning light or pressure gauge functions properly. Whenever you service either the lubrication system or the cooling system, test the warning system before returning the vehicle to the customer.

Engine Oil Pressure Warning System

Every vehicle has a means of communicating low oil pressure to the driver. Most vehicles use a warning light, either with or without an oil pressure gauge (Figure 4-60). Test to be sure that the circuit is functional by observing the indicator with the key on, engine off. If it does not light, check the oil pressure sending unit wiring, instrument panel bulb, and power supply before starting the vehicle. If the oil pressure warning light is on and no engine noise is heard, check the circuit electrically first. The oil pressure sending unit is typically located near the oil filter housing. Most often you can remove the single wire from the unit and the light will go off, and ground the wire to clean, bare metal to make the light come on or the gauge rise. Be certain to check the manufacturers' wiring diagrams and service information first; this procedure could cause damage on some systems! If the circuit appears to function as designed, run an oil pressure test to check your findings.

To replace an oil pressure switch, sending unit, or temperature sensor, unscrew it from its bore. If it is physically damaged, be certain you extract all the pieces from the engine block. If specified, use a small amount of thread tape or pipe sealant to prevent leakage. Tighten the unit to specification. Always check for proper operation of the system after replacing a component. Also check for and correct any leaks from the new part.

Cooling System Temperature Warning System

Vehicles have either a cooling system warning light or a temperature gauge with a warning light that indicates excessive cooling system temperature to the driver. To test that the warning light is functional, turn the key to the crank position and observe the light. If it does

FIGURE 4-60 This oil light comes on when the key is on and the engine is off as a bulb check, and when oil pressure falls below about 3 psi.

FIGURE 4-61 Aim the infrared pyrometer at the thermostat housing and compare the reading to the coolant temperature on the scan tool.

not come on during the bulb check prove out test, check the bulb and the instrument panel wiring. Be sure to check the manufacturer's test procedures; the following is an example of how to test the coolant sensor and gauge. One way to test the cooling system temperature sensor is to disconnect the electrical connector. Observe the gauge; it should fall to the minimum temperature or the light should go off. Next, short the terminals together. The light should come on or the gauge should rise to maximum temperature. If it does not, check the gauge assembly and the wiring to it. On most modern vehicles, you can check the accuracy of the temperature gauge by comparing the actual coolant temperature to the temperature that the PCM is displaying. Check the actual temperature by using an infrared temperature gauge (Figure 4-61). Compare that to the temperature displayed on a scan tool. Discrepancies in the readings are usually caused by a faulty temperature sensor or wiring.

Scan Tool Diagnostics

Using a Scan Tool to Check for DTC's. A scan tool can be a very powerful and useful tool. Understanding the basics of scan tool diagnostics is necessary when diagnosing basic engine problems. The different types of scan tools and computer systems are discussed in Chapter 3. A DTC may be set in the PCMs memory even if the MIL is not illuminated. If an engine is running rough, these DTCs may help you diagnose the problem quicker, maybe even saving money on major engine repairs. For example, if there are overheating and misfire codes set in the computer, you may be able to detect a slightly blown head gasket. You may be able to repair the condition before the engine further overheated and cracked the cylinder head, and possibly damaged other expensive engine components.

The first step is to connect the scan tool cable to the DLC (Figure 4-62). Most scan tools are powered by the DLC, but some are self-powered (meaning they do not have to be plugged into the DLC to power up). All OBD II DLCs have pins for battery power and ground; this powers up a scan tool. Once the scan tool is plugged into the DLC, you have to get through the menu screens by selecting the year, make, model and sometimes parts of the VIN. Turn the key to the run position (do not start the engine). Once in the menu screen you should be able to maneuver around to read the codes. The description of each code can be looked up in the service manual. The manufacturer's service information will have a diagnostic troubleshooting tree for each code. Refer to Photo Sequence 7 for the typical steps for using a scan tool to retrieve codes.

FIGURE 4-62 Connect the scan tool cable to the DLC.

Understanding these codes and their relationship to how the engine is running is important. Depending on the code, you can then further diagnose the engine by choosing a secondary test. These secondary tests may include compression tests, cylinder leakage tests, power balance tests, and others. These tests are further explained in Chapter 7.

CASE STUDY

A customer brought his 2002 Ford Focus in because the temperature gauge went into the red zone twice while driving on the highway. The technician asked him for a few more details about the events and found out that less than a minute after the temperature gauge rose into the red it returned back to a normal temperature. The technician brought the vehicle in and inspected the coolant level and condition; it was fine. Next, she checked the water pump drive belt for damage and appropriate tension; it too was in good shape. Suspicious of a sticking thermostat, she started the engine while monitoring the temperature of the upper radiator hose. It did get hot when the engine temperature came up to normal. The fan also operated properly. The technician replaced the thermostat. She let the vehicle run in the shop until the fan cycled on a few times, and then she took the vehicle for a good road test. The temperature stayed in the normal range. She returned the vehicle to the customer and then followed up with a telephone call a few days later to be sure she had corrected the problem. The customer reported that all was well and that he appreciated the call.

TERMS TO KNOW

Aluminum hydroxide deposits
Battery terminal voltage drop test
Bearingst
Blowby
Capacity (load) test
Conductance test
Current draw test
Flex fons
Hydrostatic lock
Jet valves
Oil pressure testing
Oil pump
Open circuit voltage test
Pickup tube
Positive crankcase ventilation (PCV) system
Radiator
Relief valve
Starter quick test

Typical Procedure for Using a Scan Tool to Retreive DTCs

P7-1 Roll the driver's side window down, turn the ignition to the off position, and block the wheels (or set the parking brake).

P7-2 Select the proper scan tool data cable for the vehicle you are working on.

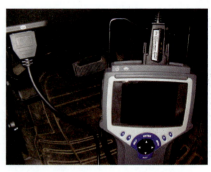

P7-3 Connect the scan tool data cable to the DLC on the vehicle.

P7-4 Turn the ignition key to the KOEO (Key On Engine Off) position. Some scan tool will automatically turn on at this point, others may need to be turned on.

P7-5 Enter the model year, make, and selected VIN components into the scan tool.

P7-6 Maneuver through the scan tool's menu and select "read/retrieve DTC."

P7-7 Record the DTC's on the repair order.

ASE-STYLE REVIEW QUESTIONS

1. *Technician A* says a stuck-open oil pump relief valve can cause higher than normal oil pressure.
   Technician B says that worn oil pump gears can cause low oil pressure.

   Who is correct?
   A. A only   C. Both A and B
   B. B only   D. Neither A nor B

2. *Technician A* says that you can use a chemical dye and a black light to pinpoint the cause of an oil leak.
   Technician B says to put 80 psi of air pressure in the dipstick tube and check for leaks.

   Who is correct?
   A. A only   C. Both A and B
   B. B only   D. Neither A nor B

3. *Technician A* says that a weak battery can cause engine performance problems.
   Technician B says that a discharged battery is a common cause of a no-start, no-crank problem.

   Who is correct?
   A. A only   C. Both A and B
   B. B only   D. Neither A nor B

4. To perform a check of the battery clamp connection:
   A. Place an ohmmeter on the battery posts.
   B. Use a VAT tester to measure current flow.
   C. Place voltmeter leads on the battery clamp and post while cranking the engine.
   D. Place the voltmeter leads on the battery post and the top of the battery.

5. While discussing proper oil change procedures:
   Technician A says to lift the vehicle to check for fluid leaks after restarting the vehicle.
   Technician B says to adjust the tire pressures and check all fluid levels.

   Who is correct?
   A. A only   C. Both A and B
   B. B only   D. Neither A nor B

6. Each of the following is a likely cause of a no-crank condition, except:
   A. A bad battery connection
   B. A faulty starter solenoid
   C. A loose connection at the S terminal of the starter
   D. A chipped starter drive gear

7. A customer says his vehicle overheats only in traffic while idling.
   Technician A says the fan may be inoperative.
   Technician B says the thermostat may be stuck closed.

   Who is correct?
   A. A only   C. Both A and B
   B. B only   D. Neither A nor B

8. A vehicle has contaminated coolant.
   Technician A says to drain and refill the coolant.
   Technician B says to flush the cooling system.

   Who is correct?
   A. A only   C. Both A and B
   B. B only   D. Neither A nor B

9. The oil pressure warning light does not come on with the key on.
   Technician A says the wire on the sending unit may be disconnected.
   Technician B says to ground the sending unit wire, and if the light comes on the sending unit is bad.

   Who is correct?
   A. A only   C. Both A and B
   B. B only   D. Neither A nor B

10. *Technician A* says that the radiator should be cleaned or replaced after an engine overhaul.
    Technician B says to replace the water pump after an engine overhaul.

    Who is correct?
    A. A only   C. Both A and B
    B. B only   D. Neither A nor B

1. A starter makes a grinding noise when it engages the flywheel teeth.
    Technician A says it could be shimmed improperly.
    Technician B says the starter drive gear could be damaged.
    Who is correct?

    A. A only   C. Both A and B
    B. B only   D. Neither A nor B

2. A technician is performing a starter current draw test because the engine will not turn over. The test results indicate that the current draw on the starter is higher than the specification. This means:

    a. That the starter solenoid is bad.
    b. The battery needs charging.
    c. The ignition switch is bad.
    d. The engine may be hydrostatically locked.

3. A cooling system is pressurized with a pressure tester to locate a leak. After 15 minutes, the tester gauge has dropped from 15 psi to 5 psi, and there are no visible leaks in the engine compartment.
    Technician A says the engine may have an internal head gasket leak.
    Technician B says the heater core may be leaking.
    Who is correct?

    A. A only   C. Both A and B
    B. B only   D. Neither A nor B

4. A customer says that his oil pressure warning light comes on while the car is idling. An oil pressure test shows low oil pressure.
    Technician A says that the engine bearings may be worn.
    Technician B says that the oil pressure relief valve may be stuck closed. Who is correct?

    A. A only   C. Both A and B
    B. B only   D. Neither A nor B

5. A coolant temperature gauge does not move from its lowest reading when the vehicle is driven.
    Technician A says the coolant temperature sensor wires may be disconnected.
    Technician B says that the thermostat could be stuck open. Who is correct?

    A. A only   C. Both A and B
    B. B only   D. Neither A nor B

Name _____ **Date** _____

PERFORMING A BATTERY TEST

Upon completion of this job sheet, you should be able to properly perform a battery capacity and conductance test.

Tools and Materials

Conductance battery tester
Capacity (load) battery tester

Describe the vehicle being worked on:

Year _____ Make _____ Model _____

VIN _____ Engine type and size _____

Procedure

Task Completed

Your instructor will provide you with a specific vehicle and engine package. Write down the information, then perform the following.

1. Locate the test procedure in the service manual.

2. What is the battery's CCA? _____   ☐

NOTE: The battery must be fully charged to 12.65 volts before testing.

3. Connect the battery to the capacity (load) test machine by placing the large red and   ☐
   black clamps on the terminals and the inductive pickup lead (amp clamp) on the
   negative lead (make sure that the arrow is pointing the right way).

4. Zero the ammeter on the tester.   ☐

5. Be prepared to observe the amperage and voltage during the test. Turn the test machine
   on and draw the energy out of the battery. If this is done manually on the tester, use 1/2
   of the CCA rating for 15 seconds. After the test, complete the section below.

   Amount of current used to test: _____
   Lowest voltage reading while testing: _____
   Is the battery good or bad? _____

6. Disconnect and remove the capacity (load) tester and connect the conductance tester.   ☐

7. Perform the battery conductance test and report the results below.

   _____
   _____
   _____

Instructor's Response _____

_____

_____

Name _____ Date _____

TESTING A STARTER

Upon completion of this job sheet, you should be able to measure the current draw of a starter motor properly and interpret the results of the test.

Tools and Materials

Battery starter tester

Describe the vehicle being worked on:

Year _____ Make _____ Model _____
VIN _____Engine type and size _____

Procedure

Task Completed

Your instructor will provide you with a specific vehicle and engine package. Write down the information, then perform the following.

1. Locate the test procedure in the service manual. ☐

2. Disable the ignition system or fuel injection system to prevent the engine from starting. Briefly describe the procedure.

   _____

   _____

   _____

3. Expected starter current draw is _____ amps.
   Voltage should not drop below _____ volts.

4. Connect the starting/charging system tester cables to the vehicle. ☐

5. Zero the ammeter on the tester. ☐

6. Be prepared to observe the amperage when the engine begins to crank and while it is being cranked. Also, note the voltage while the engine is cranking and again after you stop cranking the engine.

   Initial current draw: _____

   Current draw while cranking: _____

   Voltage while cranking: _____

   Voltage after cranking: _____

7. What is indicated by the test results? Compare test measurements with specifications.

   _____

   _____

   _____

   _____

   _____

☐   **8.** Reconnect the ignition or fuel injection system.

Instructor's Response _____

_____

_____

_____

JOB SHEET

Name _____ Date _____

PERFORMING AN OIL AND FILTER CHANGE

Upon completion of this job sheet, you should be able to properly perform an oil and filter change.

ASE Correlation

This job sheet is related to ASE Engine Repair Test content area: Lubrication and Cooling Systems Diagnosis and Repair: Perform oil and filter change.

Tools and Materials

Technician's tool set
Shop rag
Correct type of engine oil
Service manual
Oil filter wrench
Used oil container
Torque wrench

Describe the vehicle being worked on:

Year _____ Make _____ Model _____

VIN _____ Engine type and size _____

Procedure

Follow the instructions below to complete the oil and filter change.

Task Completed

1. Pull the oil fill level indicator (dipstick) and note the level and condition of the oil.

   _____

   _____

   _____

2. What are the API and SAE ratings for the oil you are going to use (consult the service manual for the correct information)?

   _____

   _____

4. Warm the engine to operating temperature. The oil will flow faster when the drain plug is removed. ☐

5. Properly raise the vehicle on a lift. ☐

6. Place the used oil container (or catch basin) under the drain plug. Remove the drain plug and let the oil flow into the container. You may have to adjust the position of the container because the flow rate and position will change as more oil flows out of the engine. ☐

7. Now remove the oil filter. You may have to use a special wrench or remove some panels. Make sure all of the oil flows into the used oil container.

☐   8. Look at the oil filter mounting surface on the engine. Was the old oil filter rubber gasket removed with the old oil filter? _____

9. Describe what could happen if the old oil filter gasket is not removed.

_____

_____

10. Check the condition of the drain plug. Are the threads or the gasket damaged?

11. Locate the new oil filter and use new engine oil to lightly lubricate the rubber gasket. Compare it to the old one to make sure it is the correct fit.

☐   12. If there is a manufacturer's specification to tighten the oil filter to, then use it. Otherwise refer to the oil filter installation instructions on the box it came in or use the accepted industry standard of tightening it 1-2 to 1 full turn past the contact point. This will allow for the gasket to remain sealed until the next oil change.

☐   13. Install the oil drain plug and tighten it to the manufacturer's torque specification.

☐   14. Lower the vehicle and refill the engine with the correct amount of oil.

☐   15. Pull the oil fill level indicator (dipstick) and check the oil level. Add oil if needed.

☐   16. Start the engine. Check to make sure that oil pressure builds immediately (or the oil light goes off).

☐   17. Check under the car. Are there any major oil leaks? _____

☐   18. Turn the engine off, wait 30 seconds, and check the oil level on the dipstick. Add if needed.

19. Dispose of the used and filter properly.

Instructor's Response _____

_____

_____

Name _____ Date _____

OIL PUMP INSPECTION AND MEASUREMENT

Upon completion of this job sheet, you should be able to properly perform an oil and filter change.

ASE Correlation

This job sheet is related to ASE Engine Repair Test content area: Lubrication and Cooling Systems Diagnosis and Repair: Inspect oil pump gears or rotors, housing, pressure relief devices, and pump drive; perform necessary action.

Tools and Materials

Technician's tool set
Service manual
Feeler gauge set
Micrometer set
Dial caliper
Oil pump removed from an engine

Describe the vehicle being worked on:

Year _____ Make _____ Model _____
VIN _____ Engine type and size _____

Procedure

Your instructor will provide you with a specific vehicle. Write down the information, and then perform the following tasks:

1. Remove, disassemble, and draw the oil pump and its parts in the space below.

2. What type of oil pump is it (gear, rotor, etc.)? _____

3. Inspect the various parts of the oil pump. Do any of the parts show signs of wear or damage?

   _____

   _____

4. Each manufacturer uses a different method for measuring the oil pump. Using the service manual, look up all the measurements and specifications that you will need.

4. Measure the oil pump and compare against the specifications. Use the chart below as a guide to measure the type of pump you have.

For gear oil pumps, measure the following:

   Gear to gear clearance _____

   Gear to gear specification _____

   Gear to housing clearance _____

   Gear to housing specification _____

For rotor oil pumps, measure the following:

Gear to rotor clearance _____

Gear to rotor specification _____

Rotor to housing clearance _____

Rotor to housing specification _____

For internal/external oil pumps, measure the following:

Gear to gear clearance _____

Gear to gear specification _____

Gear to crescent clearance _____

Gear to crescent specification _____

Gear to housing clearance _____

Gear to housing specification _____

5. According to your measurements and inspection, could the oil pump have caused any problems with oil pressure and it is reusable?

_____

_____

Instructor's Response _____

_____

_____

_____

JOB SHEET

Name _____ Date _____

OIL PRESSURE TESTING

Upon completion of this job sheet, you should be able to perform an oil pressure test properly and interpret the test results.

ASE Correlation

This job sheet is related to ASE Engine Repair Test content area: Lubrication and Cooling Systems Diagnosis and Repair; task: Perform oil pressure tests; determine necessary action.

Tools and Materials

Oil pressure gauge
Technician's tool set

Describe the vehicle being worked on:

Year _____ Make _____ Model _____

VIN _____ Engine type and size _____

Procedure

Task Completed

Your instructor will provide you with a specific vehicle and engine package. Write down the information, then perform the following.

1. Using the service manual, locate the oil pressure specifications. Record the specifications below.

   _____

   _____

   _____

2. Check the oil level.   FULL   LOW   OVERFULL
   If low, fill to proper level.
   If overfull, drain the oil to the proper level. Check for gas or coolant in the oil.

(**NOTE:** Be sure the proper dipstick is installed.)

3. Remove the oil pressure sending unit from the engine.   ☐

4. Use the required adapters to connect the oil pressure test gauge to the engine properly.   ☐

5. Start the engine and observe the gauge as the engine is idling; record your reading.

   _____

6. Record the gauge reading after the engine reaches normal operating temperatures.

   _____

7. If oil pressure is present, increase the engine speed to 2,000 rpm, and record the oil pressure.

   _____

8. Shut off the engine and compare test results to specifications. What are your recommendations?

_____

_____

_____

_____

_____

☐   **9.** Remove the test gauge and reinstall the oil pressure sending unit.

☐   **10.** Start the engine and look for leaks from the sending unit. Shut engine off.

Instructor's Response _____

_____

_____

_____

JOB SHEET

Name _____ **Date** _____

OIL LEAK DIAGNOSIS

Upon completion of this job sheet, you should be able to determine the location of oil leaks properly and recommend proper repairs.

ASE Correlation

This job sheet is related to ASE Engine Repair Test content area: General Engine Diagnosis; task: Inspect engine assembly for fuel, oil, coolant, and other leaks; determine necessary action.

Tools and Materials

Oil leak dye

Describe the vehicle being worked on:

Year _____ Make _____ Model _____

VIN _____ Engine type and size _____

Procedure

Your instructor will provide you with a specific vehicle and engine package. Write down the information, then perform the following.

1. Check the engine oil level and correct to proper level if needed.

2. Perform a visual inspection of the engine. Record any suspect areas you find.

   _____

   _____

   _____

3. Determine what method you are going to use to locate the oil leak (dye, air pressure, etc.).

4. Briefly describe the procedure you have chosen to use.

   _____

   _____

   _____

   _____

   _____

5. Based on your test results, what is your repair recommendation?

   _____

   _____

   _____

   _____

   _____

_____

Task Completed

☐

Instructor's Response _____

_____

_____

_____

TESTING AN OIL PRESSURE SWITCH AND SENSOR

Name _____ Date _____

Upon completion of this job sheet, you should be able to properly test an oil pressure sensor and switch.

ASE Correlation

This job sheet is related to ASE Engine Repair Test content area: Lubrication and Cooling Systems Diagnosis and Repair: Inspect, test, and replace oil temperature and pressure switches and sensors.

Tools and Materials

Service manual
Technician's tool set
Digital multimeter (DMM)

Describe the vehicle being worked on:

Year _____ Make _____ Model _____
VIN _____ Engine type and size _____

Procedure

Task Completed

Your instructor will provide you with a specific vehicle and engine package. Write down the information, then perform the following.

1. Check the level and condition of the engine oil. If the oil level is low, add oil as needed. Change the oil and oil filter if necessary. ☐

2. In order to verify that the problem is electrical, you should first perform an oil pressure test. This will rule out the possibility that the problem is with the engine. ☐

Note: Steps 3–10 are for diagnosing a vehicle with an oil pressure switch.

3. How much engine oil pressure does it take to turn the dash warning light off? _____

4. Is the oil pressure switch a normally open or closed switch? _____

5. Turn the key to the ON position. With the engine off, does the warning light come on? _____

6. If the warning light does come on, does this mean that the circuit is working? _____

7. Disconnect the oil pressure switch connector and start the engine. What happened to the light?
_____

8. Should that have happened? _____

6. Position the coolant catch basin under the radiator drain valve (petcock).

☐ 7. Loose the radiator drain valve, let the coolant drain, and then remove the valve.

☐ 8. After the coolant is done draining install the drain valve and lower the vehicle.

☐ 9. What type of coolant and mixture should this vehicle use?

10. Add the proper coolant type and mixture to the system until it is full. You may need to add more fluid after a few minutes.

☐ 11. Start the engine with the cap off and let it run for a few minutes. Add coolant as needed to the radiator or reservoir.

☐ 12. Install the coolant cap correctly and start the engine again. Locate the cooling system bleeders (if applicable). Describe where they are located.

_____

☐ 13. With the engine running and at operating temperature, slowly open the bleeder screw. Keep it open until only liquid comes out (no bubbles).

Instructor's Response _____

_____

_____

_____

Name _____ Date _____

TESTING AN OIL PRESSURE SWITCH AND SENSOR

Upon completion of this job sheet, you should be able to properly test an oil pressure sensor and switch.

ASE Correlation

This job sheet is related to ASE Engine Repair Test content area: Lubrication and Cooling Systems Diagnosis and Repair: Inspect, test, and replace oil temperature and pressure switches and sensors.

Tools and Materials

Service manual
Technician's tool set
Digital multimeter (DMM)

Describe the vehicle being worked on:

Year _____ Make _____ Model _____
VIN _____ Engine type and size _____

Procedure

Task Completed

Your instructor will provide you with a specific vehicle and engine package. Write down the information, then perform the following.

1. Check the level and condition of the engine oil. If the oil level is low, add oil as needed. Change the oil and oil filter if necessary.   ☐

2. In order to verify that the problem is electrical, you should first perform an oil pressure test. This will rule out the possibility that the problem is with the engine.   ☐

Note: Steps 3–10 are for diagnosing a vehicle with an oil pressure switch.

3. How much engine oil pressure does it take to turn the dash warning light off? _____

4. Is the oil pressure switch a normally open or closed switch? _____

5. Turn the key to the ON position. With the engine off, does the warning light come on? _____

6. If the warning light does come on, does this mean that the circuit is working? _____

7. Disconnect the oil pressure switch connector and start the engine. What happened to the light?

_____

8. Should that have happened? _____

9. Turn the engine off, but leave the key in the ON position, and check for voltage at the wire connector terminal (coming from the dash). Is there voltage present?

   _____

10. Based on your testing, what can you conclude about the oil pressure switch?

    _____

    _____

Note: Steps 11–15 are for diagnosing a vehicle with an oil pressure sensor.

11. Describe how the sensor's resistance readings relate to the engine oil pressure readings.

    _____

    _____

12. With the engine running, carefully measure the sensor's resistance at idle.

    _____

13. With the engine running, carefully measure the sensor's resistance at 2,500 rpm.

    _____

14. Describe if the resistance readings are within specifications and if the sensor could be the possible cause of an incorrect dash reading.

    _____

    _____

15. According to the sensor's resistance range, what should happen to the gauge when you disconnect the wire from the sensor?

    _____

☐ 16. Disconnect the sensor's connector and verify your answer.

17. Explain why the gauge moved when the sensor was disconnected.

    _____

    _____

18. Based on your test results, what is your repair recommendation?

    _____

    _____

    _____

    _____

    _____

Instructor's Response _____

_____

_____

Name _____ **Date** _____

TEST, DRAIN, AND REFILL COOLANT

Upon completion of this job sheet, you should be able to properly perform test, drain, and refill the coolant and recommend repairs.

ASE Correlation

This job sheet is related to ASE Engine Repair Test content area: Lubrication and Cooling Systems Diagnosis and Repair: Test coolant; drain and recover coolant; flush and refill cooling system with recommended coolant; bleed as required.

Tools and Materials

Technician's tool set
Coolant test strips
Refractometer
Voltmeter
Coolant recovery container

Describe the vehicle being worked on:

Year _____ Make _____ Model _____

VIN _____ Engine type and size _____

Procedure

Task Completed

Your instructor will provide you with a specific vehicle and engine package. Write down the information, then perform the following.

1. Place a shop rag over the radiator cap and using caution, slowly open the cap.   ☐

2. Place a coolant test strip in the coolant and quickly pull it out. Shake the strip for 30 seconds and compare it to the bottle's rating system. Describe what you found.

   _____

   _____

3. Using the eye dropper from the refractometer kit, place a small drop of coolant on the lens of the refractometer. Hold it up to some light and describe what you found.

   _____

   _____

4. Set the voltmeter to read DC volts. Connect the test leads to the meter. Place the negative lead on the battery negative terminal and the positive lead in the coolant itself (do not touch the radiator or any other solid pieces). Describe your findings.

   _____

   _____

5. Leave the radiator cap off and raise the vehicle up.   ☐

6. Position the coolant catch basin under the radiator drain valve (petcock).

☐ 7. Loose the radiator drain valve, let the coolant drain, and then remove the valve.

☐ 8. After the coolant is done draining install the drain valve and lower the vehicle.

☐ 9. What type of coolant and mixture should this vehicle use?

10. Add the proper coolant type and mixture to the system until it is full. You may need to add more fluid after a few minutes.

☐ 11. Start the engine with the cap off and let it run for a few minutes. Add coolant as needed to the radiator or reservoir.

☐ 12. Install the coolant cap correctly and start the engine again. Locate the cooling system bleeders (if applicable). Describe where they are located.

_____

☐ 13. With the engine running and at operating temperature, slowly open the bleeder screw. Keep it open until only liquid comes out (no bubbles).

Instructor's Response _____

_____

_____

_____

JOB SHEET

Name _____ Date _____

RADIATOR REPLACEMENT

Upon completion of this job sheet, you should be able to properly replace a radiator.

ASE Correlation

This job sheet is related to ASE Engine Repair Test content area: Lubrication and Cooling Systems Diagnosis and Repair: Remove and replace radiator, and inspect and replace engine cooling and heater system hoses.

Tools and Materials

Technician's tool set
Coolant recovery container

Describe the vehicle being worked on:

Year _____ Make _____ Model _____
VIN _____ Engine type and size _____

Procedure

Your instructor will provide you with a specific vehicle and engine package. Write down the information, then perform the following.

Task Completed

1. Let the engine cool down. Then carefully remove the radiator cap (or coolant recovery cap). ☐

2. Open the radiator petcock and drain the coolant from the radiator into the catch basin. ☐

3. Remove the upper and lower radiator hoses, the fan shroud, and any electrical connections on the radiator (which may include the electrical fan). ☐

4. Remove the transmission cooler lines with a flare wrench (if applicable). ☐

5. Remove the radiator brackets and lift the radiator out. ☐

6. Before reinstalling the radiator, check the mounts and surrounding areas for cracks, rust, and damage. If you are installing a new radiator, check the size of the hoses, petcock locations, and other dimensions to make sure it fits. ☐

7. Reinstall the radiator in the reverse process of removal. ☐

8. Add the proper coolant type and mixture to the system until it is full. You may need to add more fluid after a few minutes. ☐

9. Start the engine with the cap off and let it run for a few minutes. Add coolant as needed to the radiator or reservoir. ☐

☐   **12.** Install the coolant cap correctly and start the engine again. Locate the cooling system bleeders (if applicable). Describe where they are located.

_____

☐   **13.** With the engine running and at operating temperature, slowly open the bleeder screw. Keep it open until only liquid comes out (no bubbles).

Instructor's Response _____

_____

_____

_____

JOB SHEET

Name _____   Date _____

Diagnosing Coolant Leaks

Upon completion of this job sheet, you should be able to determine the location of coolant leaks properly and recommend proper repairs.

ASE Correlation

This job sheet is related to ASE Engine Repair Test content area: General Engine Diagnosis; task: Inspect engine assembly for fuel, oil, coolant, and other leaks; determine necessary action.

Tools and Materials

Cooling system pressure tester

Describe the vehicle being worked on:

Year _____ Make _____ Model _____

VIN _____ Engine type and size _____

Procedure

Your instructor will provide you with a specific vehicle and engine package. Write down the information, then perform the following.

Task Completed

1. Check the coolant level and correct to proper level if needed.   ☐

2. Visually inspect the engine and cooling system for indications of a leak. Record your findings.

   _____

   _____

   _____

3. Using the proper service manual, locate the cooling system pressure specifications and record them.

   _____

4. Remove the radiator cap (cold engine only) and inspect it.

   _____

5. Install the radiator cap to the pressure tester, and pump the pressure to the minimum holding pressure. Does the cap pass the test?   ☐ Yes   ☐ No

6. Install the pressure tester to the radiator fill neck, and pump the pressure to specifications.

   Does the system hold the pressure for 15 minutes?   ☐ Yes   ☐ No

   If the pressure drops below specifications, visually inspect all areas for external leaks.

Remove the dipstick and look for evidence of an internal leak.

What are the results of your inspections?

_____

_____

_____

☐

7. Apply 5 psi of pressure to the radiator, and start the engine. With the engine idling, observe the pressure gauge. Does the reading indicate a cylinder head gasket leak? ☐ Yes ☐ No

8. If the cause of the leak is not found, determine what method you are going to use to locate the coolant leak (dye, infrared exhaust gas analyzer, etc.).

9. Briefly describe the procedure you have chosen to use.

_____

_____

_____

_____

_____

10. Based on your test results, what is your repair recommendation?

_____

_____

_____

_____

_____

Instructor's Response _____

_____

_____

_____

JOB SHEET

Name _____ **Date** _____

REPLACING A WATER PUMP

Upon completion of this job sheet, you should be able to replace a water pump and properly drain and refill the cooling system.

Tools and Materials

Basic tool set
Drain pan

Describe the vehicle being worked on:

Year _____ Make _____ Model _____
VIN _____ Engine type and size _____

Procedure

Your instructor will provide you with a specific vehicle. Write down the information, then perform the following.

1. Locate the procedure to remove and replace the water pump in the service information. Briefly outline the steps: _____
   _____
   _____
   _____
   _____

2. Loosen the radiator petcock to drain the coolant into a drain pan. How do you handle the used coolant? _____
   _____

3. Remove the lower radiator hose and inspect it. Remove the water pump drive belt and inspect it. Describe the results of your inspections: _____
   _____
   _____

4. Follow the manufacturer's procedure to remove the water pump. Clean the engine block mating surface thoroughly. Install the new gasket, apply sealant if recommended, and torque the new water pump into place. Recommended sealant: _____

   Water pump bolts' torque specification: _____
   Reassemble as needed.

5. Install the water pump drive belt and adjust the tension as needed. Describe the procedure to adjust the tension: _____
   _____
   _____
   _____

6. Refill the cooling system, bleed it as required, and pressure test the system for leaks.
   Results: _____

   Instructor's Response _____
   _____
   _____
   _____

JOB SHEET

Name _____ Date _____

DIAGNOSING AND REPLACING A THERMOSTAT

Upon completion of this job sheet, you should be able to properly diagnose and replace a thermostat. The thermostat is often the most replaced component in the cooling system.

ASE Correlation

This job sheet is related to ASE Engine Repair Test content area: Lubrication and Cooling Systems Diagnosis and Repair: Inspect, test, and replace thermostat and gasket/seal and identify causes of engine overheating.

Tools and Materials

Technician's tool set
Coolant recovery container
Infrared temperature gun

Describe the vehicle being worked on:

Year _____ Make _____ Model _____

VIN _____ Engine type and size _____

Procedure

Your instructor will provide you with a specific vehicle. Write down the information, then perform the following.

Task Completed

1. Find the correct operating temperature using the service manual. _____

2. Let the engine warm up to operating temperature and check the temperature of the engine using the dash gauge, and infrared gun, or a scan tool. ☐

3. Check the operation of the thermostat by watching the scan tool data or using the IR gun on the upper radiator hose. Is the thermostat opening and closing properly? _____

4. Let the engine cool down, then carefully remove the radiator cap (or coolant recovery reservoir cap). ☐

5. Open the radiator petcock and drain the coolant from the radiator into the catch basin until the level of coolant is below the thermostat. ☐

6. Remove the correct radiator hose and the thermostat housing. ☐

7. Remove the thermostat. How does it look? _____

8. Clean the gasket mating surface. ☐

9. After cleaning the thermostat housing and gasket surface, install the new thermostat and new gasket (or sealing ring). ☐

10. Reinstall the housing and radiator hose. Torque the thermostat housing bolts to the manufacturer's specification.

11. Refill the cooling system, bleed it as required, and pressure test the system for leaks. Results:

_____

_____

Instructor's Response _____

_____

_____

JOB SHEET

Name _____ **Date** _____

ENGINE BELT AND SYSTEM INSPECTION

Upon completion of this job sheet, you should be able to properly diagnose and inspect the engine belts, pulleys, and tensioners.

ASE Correlation

This job sheet is related to ASE Engine Repair Test content area: Lubrication and Cooling Systems Diagnosis and Repair: Inspect, replace, and adjust drive belts, tensioners, and pulleys; check pulley and belt alignment.

Tools and Materials

Technician's tool set
Serpentine belt tensioner tool
Cricket belt tensioner tool
Pulley alignment laser (or straightedge)

Describe the vehicle being worked on:

Year _____ Make _____ Model _____
VIN _____ Engine type and size _____

Procedure

Task Completed

Your instructor will provide you with a specific vehicle. Write down the information, then perform the following.

1. Describe the method outlined in the service manual for removing the tension on the engine belt.

   _____

   _____

2. Using the proper tools, release the tension on the belt and remove it.   ☐

3. If the system has an automatic tensioner, attempt to move it back and forth. Does it move smoothly and appear to be in reusable condition? _____

4. Check the belt. If it has more than three cracks in 3 inches, it should be replaced. How does the belt you are working with look? _____

5. Use a pulley alignment laser (or a straightedge) to check the pulleys for alignment. Also look at the belt for signs that the pulleys are misaligned (frayed on the edges). Describe your findings.

   _____

   _____

6. Rotate and spin each pulley on the system. Describe your findings.

_____

_____

☐ 7. Reinstall the old belt if it is reusable, or install a new belt.

Instructor's Response _____

_____

_____

JOB SHEET

Name _____ **Date** _____

COOLING SYSTEM FAN INSPECTION

Upon completion of this job sheet, you should be able to properly diagnose and inspect the engine cooling fan (electrical or mechanical).

ASE Correlation

This job sheet is related to ASE Engine Repair Test content area: Lubrication and Cooling Systems Diagnosis and Repair: Inspect, and test fan(s) (electrical or mechanical), fan clutch, fan shroud, and air dams.

Tools and Materials

Technician's tool set
Digital multimeter (DMM)
Scan tool
Infrared temperature gun

Describe the vehicle being worked on:

Year _____ Make _____ Model _____

VIN _____ Engine type and size _____

Procedure

Task Completed

Your instructor will provide you with a specific vehicle. Write down the information, then perform the following.

1. Visually inspect the air shroud and air dams. Do they appear to be cracked or damaged?

   _____

2. If you have a fan clutch, visually inspect it for any signs of leakage and blade damage. Describe your findings.

   _____

3. Start the engine and let it warm up. Does the fan clutch turn? _____

4. Increase the engine speed to 2,500 rpm. Does the fan change speed or sound? _____

5. For vehicles with electric fans, visually inspect the fan and its electrical connections.

6. Turn the engine on and let it warm up. Using the IR gun, a scan tool, or the dash gauge watch the temperature. What temperature should the fan come on at? _____

7. Turn the engine off and disconnect the electrical connector to the electric fan. ☐

8. Using the DMM, test the fan for continuity (resistance) by placing both test leads on the two motor connector terminals. What is your reading? _____

9. Based on your testing and inspections, what are your recommendations?

_____

_____

_____

Instructor's Response _____

_____

_____

JOB SHEET

Name _____ **Date** _____

USING A SCAN TOOL TO RETRIEVE ENGINE CODES

Upon completion of this job sheet, you should be able to properly use a scan tool to retrieve DTCs on an OBD vehicle.

Tools and Materials

Scan tool

Describe the vehicle being worked on:

Year _____ Make _____ Model _____

VIN _____ Engine type and size _____

Procedure

Task Completed

Your instructor will provide you with a specific vehicle. Write down the information, then perform the following.

1. Apply the parking brake or block the drive wheels and roll the driver's side window down. ☐

2. Describe the manufacturer, model, and version of the scan tool that you can use for this vehicle.

   _____

3. Is the scan tool you are using a generic or manufacturer-specific scan tool? _____

4. Make sure the key is off and connect the scan tool to the diagnostic link connector. ☐

5. Turn the ignition to the run (not start) position and maneuver through the scan tool menu to retrieve the codes. List all the codes here.

   _____

   _____

   _____

   _____

6. Describe the process for clearing the codes (do not actually clear them).

   _____

   _____

Instructor's Response _____

   _____

   _____

Chapter 5

REPAIR AND REPLACEMENT OF ENGINE FASTENERS, GASKETS, AND SEALS

UPON COMPLETION AND REVIEW OF THIS CHAPTER, YOU SHOULD BE ABLE TO:

- Inspect internal and external threads; repair as needed.
- Install a thread insert.
- Replace engine gaskets.
- Replace engine seals.
- Choose and use sealants properly.

During the processes of removing an engine or performing major repairs, you are likely to experience broken bolts or damaged threads. Each fastener is important to proper vehicle and engine operation or it would not be used. It is important that each fastener have proper threads to hold securely. Also, gaskets, seals, and sealants are used to contain engine fluids. Proper replacement of gaskets and seals is an integral part of engine repair.

This chapter discusses the various methods of fastener removal and thread repair. You will learn how to install a thread insert. We will also discuss the proper replacement techniques for engine gaskets and seals and cover the use of sealants.

THREAD REPAIR

SPECIAL TOOLS
Stud remover
Arc welder
Nut splitter
Left-hand twist
drill bits
Reversible drill
Extractor set
Tap and die set
Thread insert
repair kit

At some time, every technician is frustrated because of a broken fastener or damaged threads. These must be repaired to provide proper clamping forces of the mating components. It is usually a lot faster to take the proper steps and precautions to prevent fastener breakage than it is to perform the repair procedures. Even with all steps taken, the fastener may still break. To help reduce the potential of fastener breakage and damage, the following offers some advice. First, spray all fasteners with a penetrating oil *before* attempting to remove them. Using a little oil now may save lots of time later. A saying that can prevent a lot of problems later is, "The best time to use penetrating oil is before you need it." The oil will work its way into the threads and dissolve the rust and corrosion. Also, the oil will lubricate the threads to aid in removal. Allow the oil to soak into the threads before attempting to remove the fastener.

Next, be sure you are using the proper tools. Use a box-end wrench or socket that fits the fastener head snug. Avoid the use of open-end wrenches during disassembly.

If the fastener will not come loose with normal force, and you performed the first two steps, then make sure of the thread direction. Although not common on most engine applications, left-handed threads may occasionally be found.

If the fastener still does not loosen, remember, never force a fastener loose. The intent is to remove the fastener, not to break it off. If the fastener will not loosen using normal force, try to tighten the fastener slightly and then back it out. Sometimes this will break the corrosion loose and allow the fastener to be removed.

If the fastener still will not loosen, use a punch on the head of the bolt and strike it with a hammer. The shock may break the corrosion loose. If the fastener still will not loosen, apply heat to the fastener. Use a minimal amount of heat; in most cases, a simple propane torch is sufficient. Heat either the fastener or the casting around the fastener, but not both. The trick is to get one component to expand more than the other; for example, if the nut on the end of a bolt will not loosen, heat the nut to get it to expand. If a bolt is stuck in the block, heat the casting to get the hole enlarged so the bolt can be removed.

WARNING: Never use penetrating oil or spray lubricants on a fastener that has been heated and is still warm. Most of these materials are flammable and could result in a flare-up.

At this point, you are running out of options. Try heating the fastener with a torch, and while it is still hot, hold a piece of paraffin wax against the fastener, allowing the wax to run down and into the threads. Allow the wax to sit in the threads for a few minutes before attempting to remove the fastener.

If you are still unable to remove the fastener, it will probably need to be cut off and drilled out. These procedures are described under removal of broken bolts.

If the existing fastener cannot be removed due to a damaged head, there are several methods available for removing the fastener:

- It may be possible to get a hold of the fastener with a stud remover (Figure 5-1).
- Use a chisel (Figure 5-2). Position the chisel off center of the fastener, then strike it with a hammer to drive the fastener in a counterclockwise direction. Do not use a chisel on a fastener seated against aluminum or sheet metal.
- Use an arc welder to attach a rod to the head of the fastener. Then use locking pliers to turn the fastener out.
- If a nut is damaged, use a nut splitter to cut off the nut (Figure 5-3).
- Use a left-hand drill bit with a reversible drill and an extractor (Figure 5-4).
- Use an **extractor** by drilling a hole in the center of the fastener, then tapping the extractor into the hole. Turn the extractor in the counterclockwise direction to remove the fastener (Figure 5-5).
- Drill out the center of the fastener using progressively larger drill bits until all that remains of the bolt is a thin skin. Peel the skin from the threads of the hole.

An **extractor** is a special tool that is used to remove a broken bolt from a bolt hole.

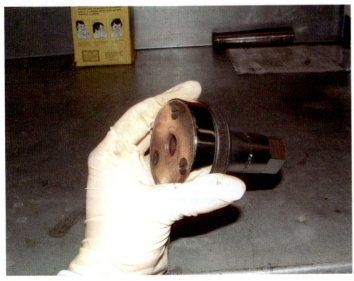

FIGURE 5-1 A stud remover.

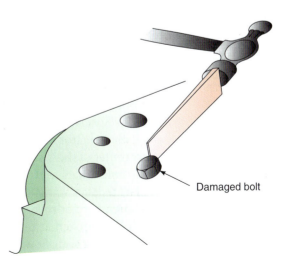

FIGURE 5-2 Using a chisel and hammer to remove a damaged bolt.

Damaged bolt

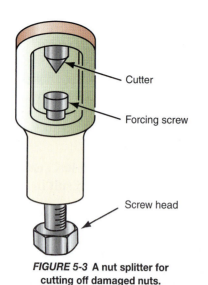

FIGURE 5-3 A nut splitter for cutting off damaged nuts.

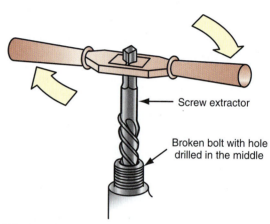

FIGURE 5-4 An extractor set.

Cutter

Forcing screw

Screw head

Screw extractor

Broken bolt with hole drilled in the middle

FIGURE 5-5 Using an extractor to remove a broken bolt.

⚠️ **CAUTION:**
Chisel fasteners that are seated on iron castings only. Attempting to chisel a fastener on aluminum or stamped steel can result in damage to these components.

If the fastener head is so badly damaged that a tool cannot be fitted onto it, and the above procedures do not work, the head must usually be cut off. This can be done with a die grinder and a cutting wheel. Be careful not to damage the casting area around the fastener. Usually, removing the head will relieve the tension against the fastener and allow it to be removed by cutting a slot across the shank and using a standard screwdriver. If the fastener still will not loosen, an extractor may be needed.

If the fastener head is broken off due to excessive torque or bottoming out in the hole, do not use an extractor to remove it. The extractor is made of very hard steel. Because of the hardness, it is very brittle and breaks easily. If the extractor should break off in the fastener, it is very difficult and time consuming to remove it. If an extractor should break off, do not attempt to drill it out. The drill bit will break off, leaving the extractor and the bit in the fastener. Special tap extractors can be used to remove some broken extractors.

If an extractor is required, follow these steps: begin by using a center punch to mark the *exact* center of the fastener. This is the most important step. If the center punch is off-center, or the hole drilled at an angle, the extractor bit will have a greater potential of breaking.

Next, determine the largest size extractor you can use to remove the fastener. The larger, the better, but not so large as to damage the threads of the casting. The intent is to make the walls of the broken fastener as thin and flexible as possible, so the fastener will lose its grip on the casting and allow the extractor to bite into the fastener better. Once the extractor size is determined, the correct size drill bit is selected. Most extractors will be stamped with the correct drill bit to use.

Using the correct size drill bit, drill a hole all the way through the center of the fastener. Use a drill guide if needed. If possible, use a left-hand drill bit, since it will sometimes draw the fastener out as the hole is drilled. For larger size fasteners, start with a small-size drill bit and work up to the correct size.

Once the hole is drilled, place the extractor into the hole and tap it lightly with a brass hammer. Make sure the extractor is straight into the hole. Tap (not pound) the extractor in just enough that the extractor will bite the wall of the drilled hole as it is turned. If you drive the extractor in too deep, the walls of the fastener will expand and cause the bond between the fastener and the casting to become even tighter. Now use a tap wrench or open-end wrench to turn the extractor counterclockwise. Some kits come with a nut that fits over the extractor and the wrench fits over the nut. Never use power tools to turn the extractor, because it will probably break.

If the extractor begins to shear off, stop! It is a lot easier to remove an extractor that is in one piece than to remove a broken one. If the extractor does break, heat the drilled hole with a torch and then cool the broken extractor quickly by dripping water onto it (paraffin wax can also be used). Use a tap extractor to try to remove the broken extractor. Never attempt to drill out a broken extractor, all you will do is break the bit.

An alternative to using extractors is to use a common set of Allen wrenches. Select an Allen wrench that is slightly smaller than the width of the broken fastener; the larger the Allen wrench, the better. Drill a hole through the bolt using a bit size the same as the Allen wrench when measured across the flats. Tap the Allen wrench into the hole using a brass hammer. Usually, the corners of the Allen wrench will grip the fastener walls and cause the fastener to follow as the wrench is turned.

It is also possible to remove a broken fastener by drilling it out. For this operation to work, the exact center of the bolt shank must be located and center punched. Starting with a small drill bit and working up, drill out the fastener until the last drill bit used is the same size as the minimum diameter of the fastener (the size bit used to tap the hole to fit the fastener). This will leave a very thin wall where the fastener used to be. Use a magnet or a pocket screwdriver to pick out the very thin wall of the fastener. If the exact center of the fastener is not drilled and the hole is off-center, this process will not work.

Internal thread damage can occur if the fastener is not properly installed or if corrosion has eaten away some of the threads. Damaged internal threads can usually be repaired using a thread chaser, **tap**, or thread inserts. Prior to making thread repairs, it is important to properly identify the pitch, size, and length of the fastener to be used.

Thread chasers can be used to clean internal and external threads. Thread files can be used to remove burrs from external threads (Figure 5-6).

Thread chasers and thread files can be used to clean and repair minor thread damage. These tools are designed to roll the threads back to their original shape, instead of cutting

SERVICE TIP:

A common trick to remove a small piece of a broken tap is to use a center punch and hammer to deliver a sharp, hard blow to the tap. This will cause the tap to shatter into pieces, which can be removed with a magnet. DO NOT attempt this on a broken extractor. Extractors are harder than taps. Attempting to shatter one in this manner will cause it to wedge tighter into the drilled hole and make matters worse. If needed, take the component to a machine shop and have them remove the broken fastener for you. Some machine shops use a process called electrical discharge machining (EDM) to remove broken tools.

A **tap** is a special tool used to thread openings in metal components.

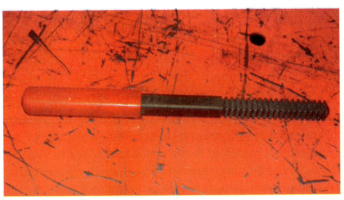

FIGURE 5-6 A thread file will refinish lightly damaged threads.

FIGURE 5-7 A master tap and die set with metric and standard equipment.

FIGURE 5-8 Use a pitch gauge to determine the correct selection of tap.

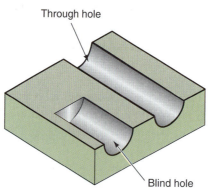

Through hole

Blind hole

FIGURE 5-9 Comparison of through and blind holes.

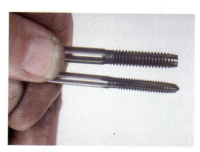

FIGURE 5-10 Two different types of taps. The tapered tap is used for through holes, and the bottoming tap is used for blind holes.

Thread chasers can be used to repair minor thread damage by rolling threads back to their original shape.

new threads. Thread chasers, taps, or **dies** should be used to clean the threads of every fastener used on the engine during the rebuilding process.

If the threads cannot be restored by use of a thread chaser, it may be possible to use a tap or die to repair the damaged threads (Figure 5-7). Proper size selection is important. Use a thread pitch gauge to determine the correct pitch (Figure 5-8). Internal threads can be located in through holes or blind holes (Figure 5-9). Through holes are threaded using a taper tap, while bottoming taps are used for blind holes (Figure 5-10).

WARNING: Wear proper eye protection whenever using a tap or die. These tools are very brittle and may shatter.

Once the correct tap or die is selected, the following guidelines should be adhered to:

- Thoroughly clean the hole or bolt.
- Lubricate the tap or die with cutting oil.
- Start the tap or die onto the existing threads, being careful to start the tool straight.
- Use the correct holding wrench for the tap.
- Chamfer the hole after the threads are repaired (Figure 5-11).

A **die** is a special tool that is used to clean threads or make new threads on a bolt.

If a fastener had to be removed and all of the threads were damaged, it may be possible to drill out the hole and tap it to fit a larger fastener. This usually works for such things as attaching bolts for the exhaust manifold to the cylinder head; however, if larger bolts are used, consider these before beginning; there must be enough casting to support the larger fastener, and the attaching component (i.e., exhaust manifold) may need to be drilled to accept the larger bolt.

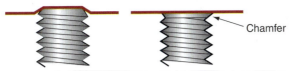

FIGURE 5-11 Chamfering the hole removes the raised edge.

If it is determined that using a larger bolt is indeed the less expensive and time-consuming method of thread repair, and there are no interferences in the casting to prevent it, the first step is to determine the size of the fastener that will be used. Once the fastener size is determined, the size of drill bit needs to be selected. Usually, the tap kit will have a tap drill size chart that indicates the correct size bit for the tap. A drill tap usually provides about 75 percent of a full thread. If a chart is not available, thread the correct size nut onto the bolt that will be used. Once the correct nut is determined, use it to select the drill bit by using the largest bit that will fit into the nut. Using too large of a bit will result in the tap not cutting threads that will be deep enough to hold the fastener properly. Too small of a bit makes cutting the threads difficult.

Once the proper size hole is drilled, it is then tapped. When tapping the hole, advance the tap a few turns, then back it off one-quarter turn. When cutting any metal other than iron, lubricate the tap with cutting oil, and keep it well lubricated throughout the operation. Iron can be tapped without the use of oil. Continue to advance and back off the tap until the hole is completely threaded. Steps to remove a broken bolt and clean the threads using a tap are outlined in Photo Sequence 8.

If it is not possible to enlarge the fastener size, the original size will have to be restored using a form of **thread insert** (Figure 5-12). There are several different types of thread inserts available:

- Helical inserts
- Self-tapping inserts
- Key-locking inserts
- Chemical thread repair

Helical inserts are made from stainless steel and look like a small spring. Installing a helical insert is a four-step process (Figure 5-13). First, the damaged hole is enlarged to a specified size. The drill bit size depends on the size of the insert and is supplied by the manufacturer. If necessary, install a stop collar on the drill bit to prevent it from drilling too deep.

> A thread insert may be called a heli-coil.

> The **thread insert** allows for major thread repairs while keeping the same size fastener.

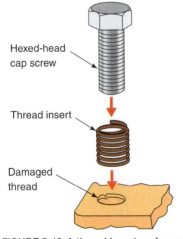

FIGURE 5-12 A thread insert replaces the damaged threads without having to change the size of the fastener.

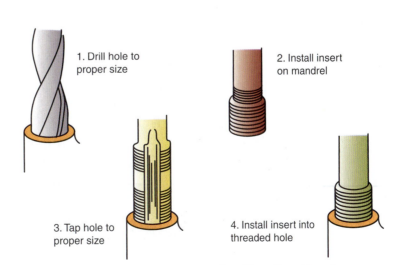

1. Drill hole to proper size
2. Install insert on mandrel
3. Tap hole to proper size
4. Install insert into threaded hole

FIGURE 5-13 The four steps to installing a thread insert.

REMOVING A BROKEN BOLT AND USING A TAP

P8-1 This bolt broke off in the block and must be extracted.

P8-2 A stud extractor was used but could not get enough of a bite to remove the broken bolt.

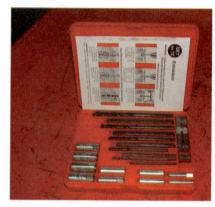

P8-3 Select the correct size drill to match the bolt extractor suitable for the broken bolt.

P8-4 Center the drill carefully and drill a hole to use the bolt extractor.

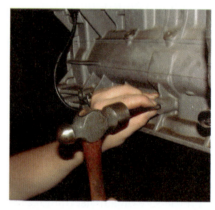

P8-5 Lightly tap the thread extractor into the drilled hole. Do not drive it in too far or you will tighten the bolt threads in the hole.

P8-6 Turn the thread extractor counterclockwise and hopefully the broken bolt will come with it.

P8-7 With the broken bolt removed, clean the threads with a tap.

P8-8 Use the correct size tap and use a light touch to clean the threads of the hole.

P8-9 Clean the bolt hole of any metal filings. Be sure you have safety glasses on!

REMOVING A BROKEN BOLT AND USING A TAP

P8-10 Lightly lubricate the threads of the new bolt.

P8-11 Start the bolt in by hand to be sure it turns freely.

Second, clean the hole out thoroughly, then tap the hole. Use the tap size specified by the manufacturer. The tap is a special tap for use with inserts. The large diameter and course pitch are not standard patterns for the fastener that will be installed. Do not attempt to substitute this tap with a regular one. Clean the hole again.

The third step is to place the insert onto the mandrel and engage the tang of the insert onto the end of the mandrel. Next, apply a light coating of thread locking compound to the external threads of the insert and thread it into the hole. Once the insert is started into the hole, spring pressure will prevent it from unscrewing. Once started, the insert cannot be removed. When correctly installed, the insert should be one-quarter to one-half turn below the surface. If the hole is deeper than the length of a standard insert, thread the insert down far enough to allow a second insert to be threaded in on top of the first. If the hole is too shallow for the length of the insert, cut the insert to the proper length prior to installing it onto the mandrel.

Remove the mandrel by unscrewing it from the insert. Use a small punch to break off the tang at the bottom of the insert.

Another variation of the threaded insert is the **self-tapping insert**. These inserts are good to use in blank holes since they cut their own thread. Cutting their own thread eliminates the need of tapping the hole after it is drilled. The process of installing self-tapping inserts is the same as the helical insert.

Solid bushing inserts are also available. These inserts use a machine thread, so a standard tap may be used. All other installation steps are the same as with helical inserts.

Some inserts are held in place by small keys. Key-locking inserts are similar to solid bushing inserts, except after they are installed, the keys are driven into place (perpendicular to the threads) to prevent the insert from coming out.

Inserts for repairing spark plug holes are also available. The installation is the same as discussed above; however, some spark plugs use a tapered seat, while others use a flat seat with a gasket. Be sure to install the correct insert that matches the spark plugs to be used.

> **CUSTOMER CARE:** If you have to install a thread insert into a spark plug hole, you should inform the customer. If something ever happens to the insert and the customer doesn't know that the hole had previously been damaged, they could become angry and distrustful of your work.

Finally, some threads can be repaired using epoxies. Epoxies cannot be used to repair threads in high stress, high torque components. Epoxy should not be used to repair the threads of critical clamping force areas such as cylinder heads, intake manifolds, and internal

SERVICE TIP:
Placing a rubber hose over the drill bit works to prevent the bit from going too deep. Cut the hose so the desired depth of the hole is equal to the amount of exposed drill bit.

A **self-tapping insert** provides a new thread when installed in an opening with damaged threads.

engine parts. Epoxy thread repair kits usually contain a two-part epoxy and a releasing agent. To use the epoxy, first clean and dry the threads of the hole and the fastener thoroughly. Apply a very thin coat of the releasing agent to the threads on the fastener, covering the entire thread surface area. Mix together the two-part epoxy (resin and hardener), making sure there will be enough mixed to complete the job. Spread the epoxy as evenly as possible on the inside of the hole. If the hole goes through the casting, use tape to shut off one end of the hole so the epoxy does not run out. Screw the fastener into the hole, but do not tighten it. Allow the epoxy to cure the required amount of time specified in the instructions. After the epoxy dries, tighten the bolt.

GASKET INSTALLATION

SPECIAL TOOLS
Torque wrench

For the gasket to perform its function, it must be installed correctly. There are a few basic steps to follow when preparing to replace a gasket:

1. Follow all instructions provided by the gasket manufacturer.
2. Thoroughly clean the mating surfaces.
3. Check the gasket for proper fit.
4. Never reuse gaskets.

Many gaskets are installed dry without additional sealants. Never use sealant unless the manufacturer specifies it; then, follow the instructions closely. Most gasket and seal failures are caused by improper installation. When installing a gasket, you need to torque the cover or component on properly. Over-tightening a cork gasket, for example, will cause the gasket to crack. Unequal torque on any gasket is likely to allow leakage. Use the recommended torque sequence. If one is not provided, tighten the bolts in a diagonal pattern starting at the inside and working your way out. Use a torque wrench and tighten the bolts to the proper specification. Do not use air tools to tighten gasket covers.

CYLINDER HEAD GASKET INSTALLATION

SERVICE TIP:
No-retorque gaskets are available for most engine applications; however, many imports and older engines require the cylinder head to be retorqued. This is required because the original set of the gasket occurs after initial engine operation and then relaxes. This causes the clamping pressure to be decreased. Retorquing the bolts will reset the proper clamping pressure.

Cylinder head gaskets must be capable of withstanding combustion pressures up to 2,700 psi (1,862 kPa) and temperatures over 2,000°F (1,100°C). Proper installation will assure the gasket will withstand these conditions. Following is a typical procedure for cylinder head gasket replacement:

1. Check the gasket for proper fit. Note that the gasket may be marked for proper installation.
2. Clean and prepare the bolts. Make sure all guides or dowel pins are in place (Figure 5-14).
3. Install the gasket and cylinder head.
4. Follow the service manual torque sequence and torque the bolts to specifications.

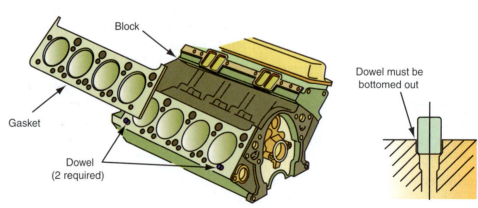

FIGURE 5-14 Some engines use dowel pins to guide the head and gasket.

FIGURE 5-15 Remove the deformation around the bolt holes of the valve cover before installing it.

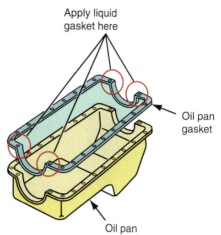

Apply liquid gasket here

Oil pan gasket

Oil pan

FIGURE 5-16 Oil pan and gasket installation.

OIL PAN GASKET INSTALLATION

When an engine is rebuilt or replaced or when the oil pan gasket is leaking, install a fresh gasket under the oil pan. When the job is done with the engine in the car, it may be necessary to remove some portions of the frame, suspension, or exhaust. One of the most important things to do when replacing a gasket is to be certain that all of the old gasket and any oil is removed from the engine block and from the oil pan. If the oil pan is stamped steel, it may be necessary to straighten deformities around the bolt holes. Use a ball peen hammer and a block of wood to flatten the bolt holes (Figure 5-15). Do not attempt this method on an aluminum oil pan; it will crack. Next, line up the new gasket on the pan and be sure it fits properly. Locate the service procedure or gasket instructions to determine if any sealant should be used (Figure 5-16). Locate and follow the torque sequence and specification to tighten the oil pan. Check for leaks after the engine has started.

VALVE COVER GASKET REPLACEMENT

A common source of oil leaks is from the valve cover gasket(s). Replacement of these gaskets is typically straightforward, although on some vehicles it may be necessary to remove a number of components to gain access to the cover. The general procedures are very similar to those for replacement of an oil pan gasket. One of the leading causes of gasket failure is improper installation, so be certain to follow the manufacturer's procedures for tightening the cover and using any sealant. Photo Sequence 9 shows typical steps involved in removing and replacing a valve cover gasket on a modern pickup truck.

SEAL INSTALLATION

Most one-piece seals are removed by prying them out of the bore (Figure 5-17). When performing this task, care must be taken not to damage the bore. A special seal-removing tool may be required to remove the seal if it is recessed into the bore (Figure 5-18).

Prior to installing a seal with a steel outer ring, coat the outer surface with a silicone sealer. The seal lip must always face toward the flow of lubricant. Use a seal driver or press of the proper size to install the seal (Figure 5-19). Support the back of the component where the seal is being installed, if possible, to prevent warpage or breakage. Do not use excessive force to install the seal. Drive the seal until it is the proper depth into the bore, as specified in the service manual. Coat the rubber seal with engine oil to protect it on initial startup.

SERVICE TIP:
If the seal is stubborn, drill two small holes 180° apart on the metal outer ring of the seal. Thread self-tapping sheet metal screws into the holes, then grip the screws with pliers or side cutters and pry the seal out.

CAUTION:
If the outer ring of the seal is coated with a silicone from the factory, do not use oil to aid in installation. The oil may not be compatible with the silicone and may cause leakage. Use soapy water to aid installation.

REPLACING A VALVE COVER GASKET

P9-1 After identifying the problem, consult the service information for instructions on how to replace the valve cover gasket properly.

Leaking valve cover gasket

P9-2 Remove the air distribution duct-work to gain access to the valve cover.

P9-3 Remove and label connectors and vacuum lines as needed.

P9-4 With the intake ducts out of the way, remove the spark plug wire loom and wires that will obstruct removal.

P9-5 Loosen the valve cover bolts using the proper sequence if one is specified.

P9-6 Maneuver the valve cover off of the cylinder head.

P9-7 Remove the old gasket material and thoroughly clean and dry the valve cover.

P9-8 Install the new valve cover gasket securely. Use sealant only if it is specified.

P9-9 Always properly torque the valve cover to the specification to ensure a quality repair.

REPLACING A VALVE COVER GASKET

P9-10 Reinstall the ductwork securely.

P9-11 Reconnect all wiring and vacuum lines. Wipe down the valve cover and surrounding components. Finally, start the vehicle and check to be sure it runs properly and has no leaks.

FIGURE 5-17 Prying out the old seal.

FIGURE 5-18 Use a seal remover to pry the old seal out.

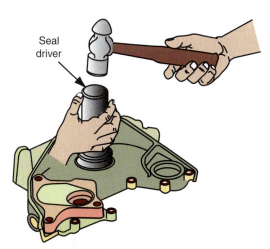

Seal driver

FIGURE 5-19 Driving in a new seal.

USING SEALANTS

There are several locations on the engine where sealants and form-in-place gaskets are used. The proper sealant must be selected for the application.

Aerobic sealers, such as room-temperature vulcanizing (RTV) sealant, are used on flexible metal flanges. Anaerobic sealers are used between two machined metallic surfaces. Before applying any type of sealer, the surface must be properly prepared. Use a flat-blade

CAUTION:
Do not use sealants unless specified in the service manual.

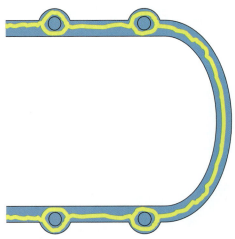
FIGURE 5-20 Apply sealant around any bolt holes or passages.

Very rusted and corroded exhaust manifold studs and nuts

FIGURE 5-21 These very rusty nuts and studs required heat to extract them from the cylinder head.

CAUTION:
Always check the expiration date on the sealant before using it. Most sealants have a shelf life of 1 year. If the sealant is too old, it may not set correctly.

putty knife, scraper, or wire brush to clean all of the gasket surfaces. Inspect the surfaces for warpage, dents, and old gasket materials. Wipe the surfaces free of dirt and oil.

Apply RTV sealant in a 1/8 in. (3 mm) diameter continuous bead. Circle any bolt holes or fluid passages (Figure 5-20). For corner sealing, place a 1/8 to 1/4 in. (3 to 6 mm) drop in the center of the contact area. Use a shop towel to wipe away any sealant that expands out of the seam. RTV begins to cure in about 10 minutes, requiring all components to be installed and torqued before the sealant cures. RTV is typically used as a substitute for a gasket, not in addition to a gasket. In some cases, it is used at the mating corners where gaskets connect, such as the ends of two-piece intake manifold gaskets on vee engines. Be careful not to use too much RTV; more is not better. Using excess RTV can allow it to be drawn into the engine and clog critical fluid passages. Make sure to carefully follow the RTV manufacturer's directions on total curing time before putting the gasket to the test.

Apply anaerobic sealants in continuous beads about 0.04 in. (1 mm) in diameter. Apply the sealer to only one gasket surface, making sure to circle any mounting hole or passages. Install the components and torque within 15 minutes.

TERMS TO KNOW

Die

Extractor

RTV

Sealants

Self-tapping insert

Tap

Thread insert

CASE STUDY

A student brought an engine into school to be overhauled. He had rebuilt the engine over the summer, but it started knocking just a few hundred miles later. Once the oil pan was removed, we could see the cause. The oil pickup screen was completely plugged with RTV. He had covered the oil pan gasket with RTV, thinking that it would provide the best seal.

CASE STUDY

When replacing a cracked exhaust manifold on a vehicle, a technician ran into some very rusty hardware (Figure 5-21). Some of the threads on the studs had actually rusted away. Rather than rethread the existing studs and risk having the nuts loosen up over time, the technician did the right thing and replaced the studs. She sprayed the studs generously with penetrating oil and let them sit before using the stud remover to extract the studs. Only one stud gave her trouble, and with a little heat to the head it did break free and turn out.

ASE-STYLE REVIEW QUESTIONS

1. *Technician A* says to use a socket or a box-end wrench to loosen tight bolts.
   Technician B says to use locking pliers to free up rusted bolts.
   Who is correct?

   A. A only   C. Both A and B
   B. B only   D. Neither A nor B

2. A bolt will not loosen:
   Technician A says to tighten the bolt slightly.
   Technician B says to rap the head of the bolt.
   Who is correct?

   A. A only   C. Both A and B
   B. B only   D. Neither A nor B

3. *Technician A* says to spray penetrating oil on rusty bolts before trying to loosen them.
   Technician B says to apply wax to the head of the bolt.
   Who is correct?

   A. A only   C. Both A and B
   B. B only   D. Neither A nor B

4. *Technician A* says to heat a rusted nut and then spray penetrating oil on it.
   Technician B says you can use a nut splitter to remove damaged nuts.
   Who is correct?

   A. A only   C. Both A and B
   B. B only   D. Neither A nor B

5. A bolt head is damaged and it won't turn out:
   Technician A says you can use a stud remover.
   Technician B says you can use a center punch.
   Who is correct?

   A. A only   C. Both A and B
   B. B only   D. Neither A nor B

6. To use an extractor:
   Technician A says to drill a hole into one side of the bolt.

7. An extractor broke inside a bolt:
   Technician A says to drill it out using a left-hand drill.
   Technician B says to heat the extractor and chisel it out.
   Who is correct?

   A. A only   C. Both A and B
   B. B only   D. Neither A nor B

8. *Technician A* says that the sealing surfaces must be perfectly clean before fitting a new gasket.
   Technician B says that most gaskets should be lubricated with oil before installation.
   Who is correct?

   A. A only   C. Both A and B
   B. B only   D. Neither A nor B

9. *Technician A* says that you should use RTV sealant between two machined surfaces.
   Technician B says that anaerobic sealant hardens within five minutes.
   Who is correct?

   A. A only   C. Both A and B
   B. B only   D. Neither A nor B

10. *Technician A* says that you should apply a thin layer of RTV to the lip of a seal.
    Technician B says you can use a chisel to get a stubborn seal out.
    Who is correct?

    A. A only   C. Both A and B
    B. B only   D. Neither A nor B

Technician B says to tap the extractor into the hole and turn it counterclockwise.
Who is correct?

A. A only   C. Both A and B
B. B only   D. Neither A nor B

ASE Challenge Questions

1. While discussing thread insert installation:
   Technician A says the first step is to use a tap to match the external threads on the heli-coil.
   Technician B says the thread insert should be installed using the correct size drill bit.
   Who is correct?
   A. A only
   B. B only
   C. Both A and B
   D. Neither A nor B

2. When using RTV for gasket applications:
   A. The gasket surface should be cleaned with acid.
   B. The RTV should be placed on the inside of the bolt holes.
   C. The RTV should be allowed to cure for 15 minutes before assembly.
   D. The RTV bead should be 1/8 in. (3 mm) in width.

3. While discussing seal installation:
   Technician A says a seal lip should face toward the flow of lubricant it is sealing.
   Technician B says to tap around the edges of the new seal to fit it into place.
   Who is correct?
   A. A only
   B. B only
   C. Both A and B
   D. Neither A nor B

4. *Technician A* says that you should spray head gaskets with copper sealer before installation.
   Technician B says that most head gaskets have a top marking on them.
   Who is correct?
   A. A only
   B. B only
   C. Both A and B
   D. Neither A nor B

5. *Technician A* says that you can use a 1/4-inch drive air ratchet to install the oil pan.
   Technician B says to tighten the corners first and work your way in.
   Who is correct?
   A. A only
   B. B only
   C. Both A and B
   D. Neither A nor B

JOB SHEET

Name _____   Date _____

IDENTIFYING ENGINE FASTENERS AND TORQUE SPECIFICATIONS

Upon completion of this job sheet, you should be able to properly identify the properties of common engine fasteners and their torque specifications.

Tools and Materials

Service manual
Technician's tool set
Torque wrench
Thread pitch gauge set
Dial caliper

Describe the vehicle being worked on:

Year _____ Make _____ Model _____

VIN _____ Engine type and size _____

Procedure

Your instructor will show you which fasteners should be removed from the engine. Remove each of the fasteners as directed. Then find the length, width, thread pitch, designation (metric or SAE), and torque specification. After you identify each fastener, reinstall them using a torque wrench to the manufacturer's specification.

1. Fastener location _____
   Thread length _____ Total length _____
   Thread pitch _____ Designation _____
   Torque specification _____

2. Fastener location _____
   Thread length _____ Total length _____
   Thread pitch _____ Designation _____
   Torque specification _____

3. Fastener location _____
   Thread length _____ Total length _____
   Thread pitch _____ Designation _____
   Torque specification _____

4. Fastener location _____
   Thread length _____ Total length _____
   Thread pitch _____ Designation _____
   Torque specification _____

5. Fastener location _____
   Thread length _____ Total length _____
   Thread pitch _____ Designation _____
   Torque specification _____

Instructor's Response _____
_____
_____

JOB SHEET

Name _____ Date _____

LOOSENING A RUSTED FASTENER

Upon completion of this job sheet, you should be able to remove a frozen fastener.

Tools and Materials

Technician's tool set

Describe the vehicle being worked on:

Year _____ Make _____ Model _____

VIN _____ Engine type and size _____

Procedure

1. Locate a damaged or seized fastener as guided by your instructor. Is your fastener a damaged nut, bolt, or stud?

   _____

2. Spray penetrating oil on the fastener and allow it to soak in.

3. Will your fastener loosen with a socket or box-end wrench? _____

4. Describe your next strategy for loosening the fastener: _____

   _____

   Did that loosen the fastener? _____

5. Describe your next strategy for loosening the fastener: _____

   _____

   Did that loosen the fastener? _____

6. Describe your next strategy for loosening the fastener: _____

   _____

   Did that loosen the fastener? _____

7. If the fastener head is damaged, use a stud remover or an extractor to remove the fastener.

   Describe your procedure and your results: _____

   _____

   _____

   _____

8. Describe any additional steps required to remove the seized fastener: _____

   _____

   _____

   _____

9. Were you successful in removing the frozen fastener? _____
_____

Instructor's Response _____
_____
_____

JOB SHEET

Name _____ **Date** _____

REPLACING AN OIL PAN GASKET

Upon completion of this job sheet, you should be able to properly replace the oil pan gasket.

ASE Correlation

This job sheet is related to ASE Engine Repair Test content area: Engine Block Assembly, Diagnosis, and Repair; task: Assemble the engine using gaskets, seals, and formed-in-place sealants, thread sealers, etc., according to manufacturer's specifications.

Tools and Materials

Technician's tool set

Describe the vehicle being worked on:

Year _____ Make _____ Model _____

VIN _____ Engine type and size _____

Procedure

Task Completed

1. Locate the manufacturer's procedure for removing and replacing the oil pan. ☐

2. Drain the oil and remove the oil pan. ☐

3. Clean the oil pan and the engine block of oil and old gasket material. ☐

4. Lay the new gasket on the oil pan, and be sure the holes line up properly. ☐

5. What type of gasket is used? _____

   _____

6. Apply the recommended sealant to the oil pan if applicable. What sealant is recommended and where for your installation? _____

   _____

7. Install the new gasket and tighten the oil pan using the appropriate sequence and torque specification. What is the torque specification for the oil pan attaching bolts?

   _____

8. If working on an engine in a vehicle, replace the oil filter and reassemble the vehicle. ☐

9. Refill the engine with the proper oil. What is the recommended oil? _____

   What is the engine oil capacity? _____

10. Start the engine, if applicable, and check for leaks. ☐

Instructor's Response _____

_____

_____

JOB SHEET

34

Name _____ Date _____

IDENTIFYING ENGINE MATERIALS

Upon completion of this job sheet, you should be able to identify various engine materials and components used on late-model vehicles.

Tools and Materials

Service Information

Procedure

1. Locate an engine that uses a cast-iron block and an aluminum cylinder head. Describe the vehicle:
   Year _____ Make _____ Model _____
   VIN _____ Engine type and size _____

2. Locate an engine that uses an aluminum block and a cast-iron cylinder head. Describe the vehicle:
   Year _____ Make _____ Model _____
   VIN _____ Engine type and size _____

3. Locate an engine that has a lost foam casting block. Describe the vehicle:
   Year _____ Make _____ Model _____
   VIN _____ Engine type and size _____

4. Locate an engine that uses an aluminum oil pan. Describe the vehicle:
   Year _____ Make _____ Model _____
   VIN _____ Engine type and size _____

5. Locate an engine that uses a composite intake manifold. Describe the vehicle:
   Year _____ Make _____ Model _____
   VIN _____ Engine type and size _____

Instructor's Response _____
_____
_____

Chapter 6

INTAKE AND EXHAUST SYSTEM DIAGNOSIS AND SERVICE

UPON COMPLETION AND REVIEW OF THIS CHAPTER, YOU SHOULD BE ABLE TO:

- Service and replace air filters.
- Diagnose driveability problems resulting from intake system vacuum leaks.
- Use a vacuum gauge to diagnose engine problems.
- Test the intake system for vacuum leaks.
- Remove, inspect, and replace intake manifolds and gaskets.
- Perform an exhaust system inspection.

- Perform an exhaust system restriction test.
- Remove and replace exhaust system components.
- Test catalytic converters.
- Inspect the turbocharger system.
- Inspect the supercharger system.
- Test for proper boost and boost control.

BASIC TOOLS

Service manual
Technician's tool set
Penetrating oil
Threadlock

The intake system must conduct the airflow into the cylinders without leaking or providing excessive airflow restriction. A small leak in the air intake duct between the air cleaner and the throttle body allows dirt particles to enter the air intake, which greatly reduces engine life. On an engine that uses a mass airflow sensor, a leak between the MAF and the intake manifold will allow air to enter the engine that is not measured. The PCM will "think" there is less air flowing into the engine and will deliver less fuel than is required. This can cause serious driveability issues, such as hesitation and misfiring during acceleration. A leak between the throttle body and the intake valves allows air to be drawn into the intake manifold, and this action causes a lean air-fuel ratio, which results in performance problems such as engine surging at low speed.

The exhaust system must conduct the exhaust gases from the cylinders to the atmosphere without excessive restriction or noise. A restricted exhaust system results in a power loss at higher speeds. Accurate intake and exhaust system diagnosis and service is essential to provide proper engine performance and economy.

When a turbocharger or supercharger system malfunctions, it can create an underboost or overboost condition. When too little boost is developed, the engine will lack its normal power and the engine will hesitate on acceleration. When the overboost control system malfunctions and too much boost is allowed, serious engine damage can occur; prompt repairs are essential.

AIR CLEANER SERVICE

If an air filter is doing its job, it will get dirty. That is why filters are made of pleated paper. The paper is the actual filter. It is pleated to increase the filtering area. By increasing the area, the amount of time it will take for dirt to plug the filter becomes longer. As a filter gets dirty, the amount of air that can flow through it is reduced. This is not a problem until less air than the engine needs flows through the filter (Figure 6-1). Without the proper amount of air, the engine will not be able to produce the power it should, nor will it be as fuel efficient as it should be.

An engine with a plugged air filter will hesitate on acceleration and may not reach cruising speeds. Usually the engine will not misfire; it will just feel as though the engine does not respond to the accelerator pedal. When it is severely plugged, the engine may not start or will stall after starting.

Included in the preventative maintenance plan for all vehicles is the periodic replacement of the air filter. This mileage or time interval is based on normal vehicle operation. If the vehicle is used or has been used in heavy dust, the life of the filter is shorter. Always use a replacement filter that is the same size and shape as the original.

Follow these steps for air filter service or replacement:

1. Remove the wing nut, clips, or screws that retain the air filter cover and remove the cover. You may also need to remove the intake ductwork (Figure 6-2).
2. Remove the air filter element from the air cleaner, and be sure that no foreign material such as small stones drop into the air cleaner duct or throttle body (Figure 6-3). If the air cleaner has an **airflow restriction indicator,** inspect the color of the indicator window (Figure 6-4). If the window appears orange and "Change Air Filter" appears, the air filter is restricted. When the window appears green, the air filter is not restricted. If the airflow restriction indicator window appears orange, press the reset button on top of the indicator to reset the indicator so green appears in the window.
3. Carefully clean the air cleaner housing to remove any small stones, bugs, and debris. Inspect the air cleaner housing for cracks, holes, and damage.

SPECIAL TOOLS

Vacuum gauge
Vacuum pump
Ultrasonic vacuum
leak detector
Smoke-type vacuum
leak detector
Slitting tool
Digital pyrometer

SERVICE TIP:
Rodents often plug up air filters and air filter housings by storing food in them. They are attracted to the warmth of the engine. Check underneath a filter for dog food, nuts, or bedding.

FIGURE 6-1 This air filter was causing a significant hesitation on acceleration.

FIGURE 6-2 You have to remove the intake duct and the MAF sensor to replace this air filter.

MAF sensor   Air filter housing   Retaining clip

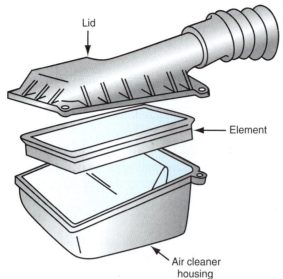

FIGURE 6-3 Air filter and air cleaner housing.

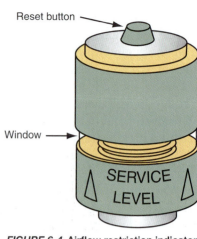

FIGURE 6-4 Airflow restriction indicator.

FIGURE 6-5 The PCV tube is sometimes connected to the air intake tube.

The **airflow restriction indicator** displays the amount of air filter element restriction.

CAUTION:

Do not use pressurized air to clean the housing unless you are sure that debris will not get into the engine or other compartments.

When servicing an air filter always service the positive crankcase ventilation (PCV) inlet filter.

4. Visually inspect the air filter for pin holes in the paper element and damage to the paper element, sealing surfaces, or metal screens on both sides of the element. If the air filter is damaged or contains pin holes, replace the filter.

5. Place a shop trouble light on the inside surface of the air filter and look through the filter toward the light. The light should be clearly visible through the filter, but there must be no pin holes in the paper element. If the light is not visible over most of the filter, or the filter is contaminated with oil, the air filter must be replaced. If the air filter is contaminated with oil, the engine has excessive blowby past the piston rings, or the positive crankcase ventilation (PCV) valve or hose is restricted (Figure 6-5). When either of these conditions exists, excessive crankcase pressure forces oil vapors out of the engine through the PCV clean air hose into the air cleaner. Clean air should normally flow from the air cleaner through the PCV clean air hose into the engine.

6. Clean or replace the PCV system filter in the air cleaner.
7. Install the replacement air filter in the air cleaner housing. Be sure the sealing surfaces on the element fit snugly on the air cleaner housing. Install the air cleaner cover and tighten the retaining clamps, screws, or wing nut.
8. Be sure the PCV hose and any other hose sensors or wiring connectors are properly connected to the air cleaner.

VACUUM SYSTEM DIAGNOSIS AND TROUBLESHOOTING

Vacuum system problems can produce or contribute to the following driveability symptoms:

1. Stalls
2. Poor fuel economy
3. Detonation, or knock or pinging
4. Rich or lean stumble
5. Rough idle
6. Poor acceleration
7. Backfire (deceleration)
8. Overheating
9. Hard start (hot soak)
10. Rotten eggs exhaust odor
11. No start (cold)

As a routine part of problem diagnosis, a technician who suspects a vacuum problem should first:

1. Inspect vacuum hoses for improper routing or disconnections (engine decal identifies hose routing).
2. Look for kinks, tears, or cuts in vacuum lines.
3. Check for vacuum hose routing and wear near hot spots, such as exhaust manifold or the EGR tubes.
4. Make sure there is no evidence of oil or transmission fluid in vacuum hose connections. (Valves can become contaminated by oil getting inside.)
5. Inspect vacuum system devices for damage (dents in cans; bypass valves; broken nipples on vacuum control valves; broken tees in vacuum lines, and so on).

Broken or disconnected hoses allow vacuum leaks that admit more air into the intake manifold than the engine is calibrated for. The most common result is a rough-running engine due to the leaner air-fuel mixture created by the excess air. Kinked hoses can cut off vacuum to a component, thereby disabling it. For example, if the vacuum hose to the EGR valve is kinked, vacuum cannot be used to move the diaphragm. Therefore, the valve will not open.

To check vacuum controls, refer to the service manual for the correct location and identification of the components. Typical locations of vacuum-controlled components are shown in Figure 6-6.

Tears and kinks in any vacuum line can affect engine operation. Any defective hoses should be replaced one at a time to avoid misrouting. Original equipment manufacturers (OEM) vacuum lines are installed in a harness consisting of 1/8-inch (3.18 mm) or larger outer diameter and 1/16-inch (1.59 mm) inner diameter nylon hose with bonded nylon or rubber connectors. Occasionally, a rubber hose might be connected to the harness. The nylon connectors have rubber inserts to provide a seal between the nylon connector and the component connection (nipple). In recent years, many domestic car manufacturers have been using ganged steel vacuum lines.

SERVICE TIP:
Do not clean air filters with shop air. This can break holes in the filter element and allow large particles into the engine, potentially causing serious damage. If the air filter is dirty, replace it; air filters are generally not terribly expensive.

Classroom Manual
Chapter 6

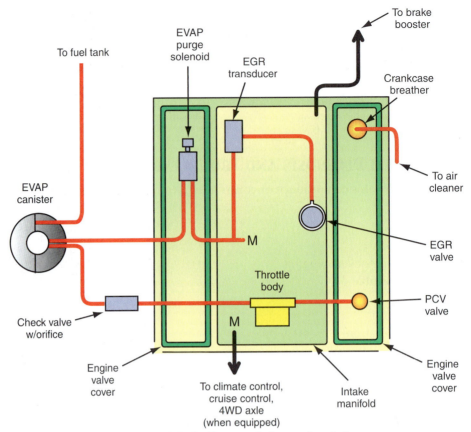

FIGURE 6-6 Typical vacuum devices and controls.

VACUUM TESTS

The **vacuum gauge** is a very important engine diagnostic tool used by technicians. With the gauge connected to the intake manifold and the engine warm and idling, watch the action of the gauge needle. A healthy engine will give a steady, constant vacuum reading between 17 and 22 in. Hg (57.5 and 74.4 kPa) (Figure 6-7). On some 4- and 6-cylinder engines, however, a reading of 15 inches (50.7 kPa) is considered acceptable. With high performance engines,

FIGURE 6-7 This engine has a healthy 18 in. Hg vacuum at idle.

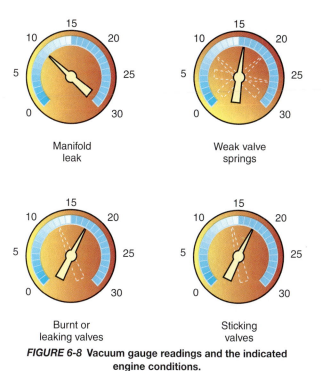

Manifold
leak

Weak valve
springs

Burnt or
leaking valves

Sticking
valves

FIGURE 6-8 Vacuum gauge readings and the indicated
engine conditions.

FIGURE 6-9 A smoke leak detector can be
used to find intake or exhaust system leaks as
well as evaporative emissions system leaks.

a slight flicker of the needle can also be expected. Keep in mind that the gauge reading will drop about 1 inch (2.54 cm) for each 1,000 feet (305 m) above sea level. Figure 6-8 shows some of the common readings and what engine malfunctions they indicate.

If your vacuum test or other symptoms indicate a vacuum leak, further testing may be needed to isolate the location of the leak. Vacuum can leak from a faulty intake manifold gasket, a cracked intake manifold, faulty injector O-rings, or from a vacuum line. The most effective method to test for a vacuum leak is with a smoke leak detector (Figure 6-9). You can connect the tester through the throttle bore and then depress the push button to introduce smoke into the intake system. If there is a leak in the system, you will see smoke emitted from the source of the leak (Figure 6-10). A flashlight or a light provided by the smoke machine's manufacture will assist you in finding smaller and hard-to-see vacuum leaks.

Smoke billowing from
the leaking intake
manifold gasket

Smoke leak
detector adaptor

FIGURE 6-10 This smoke test shows a serious leak at the intake
manifold gasket.

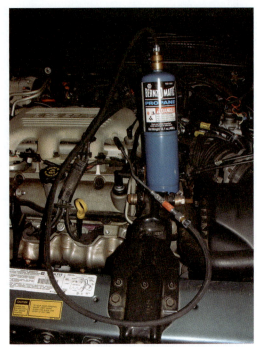

FIGURE 6-11 A small propane tank with a pressure control nozzle makes an excellent and inexpensive vacuum leak detection tool.

SPECIAL TOOLS

Vacuum pump
Smoke-type vacuum leak detector

The **intake manifold** distributes clear air to each cylinder of the engine.

SERVICE TIP:
The throttle body, fuel rail, and fuel injectors may need to be removed as an assembly when removing the intake manifold.

Another method of testing for vacuum leaks is to use a propane tank with a low-pressure outlet line (Figure 6-11). First verify that there are no sources of ignition in or around the engine compartment. Then, with the engine running, direct a small amount of propane at potential areas of leakage. When the propane is introduced toward a vacuum leak, the engine will pull the added fuel in and the idle quality will change. Turn the propane on and off at the site of the leak to confirm your suspicion. Be careful to perform this procedure in a well-ventilated area, and avoid using any more propane than is needed for diagnosis.

INTAKE MANIFOLD SERVICE

There are few reasons why an **intake manifold** would need to be replaced. Obviously, if the manifold is cracked or the sealing surfaces severely damaged, it should be replaced. The sealing surfaces should also be checked for flatness with a straightedge and feeler gauge. Minor imperfections on the surface can be filed away; however, do not attempt to clean up any serious damage.

Another common problem is a leaking intake manifold gasket. This will create a vacuum leak that can cause stalling, rough running, misfire, and detonation from a lean air-fuel mixture. The intake manifold gasket can also allow coolant or oil leaks. You will need to remove the intake manifold and inspect the sealing surfaces to replace the intake manifold gasket.

The following is a typical intake manifold removal procedure:

1. Connect a 12V power supply recommended by the vehicle manufacturer to the cigarette lighter socket, and disconnect the negative battery cable. If the vehicle is air bag–equipped, wait for the time period specified by the vehicle manufacturer before working on the vehicle.
2. Remove the sight cover bolts and the sight cover (Figure 6-12), and remove the duct between the air cleaner and the throttle body.

FIGURE 6-12 Remove the sight cover from the intake manifold.

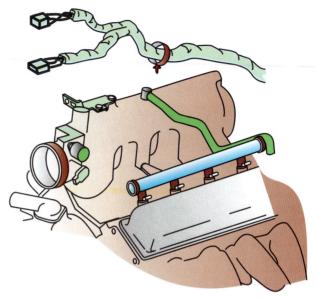

FIGURE 6-13 Remove the engine wiring harness retainer.

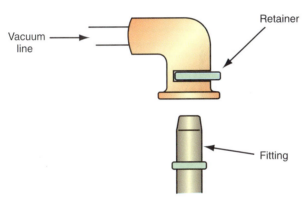

Vacuum line

Retainer

Fitting

FIGURE 6-14 Remove the purge solenoid vent hose.

3. Open the large electrical harness retainer and then remove the retainer bolt, in order to separate the engine wiring harness from the intake manifold (Figure 6-13).
4. Disconnect the following electrical connectors:
   (a) eight fuel injector connectors
   (b) idle air control motor connector
   (c) throttle position sensor (TPS) connector
   (d) evaporative emission canister purge solenoid connector
   (e) manifold absolute pressure (MAP) sensor connector

Position the electrical connectors so they do not interfere with the intake manifold removal.

5. Squeeze the purge solenoid vent tube retainer, and pull the tube from the solenoid (Figure 6-14). Remove the inlet coolant hose from the throttle body.

The metal shield around the exhaust manifold from which a heated air inlet system picks up hot air may be called a heat stove.

WARNING: Never disconnect any coolant hose or loosen the cooling system pressure cap if the engine coolant is warm or hot. The sudden decrease in cooling system pressure may cause the coolant to boil, resulting in personal injury.

FIGURE 6-15 Fuel lines with quick disconnect fittings.

FIGURE 6-16 The special quick disconnect fitting tools slide over the fitting to depress the locking tabs and allow the line to pull a part.

FIGURE 6-17 Remove the PCV valve and hose.

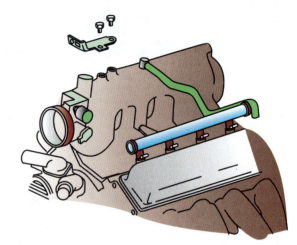

FIGURE 6-18 Remove the throttle cable and bracket.

6. Connect a fuel pressure gauge to the Schrader valve on the fuel rail, and place the gauge fuel drain hose in an approved gasoline container. Press the pressure relief valve on the gauge to relieve the fuel pressure.

7. Remove the dust covers from the quick disconnect fuel pipe fittings. Disconnect the quick disconnect fittings on the fuel supply and return lines. Use an air gun and shop air to blow any dirt from the **quick disconnect fittings** (Figure 6-15). Use a special tool on the quick disconnect fitting and pull the connector apart (Figure 6-16).

8. Disconnect the vacuum brake booster hose from the intake manifold.

9. Remove the PCV valve and hose (Figure 6-17).

10. Remove the accelerator control cable bracket and retaining bolt, and disconnect the accelerator cable (Figure 6-18). If the vehicle is equipped with cruise control, disconnect the cruise control cable.

11. Remove the intake manifold retaining bolts, and remove the intake manifold and gaskets (Figure 6-19).

Inspect the intake manifold and cylinder head mating surfaces for nicks, scratches, cracks, and damage. Minor scratches may be polished out with crocus cloth. Always plug the cylinder head and intake manifold intake air passages with clean shop towels when the intake manifold is removed to prevent any dirt particles or other objects from entering these passages. Clean the mating surfaces on the intake manifold and cylinder heads with a plastic scraper. Refer to Photo Sequence 10 for step by step pictures for removing and replacing an intake manifold gasket on a typical V6 engine.

Quick disconnect fittings are fuel line fittings that are removable with a special tool rather than threaded fittings.

TYPICAL PROCEDURE FOR REMOVING AND REPLACING AN INTAKE MANIFOLD GASKET

P10-1 A smoke leak detector reveals leakage from the intake manifold gasket.

P10-2 This manifold looks fairly simple to remove but check the service information for tips and torque specifications and sequence.

P10-3 Disconnect the necessary vacuum lines and electrical connectors.

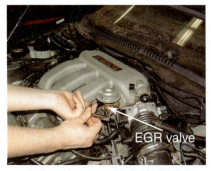

P10-4 This EGR valve needs to be unbolted from the manifold.

P10-5 Disconnect the throttle linkage.

P10-6 Remove the intake ducting.

P10-7 Loosen the intake manifold bolts in the opposite of the tightening sequence.

P10-8 Carefully lift the intake manifold off of the engine.

P10-9 Thoroughly clean the sealing surfaces and install the new gaskets securely. These are installed dry.

P10-10 Place the intake manifold back into position without disturbing the gaskets.

P10-11 Torque the intake manifold in the correct sequence to the proper specification.

P10-12 Reconnect the intake ducting, thr-ottle cable, EGR valve and vacuum lines.

TYPICAL PROCEDURE FOR REMOVING AND REPLACING AN INTAKE MANIFOLD GASKET

P10-13 Double check your work to be sure you have reconnected all the electrical connectors and vacuum lines properly. Start the engine, listen for a smooth idle and check for adequate vacuum.

CAUTION:
Never use a metal scraper or wire brush to clean surfaces on aluminum components. This action may scratch these components, so they would have to be replaced.

CAUTION:
If any dirt particles or foreign objects enter the intake manifold or cylinder head air intake passages, they will likely be pulled into the combustion chambers when the engine is started, resulting in severe engine damage.

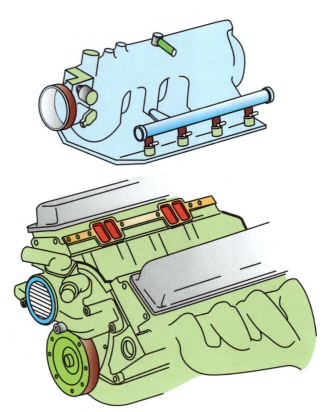

FIGURE 6-19 Remove the intake manifold.

CAUTION:
Never reuse old intake manifold gaskets. This action may cause vacuum leaks and driveability problems.

If the cylinder head or intake manifold mating surfaces are cracked or damaged, replace the necessary component(s). Inspect the intake manifold surface for cracks.

Follow these steps to install the intake manifold:

1. Install new intake manifold gaskets on the intake manifold.
2. Place a 0.20-in. (5 mm) band of threadlock GM part no. 12345382 or equivalent to the threads on the intake manifold bolts.
3. Install the intake manifold and the retaining bolts. Start each bolt into its threaded opening by hand, and then tighten each bolt to the specified torque in the sequence in Figure 6-20.
4. Install the PCV valve and hose.

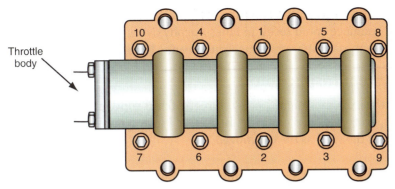

Throttle body

FIGURE 6-20 Tighten the intake manifold using the proper tightening sequence.

5. Install the accelerator cable, bracket, and retaining bolt. Tighten this bolt to the specified torque.
6. Install the cruise control cable if equipped.
7. Install the vacuum brake booster hose.
8. Place a few drops of engine oil on the male ends of the fuel supply and return pipes. Push the quick disconnect fitting together. After these fittings are pushed together, pull on both sides of the quick disconnect fitting to be sure the fitting does not pull apart. Install the dust covers over the quick disconnect fittings.
9. Install the purge solenoid vent tube.
10. Disconnect the fuel pressure gauge from the Schrader valve on the fuel rail.
11. Connect all the electrical connectors to sensors and components mounted on the intake manifold.
12. Install the engine wiring harness retainer attaching bolt, and install this harness in the retainer.
13. Install the sight cover and bolts.
14. Reconnect the negative battery cable and disconnect the 12V power supply from the cigarette lighter socket.
15. Start the engine and be sure engine operation and idle speed are normal.
16. Use a vacuum gauge, smoke machine, or scan tool to verify your work has been completed and works fine before you close the hood.
17. Test-drive the vehicle to make sure the customer complaint was repaired.

CUSTOMER CARE: Never sell customers automotive service that is not required on their cars. Selling preventative maintenance is a sound business practice and may save customers some future problems. An example of preventative maintenance is selling a cooling system flush when the cooling system is not leaking but the manufacturer's recommended service interval has elapsed. If customers find out they were sold some unnecessary service, and some will find out, they will probably never return to the shop. They will likely tell their friends about their experience, and that kind of advertising the shop can do without.

EXHAUST SYSTEM SERVICE

Exhaust system components are subject to both physical and chemical damage. Any physical damage to an exhaust system part that causes a partially restricted or blocked exhaust system usually results in loss of power or backfire up through the throttle plate(s). In addition to improper engine operation, a blocked or restricted exhaust system causes

Classroom Manual

Chapter 6

increased noise and air pollution. Leaks in the exhaust system caused by either physical or chemical (rust) damage could result in illness, asphyxiation, or even death. Remember that vehicle exhaust fumes can be very dangerous to one's health.

Exhaust System Inspection

The **front pipe** connects the exhaust manifolds to the catalytic converter or muffler.

Most parts of the exhaust system, particularly the **front pipe, muffler,** and **tailpipe,** are subject to rust, corrosion, and cracking. Broken or loose clamps and hangers can allow parts to separate or hit the road as the car moves.

WARNING: If the engine has been running, exhaust system components may be extremely hot! Wear protective gloves when working on these components.

WARNING: Exhaust gas contains poisonous carbon monoxide (CO) gas. This gas can cause illness and death by asphyxiation. Exhaust system leaks are dangerous for customers and technicians.

The **muffler** is a device, mounted to the exhaust system behind the catalytic converter, that reduces engine noise.

Any exhaust system inspection should include listening for hissing or rumbling that would result from a leak in the system. An on-lift inspection should pinpoint any of the following types of damage:

1. Holes, road damage, separated connections, and bulging muffler seams
2. Kinks and dents
3. Discoloration, rust, soft corroded metal, and so forth
4. Torn, broken, or missing hangers and clamps
5. Loose tailpipes or other components
6. Bluish or brownish catalytic converter shell, which indicates overheating

Replacing Exhaust System Components

The **tailpipe** conducts exhaust gases from the muffler to the rear of the vehicle.

Before beginning work on an exhaust system, make sure it is cool to the touch. Some technicians disconnect the battery's negative cable before starting to work to avoid short-circuiting the electrical system. Soak all rusted bolts, nuts, and other removable parts with a good penetrating oil. Finally, check the system for critical clearance points, so they can be maintained when new components are installed.

Most exhaust work involves the replacement of parts. When replacing exhaust parts, make sure the new parts are exact replacements for the original parts. Doing this will ensure proper fit and alignment, as well as ensure acceptable noise levels. Exhaust system component replacement might require the use of special tools (Figure 6-21).

Many technicians will use oxyacetylene torches to heat or cut exhaust connections and components. With practice, it is possible to cut a leaky pipe off of a good one without damaging the good pipe. This process is often faster than trying to use an air chisel or muffler cutter to remove a bad pipe. You can also use torches to apply heat. Heating the outer pipe of a connection will usually cause the metal to expand enough to allow you to pull the pipes apart. Similarly, heating rusty nuts in the exhaust system will greatly help removal.

Exhaust Manifold and Exhaust Pipe Servicing

As mentioned, the manifold itself rarely causes any problems. On occasion, an exhaust manifold will warp because of excess heat. A straightedge and feeler gauge can be used to check the machined surface of the manifold.

Another problem, also the result of high temperatures generated by the engine, is a cracked manifold. This usually occurs after the car passes through a large puddle and cold

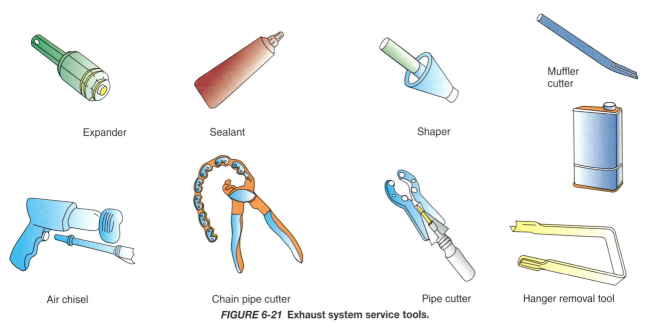

Expander   Sealant   Shaper   Muffler cutter

Air chisel   Chain pipe cutter   Pipe cutter   Hanger removal tool

FIGURE 6-21 Exhaust system service tools.

water splashes on the manifold's hot surface. If the manifold is warped beyond manufacturer's specifications or is cracked, it must be replaced. Also, check the exhaust pipe for signs of collapse. If there is damage, repair it. These repairs should be done as directed in the vehicle's service manual.

Replacing Exhaust System Gaskets and Seals

The most likely spot to find leaking gaskets and seals is between the exhaust manifold and the front pipe (Figure 6-22). When installing exhaust gaskets, carefully follow the recommendations on the gasket package label and instruction forms. Read through all installation steps before beginning. Take note of any of the original equipment manufacturer's recommendations in service manuals that could affect engine sealing. Manifolds warp more easily if an attempt is made to remove them while they are still hot. Remember, heat expands metal, making assembly bolts more difficult to remove and easier to break.

SERVICE TIP:
If a heated oxygen sensor (HO_2S) is mounted in the exhaust manifold, a cracked manifold upstream from the HO_2S allows air to enter the exhaust manifold. This air affects the HO_2S sensor signal and causes the powertrain control module (PCM) to supply a rich air-fuel ratio.

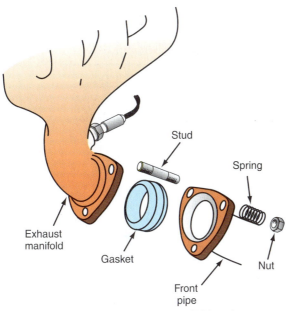

Stud

Spring

Exhaust manifold

Gasket

Nut

Front pipe

FIGURE 6-22 Front pipe to manifold gasket.

To replace an exhaust manifold gasket, follow the torque sequence in reverse to loosen each bolt. Repeat the process to remove the bolts. This minimizes the chance that components will warp. Some exhaust manifolds do not come with a gasket from the factory because the surfaces can be machined very smooth. Aftermarket and factory gaskets are available for these. Exhaust manifold studs and nuts are often very rusty due to the high heat they are subjected to (Figure 6-23). Heat the nuts to prevent breaking the studs off in the cylinder head.

Any debris left on the sealing surfaces increases the chance of leaks. A gasket remover solvent applied to the old gasket will quickly soften it and the old adhesive for quick removal with a hand scraper. Carefully remove the softened pieces with a scraper and a wire brush. Be sure to use a nonmetallic scraper when attempting to remove gasket material from aluminum surfaces.

Inspect the manifold for irregularities that might cause leaks, such as gouges, scratches, or cracks. Replace it if it is cracked or badly warped. You can use a feeler gauge set and a straightedge to check for warpage. File down any imperfections to ensure proper sealing of the manifold.

Due to high heat conditions, it is important to retap and redie all threaded bolt holes, studs, and mounting bolts. This procedure ensures tight, balanced clamping forces on the gasket. Lubricate the threads with a good high temperature antiseize lubricant. Use a small amount of contact adhesive to hold the gasket in place. Align the gasket properly before the adhesive dries. Allow the adhesive to dry completely before proceeding with manifold installation.

Install the bolts finger-tight. Tighten the bolts in three steps, one-half, three-quarters, and full torque, following the torque tables in the service manual or gasket manufacturer's instructions. Torquing is usually begun in the center of the manifold, working outward in an X pattern.

To replace a damaged front pipe, begin by supporting the converter to keep it from falling. Carefully remove the oxygen sensor if there is one. Remove any hangers or clamps holding the exhaust pipe to the frame. Unbolt the flange holding the exhaust pipe to the exhaust manifold. When removing the exhaust pipe, check to see if there is a gasket. If so, discard it and replace it with a new one. Once the joint has been taken apart, the gasket loses its effectiveness. Disconnect the pipe from the converter and pull the front exhaust pipe loose and remove it.

CAUTION:
Be sure no part of the exhaust system contacts any part of the chassis, fuel lines, fuel tank, or brake lines. This condition causes a rattling noise and/ or damage to other components.

FIGURE 6-23 It will be necessary to heat these rusty nuts before trying to loosen them, to avoid breaking the studs off in the cylinder head.

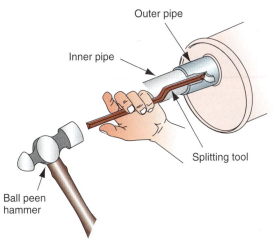

FIGURE 6-24 Using an exhaust pipe splitting tool.

Outer pipe

Inner pipe

Splitting tool

Ball peen
hammer

FIGURE 6-25 Install the new pipe and secure it with a U-clamp.

Although most exhaust systems use flanges or a slip joint and clamps to fasten the pipe to the muffler, a few use a welded connection. If the vehicle's system is welded, cut the pipe at the joint with a hacksaw or pipe cutter. The new pipe need not be welded to the muffler. An adapter, available with the pipe, can be used instead. When measuring the length for the new pipe, allow at least 2 inches (50.8 mm) for the adapter to enter the muffler.

WARNING: Wear safety goggles to protect your eyes and work gloves to protect your hands from burns and cuts.

When trying to replace a part in the exhaust system, you may run into parts that are rusted together. This is especially a problem when a pipe slips into another pipe or the muffler. If you are trying to reuse one of the parts, you should carefully use a cold chisel or slitting tool (Figure 6-24) on the outer pipe of the rusted union. You must be careful when doing this, because you can easily damage the inner pipe. It must be perfectly round to form a seal with a new pipe.

Slide the new pipe over the old. Position the rest of the exhaust system so that all clearances are evident and the parts aligned, then put a U-clamp over the new outer pipe to secure the connection (Figure 6-25).

Catalytic Converter Testing

The shell of a **catalytic converter** can become damaged like any other part of the exhaust system and require replacement (Figure 6-26). The converter can also be damaged internally and fail to perform as designed. A converter can fail to reduce emissions adequately. On an OBD II vehicle, the malfunction indicator light (MIL) will illuminate and a diagnostic trouble code (DTC) will be stored if the catalytic converter is not functioning properly. If the MIL is flashing during engine operation, it is a warning that if the vehicle continues to operate in that condition, catalytic converter damage may occur. A flashing MIL, usually caused by a severe misfire, will set a DTC or set continuing of misfires. The DTC will usually be P0420, low catalyst efficiency. The on-board diagnostic (OBD) system checks the efficiency of the catalytic converter by comparing the signals from the pre- and post-catalytic converter oxygen sensors. When the OBD system has determined that the catalyst is no longer 60% efficient, it must be replaced.

On an older vehicle, you can also test that the catalytic converter is functioning by using a **digital pyrometer.** Be sure the vehicle is thoroughly warmed up before testing. Measure the temperature of the pipe going into the converter and the temperature of the pipe

SPECIAL TOOLS
Slitting tool

SPECIAL TOOLS
Digital pyrometer

A **catalytic converter** reduces tailpipe emissions of carbon monoxide, unburned hydrocarbons, and nitrogen oxides.

A **digital pyrometer** is an electronic device that measures heat.

FIGURE 6-26 A catalytic converter.

coming out. The outlet pipe should be at least 100°F (37.7°C) hotter than the inlet pipe if the converter is working. If the outlet temperature is the same or lower, nothing is happening inside the converter.

A common source of driveability problems related to the catalytic converter is a **restricted exhaust.** The catalytic converter can often become partially or fully plugged, particularly after it has been overheated by extended periods of misfire or rich running. When this happens, the engine will lose power. It will accelerate slowly and have a hard time reaching higher speeds. In severe cases, the engine may stall or fail to start. If you suspect a restricted exhaust, give a light tap to the converter and to the muffler. Rattling in either component indicates internal damage that may be causing a restriction. There are several ways to test for a restricted exhaust. One way is to use the vacuum test we discussed previously. Place a vacuum gauge on an intake manifold vacuum port, and watch the gauge while you rev the engine to 2,500 rpm. Hold the rpm up for a couple of minutes and monitor the gauge. If the vacuum falls, the exhaust is restricted. The vacuum falls because the high volume of air trying to pass through the exhaust starts to back up in the system and leaves residual pressure in the cylinder on the intake stroke. This reduces engine vacuum and power.

You can also use a backpressure gauge to test the exhaust system (Figure 6-27). Install the gauge into the exhaust ahead of the converter. The gauge will typically read from 0 to 10 psi in half-psi increments. Often the gauge has an adapter that screws into the front oxygen sensor port. Other setups may require that you drill or punch a small hole in the front exhaust pipe and screw the gauge in. Always seal that hole before returning the vehicle to the customer. With the gauge installed, check the pressure at idle, 2,500 rpm, and under a hard, quick snap of the throttle. The pressure at idle should be near 0 psi, and it should stay below 1.5 psi at 2,500 rpm. During a snap of the throttle, the pressure should not exceed 4 psi. If the pressure is too high, remove the converter, and inspect it for damage. It is more common for a converter to fail than a muffler, but you may also have to remove and check the muffler or mufflers. A converter is expensive, so be sure that replacing it will fix the problem!

After replacing the catalytic converter, you should carefully verify that the engine is running properly. Most converters fail due to overheating from burning excess fuel. An ignition misfire or an excessively rich mixture can damage a catalytic converter in a relatively short time. Be sure that you cure the cause of the problem, to prevent the vehicle from returning with another damaged converter.

Restricted exhaust is an exhaust system that offers more- than-normal restriction to the exhaust flow.

Before purchasing a new catalytic converter, check with the manufacturer for a warranty. Catalytic converters are covered by the emission control components of federal law. They are covered for 8 years/80,000 miles. New aftermarket converters must have a 5 year/50,000 miles warranty on the shell, casing, and end pipes; and a 25,000-mile warranty on the substrate and inside components.

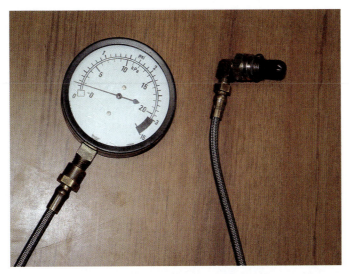

FIGURE 6-27 An exhaust pressure gauge.

TURBOCHARGER DIAGNOSIS

A faulty turbocharger can cause a serious lack of power. If the turbocharger fails to produce adequate boost pressure, the engine will be sluggish on acceleration. The turbocharger itself could be faulty, but there are other possible causes of low boost pressure. The whole system must be inspected to correctly diagnose the cause. Other symptoms caused by problems with the turbocharging system include:

1. Whining on acceleration from worn turbocharger bearings
2. Blue smoke from worn turbocharger seals
3. Misfire and hesitation on acceleration from a leak in the intake ductwork
4. Engine detonation or overheating from overboosting

WARNING: **If the engine has been running, turbochargers and related components are extremely hot. Use caution and wear protective gloves to avoid burns when servicing these components.**

The first step in **turbocharger** diagnosis is to check all linkages and hoses connected to the turbocharger (Figure 6-28). Inspect the wastegate diaphragm linkage for looseness and binding, and check the hose from the wastegate diaphragm to the boost control solenoid for cracks, kinks, and restrictions. Also, check the coolant hoses and oil line connected to the turbocharger for leaks.

Excessive blue smoke in the exhaust may indicate worn turbocharger seals. The technician must remember that worn valve guide seals or piston rings also cause oil consumption and blue smoke in the exhaust. When oil leakage is noted at the turbine end of the turbocharger, always check the oil drain tube and engine crankcase breathers for restrictions. If sludge is found in the crankcase, the engine's oil and filter must be changed.

Check all turbocharger mounting bolts for looseness. A rattling noise may be caused by loose turbocharger mounting bolts. Some whirring noise is normal when the turbocharger shaft is spinning at high speed. Excessive internal turbocharger noise may be caused by too much shaft end play, which allows the blades to strike the housings.

Check for exhaust leaks in the turbine housing and related pipe connections. If exhaust gas is escaping before it reaches the turbine wheel, turbocharger effectiveness is reduced. Check for intake system leaks. If there is a leak in the intake system before the compressor

A **turbocharger** is a device that is driven by exhaust pressure and that forces air into the cylinders.

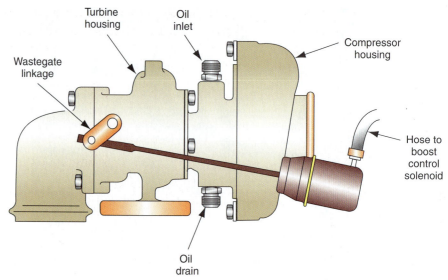

Turbine housing

Oil inlet

Compressor housing

Wastegate linkage

Hose to boost control solenoid

Oil drain

FIGURE 6-28 Typical turbocharger assembly.

housing, dirt may enter the turbocharger and damage the compressor or turbine wheel blades. When a leak is present in the intake system between the compressor wheel housing and the cylinders, turbocharger pressure is reduced.

If the turbocharger's boost is controlled by the engine computer, a diagnostic trouble code (DTC) is stored in the powertrain control module (PCM) memory if a fault is present in the boost control solenoid or solenoid-to-PCM wiring.

The following is a guide for turbocharger diagnosis and service:

1. A basic turbocharger inspection involves inspecting all turbocharger linkages, hoses, and lines, and checking for intake system and exhaust leaks.
2. Leaks in the air intake system may allow dirt particles to enter the turbocharger and damage the blades.
3. Exhaust leaks between the cylinders and the turbocharger decrease turbocharger efficiency.
4. Blue smoke in the exhaust may be caused by worn turbocharger seals, worn valve stem seals and guides, or worn piston rings.
5. Low cylinder compression reduces airflow through the engine and decreases turbocharger shaft speed and boost pressure.
6. Higher-than-specified boost pressure may be caused by a defective wastegate system.
7. Lower-than-specified boost pressure may be caused by a faulty wastegate system, worn turbocharger bearings or blades, or low cylinder compression.
8. When the axial movement on the turbocharger shaft is more than specified, the blades may strike the end housings.
9. Premature turbocharger bearing failure may be caused by lack of lubrication, contaminated oil, or a contaminated cooling system.
10. After the turbocharger has been replaced, it must be prelubricated before starting the engine.

Common Turbocharger Failures

The turbocharger is designed to provide years of useful service. Most premature failures of the turbocharger are due to lack of lubrication, contamination of the lubricant, or ingestion of foreign objects. Figure 6-29 provides information concerning troubleshooting the turbocharger and recommended remedies.

Condition	Possible Causes Code Numbers	Remedy Description by Code Numbers
Engine lacks power	1, 4, 5, 6, 7, 8, 9, 10, 11, 18, 20, 21, 22, 25, 26, 27, 28, 29, 30, 37, 38, 39, 40, 41, 42, 43	1. Dirty air cleaner element 2. Plugged crankcase breathers 3. Air cleaner element missing, leaking, not sealing correctly; loose connections to turbocharger 4. Collapsed or restricted air tube before turbocharger 5. Restricted-damaged crossover pipe, turbocharger to inlet manifold
Black smoke	1, 4, 5, 6, 7, 8, 9, 10, 11, 18, 20, 21, 22, 25, 26, 27, 28, 29, 30, 37, 38, 39, 40, 41, 43	6. Foreign object between air cleaner and turbocharger 7. Foreign object in exhaust system (from engine, check engine) 8. Turbocharger flanges, clamps, or bolts loose 9. Inlet manifold cracked; gaskets loose or missing; connections loose
Blue smoke	1, 2, 4, 6, 8, 9, 17, 19, 20, 21, 22, 32, 33, 34, 37, 45	10. Exhaust manifold cracked, burned; gaskets loose, blown, or missing 11. Restricted exhaust system 12. Oil lag (oil delay to turbocharger at startup) 13. Insufficient lubrication 14. Lubricating oil contaminated with dirt or other material
Excessive oil consumption	2, 8, 15, 17, 19, 20, 29, 30, 31, 33, 34, 37, 45	15. Improper type lubricating oil used 16. Restricted oil feed line 17. Restricted oil drain line 18. Turbine housing damaged or restricted 19. Turbocharger seal leakage
Excessive oil turbine end	2, 7, 8, 17, 19, 20, 22, 29, 30, 32, 33, 34, 45	20. Worn journal bearings 21. Excessive dirt buildup in compressor housing 22. Excessive carbon buildup behind turbine wheel 23. Too fast acceleration at initial start (oil lag)
Excessive oil compressor end	1, 2, 4, 5, 6, 8, 19, 20, 21, 29, 30, 33, 34, 45	24. Too little warm-up time 25. Fuel pump malfunction 26. Worn or damaged injectors 27. Valve timing 28. Burned valves
Insufficient lubrication	8, 12, 14, 15, 16, 23, 24, 31, 34, 35, 36, 44, 46	29. Worn piston rings 30. Burned pistons 31. Leaking oil feed line 32. Excessive engine pre-oil 33. Excessive engine idle
Oil in exhaust manifold	2, 7, 17, 18, 19, 20, 22, 29, 30, 33, 34, 45	34. Coked or sludged center housing 35. Oil pump malfunction 36. Oil filter plugged 37. Oil-bath-type air cleaner:
Damaged compressor wheel	3, 4, 6, 8, 12, 15, 16, 20, 21, 23, 24, 31, 34, 35, 36, 44, 46	◆ Air inlet screen restricted ◆ Oil pullover ◆ Dirty air cleaner ◆ Oil viscosity low ◆ Oil viscosity high
Damaged turbine wheel	7, 8, 12, 13, 14, 15, 16, 18, 20, 22, 23, 24, 25, 28, 30, 31, 34, 35, 36, 44, 46	38. Actuator damaged or defective 39. Wastegate binding 40. Electronic control module or connector(s) defective 41. Wastegate actuator solenoid or connector defective
Drag or bind in rotating assembly	3, 6, 7, 8, 12, 13, 14, 15, 16, 18, 20, 21, 22, 23, 24, 31, 34, 35, 36, 44, 46	42. EGR valve defective 43. Alternator voltage incorrect 44. Engine shut off without adequate cool-down time 45. Leaking valve guide seals 46. Low oil level
Worn bearings, journals, bearing bores	6, 7, 8, 12, 13, 14, 15, 16, 23, 24, 31, 35, 36, 44, 46	
Noisy	1, 3, 4, 5, 6, 7, 8, 9, 10, 11, 12, 13, 14, 15, 16, 18, 20, 21, 22, 23, 24, 31, 34, 35, 36, 37, 44, 46	
Sludged or coked center housing	2, 11, 13, 14, 15, 17, 18, 24, 31, 35, 36, 44, 46	

FIGURE 6-29 Turbocharger troubleshooting guide.

Turbocharger Inspection

To inspect the turbocharger, start the engine and listen for noises being generated within the turbocharger. A high-pitched sound may indicate an air leak between the compressor wheel and the intake manifold. If the sound changes in intensity, the likely causes are a plugged air filter element, loose material in the inlet ducts, or dirt buildup on the wheels and housing.

Check for loose clamps on all ducting associated with the turbocharger system (including the intercooler). Visually inspect all hoses, gaskets, and tubing for proper fit, damage, and wear. If there is no problem found in this area, remove the air cleaner and the ducting from the air cleaner to the turbo. Look for dirt buildup and damage from foreign objects.

Once the turbocharger has fully cooled, inspect the wheels for free rotation. Also, look into the housing and visually inspect for signs of wheel-to-housing interference.

The high-pressure side of the turbocharger system can be checked for leaks with the use of soapy water. While applying the solution to the connections, tubing, and the intercooler, watch for the presence of bubbles to pinpoint the leak.

Leakage or restrictions in the exhaust system above the turbine housing will reduce turbocharger efficiency. Also, check for free operation of the wastegate actuator. Be sure to check the hose that is attached to the actuator for damage. Photo Sequence 11 shows a typical procedure for inspecting turbochargers and testing boost pressure.

Testing Boost Pressure

To test **boost pressure,** begin by connecting a pressure gauge to the intake manifold. The pressure gauge hose should be long enough so the gauge may be positioned in the passenger compartment. Road test the vehicle at the speed specified by the vehicle manufacturer, and observe the boost pressure. Some vehicle manufacturers recommend accelerating from a stop to 60 mph (96 kph) at wide-open throttle while observing the boost pressure.

Higher-than-specified boost pressure may be caused by a defective wastegate system. Low boost pressure may be caused by the wastegate system or turbocharger defects such as damaged wheel blades or worn bearings. An engine with low cylinder compression will usually have low boost pressure.

Wastegate malfunctions can often be traced to carbon buildup that prevents bypass valve closing or causes it to bind. Also, check the linkage for bends that would prevent free movement. Usually, adjustment of the actuator rod is not necessary (unless someone misadjusted it previously). To test the wastegate on a PCM-controlled system, install a vacuum pump on the inlet hose. Pull a vacuum while monitoring the wastegate actuator rod. It should move freely and hold the wastegate open while vacuum is applied. If it fails to move, be sure that the linkage is not seized. If the linkage moves freely, replace the faulty wastegate. On an older system that relies on boost pressure to operate the wastegate, you'll apply pressure to verify wastegate operation. Locate the pressure at which the wastegate is supposed to open. Apply that pressure to the wastegate hose to check actuator rod movement.

On a PCM-controlled turbocharging system, be sure that there are no DTCs and that the fuel and ignition systems are functioning properly. A fault in the intake air temperature sensor, for example, could cause the PCM to limit boost pressure. Repair all related malfunctions before condemning the turbocharger.

Turbocharger Removal

The turbocharger removal procedure varies depending on the engine; for example, on some cars, such as a Nissan 300 ZX, the manufacturer recommends the engine be removed to gain access to the turbocharger. On other applications, the turbocharger may be removed with the engine in the vehicle. Always follow the turbocharger removal procedure in the

SERVICE TIP:
If the engine has low cylinder compression, there is reduced airflow through the cylinders, which results in lower turbocharger shaft speed and boost pressure.

SERVICE TIP:
Excessive boost pressure causes engine detonation and possible engine damage.

Boost pressure is the amount of pressure supplied by the turbocharger to the intake manifold.

Low turbocharger boost pressure causes reduced engine performance.

A **wastegate** is a valve that opens to bypass some exhaust around the turbine wheel to limit turbocharger boost pressure.

TYPICAL PROCEDURE FOR INSPECTING TURBOCHARGERS AND TESTING BOOST PRESSURE

P11-1 Check all turbocharger linkages for looseness, and check all turbocharger hoses for leaks, cracks, kinks, and restrictions.

P11-2 Check the level and condition of the engine oil on the dipstick.

P11-3 Check the exhaust for evidence of blue smoke when the engine is accelerated.

P11-4 Check all turbocharger mounting bolts for looseness.

P11-5 Check for exhaust leaks between the engine and the turbocharger, and use a stethoscope to listen for excessive turbocharger noise.

P11-6 Use a propane cylinder, metering valve, and hose to check for intake man-ifold vacuum leaks.

P11-7 Connect a pressure gauge to the intake manifold and locate the gauge in the passenger compartment where it can be seen easily by the driver.

P11-8 Road test the car and accelerate from 0 to 60 mph at wide-open throttle while observing the pressure gauge.

P11-9 Disconnect the pressure gauge from the intake manifold and remove the gauge from the passenger compartment.

vehicle manufacturer's service manual. The following is a typical turbocharger removal procedure:

1. Disconnect the negative battery cable, and drain the cooling system.
2. Disconnect the exhaust pipe from the turbocharger.
3. Remove the support bracket between the turbocharger and the engine block.
4. Remove the bolts from the oil drain back housing on the turbocharger.
5. Disconnect the turbocharger coolant inlet tube nut at the block outlet, and remove the tube-support bracket.
6. Remove the air cleaner element, air cleaner box, bracket, and related components.
7. Disconnect the accelerator linkage, throttle body electrical connector, and vacuum hoses.
8. Loosen the throttle body–to–turbocharger inlet hose clamps, and remove the three throttle body–to–intake manifold attaching screws. Remove the throttle body.
9. Loosen the lower turbocharger discharge hose clamp on the compressor wheel housing.
10. Remove the fuel rail–to–intake manifold screws and the fuel line bracket screw. Remove the two fuel rail bracket–to–heat shield retaining clips, and pull the fuel rail and injectors upward out of the way. Tie the fuel rail in this position with a piece of wire.
11. Disconnect the oil supply line from the turbocharger housing.
12. Remove the intake manifold heat shield.
13. Disconnect the coolant return line from the turbocharger and the water box. Remove the line-support bracket from the cylinder head and remove the line.
14. Remove the four nuts retaining the turbocharger to the exhaust manifold, and remove the turbocharger from the manifold studs. Move the turbocharger downward toward the passenger side of the vehicle, and then lift the unit up and out of the engine compartment.

SPECIAL TOOLS

Pressure gauge
Hand pressure pump
Dial indicator

CUSTOMER CARE: When returning vehicles to customers after a turbocharger replacement, be sure to discuss proper care and maintenance with them. Remind them that they should allow the turbo to wind down before shutting the vehicle off. This simply means that they should let the engine idle for a minute after driving before turning the ignition off. You should also remind them that regular oil changes are essential to turbocharger life. They can help ensure that the customer won't be back any time soon for turbocharger service.

Turbocharger Component Inspection

If the vehicle manufacturer recommends turbocharger disassembly, inspect the wheels and shaft after the end housings are removed. Lack of lubricant or lubrication with contaminated oil results in bearing failure, which leads to wheel rub on the end housings. A contaminated cooling system may provide reduced turbocharger bearing cooling and premature bearing failure. Bearing failure will likely lead to seal damage. Inspect the shaft and bearings for a burned condition (Figure 6-30). If the shaft and bearings are burned, replace the complete center housing assembly or individual parts as recommended by the manufacturer.

If the shaft and bearings are in satisfactory condition, but the blades are damaged, check the air intake system for leaks or a faulty air cleaner element. When the blades or shaft and bearings must be replaced, always check the end housings for damage (Figure 6-31). When

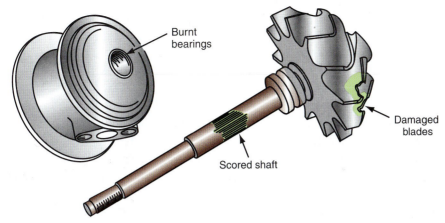

FIGURE 6-30 Inspect the turbocharger components for damage.

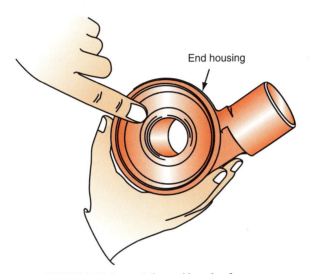

FIGURE 6-31 Inspect the end housing for wear.

these housings are marked or scored, replacement is necessary. Since turbocharger components are subjected to extreme heat, use a straightedge to check all mating surfaces for a warped condition. Replace warped components as necessary.

Turbocharger Installation and Prelubrication

Prior to reinstalling the turbocharger, be sure the engine oil and filter are in satisfactory condition. Change the oil and filter as required, and be sure the proper oil level is indicated on the dipstick. Check the coolant for contamination. Flush the cooling system if contamination is present. Reverse the turbocharger removal procedure explained previously to install the turbocharger. Replace all gaskets and be sure all fasteners are tightened to the specified torque. Before installing the new turbocharger, carefully check the oil supply and return lines for restrictions. If there is any sludge buildup in the hoses, clean them thoroughly or replace them. Restricted oil lines can cause premature failure of the new turbocharger.

Follow the vehicle manufacturer's recommended procedure for filling the cooling system. On some DaimlerChrysler engines, this involves removing a plug on top of the water box and pouring coolant into the radiator filler neck until coolant runs out the hole in the top of the water box (Figure 6-32). Install the plug in the top of the water box, and tighten the plug to the specified torque. Continue filling the cooling system to the maximum-level mark on the reserve tank.

CAUTION:
Failure to prelubricate turbocharger bearings may result in premature bearing failure.

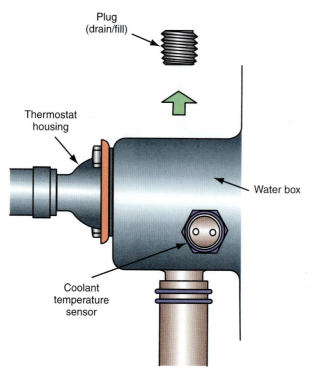

Plug
(drain/fill)

Thermostat
housing

Water box

Coolant
temperature
sensor

FIGURE 6-32 Remove the plug at the top of the water box
while filling the cooling system.

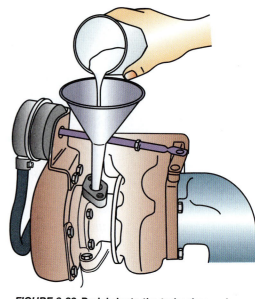

FIGURE 6-33 Prelubricate the turbocharger bearings by pouring oil in the oil inlet.

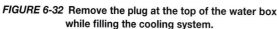

Prelubrication
entails lubricating
turbocharger
bearings or other
components before
starting the engine.

The turbocharger bearings must be prelubricated before starting the engine to prevent bearing damage. Some vehicle manufacturers recommend removing the turbocharger oil supply pipe and pouring a half pint of the specified engine oil into the turbocharger to **prelubricate** the bearings (Figure 6-33).

Other vehicle manufacturers recommend disabling the ignition system by disconnecting the positive primary coil wire and cranking the engine for 10 to 15 seconds to allow the engine lubrication system to lubricate the turbocharger bearings. Always follow the turbocharger prelubrication instructions in the vehicle manufacturer's service manual.

SUPERCHARGER DIAGNOSIS AND SERVICE

**Classroom
Manual**

Chapter 6

The supercharger is not subjected to high speeds and temperatures like the turbocharger; however, it does operate at a high speed with very close tolerances. Proper lubrication is essential for long supercharger life.

The fluid in the front supercharger housing lubricates the rotor drive gears. This fluid does not require changing for the life of the vehicle; however, this fluid level should be checked at 30,000-mile (48,000-km) intervals. To check the fluid level in the front supercharger housing, remove the Allen head plug in the top of this housing. The fluid should be level with the bottom of the threads in the plug opening. If the fluid level is low, add the required amount of synthetic fluid that meets the vehicle manufacturer's specifications.

Supercharger Diagnosis

A **supercharger** is
a belt-driven device
that forces more air
into the cylinders.

The **supercharger** on most vehicles is serviced as an assembly only. Specialty shops with proper equipment may rebuild some superchargers if parts are available. Therefore, the technician must diagnose supercharger problems and replace the supercharger only if necessary. Supercharger problems include low boost, high boost, reduced vehicle response and/or fuel economy, noise, and oil leaks.

Supercharger Removal

Follow these steps for supercharger removal:

1. Disconnect the negative battery cable.
2. Remove the air inlet tube from the throttle body.
3. Remove the cowl vent screens.
4. Drain the coolant from the radiator.
5. Disconnect the spark plug wires from the spark plugs in the right cylinder head and position them out of the way.
6. Remove the electrical connections from the air charge temperature sensor, throttle position sensor, and idle air control valve.
7. Disconnect the vacuum hoses from the supercharger air inlet plenum.
8. Remove the exhaust gas recirculation (EGR) transducer from the bracket, and disconnect the vacuum hose to this component.
9. Disconnect the positive crankcase ventilation (PCV) tube from the supercharger air inlet plenum.
10. Disconnect the throttle linkage, and remove the throttle linkage bracket. Position this linkage and bracket out of the way. Remove the cruise control linkage if equipped.
11. Remove the two EGR valve–attaching bolts, and place this valve out of the way.
12. Remove the coolant hoses from the throttle body.
13. Remove the supercharger drive belt.
14. Remove the inlet and outlet tubes from the intercooler (Figure 6-34).
15. Remove three intercooler adapter–attaching bolts from the intake manifold.
16. Remove three supercharger mounting bolts (Figure 6-35).
17. Remove the supercharger and air intake plenum as an assembly.

Follow these steps to install the supercharger:

1. Clean all gasket surfaces, and inspect these surfaces for scratches and metal burrs. Remove metal burrs as required.
2. Place a new gasket on the intake manifold surface that mates with the supercharger intercooler adapter.

A slipping drive belt is often the cause of low supercharger boost pressure.

Changing supercharger pulley size alters the supercharger rpm in relation to engine rpm, and this may cause improper boost pressure and engine performance problems.

A blown engine may refer to a supercharged engine. This term may also refer to an engine with serous mechanical problems.

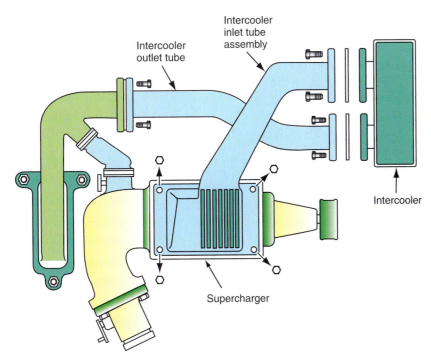

FIGURE 6-34 Intercooler inlet and outlet tubes.

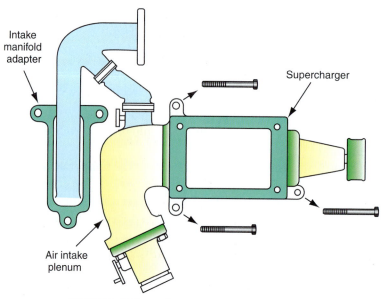

FIGURE 6-35 Supercharger mounting bolt location.

3. Install the supercharger, throttle body, and air intake plenum as an assembly.
4. Install three supercharger mounting bolts, and tighten these bolts to the specified torque.
5. Install three bolts that retain the intercooler adapter to the intake manifold, and tighten these bolts to the specified torque.
6. Install the intercooler inlet and outlet tubes, and tighten all mounting bolts to the specified torque.
7. Install the coolant hoses on the throttle body, and tighten these hose clamps.
8. Install a new EGR valve gasket, and install the EGR valve. Tighten the EGR valve–mounting bolts to the specified torque.
9. Install the throttle linkage and bracket, and tighten the bracket bolts to the specified torque.
10. Connect all the vacuum hoses to the original locations on the air intake plenum, EGR valve, and EGR transducer.
11. Install the spark plug wires on the spark plugs in the right cylinder head, and connect the electrical connectors to the throttle position sensor, air charge temperature sensor, and idle air control valve.
12. Install the cowl covers.
13. Install and tighten the air inlet tube on the throttle body.
14. Install the supercharger drive belt.
15. Refill the cooling system to the proper level.
16. Connect the battery ground cable.
17. Start the engine, and check for air leaks in the supercharger system.

Some German manufactures call a supercharger a compressor because it compresses the incoming air.

CASE STUDY

A customer complained about loss of power and a top speed of 55 mph (88 kmh) on a 2000 Oldsmobile Intrigue with a 3.5L V6 engine. The technician road tested the car to verify the customer's complaint. The car did have a maximum top speed of 55 mph (88 km/h), but the engine ran smoothly at idle speed. During the road test, the technician observed the malfunction indicator light (MIL) in the instrument panel. This light was not illuminated with the engine running, indicating there were no diagnostic trouble codes (DTCs) in the PCM. When the technician returned to the shop, he performed a visual under-hood and under-car inspection of the wiring harness, fuel lines, exhaust system, and vacuum system. The visual inspection did not reveal any damaged, loose, disconnected, or defective components.

The technician tested the fuel pressure and flow and found these values were within specifications. Next the technician connected a vacuum gauge to the intake manifold. With the engine idling, the vacuum was a steady 18 in. Hg (60.8 kPa). When the technician held the engine at 2,500 rpm for 3 minutes, the vacuum slowly decreased to 12 in. Hg (40.6 kPa), indicating a restricted exhaust system. When the front exhaust pipe was disconnected, the technician discovered the inside wall of this double-walled pipe had collapsed and severely restricted the exhaust passage through the pipe. After replacing this exhaust pipe, normal vehicle operation was restored.

TERMS TO KNOW

Airflow restriction indicator
Boost pressure
Catalytic converter
Digital pyrometer
Front pipe
Intake manifold
Muffler
Prelubrication
Quick disconnect fittings
Restricted exhaust
Supercharger
Tailpipe
Turbocharger
Wastegate
Vacuum gauge

ASE-STYLE REVIEW QUESTIONS

1. Airflow restriction indicators are being discussed.
   Technician A says if the restrictor window appears green, the air filter is restricted.
   Technician B says if the restrictor window appears green, the airflow restriction indicator must be replaced.
   Who is correct?
   A. A only
   B. B only
   C. Both A and B
   D. Neither A nor B

2. Air filter service is being discussed.
   Technician A says the air gun should be held tight against the air filter element when blowing dirt out of the filter.
   Technician B says when the outside of the air filter is observed with a trouble light inside the filter, small holes in the filter indicate filter replacement is required.
   Who is correct?
   A. A only
   B. B only
   C. Both A and B
   D. Neither A nor B

3. All of these driveability and engine problems may be caused by an intake manifold vacuum leak EXCEPT:
   A. Rough idle operation.
   B. Low engine operating temperature.
   C. Acceleration stumbles.
   D. Engine detonation.

4. A low steady vacuum gauge reading with the gauge connected to the intake manifold may be caused by:
   A. A vacuum leak.
   B. A burned exhaust valve.
   C. A defective spark plug.
   D. Weak valve springs.

5. An engine has 19 in. Hg (483 mm. Hg) of vacuum in the intake manifold with the engine idling. With the engine running at 2,500 rpm for 3 minutes, the vacuum slowly decreases to 10 in. Hg (254 mm. Hg). This problem could be caused by:
   A. A restriction in the catalytic converter or muffler.
   B. A leak in the exhaust pipe.
   C. A leak in the intake manifold gaskets.
   D. A restricted vacuum hose connected to the brake booster.

6. Removing and replacing an aluminum intake manifold is being discussed.
   Technician A says the old intake manifold gaskets may be reused.
   Technician B says a steel scraper should be used to remove old gasket material from an aluminum intake manifold.
   Who is correct?
   A. A only
   B. B only
   C. Both A and B
   D. Neither A nor B

7. An exhaust manifold with an HO$_2$S sensor is cracked upstream from the sensor.
Technician A says this condition does not affect the HO$_2$S sensor signal.
Technician B says this condition will damage the HO$_2$S sensor.
Who is correct?
A. A only C. Both A and B
B. B only D. Neither A nor B

8. When diagnosing a catalytic converter with the engine running at normal operating temperature:
A. The converter inlet should be 150°F (65°C) hotter than the outlet.
B. The converter outlet should be 100°F (38°C) hotter than the inlet.
C. The converter outlet should be 20°F (6.6°C) cooler than the inlet.
D. The converter inlet should be 50°F (19°C) hotter than the outlet.

9. While discussing turbocharger inspection and diagnosis:
Technician A says excessive turbo shaft endplay can cause a rattling noise from the turbocharger.
Technician B says that an intake system air leak can damage the turbocharger.
Who is correct?
A. A only C. Both A and B
B. B only D. Neither A nor B

10. While discussing turbocharger performance:
Technician A says that worn turbocharger seals can cause blue smoke from the exhaust.
Technician B says that a stuck-open wastegate can cause blue smoke from the exhaust.
Who is correct?
A. A only C. Both A and B
B. B only D. Neither A nor B

ASE CHALLENGE QUESTIONS

1. With a vacuum gauge connected to the intake manifold, the vacuum gauge reading is a steady 18 in. Hg (457 mm. Hg) at idle. When the engine speed is increased to 2,500 rpm, the reading increases to 25 in. Hg (635 mm. Hg).
Technician A says the engine may have a restricted catalytic converter.
Technician B says the EGR valve may be stuck open.
Who is correct?
A. A only C. Both A and B
B. B only D. Neither A nor B

2. When a vacuum gauge is connected to the intake manifold, the gauge needle fluctuates between 15 and 20 in. Hg (381 mm. Hg) The cause of this problem could be:
A. Late ignition timing.
B. Intake manifold gaskets leaking.
C. A restricted exhaust system.
D. Sticking valve stems and guides.

3. *Technician A* says an intake manifold vacuum leak may cause a cylinder misfire with the engine idling.
Technician B says an intake manifold vacuum leak may cause a cylinder misfire during hard acceleration.
Who is correct?
A. A only C. Both A and B
B. B only D. Neither A nor B

4. Reduced turbocharger boost pressure can be caused by:
A. A stuck-closed wastegate.
B. A stuck-open wastegate.
C. A leaking wastegate diaphragm.
D. Both A and C.

5. While discussing turbocharger performance:
Technician A says that reduced engine compression can cause reduced boost pressure.
Technician B says that an air leak between the turbocharger and the throttle can cause misfire.
Who is correct?
A. A only C. Both A and B
B. B only D. Neither A nor B

Name _____ Date _____

INSPECT AND SERVICE AIR CLEANER

Upon completion of this job sheet, you should be able to inspect and service air cleaners.

ASE Correlation

This job sheet is related to ASE Engine Repair Test's content area: Fuel, electrical, ignition, and exhaust system inspection and service; task: Inspect, service, or replace air filters, filter housings, and intake ductwork.

Tools and Materials

Shop towels
Air gun
Shop light
Vehicle with an air cleaner and airflow restriction indicator

Describe the vehicle being worked on.

Year _____ Make _____ Model _____

VIN _____ Engine type and size _____

Describe the type of air cleaner mounting and type of filter element.

_____

_____

Procedure

Task Completed

1. Remove the air cleaner cover wing nut, clips, or screws, and remove the cover and air filter.

   ☐

2. Place a shop light inside the air filter, and observe the air filter condition.
   Air filter satisfactory _____ Requires cleaning _____
   Requires replacement _____

3. Inspect the air flow restriction indicator.
   Window color _____
   Does "Change Air Filter" appear in window?   ☐ Yes   ☐ No

4. Inspect air cleaner housing for cracks, holes, leaks.
   Air cleaner housing, satisfactory _____ unsatisfactory _____
   If air cleaner housing is unsatisfactory, state conditions and necessary repairs.

   _____

5. Clean the inside of the air cleaner housing with a shop towel, being sure that nothing falls into the air cleaner duct or throttle body.
   Air cleaner housing properly cleaned _____   ☐ Yes

6. Inspect all air intake ducts for cracks, holes, leaks, damage, and deterioration.
   Air intake ducts, satisfactory _____ unsatisfactory _____
   If air intake ducts are unsatisfactory, state necessary repairs.

   _____

7. If the air filter requires cleaning, place an air gun with 30 psi (207 kPa) shop air pressure six inches from the inside of the filter and blow the dirt out of the filter.
   Reinspect the air filter with a shop light.
   Air filter, satisfactory _____ Requires replacement _____

8. Inspect the sealing surfaces of the air cleaner housing and cover.
   Do the sealing surfaces of the housing and cover fit squarely against the air filter seals?   ☐ Yes   ☐ No
   If the answer above is no, state necessary repairs.

   _____

☐

9. Install the air filter, air filter cover, and retaining clips, screws, or wing nut.

   Instructor's Response _____
   _____
   _____
   _____

JOB SHEET

Name _____ Date _____

DIAGNOSING VACUUM LEAKS

Upon completion of this job sheet, you should be able to test a vehicle for proper vacuum and vacuum leaks, recognize symptoms of a vacuum leak or other problems, and recommend appropriate repairs.

Tools and Materials

Vacuum gauge
Propane tank
Smoke leak detector

Describe the vehicle being worked on.

Year _____ Make _____ Model _____

VIN _____ Engine type and size _____

Procedure

Your instructor will provide you with a specific vehicle. Write down the information, then perform the following tasks:

1. Start the vehicle and observe the engine idling condition and rpm. Describe your results:

   _____

   _____

   _____

2. Attach a vacuum gauge and observe the readings at idle and at 2,500 rpm. Describe the readings and the needle movement during the tests:

   _____

   _____

3. With the vacuum gauge still installed, hold the engine rpm at 2,500 for one minute and observe the vacuum gauge. Describe your results:

   _____

   _____

4. Do your tests so far indicate any concerns with the vehicle you are working on?   ☐ Yes   ☐ No
   If there are concerns, describe the possible causes:

   _____

   _____

5. Next, check the engine for vacuum leaks. Use a smoke leak detector if available or a propane tester if not. Follow the equipment manufacturer's information for connecting the smoke leak detector to the throttle bore and apply smoke to the intake. If you

are using propane, be certain there are no sources of ignition, and direct the propane at all possible sources of vacuum leaks.

Does your vehicle have any vacuum leaks?　　☐ Yes　☐ No

If your vehicle does have a vacuum leak, describe the source of the problem and the recommended repair:

☐

_____

_____

6. Attach a vacuum gauge to a port on the intake manifold or another source of manifold vacuum. Start the vehicle and disconnect a vacuum line from the intake manifold. Describe the change in the idle condition and vacuum reading from your previous test:

_____

_____

_____

7. Turn the engine off, replace the vacuum line, and remove all test equipment. Start the engine, and be sure that it runs as smoothly as it did when you began your testing.

_____

8. Based on your tests and results, what are your recommendations for further testing and repairs?

_____

_____

_____

Instructor's Response _____

_____

_____

_____

JOB SHEET

Name _____ Date _____

EXHAUST SYSTEM INSPECTION AND TEST

Upon completion of this job sheet, you should be able to inspect an exhaust system and determine the necessary repairs.

ASE Correlation

This job sheet is related to the ASE Engine Repair Test's content area: Fuel, electrical, ignition, and exhaust systems inspection and service; task: Inspect, service, and replace exhaust manifold.

Tools and Materials

Vehicle with an exhaust system
Digital pyrometer
Vacuum gauge

Describe the vehicle being worked on.

Year _____ Make _____ Model _____

VIN _____ Engine type and size _____

Describe the type of exhaust system on the vehicle.

_____

Procedure

1. Inspect the HO_2S sensor wiring.
   HO_2S sensor wiring, satisfactory _____ unsatisfactory _____
   If the HO_2S wiring is unsatisfactory, state necessary repairs.

   _____

2. Inspect exhaust manifold(s) for cracks and exhaust leaks.
   Exhaust manifold(s), satisfactory _____ unsatisfactory _____
   If exhaust manifold(s) are unsatisfactory, state necessary repairs.

   _____

3. Inspect exhaust pipe, muffler, and tailpipe for holes, cracks, leaks, kinks, dents, separated connections, and excessive rust.
   Exhaust pipe, satisfactory _____ unsatisfactory _____
   Muffler, satisfactory _____ unsatisfactory _____
   Tailpipe, satisfactory _____ unsatisfactory _____
   If the exhaust pipe, muffler, and/or tailpipe are unsatisfactory, state necessary repairs.

   _____

4. Inspect all exhaust system hangers and clamps.
   Hangers and clamps, satisfactory _____ unsatisfactory _____
   If the hangers and/or clamps are unsatisfactory, state necessary repairs.

   _____

5. Perform exhaust system restriction test.
   Intake manifold vacuum at idle speed _____
   Intake manifold vacuum at 2,500 rpm after 3 minutes _____
   Exhaust system restriction, satisfactory _____ unsatisfactory _____
   If exhaust system restriction is unsatisfactory, state necessary corrective action.

   _____

6. Perform catalytic converter test with engine at normal operating temperature. You
   must drive the vehicle under a load to properly precondition the catalytic converter
   and ensure that it is hot enough to be functional. An extended period of idling may
   allow the converter to cool down enough to become inefficient.
   Converter inlet temperature _____ Converter outlet temperature _____
   Converter, satisfactory _____ unsatisfactory _____
   If the converter is unsatisfactory, state necessary repairs.

   _____

   Instructor's Response _____
   _____
   _____
   _____

Name _____ Date _____

TURBOCHARGER SYSTEM INSPECTION

Upon completion of this job sheet, you should be able to perform a proper inspection of the turbocharger system and determine needed repairs.

ASE Correlation

This job sheet is related to the ASE Engine Repair Test's content area: Fuel, Electrical, Ignition, and Exhaust Systems Inspection and Service; tasks: Inspect turbocharger/super-charger; determine needed action.

Tools and Materials

Vehicle with a turbocharger

Describe the vehicle being worked on.

Year _____ Make _____ Model _____

VIN _____ Engine type and size _____

Procedure

Your instructor will assign you (or your group) to a turbocharger-equipped vehicle. Using the proper service manual, you will identify the required steps to inspect the turbocharger system.

WARNING: If the engine has been running, turbochargers and related components are extremely hot. Use caution and wear protective gloves to avoid burns when servicing these components.

1. Inspect the wastegate diaphragm linkage for looseness and binding. Record your results.

   _____

   _____

2. Check the hose from the wastegate diaphragm to the intake manifold for cracks, kinks, and restrictions. Record your results.

   _____

   _____

3. Check the coolant hoses and oil line connected to the turbocharger for leaks. Record your results.

   _____

   _____

4. Check the engine oil level for proper fill and condition. Record your results.

   _____

   _____

5. Check for oil leakage at the end housings of the turbocharger. Is oil leaking?
   ☐ Yes   ☐ No

6. Check the oil drain tube and engine crankcase breathers for restrictions. Record your results.

   _____

   _____

☐

7. Check all turbocharger mounting bolts for looseness, and tighten to specifications as needed.

8. Start the engine. After oil pressure is obtained, accelerate the engine while observing the tailpipe for excessive blue smoke. Is blue smoke present?
   ☐ Yes   ☐ No

9. While the engine is running, listen for any unusual turbocharger noises. Record your results.

   _____

   _____

10. Listen for any exhaust leaks. Record your results.

    _____

    _____

11. Check for intake system leaks. Record your results.

    _____

    _____

12. Check the high pressure side of the turbocharger system for leaks using soapy water. Record your results.

    _____

    _____

13. If the boost pressure is computer controlled, check and record any diagnostic trouble codes (DTCs).

    _____

    _____

14. Shut off the engine, and remove the air cleaner and the ducting from the air cleaner to the turbo. Look for dirt buildup and damage from foreign objects. Record your results.

    _____

    _____

15. Once the turbocharger has fully cooled, inspect the wheels for free rotation. Also, look into the housing and visually inspect for signs of wheel-to-housing interference. Record your results.

    _____

    _____

16. Based on your results, what is your recommendation?

_____

_____

Instructor's Response _____

_____

_____

_____

JOB SHEET

39

Name _____   Date _____

TESTING BOOST PRESSURE

Upon completion of this job sheet, you should be able to perform a boost pressure test of the turbocharger system and determine needed repairs.

ASE Correlation

This job sheet is related to the ASE Engine Repair Test's content area: Fuel, Electrical, Ignition, and Exhaust Systems Inspection and Service; tasks: Inspect turbocharger/supercharger; determine needed action.

Tools and Materials

Vehicle with a turbocharger
Vacuum/pressure hand pump
Pressure gauge
Dial indicator

Describe the vehicle being worked on.

Year _____ Make _____ Model _____

VIN _____ Engine type and size _____

Procedure

Task Completed

Your instructor will assign you (or your group) to a turbocharger-equipped vehicle. Using the proper service manual, you will identify the required steps to test the boost pressure.

1. What is the specification for boost pressure?

2. Connect a pressure gauge to the intake manifold. The pressure gauge hose should be long enough so the gauge may be positioned in the passenger compartment. ☐

3. Road test the vehicle at the speed specified by the vehicle manufacturer and observe the boost pressure. Record the amount of boost indicated on the pressure gauge.

   _____

4. Is the boost pressure within specification?   ☐ Yes   ☐ No
   If NO, continue to step 6.

5. Check the actuator rod for bends that would prevent free movement. Record your results.

   _____

6. Check for carbon buildup that prevents the valve from closing or causes it to bind. Record your results.

   _____

7. Before condemning the wastegate, perform some basic checks of the ignition and fuel systems. Check ignition timing, the spark retard system, vacuum hoses, the knock

sensor, the O_2 sensor, and fuel delivery, to assure all are operating properly. Record your results.

_____

_____

_____

_____

_____

8. Test the stroke of the wastegate actuator rod. Connect a hand pressure pump and a pressure gauge to the wastegate diaphragm. Position a dial indicator on the outer end of the actuating rod so it can measure rod movement. While supplying the specified amount of pressure with the hand pump, read the dial indicator and note the amount of movement.

Specifications _____

At what pressure is the specification? _____

Actual rod movement measurement _____

9. Will an adjustment of the rod length correct the amount of rod movement?
   ☐ Yes   ☐ No

10. If NO, what needs to be done to correct the problem?

_____

_____

_____

Instructor's Response _____

_____

_____

_____

Chapter 7

DIAGNOSING ENGINE PERFORMANCE CONCERNS

UPON COMPLETION OF THIS CHAPTER, YOU SHOULD BE ABLE TO:

- Replace and evaluate the older-type spark plugs, with nickel alloy steel electrodes every 30,000–60,000 miles, and the newer, platinum spark plugs that have a longer service interval, typically 60,000–100,000 miles.

- Keep the spark plugs in order when removing them, so you can trace any problems with the plug to the correct cylinder.

- Diagnose engine operating concerns by "reading" the spark plugs.

- Use a power balance test to find a cylinder that is not contributing equally to the power of the engine.

- Identify and diagnose the reasons for blue smoke being emitted from the tailpipe.

- Identify black smoke caused by too rich an air-fuel mixture or incomplete combustion.

- Diagnose the causes of excessive oil consumption.

- Use a cranking compression test to assess the condition of the rings, pistons, valves, and head gasket(s).

- Use a wet compression test to help determine whether low readings are caused by worn rings or by a burned valve or piston.

- Perform a running compression test to help you evaluate the condition of the valvetrain.

- Use a cylinder leakage test to pinpoint the cause of improper combustion chamber sealing and interpret the test results.

- Listen closely to engine noises to help you diagnose engine problems correctly.

- Describe and recognize engine noises, their possible causes, and methods of diagnosis.

INTRODUCTION

You should spend adequate time diagnosing the causes of engine performance problems before jumping into engine mechanical repairs. In this chapter you will learn several methods of determining why an engine is performing poorly. Some of these will lead you to determine that the engine requires engine mechanical repairs or even engine overhaul or replacement. Others will lead you to a less dramatic repair that may cost the customer much less. You need to isolate the cause of a performance problem as much as possible to help guide your repairs and ensure that you are able to make a successful repair on the first try.

SPARK PLUG EVALUATION AND POWER BALANCE TESTING

Spark plug reading and **power balance testing** are two methods you will use to gather information about how the engine is operating. You may use these tests during a routine tune up, when there is a driveability concern, and as part of gathering information about an engine mechanical failure. Spark plugs can help identify if one or more cylinders are acting up or if all cylinders are affected. They can also help guide your diagnosis toward a fuel, ignition, or mechanical issue for example. Similarly, a power balance test can determine how much each cylinder is contributing and pick out the malfunctioning cylinder(s).

A **power balance test** evaluates approximately how much each cylinder is contributing to the overall engine power. It is used to help identify a weak cylinder.

SPARK PLUGS

The spark plug must be in good condition to allow the spark to jump the gap. If the electrodes are worn rounded, covered with oil or gas, or if the gap is too large, the spark will not be strong enough to ignite a flame front, if it sparks at all. Spark plugs are generally one of two basic types: the older type plug with nickel alloy steel electrodes and the newer, more common, platinum tip electrode plugs. Most new vehicles use the platinum plugs as original equipment (Figure 7-1). The platinum tip plugs have a service interval of 60,000 miles or longer. The older style plugs must generally be replaced every 30,000 miles.

Many times customers do not bring their cars in for service until there is a driveability concern. Spark plugs have a high failure rate; check them early on in your diagnosis. As a spark plug wears, the gap between the electrodes becomes so wide it may not allow the spark to jump the gap. This can cause partial or total misfiring on that cylinder. Ignition misfire is usually noticeable at low rpm under heavy acceleration. The engine may buck and shudder when the cylinder(s) misfires. It is often difficult to see wear on a platinum plug; the electrodes are small and do not show much visible degradation. Many technicians will replace rather than inspect and reinstall platinum plugs with over 30,000 miles on them. Given today's labor rates, this is often more cost effective for the consumer. If the engine is running poorly and you have the spark plugs out, it may be wise to perform a compression test, which is described later in this chapter.

Spark Plug Removal and Installation

While spark plug removal is a very common task in the automotive repair shop, it is not always a simple one. Many times you will have to disassemble parts of the engine to access the spark plugs. There are vehicles that require you to remove an engine mount or remove the turbocharger just to access the plugs. Every time you replace a set of spark plugs you want to be sure you do not damage the plugs or the threads while installing them. Always be sure to replace the whole set of plugs. If one has failed, the others are likely not far behind.

If access to the plugs is not simple, read through the service information to pick up any steps that will make the job easier. This task often requires a universal joint and an assortment of extensions (Figure 7-2).

CAUTION:
Do not remove spark plugs from a hot aluminum cylinder head; this can easily damage the threads in the head. Allow the engine to cool down adequately before removal of spark plugs.

FIGURE 7-1 A platinum tip spark plug.

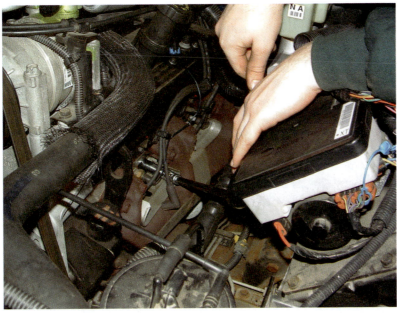

FIGURE 7-2 This spark plug is readily accessible; not every engine makes it this easy.

Blow away any sand and grit from around the spark plug holes before removing the plugs, to prevent this from falling into the cylinder head. Remove all plugs and keep them in order, so you can evaluate the condition of each plug and correlate it to the correct cylinder. Keep the spark plug wires (if applicable) organized, so you return them to their correct positions. Carefully inspect the ends of the plug wires. They can easily be damaged during removal and may require replacement. Also look closely at the plug wires for signs of electrical leakage or arcing to a ground path. If the wire's insulation is damaged, you can often see a light gray or white residue on the wire where the high voltage arcs to a ground—on the valve cover for example, rather than across the spark plug gap.

> **CUSTOMER CARE:** In one semester three students have come to me with a misfiring problem. In each case, they had recently replaced spark plugs using a popular, low-cost brand of spark plugs. Installing higher quality or OEM spark plugs repaired each of the three late-model vehicles. Use OEM or high-quality spark plugs on your customers' cars to prevent comebacks.

Check the spark plug gap before installing the new set of plugs. Sometimes the gap gets closed if the plugs have been dropped. Note that some platinum spark plugs do not have adjustable gaps. Clean the area around the spark plug hole. If you get grease on the electrodes during installation, you will likely wind up with a misfiring spark plug. Install the new plugs by hand; **do not** use air tools. If you can't reach the plug itself, turn it in a few turns using only the extension.

Finish tightening the plugs to the correct torque specification using a torque wrench. After your work is complete, always give the vehicle a thorough road test to be sure the engine runs perfectly.

Spark Plug Reading

Each spark plug can tell you a story about what is occurring in the cylinder. During a routine service, you can use this information to help determine if the engine is mechanically sound. If one spark plug comes out caked with oil deposits from oil leaking past the rings or valve

CAUTION:
Never use air tools to remove or install spark plugs. It takes a lot longer to repair a spark plug hole than it does to loosen plugs with a ratchet!

SERVICE TIP:
When the plugs are very difficult to reach, it is helpful to put a piece of vacuum line on the end of the spark plug. This flexible connection allows you to reach the spark plug hole. Turn the plug in a few turns using the end of the vacuum line. If you start to cross thread the plug, the vacuum line will spin on the spark plug and prevent thread damage.

CAUTION:
No technician has a torque wrench built into his elbow, no matter how adamantly he insists he can torque something properly. Spark plug torque specifications are very light on many new engines, and a stiff elbow can easily strip the spark plug threads on an aluminum cylinder head.

seals, you will be able to let the customer know that the engine is in need of more serious work. When performing driveability diagnosis, the spark plug can help guide your diagnosis. If one spark plug were wet with fuel, you would first confirm that the ignition system is delivering adequate spark to the plug. And whenever you are trying to narrow down the cause of an engine failure, analyze the spark plugs as part of your pre-teardown investigation. The more information you have going into the job, the more confident you can be that you will find the real cause of the problem.

Spark plugs that are wearing normally should show a light tan to almost white color on the electrodes, with no deposits (Figure 7-3). The electrodes should be square and the gap should be at or near at the specified measurement. You will see plugs that have a red or orange tint on them. This is the result of fuel additives found in certain fuels or in fuel system cleaning additives.

A spark plug that has many miles on it but is wearing normally will still be light tan in color. There should be only very light deposits, if any, on the plug (Figure 7-4). On nonplatinum plugs, you will be able to see that the corners of the electrodes are rounded. The gap will usually be wider than specified. On platinum plugs, the tip may look brand new even when the plug is not firing well. Often you will have to judge wear by the miles since the last change. If the plug is not firing at all, it will be gas-soaked and smell like fuel.

> **CUSTOMER CARE:** Some spark plugs are advertised to last for over 100,000 miles. This doesn't guarantee they'll last that long; these platinum plugs have a high failure rate over 50,000 miles and often become welded into the cylinder head by 75,000 miles. A 60,000-mile service interval is a reasonable compromise that is likely to serve the customer better.

Fuel-fouled plugs will be wet with gas when you remove them from the engine. They will smell like fuel and may have a varnish on the ceramic insulator from heated fuel. The most common cause of this is inadequate spark, though a leaking fuel injector will also flood the plug. If the valves are burned or the compression rings are badly worn, the engine may lack adequate compression to make the air-fuel mixture combustible. A compression check will either verify or rule out that problem.

When a vehicle is burning oil, the spark plugs will develop tan to dark-tan deposits on and around the electrode. As the oil is heated during combustion, it bakes onto the plug, leaving a residue. You will have to rule out any other possible causes of oil consumption, such as a plugged PCV system, badly worn valve seals, or a faulty turbocharger, but this is usually

FIGURE 7-3 This spark plug is wearing nicely; it has no deposits and is a light tan color.

FIGURE 7-4 Notice the wide gap and rounded electrodes on these worn spark plugs.

Engine coolant temperature sensor

FIGURE 7-5 A faulty engine coolant temperature sensor can cause the engine to run rich.

a telltale sign that the engine is mechanically worn. If you find this problem during a routine service, it is important to warn the customer about the failing condition of his engine. If the oil is wet and thick on the plug, look for a cause of heavy oil leakage into the cylinder. On rare occasions a head gasket will crack between an oil passage and the cylinder, allowing oil into just one cylinder. This can cause an oil-soaked plug. Similarly, a hole blown in a piston can cover a plug with oil, but you will have other indicators of a serious mechanical failure.

Black-colored, soot-covered spark plugs are found when the air-fuel mixture is too rich. This means there is too much fuel and not enough air. This can be caused by something as simple as a plugged air filter or as involved in the powertrain control system as an inaccurate ECT sensor (Figure 7-5). An engine that does not have good compression can also create carbon because combustion will not be complete. The key is to notice the condition. Locate the cause before considering a maintenance service complete. Write down your observations if you are evaluating an engine for serious mechanical repairs.

Blistered or overheated spark plugs are a sign of potentially serious trouble in the combustion chamber. The porcelain insulator around the center electrode will actually have blisters in it, or there may be pieces of metal welded onto the electrode. It is possible that the wrong spark plugs were installed in the engine. Otherwise it is likely that pinging or detonation is occurring. If there is no engine damage yet, it is critical to find the cause of the problem before returning the vehicle to the customer. If you find this as you are diagnosing an engine mechanical failure, it is essential to record this information and find the source of the problem during your repair work. A cooling system passage with deposits in it could be causing one corner of a combustion chamber to be getting so hot that detonation occurs. If you were to replace the damaged piston and rings but not repair the cause of the problem, the customer would be back soon with a repeat failure. Your thorough evaluation of the engine and analysis of the causes of the symptoms will ensure customer satisfaction.

POWER BALANCE TESTING

On a rough-running or misfiring engine, you can perform a power balance test to identify a cylinder that is not equally contributing power to the engine. To perform the test, you disable one cylinder at a time while noting the change in engine rpm and idling condition. The more the rpm drops or the rougher the engine runs, the more the cylinder is contributing. If a cylinder is producing little or no power, the change in rpm and engine running will barely be noticeable. You must note the rpm drop quickly as the cylinder is initially disabled; today's PCMs will boost the idle almost immediately to compensate for the drop. You will also be able to feel and see the roughness of the engine as a functional cylinder is disabled. We'll discuss a few different methods of performing a power balance test.

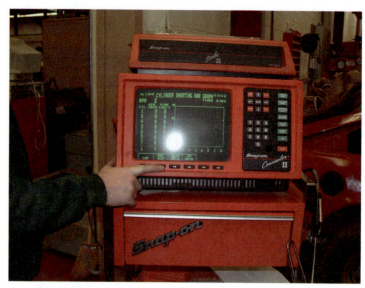

FIGURE 7-6 This engine analyzer shorts one cylinder at a time and displays the engine rpm drop.

Power Balance Testing Using an Engine Analyzer or Scan Tool

Many engine analyzers can perform a manual or automated power balance or cylinder efficiency test. With the analyzer leads connected to the ignition system as instructed in the user's manual, select the power balance test from the menu (Figure 7-6). The tester will automatically disable the ignition to one cylinder at a time for just a few seconds and display the rpm drop. The analyzer disables spark in the firing order; an engine with the firing order 1-3-4-2 would be tested on cylinder number 1 first, then number 3, number 4, and number 2. On some analyzers, the power balance results will be displayed in bar-graph format for easy comparison of cylinder contribution.

On some vehicles, the PCM is programmed to be able to perform a power balance test. The manufacturers and some aftermarket scan tools will be able to access this test. The only hookup required is to the diagnostic link connector (DLC). Then select the power balance test from the menu system. At the prompt, the scan tool will communicate with the PCM to initiate its disabling of cylinders one by one. The rpm drop for each cylinder will be displayed after the test.

Manual Power Balance Testing

Even without an engine analyzer or a capable scan tool, you can perform a power balance test. The procedure will require that you manually disable either fuel or spark to each cylinder. It is also important that you check the vehicle for diagnostic trouble codes (DTCs) after performing this test and clear the codes if any are present. You can access DTCs using a scan tool; the code can help you locate the source of the problem. If the PCM detects a misfire, it is likely to set a DTC. When the malfunction indicator light comes on, a DTC has been set. Using the menu on a scan tool, you can erase any DTCs when your testing is complete.

On vehicles that use spark plug wires to deliver the spark to the plug, you can short the spark to ground on each cylinder in turn to test the power contribution of cylinders. One way to do this is to use an insulated pair of pliers (Figure 7-7). Carefully pull the wire off the plug and hold the wire to the block or head to allow it to find a ground path. It is important that the spark can find an easy path to ground; do not hold the wire more than a half an inch away from a good metal conductor. The coil can be damaged if it uses all its possible power to find a path to ground. Sometimes the spark will track through the insulation of the

CAUTION:
You must be careful not to disable spark for long periods of time and must allow the engine to run on all cylinders for one minute between tests. Failure to follow these guidelines can overload the catalytic converter with raw fuel and damage it from overheating.

FIGURE 7-7 Use insulated pliers to remove a live plug wire. Ground it to a metal component immediately to prevent harming the coil.

FIGURE 7-8 Disconnect the low voltage primary wiring to the individual coils and record rpm drop. Be sure to clear any DTCs after your testing.

coil and destroy it. As you remove the spark plug wire, you should notice a clear change in rpm and idle quality if that cylinder is contributing adequate power.

Many newer engines use coil-on-plug ignition systems. Each spark plug has its own coil and you cannot safely remove the coil and ground it. Instead, locate the connector going into the coil (Figure 7-8). While the engine is running, you can briefly remove the connector to prevent the coil from firing. As you unplug the coil, monitor the rpm drop on a tachometer. Repeat this for each cylinder and compare your results. This testing is very likely to trigger the MIL and set a diagnostic trouble code, so be sure to clear the DTCs after your testing.

Another way to perform a manual power balance test is to unplug the connectors to the fuel injectors one by one while listening to and watching the engine rpm. By preventing fuel from entering the cylinder, you can momentarily disable the cylinder. This allows you to note how much the cylinder is contributing to the overall power of the engine.

Analyzing the Power Balance Test Results

When an engine is running properly, each cylinder should cause very close to the same rpm drop when disabled. Cylinders that are lower by just 50 rpm should be analyzed. Look at the results below:

Cyl. #1	Cyl. #3	Cyl. #4	Cyl. #2
125rpm	150rpm	50rpm	125rpm

Cylinder number four is definitely not contributing equally to the power of the engine. The rpm changed very little when it was disabled, meaning that the engine speed is barely affected by that cylinder. There could be many causes of the problem, but you now know which cylinder to investigate. The problems could relate to fuel delivery; perhaps the fuel injector was not delivering adequate fuel to support combustion. An ignition system problem, something as simple as a spark plug with a very wide gap from worn electrodes, could also cause this. A cylinder with low compression will also produce less power and be suspect during a power balance test. Compression testing and cylinder leakage testing will help you determine whether the weak cylinder has an engine mechanical problem. They can also provide information about what the specific cause may be.

ENGINE SMOKING

Engine smoking can indicate a very serious engine problem. It may also be caused by a relatively minor problem. This makes your correct diagnosis essential. Some likely causes of engine smoking and methods of diagnosis are described next.

> **CUSTOMER CARE:** A customer came in distraught, convinced that the "engine is blown" because it started blowing blue smoke out of the tailpipe. Careful checking of the specific times when the smoking occurred pointed to worn valve seals. Less than $200 later, the customer was relieved to pick up his fine-running vehicle (Figure 7-9).

Blue Smoke

There are several possible causes of blue smoke:

- Worn rings or cylinders
- A plugged PCV system
- Worn valve seals or guides
- A blown turbocharger
- A leaking head gasket

Obviously the difference in parts and labor to repair a PCV system, as opposed to replacing faulty rings, is dramatic. Be very careful to check each of these possible causes before disassembling or replacing an engine.

A clogged or frozen PCV valve or hose can cause a tremendous amount of blue smoke to pour out of the tailpipe at all times when the engine is running. The more you accelerate, the greater the smoking. Shake the PCV valve; it should rattle (Figure 7-10). Replace it if there is any possibility that it could be faulty. Blow through the larger diameter hose attached to the PCV valve that connects to the air filter housing. If it is clogged, clean it thoroughly, and road test the vehicle again to confirm your repair.

Worn valve seals will cause smoking at predictable times. At startup after the engine has been sitting for a period, you will notice a cloud of blue smoke. This occurs because when the engine is shut off the oil sitting under the valve cover can leak past the worn seals into the combustion chamber. When the engine starts, it burns this oil that produces blue smoke. It will clear up significantly after running for a few minutes. Once the engine is running, the oil is constantly circulating through the engine and not pooling up above the seals,

FIGURE 7-9 Replacing these worn valve seals cured the engine smoking concern.

FIGURE 7-10 Shake the PCV valve to be sure it rattles and blow through the hose to be sure it is clear.

so the smoke will dissipate. Also, when the engine is at idle or decelerating for a period of time, the engine vacuum is high. This can draw oil into the combustion chamber past the weak seals. When you accelerate after idling or decelerating, you may notice a puff or cloud of blue smoke. Valve seals can be replaced relatively inexpensively; the cylinder head does not need to be removed (Figure 7-11).

A turbocharger with blown seals can produce huge clouds of smoke on acceleration. This could easily be confused with worn rings, but an engine overhaul would not cure the problem. Whenever a turbocharged engine is smoking, check the turbo seals first. Remove the boot on the intake side of the compressor, and look for oil. If plenty of oil is sitting in the turbo housing and the boot, replace the faulty turbocharger. Shake the impeller up and down to help confirm this diagnosis. Usually it is wear in the bearings that allows the shaft to rock and destroy the turbo seals.

While this is not a common problem, it is possible for a head gasket to split by an oil passage and allow oil into the combustion chamber. Check the spark plugs; if one is drenched in oil, literally dripping, suspect a blown head gasket. Perform a compression test. If the readings are normal or just slightly low, they point to a head gasket. If the rings were badly worn, you would usually get low compression results. Look very closely at the head gasket once the head is off to be sure that the head gasket is the cause of the blue smoke. You should be able to see a crack or path in the gasket from an oil passage into the combustion chamber.

Worn rings are a common cause of oil consumption. The engine may smoke all the time, but it will usually emit more blue smoke while accelerating. Perform a cylinder leakage test to confirm your suspicions. When the rings are badly worn, you will need to rebuild or replace the engine. Fully evaluate the condition of the engine, including valve seating, oil pressure, noises, cylinder leakage, and oil consumption. This will help you determine whether it would be more cost effective to rebuild an engine or replace it with a rebuilt unit. If the engine has valve problems, knocking noises, and excessive cylinder leakage, for example, it will likely be more efficient and inexpensive to replace the engine. In other cases, you may conclude that a set of rings and engine bearings will cure the problems and repairing the engine will be the best option.

The combustion chamber can produce smoke with confusing different colors. Blue smoke typically indicates oil burning. White smoke typically indicates burning coolant (steam). Black smoke indicates a rich fuel mixture.

SERVICE TIP:
Many manufacturers are having problems with head gasket sealing. Sometimes replacing the gasket *is* performing a complete repair.

SERVICE TIP:
Check the manufacturers' service information, including technical service bulletins (TSBs). TSBs are bulletins that the manufacturer publishes when it has a common problem with its vehicles and has identified an effective correction. Some manufacturers are recommending that a liquid "stop leak" chemical be added to the coolant to try to seal the head gasket rather than replacing the gasket. Use only the chemical recommended by the manufacturer.

FIGURE 7-11 Replacing valve seals with the head installed.

White Smoke

When clouds of white smoke are pouring out of the tailpipe, it usually indicates that the head gasket is blown; otherwise, there is a significant crack in the block or head. The leak is most commonly past the head gasket, but a crack in the cylinder head or block will also cause coolant consumption. A distinct puff or small cloud of white smoke at startup, coupled with rough running, may be the result of a small crack in the head gasket or the cylinder head. A faulty head gasket or a cracked cylinder head or block can cause a number of different symptoms. The most obvious sign of one of these problems is excessive white, sweet-smelling smoke coming from the tailpipe. Other times the symptoms may include: erratic heat and engine temperature; consistent or intermittent overheating; rough running; misfiring, especially at startup; excessive pressure in the cooling system; coolant overflow; coolant loss without an external leak; low compression on two adjacent cylinders; and coolant mixed in the oil, creating a fluid in the lubrication system that closely resembles a coffee milkshake (its lubricating quality is about as effective as a coffee milkshake too). Prompt repair of a vehicle exhibiting any of these symptoms is essential, to prevent more severe engine damage. Overheating typically causes failures of the head gasket, head, or block. Do not simply replace the faulty part. Thoroughly evaluate the cooling system to find the cause of overheating before calling the job complete.

These failures will very often cause the telltale symptom of white smoke blowing out of the tailpipe as coolant leaks into the combustion chamber and burns. You may have witnessed a head gasket failing while you were driving; all of a sudden, a cloud of white smoke comes from the tailpipe of the unlucky person's vehicle.

Occasionally, the crack in the head gasket, head, or block is very small and will cause less-dramatic symptoms. When the engine is cool, the metals in either of the suspect

FIGURE 7-12 Pressurize the cooling system and then check for leaks into the combustion chambers.

parts contract and enlarge the crack. Coolant may leak into the affected cylinder and foul the spark plug or cause small amounts of white smoke to exit the tailpipe. Symptoms may occur for a short time after startup; then the engine heat can cause enough expansion in the metal to seal the gap. To confirm your diagnosis, remove the spark plugs, allow the engine to cool *thoroughly,* and place rated system pressure on the coolant using a pressure tester (Figure 7-12). Let the vehicle sit for as long as is practically possible, but at least one half hour. Disable the fuel system; have an assistant crank the engine over and watch closely for coolant spray out of a cylinder. Sometimes it is helpful to place a paper towel in front of each plug hole, so you are able to check each cylinder. Any coolant from a cylinder indicates a sealing failure. If you are uncertain of the result from a cylinder, perform a cylinder leakage test to confirm your suspicion.

Combustion gases leaking into the cooling system may cause erratic temperatures, overheating issues, and problems of excess pressure and boiling over. The gases can form air pockets in the engine, preventing the thermostat from operating properly. When the leak is larger, the gases produce enough pressure in the system to force the cap to relieve pressure and coolant with it. One way to check for the presence of exhaust gases in the cooling system is to install a pressure tester on the filler neck. Do not pump any pressure on the tester. If compression pressure is leaking into the coolant, the pressure on the tester gauge will rise rapidly. Sometimes the leak is large enough to diagnose by observation. Be certain that the engine is cool enough to run without the pressure cap on for a few minutes. Remove the cap, and start the engine. If engine coolant bubbles violently and sprays out the top of the filler, then the head gasket, or head or block, has failed. When the problem is not this severe, use a **chemical block tester** to confirm your suspicions of leakage (Figure 7-13). This tester uses a chemical solution that changes color when exposed to combustion gases. Remove the pressure cap and start the engine. Place the tester on the top of the reservoir or radiator. Use the bulb to pull air (not coolant) from the top of the fluid. Pump the bulb several times with the engine running, while watching the color of the liquid in the chamber. If it changes color, usually from blue to yellow, you have diagnosed a head gasket, head, or block leak.

On an older vehicle with an automatic transmission, it is possible that a leaky vacuum modulator could cause these symptoms. On transmissions with a vacuum modulator, it is possible for the diaphragm inside to rip. This allows the vacuum line attached from the

CAUTION:

If the engine does not crank over at all or at a normal rate of speed, stop your testing because the engine could be hydrolocked. This means that a cylinder(s) is so full of coolant that the engine cannot displace the coolant; nor can it compress it. Putting a battery charger on the system and continuing to crank could crack a piston, cylinder head, or even the block. Removing all the spark plugs will minimize the risk of this during testing.

A **chemical block tester** uses a chemical that reacts to combustion gases. When air is passed across the solution from on top of the radiator, no combustion gases should be present. If they are, a problem with the head or head gasket is indicated.

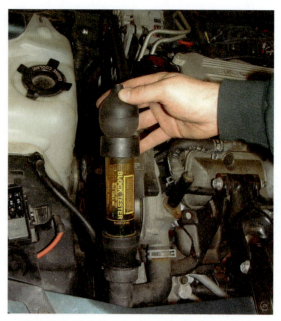

FIGURE 7-13 Pull gases from above the coolant through the chemical in the block tester.

intake manifold to the modulator to pull in high quantities of automatic transmission fluid (ATF). The ATF introduced into the intake manifold is then burned in the cylinders and can also cause excessive white smoke from the tailpipe. A simple test to eliminate the modulator as the problem is to remove the vacuum line from the modulator and check to be sure there is no ATF present in the line. If the modulator is faulty, the smoking will stop with the vacuum line disconnected.

When coolant has gotten into the combustion chamber, you must also check the condition of the oil. Coolant that mixes with the oil produces a liquid very similar to a coffee milkshake. Its lubricating qualities are less effective than a good frappe. The very thick mixture of oil and coolant will block oil passages and quickly destroy the engine bearings. If the oil has been contaminated with coolant, warn the customer that major bottom-end damage may have occurred (Figure 7-14). If the engine has been run for any length of time with this "lubrication," you will

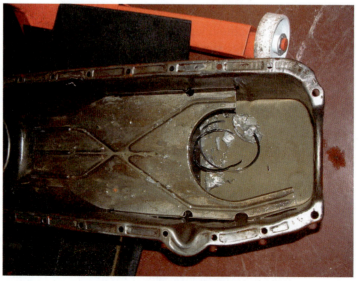

FIGURE 7-14 This coolant contaminated oil wrecked havoc in this engine.

likely hear knocking from the bottom end of the engine. When the engine has high mileage, it may be advisable to check the price and availability of a rebuilt engine to repair the extensive engine damage this problem can cause.

Black Smoke

An engine running too rich causes black smoke. Eliminate engine breathing problems first. Check the air filter and the exhaust system for restrictions. You may also need to check the fuel delivery system for faults. Possible problems include a faulty mass airflow sensor or **fuel pressure regulator,** a leaking or stuck-open injector(s), an inaccurate engine coolant temperature sensor or **oxygen sensor (O_2S)** (Figure 7-15) or faulty engine control wiring. You can use a scan tool to help verify proper operation of the sensors. Follow the manufacturer's diagnostic troubleshooting chart to locate the cause of the problem. A cylinder(s) with low compression can cause less air to be pulled into the cylinder and for the air-fuel mixture to be less combustible. This can result in imperfect combustion and the emission of black smoke from the tailpipe. Check compression to verify that the engine is mechanically sound. The catalytic converter suffers from rich running and could become partially restricted or plugged. It could also trigger the MIL and set a DTC for low-catalyst efficiency on OBD II vehicles.

Oil Consumption Diagnosis

Oil consumption can vary from engine to engine greatly. Oil consumption is defined as oil burning in the combustion chamber. Remember that some oil will normally burn in the combustion chamber due to the design and tolerances of the piston and rings. Some manufacturers use higher quality piston rings and allow for less tolerance and clearance in the combustion chamber, while others do not.

Oil consumption is considered excessive when it affects the customer's ability to drive the car normally in-between oil changes or when it surpasses the manufacturers' expectations. Oil consumption has many variables that can be controlled. Most manufacturers will provide a technical service bulletin that describes how to evaluate if a vehicle is consuming

A **fuel pressure regulator** is typically a mechanical device that maintains the correct fuel pressure in the fuel rail the injectors sit in.

An **oxygen sensor** sits in the exhaust stream and measures the amount of oxygen present in the exhaust. One use of this sensor is to help the computer to determine whether the engine is running rich or lean.

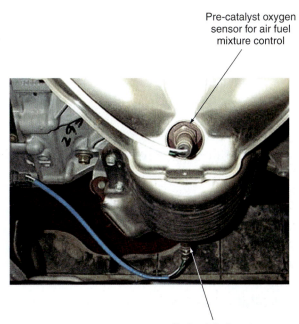

Pre-catalyst oxygen sensor for air fuel mixture control

Post-catalyst oxygen sensor for testing converter efficiency

FIGURE 7-15 **This engine uses an oxygen sensor before the catalytic converter to adjust the air-fuel ratio and another one after the converter to evaluate the converter's efficiency.**

oil excessively or not. It is sometimes normal for some vehicles to consume 1 quart of oil in 2,000 miles. Some of the following items are conditions that may help a technician indicate when it is excessive oil consumption:

- Improper SAE oil viscosity for the temperature range used.
- Improperly read oil level indicator.
  - Check the oil while the car is standing on a level surface.
  - Allow adequate drain-back time.
- Continuous high-speed driving.
- Broken, improperly installed, worn, or unseated piston rings.
- Piston improperly installed or improperly fitted.
- Heavy trailer towing (overloading) and excessive engine acceleration.
- Continuous driving in the incorrect transmission range.
- Malfunctioning PCV (positive crankcase ventilation) system.
- Worn valve stem seals and/or valve guides.
- Plugged cylinder head gasket oil drain holes.

IMPROPER COMBUSTION

Improper combustion can cause serious engine problems if it is left unrepaired. It can also be the cause of serious engine damage, and it will be absolutely critical that you recognize the problem during your engine overhaul. **Preignition** can cause a loss of power, but left uncontrolled it can lead to more damaging engine **detonation.** Detonation causes piston and ring damage, bent connecting rods and worn bearings, top ring groove wear, blown head gaskets, and possibly complete engine failure. Misfire will cause power loss, contamination of the engine oil, and catalytic converter failure.

Common causes of preignition and detonation are:

- Deposits in the cooling system around the combustion chamber
- Engine overheating
- Too hot a spark plug
- An edge of metal or gasket hanging into the combustion chamber
- Fuel with too low an octane rating
- A faulty exhaust gas recirculation (EGR) system
- Improper ignition timing
- Lean air-fuel mixtures (when there is less-than-desired fuel mixed with air)
- Carbon buildup in the combustion chamber
- A faulty **knock sensor**
- Excessive boost pressure from a turbocharger or supercharger

Each of these areas should be investigated when a concern of preignition or detonation exists.

> **CUSTOMER CARE: Communication with the customer is essential to excellent repair work. In the case of engine preignition or detonation, it is very important to discuss any recent changes in driving habits or repair work with the customer. Did the customer recently change fuel brand or octane level? Did he just have his spark plugs changed? Several questions may lead you to a simple answer to a problem that is sometimes difficult to diagnose.**

Check the engine for diagnostic trouble codes. Often the PCM will offer a hint about a problem. If the EGR system is not functioning, for example, combustion will be too hot and could easily lead to detonation. Evaluate the cooling system to be sure it is operating properly (Figure 7-16). If the engine is turbocharged or supercharged, verify that boost pressures are being controlled properly. Verify that the proper spark plugs are installed in the engine.

Preignition means that a flame starts in the combustion chamber before the spark plug ignition.

Detonation is a dangerous explosion that occurs when two flame fronts collide in the combustion chamber.

A **knock sensor (KS)** is bolted into the engine and senses engine vibrations that resemble knock. It relays this information to the PCM so it can adjust ignition timing to try to reduce knocking.

FIGURE 7-16 You can monitor areas of the engine with an infrared pyrometer.

MIL light

FIGURE 7-17 The malfunction indicator light (MIL) will come on when the PCM has detected a problem that could allow emissions to increase; this certainly includes misfire.

An air-fuel mixture that is too rich or too lean, a faulty ignition system component, a faulty fuel injector, or inadequate cylinder compression can cause engine misfire. Most OBD II vehicles will identify an engine misfire and illuminate the malfunction indicator light and set a DTC (Figure 7-17). Use a scan tool to identify the cylinder(s) whenever possible. You may also use a power balance test or a compression test to identify the faulty cylinder(s). Mechanical failures such as weak compression, burned valves, engine vacuum leaks, worn lifters, or rocker arms can all cause an engine misfire.

COMPRESSION TESTING

A **compression test** is an excellent way to determine if an engine has serious mechanical problems. When an engine is running poorly, it is important to establish that the mechanical function of the engine is sound before delving into testing the myriad of systems and components that could also cause driveability issues. Many technicians will perform a compression test on an engine that is running quite poorly before performing a major maintenance service. It is important to perform a compression test before undertaking any major engine work. You should know whether you will be performing work on the cylinder head and valves only or if the entire engine needs to be overhauled or replaced. A valve job, replacing

A cranking **compression test** measures the compression pressure in a cylinder while the engine is cranking. A running test checks compression with the engine running.

or machining the valves and seats to reestablish proper sealing, may cost upwards of $1,000 on many engines. A complete engine overhaul of the head and block or installing a crate engine is likely to set the customer back at least twice that amount. Compression testing can help provide you and the customer with information to make an appropriate repair choice.

Cranking Compression Test

A cranking compression test is the most commonly used type of compression testing. It can clearly determine if the valves, rings, and head gasket(s) are sealing the combustion chamber properly (Figure 7-18). When any of those components are faulty, the engine may not be able to develop adequate compression. The manufacturer will provide a specification for the appropriate cranking compression pressure. Low compression will cause less productive combustion. You and the customer will notice that the engine lacks power or misfires on a cylinder(s).

To perform a cranking compression test: (Photo Sequence 12)

1. Let the engine warm up to normal operating temperature. (This assures that the pistons have expanded to properly fit and seal the cylinder.)
2. Let the engine cool for ten minutes so you don't strip any threads; then remove the spark plugs. Remember to keep them in order so you can read them.
3. Block the throttle open at least part way open so the engine can breathe.
4. Disable the fuel system by removing the fuel pump fuse or relay to prevent contamination of the engine oil.
5. Disable the ignition system so you do not damage the coils. You can either remove the ignition fuse or remove the low-voltage connector to the coil(s). Do not simply remove the plug or coil wires and let them dangle. The coil can be damaged as it puts out maximum voltage trying to find a ground path for the wires.
6. Install a remote starter or have an assistant help you.
7. Carefully thread the compression tester hose into the spark plug hole and attach the pressure gauge to the adaptor hose (Figure 7-19).
8. Crank the engine over through five full compression cycles. Each cycle will produce a puffing sound as engine compression escapes through the spark plug ports.
9. Record the final pressure.
10. Repeat the test on each of the other cylinders.

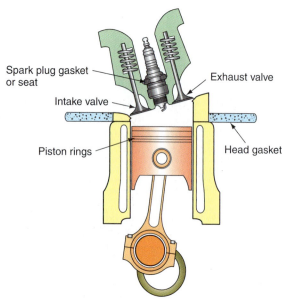

FIGURE 7-18 A cranking compression test will detect leaks at the valves, rings, pistons and head gasket.

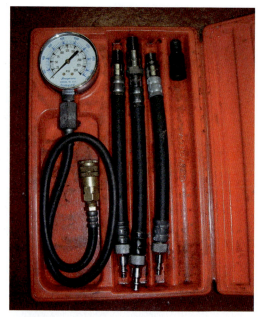

FIGURE 7-19 This style compression tester has adaptors that thread into the spark plug hole.

TYPICAL PROCEDURE FOR PERFORMING A CRANKING COMPRESSION TEST

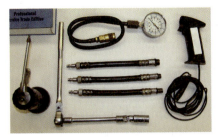

P12-1 Tools required to perform this task: compression tester and adapters, spark plug socket, ratchet, extensions, universal joint, remote starter button, battery charger, service manual, and squirt can of oil.

P12-2 Follow the service manual procedures for disabling the ignition and fuel systems.

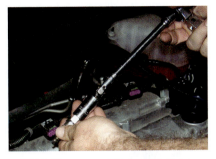

P12-3 Clean around the spark plugs with a low-pressure air gun and remove all of the spark plugs from the engine.

P12-4 Install the remote starter button.

P12-5 Carefully install the compression tester adapter into the first cylinder's spark plug hole.

P12-6 Connect the battery charger across the battery and adjust to maintain 13–14.5 volts. This is needed to allow the engine to crank at constant speeds throughout the test.

P12-7 With the throttle plate held or locked into the wide-open throttle position, crank the engine while observing the compression reading.

P12-8 Continue to rotate the engine through four compression strokes as indicated by the jumps of the needle.

P12-9 Record the reading after the fourth compression stroke. Continue until all cylinders have been tested, and compare the test results. Compare test results with the manufacturer's specifications.

Analyzing the Results

With all the pressure readings in front of you, compare them to the manufacturer's specifications. Typical compression pressures on a gas engine range between 125 and 200 psi. The readings should be close to the specification and within 15–20 percent of each other. Some manufacturers specify only that the compression be over 100 psi and that each reading is within 20 percent of the others. To calculate 20 percent easily, take the normal reading (let's say 160 psi) and drop the last digit (16 psi). Multiply by 2 (32 psi) and that is 20 percent. If the readings were 160, 150, 100, and 155, the third cylinder is more than 32 psi different than the other three. That indicates a significant problem with cylinder number three that must be identified and corrected.

If one or more cylinders are below specification, it is likely caused by worn rings, burned valves, or valves sticking open from carbon on the seats, a faulty head gasket, or a worn or broken piston. Perform a wet compression test to help find the probable cause. When all the cylinders are slightly low, it usually indicates a high mileage engine with worn rings. Two adjacent cylinders with low compression readings may indicate a blown head gasket leaking compression between two cylinders. If all the cylinders are near zero or very low, suspect improper valve timing. A timing belt, chain, or gears that have "jumped" time by slipping a tooth will allow the valves to be open at the wrong time, causing low compression readings.

WET COMPRESSION TEST

To help determine the cause of a weak cylinder, perform a wet compression test (Figure 7-20). Use an oil can and put two squirts of oil into the weak cylinder. Reinstall the compression gauge, crank the engine over five times, and record the reading. If the low reading cylinder increases to almost normal compression pressure it is likely worn rings that are causing the cylinder leakage. The extra oil in the cylinder helps the rings to seal better during the test. If a valve is burned or being held open by carbon or if there is a hole in the piston, a film of oil will not dramatically improve the compression pressure. Note that a wet compression test may not be as effective on horizontally opposed engines because the oil will only seal the lower half of the cylinder wall.

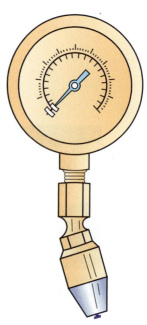

FIGURE 7-20 **Hold this compression gauge firmly against the spark plug hole to measure compression.**

Look at the examples that follow:

Compression test results

Cyl. #1	Cyl. #2	Cyl. #3	Cyl. #4	Cyl. #5	Cyl. #6
175 psi	165 psi	170 psi	80 psi	170 psi	170 psi

Wet compression test

Cyl. #4

160 psi

This is a clear indication that the rings are severely worn on cylinder number 4. If the wet compression test had shown a rise to only 105 psi, you would suspect a burned valve or leaking valve. To definitively determine the cause of low cranking compression, perform a cylinder leakage test (Figure 7-21).

RUNNING COMPRESSION TEST

It is possible that the cranking compression shows good results but the engine is still not mechanically sound. A cranking compression test does not do a good job of testing the valvetrain action. There is so much time for the cylinder to fill with air when the engine is only spinning at cranking speeds of about 150–250 rpm. A more accurate way to test the loaded operation of the valvetrain is to perform a running compression test. Running compression readings will be quite low compared to the cranking compression pressures because there is so much less time to fill the cylinders with air.

If you have identified a weak cylinder through power balance testing, check that cylinder and a few others to gain comparative information. In most cases, it is worth the time to check all the cylinders. To perform the running compression test:

1. Remove the spark plug only from the test cylinder.
2. Disable the spark to that cylinder by grounding the plug wire or removing the primary wiring connector to the individual coil.

FIGURE 7-21 This cylinder showed low compression of only 100 psi while cranking dry and wet. A leakdown test will verify a leaking valve.

3. Disable the fuel to the test cylinder by disconnecting the fuel injector connector.
4. Install the compression gauge into the spark plug port.
5. Start the engine and let it idle.
6. Release the pressure from the gauge that reading reflects cranking compression. Record the new pressure that develops at idle.
7. Rev the engine to 2,500 rpm. Release the idle pressure, and record the new pressure that develops at 2,500 rpm.
8. Remove the gauge and reinstall the spark plug. Remember to reattach the plug wire or coil and the fuel injector connector.
9. Repeat for several or all cylinders.

Analyzing the Results

Generally, the running compression will be between 60 and 90 psi at idle and 30–60 psi at 2,500 rpm (Figure 7-22). The most important indication of problems however is how the readings compare to each other.

Look at the following readings:

	Cyl. #1	Cyl. #2	Cyl. #3	Cyl. #4
Idle	70 psi	75 psi	45 psi	70 psi
2,500 rpm	40 psi	40 psi	20 psi	45 psi

Cylinder number 3 is barely below the general specifications, but it is clearly weaker than the other cylinders. These results warrant investigation of the problem in cylinder number 3. The most likely causes of low running compression readings are:

- Worn camshaft lobes
- Faulty valve lifters
- Excessive carbon buildup on the back of intake valves (restricting air flow) (Figure 7-23)
- Broken valve springs
- Worn valve guides
- Bent pushrods

FIGURE 7-22 The same cylinder showed low running compression at idle.

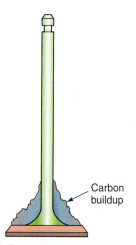

Carbon buildup

FIGURE 7-23 Excessive carbon buildup on the back of intake valves caused by a bad valve seal or guide leakage prevents the engine from pulling in enough air to make adequate power. The added weight can also cause the valve to float at higher rpms.

Use a stethoscope, vacuum testing, and visual inspection under the valve covers to help locate the reason for the low reading.

CYLINDER LEAKAGE TESTING

Once you have identified a cylinder(s) with low cranking compression, you can use a **cylinder leakage test** to pinpoint the cause of the problem. To perform a cylinder leakage test, you put pressurized air in the weak cylinder through the spark plug hole when the valves are closed. You can then listen for leakage of air from the combustion chamber at various engine ports to determine the source of the leak. This is an excellent way to gather more information before disassembling the engine or when deciding whether a replacement engine should be installed. A cylinder leakage test is also a very thorough way of evaluating an engine when someone is considering purchasing a vehicle. You can even perform a cylinder leakage test on a junkyard engine to assess its value before purchasing it.

> **CUSTOMER CARE:** A customer was anxious to buy a good-looking, performance-modified Honda. He was finally convinced to allow us to perform a cylinder leakage test after we noticed a bit of blue smoke coming from the tailpipe. A leakdown test showed very serious ring wear on two cylinders. While he still bought the car, he paid $1,500 less for it than he had planned.

Cylinder Leakage Test

The cylinder leakage test can help you evaluate the severity of an engine mechanical problem and pinpoint the cause. You can test one low cylinder to find a problem, or you can test each cylinder to gauge the condition of the engine. Like a compression test, the cylinder leakage test will detect leaks in the combustion chamber from the valves, the rings, or pistons and the head gasket. From this test, however, you can clearly identify which is the cause of the leak.

To perform a cylinder leakage test:

1. Let the engine warm up to normal operating temperature.
2. Let the engine cool for 10 minutes, and then remove the spark plug from the cylinder to be tested.
3. Remove the radiator cap and the PCV valve. This prevents damage to the radiator or engine seals if excessive pressure leaks into the radiator or the engine's crankcase.
4. Turn the engine over by hand until the cylinder is at TDC on the compression stroke. This assures that the valves will be closed and the rings will be at their highest point of travel. The cylinders wear the most at the top because of the heat and pressure of combustion, so if the rings are leaking, they will show the greatest evidence at the top. There are a few different methods of getting a cylinder to TDC compression, depending on the engine.
   a. Have an assistant turn the engine over slowly while watching the piston come up to top dead center. You should be able to feel and hear blowing out of the spark plug hole if it is coming up on the compression stroke. If you begin the leakage, test and show 100 percent leakage, you are on TDC exhaust. Rotate the engine 360 degrees and retest.
   b. Remove the valve cover, and watch the rockers or camshaft as you turn the engine over. On the intake stroke, the intake valve will open and then close. Then both the valves will remain closed while the piston comes up to TDC. If the engine has

A **cylinder leakage test** places air pressure into a combustion chamber so that you can listen for leaks from possible faulty components.

SERVICE TIP:
If the piston and connecting rod are not exactly at TDC, the engine may actually rotate as you apply air pressure. Inline engines need to be almost exactly at TDC, while vee engines can be a little off. If the engine starts to turn while you apply air pressure, try to move the cylinder closer to TDC. If you still cannot get the cylinder to stop moving, then try other things like holding the ratchet (watch your hand), using a piston position locater or checking the timing marks on the crankshaft pulley. Each vehicle is very different, but you may be able to use a bar in a location where it stops a pulley from moving.

TYPICAL PROCEDURE FOR PERFORMING A CYLINDER LEAKAGE TEST

P13-1 Loosen the ignition cassette to access the spark plugs.

P13-2 Disconnect power to the ignition cassette to avoid damaging the coils.

P13-3 Remove the ignition cassette.

P13-4 Remove the spark plugs.

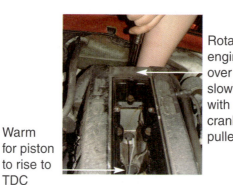

Warm for piston to rise to TDC

Rotate engine over slowly with crank pulley

P13-5 Rotate the engine to TDC compression for the test cylinder so that the piston is at TDC compression.

P13-6 Hook up shop pressure to the leakage tester and set to 0 percent leakage.

P13-7 Connect adaptor to test cylinder and read leakage. This cylinder is leaking slightly over 30%.

P13-8 Listen for air at the intake manifold for leakage past an intake valve.

P13-9 Listen for air at the oil fill to detect leakage past the rings. In this case you could feel the air blowing out the fill pipe.

TYPICAL PROCEDURE FOR PERFORMING A CYLINDER LEAKAGE TEST

P13-10 Listen for air in the coolant reservoir for indications of a leaking head-gasket.

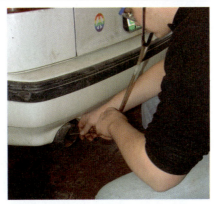

P13-11 Listen for air at the tailpipe to find leakage past an exhaust valve.

P13-12 The moderate blue smoke from the tailpipe substantiates the diagnosis of worn rings on this high mileage vehicle.

rocker arms, move them up and down to be sure they both have lash (clearance) between the arm and the pushrod. If the engine has an overhead camshaft(s), the followers will be on the base circle not the lobes of the camshaft.

c. If the vehicle has a distributor, remove the cap and rotate the engine until the rotor points at the test cylinder's firing point.

5. With the cylinder at TDC compression, thread the cylinder leakage tester adaptor hose into the spark plug hole.

6. Apply air pressure to the leakage tester and calibrate it according to the equipment instructions. On the tester shown, you turn the adjusting knob until the gauge reads 0 percent leakage without it attached to the cylinder.

7. Connect the leakage tester to the adaptor hose in the cylinder while watching the crankshaft pulley. If the engine rotates at all, reset it to TDC. When the engine is right at TDC, it will not rotate with air pressure applied.

8. Read the percentage of leakage on the tester and gauge as follows:
   - Up to 10 percent leakage is excellent; the engine is in fine condition.
   - Up to 20 percent leakage is acceptable; the engine is showing some wear but should still provide reliable service.
   - Up to 30 percent leakage is borderline; the engine has distinct wear but may perform reasonably well.
   - Over 30 percent leakage shows a significant concern that warrants repair.

9. With the air pressure still applied, use a stethoscope and listen at the following ports:
   - At the tailpipe: Leakage here indicates a burned or leaking exhaust valve.
   - At the throttle or a vacuum port on the intake manifold: Leakage here indicates a burned or leaking intake valve.
   - At the radiator or reservoir cap: Leakage here proves that the head gasket is leaking.
   - At the oil fill cap or dipstick tube: Leakage here indicates worn rings.

   If the leakage at the oil fill is near 100 percent suspect damage to a piston (Figure 7-24).

10. Release the pressure; remove the adaptor, and rotate the engine over to the next test cylinder. Repeat as indicated. . Refer to Photo Sequence 13 for detailed pictures and directions on performing a cylinder leakage test.

SERVICE TIP:
There will always be some leakage past the rings; a light noise is to be expected. Excessive ring wear will cause distinct blowing out of the cap or dipstick hole. You will be able to hear the noise and feel the pressure.

FIGURE 7-24 This destroyed piston showed 100 percent leakage.

CUSTOMER CARE: While a cylinder leakage test can identify the primary cause of low compression readings, remember to keep your mind and eyes open to other problems when repairing the engine. Do not automatically inform a customer that an exhaust valve is leaking and a valve job will cure the problem. Consider the engine mileage, oil condition, pressure and usage, and degree of leakage past the rings. If the engine has high miles, it is very likely that the rings are significantly worn. By replacing the faulty valve(s) and restoring the others to like-new condition, you will increase the compression and combustion pressures. The change may be significant enough that the old rings will not seal as well as they had been before the valve job. In this case, the customer would be back complaining about blue smoke from oil consumption or about the engine still lacking adequate power. This may very well be an example of when a replacement engine would be the most cost-effective repair for the customer. Evaluate the whole situation, and make your recommendation to the customer. If you offer the customer repair options, be very clear about explaining the possible consequences.

ENGINE NOISE DIAGNOSIS

Irregular engine noises may be caused by major or minor engine problems. Your job is to diagnose the noise as accurately as possible to be able to repair the engine in the most effective and efficient way. Low oil pressure can cause or increase the level of engine noises, so check that the engine has adequate oil and pressure before drawing conclusions. In many cases, you should consider changing the oil before performing an engine noise diagnosis. This allows you to use the correct oil and filter. Some bottom-end noises clearly indicate an engine that requires a thorough overhaul or replacement. Other noises may be resolved by much less drastic measures. While it is difficult to describe noises on paper, a few guidelines should help you narrow down the possible problem(s). A deep knock within the engine will require engine disassembly or replacement.

SPECIAL TOOLS

Stethoscope

If you disassemble the engine, be sure to look for all potential problems. Noise diagnosis should help you identify problems but not limit your investigation once the engine is apart.

It is important to remember the engine's operation, how the various engine components are manufactured, and how much each one might weigh. These items may help clue you into what component is making noise and what has to be done to repair it. For example, connecting rods and the crankshaft all rotate at the same speed and are located in the lower part of the engine. Valvetrain components rotate half as fast, weigh less, and are located in the upper part of the engine. Valvetrain components will also have a different frequency because they rotate at half the speed of the crankshaft. This is important because as the engine speed increases, the difference in frequency will be greater. Body and exhaust components can also create noises that will echo under the hood and seem like an engine noise. Make sure to check these and eliminate them before going too far.

Bottom-End Knock

A loud knocking noise from the bottom end of the engine, a **bottom-end knock** portends imminent failure of the engine. This is a deep knocking sound caused by excessive clearance between the main or rod bearings and the crankshaft (Figure 7-25). Place a stethoscope on the block just above the oil pan, and listen while the engine idles and when you open the throttle. A loud rapping noise indicates excessive bearing wear. Check the oil pressure at idle; low readings confirm your diagnosis. Excess bearing clearances allow oil to leak through the bearings, lowering pressures and allowing the crankshaft journals to crash against the bearing face without a soft cushion of oil. Engine knocking that is repaired promptly may require only engine bearing and ring replacement. If the engine is allowed to run for extended periods of time with worn bearings, the crankshaft and connecting rods will likely be damaged. This will often warrant a replacement engine rather than an overhaul. Be certain that the flexplate or flywheel is not cracked or loose and that the catalytic converter is not rattling; these problems can cause knocking that sounds very much like a bearing knock. Use your stethoscope on the **bell housing** and on the catalytic converter to clearly identify the source of the noise.

Crankshaft End Play

When an engine knocks or clunks loudly on acceleration, suspect excessive **crankshaft end play.** The repair requires replacement of the engine main bearings, but while you are disassembling the engine, you must pay particular attention to the crankshaft and block. If the crankshaft has been slamming against the block without the benefit of a cushion from the **thrust bearing,** significant damage may have occurred. The thrust bearing is designed

A **bottom-end knock** is a loud, deep knocking noise heard from the lower end of the engine. It is usually an indication of serious engine problems, such as worn main or rod bearings or crankshaft, or low oil pressure.

The **bell housing** is the portion of the transmission that bolts to the back of the engine.

A small amount of excessive **crankshaft end play** is required. Excessive end play is caused by bearing wear and can cause serious engine damage.

FIGURE 7-25 These rod journals were destroyed when a bearing spun; the knocking sounded as serious as the fault.

FIGURE 7-26 This thrust bearing was worn down to the copper causing a clunking on acceleration. It did its job of protecting the crank and block; a new set of bearings cured the problem.

to minimize forward and backward movement of the crankshaft to protect the crankshaft and block from wear, but look for scoring along the thrust surfaces of the block and crankshaft (Figure 7-26). Minor damage can be corrected by a different thrust bearing, but serious damage requires component replacement. If left unrepaired, excessive crankshaft end play can result in the crankshaft breaking through the block.

Many other drivetrain problems can cause clunking on acceleration. Make sure that the noise is coming from the engine rather than the differential or a U-joint, for example. You can often feel excessive crankshaft end play with the engine installed. Put the engine on a lift and pry and push the crankshaft pulley forward and backward in the engine block. Movement and noise should be minimal. If the crankshaft moves significantly (more than a few thousandths of an inch) and you can hear a knocking noise, the engine will require repairs.

Piston Pin Knock

When a piston pin develops clearance within the piston pin bore, you will hear a higher pitch, double-time knock at the top of the block. The noise is deeper than **valvetrain clatter** but not as deep as bottom-end knocking. Place a stethoscope near the top of the block and listen for two knocks close together. The pin will knock as the rod pushes the piston up and then again as combustion forces the piston down. It is important to repair this problem immediately. Excessive pin clearance will often break the piston off the rod (Figure 7-27). This can cause extensive damage to the block and cylinder head.

Piston Slap

Piston slap occurs when there is too much clearance between the piston and the cylinder. It usually begins as a knocking noise when the engine is cold and goes away as the engine fully warms up. Prompt repair can prevent piston breakage and more serious engine damage. Once the piston is slapping continuously, even when the engine is warm the engine is at risk of a catastrophic failure. If you rebuild the engine, carefully measure the cylinder bore diameter, out-of-round, and taper. Often the repair for piston slap includes boring the cylinders out oversize and installing new pistons. In many cases, these extensive repairs will cost more than a crate engine, so a rebuilt engine assembly will be installed. Some manufacturers have been having problems with excessive piston slap on very low mileage engines. These engines are being replaced under warranty.

FIGURE 7-27 The piston let loose due to excessive pin wear. It caused extensive cylinder and head damage.

FIGURE 7-28 This badly worn lifter was causing serious valve train clatter. The other lifters were not much better.

Timing Chain Noise

A loose timing chain or worn gears will make a clacking or rattling noise, particularly when you allow the engine to decelerate after giving it some throttle. A chain will often slap on the timing chain cover when it wears. You will learn to identify this noise, but you can also verify it with a stethoscope on the timing cover. If the engine has gears, the noise will be less of a rattle and more of a knock. It will still be worst on deceleration.

Valvetrain Clatter

When there is excessive clearance between components in the valvetrain, you will hear a tinny clattering sound from the top of the engine. A common cause of valvetrain clatter is a collapsed hydraulic lifter. You may hear this in the morning, after a car has been sitting for an extended period, or after an oil change. The noise should go away within a few seconds. If the noise persists, the lifters are probably worn. Be sure that the engine is properly filled with clean oil before dismantling the engine to replace the lifters. Metallic clicking noises can also be caused by valves out of adjustment (too loose), a worn camshaft, worn lifters, or worn or broken rocker arms or rocker arm shafts. Remove the valve cover(s) to inspect the valvetrain components and look closely for signs of wear (Figure 7-28). Again, a stethoscope can help point you to the right area of the cylinder head or even to the affected cylinder.

Always check for TSBs when diagnosing engine noises.

CASE STUDY

A vehicle that was running very poorly with low power had been to two shops that tried unsuccessfully to locate the cause of the problem. A full maintenance service had been performed, as well as a new PCM, ECT, MAF, and turbocharger installed, to no avail! A cranking compression test showed adequate compression on each cylinder. After double-checking some of the previous work, the young technician at the new shop decided to perform a running compression test. The test showed very low readings on two cylinders. Careful inspection of the overhead camshaft showed well-worn lobes preventing adequate airflow into the engine. A new camshaft made the vehicle perform beautifully and made the frustrated customer very happy.

ASE-STYLE REVIEW QUESTIONS

1. A spark plug gap that is too wide can cause:
   A. Detonation
   B. Low compression
   C. Valvetrain clatter
   D. Misfire

2. *Technician A* says that black flaky carbon on a spark plug is caused by detonation.
   Technician B says that heavy tan deposits are caused by oil consumption.
   Who is correct?
   A. A only
   B. B only
   C. Both A and B
   D. Neither A nor B

3. *Technician A* says that some platinum spark plugs may last up to 100,000 miles.
   Technician B says that it is responsible to replace them at 60,000 miles.
   Who is correct?
   A. A only
   B. B only
   C. Both A and B
   D. Neither A nor B

4. One weak cylinder is found during a power balance test.
   Technician A says it could be caused by a plugged fuel filter.
   Technician B says it could be caused by a bad fuel injector.
   Who is correct?
   A. A only
   B. B only
   C. Both A and B
   D. Neither A nor B

5. *Technician A* says to check compression before disassembling an engine.
   Technician B recommends performing a compression test before completing a major maintenance service if the engine is running very badly.
   Who is correct?
   A. A only
   B. B only
   C. Both A and B
   D. Neither A nor B

6. *Technician A* says that cranking compression pressures are usually between 200 and 250 psi.
   Technician B says running compression pressures are lower than cranking compression pressures.
   Who is correct?
   A. A only
   B. B only
   C. Both A and B
   D. Neither A nor B

7. A compression test is performed on an engine that is misfiring. All the pressures are between 175 and 185 psi except cylinder number 5; it is 80 psi.
   Technician A says a valve seal could be bad.
   Technician B says the rings could be worn or broken.
   Who is correct?
   A. A only
   B. B only
   C. Both A and B
   D. Neither A nor B

8. An engine produced the following compression pressures: cylinder number 1, 150 psi; cylinder number 2, 55 psi; cylinder number 3, 45 psi; cylinder number 4, 145 psi. Which is the most likely cause?
   A. A hole in the pistons
   B. A worn camshaft
   C. Bad rings
   D. A blown head gasket

9. An engine produces the following compression pressures: cylinder number 1, 95 psi; cylinder number 2, 140 psi; cylinder number 3, 145 psi; cylinder number 4, 135 psi. When a wet test is performed on cylinder number 1 the pressure rises to 160 psi. The most likely cause of the problem with cylinder number 1 is:
   A. A blown head gasket
   B. Worn rings
   C. A bad oil pump
   D. A burned valve

10. An engine has 45 percent leakage and air is heard at the intake manifold.
    Technician A says the rings are bad.
    Technician B says the engine should run fine.
    Who is correct?
    A. A only
    B. B only
    C. Both A and B
    D. Neither A nor B

ASE Challenge Questions

1. An engine has low-running compression on two out of four cylinders; the cranking compression was normal.

   Technician A says the camshaft lobes may be worn.

   Technician B says the head gasket may be blown.

   Who is correct?

   A. A only   C. Both A and B

   B. B only   D. Neither A nor B

2. A loud clattering is heard from the top of the engine.

   Technician A says it may be caused by collapsed lifters.

   Technician B says the rocker arms could be worn.

   Who is correct?

   A. A only   C. Both A and B

   B. B only   D. Neither A nor B

3. *Technician A* says that a loud clunk on acceleration may be caused by a loose timing chain.

   Technician B says that metallic clacking noise on deceleration is usually caused by worn camshaft bearings.

   Who is correct?

   A. A only   C. Both A and B

   B. B only   D. Neither A nor B

4. A customer started up her car one extremely cold morning, and within a few minutes, a cloud of blue smoke was coming from the tailpipe. She stated that the engine had been running perfectly, with no smoking at all the day before. Which is the most likely cause?

   A. A blown head gasket

   B. Worn rings

   C. Bad main bearings

   D. Plugged PCV hose

5. An engine with 135,000 miles is running poorly. The cause of low compression on the third cylinder is being diagnosed using a cylinder leakage test. The results show 55 percent leakage. Most of the leakage is past an intake valve, but there is significant air coming out the oil fill port.

   Technician A says that he would replace the bad valve.

   Technician B says that he would recommend an engine overhaul or replacement.

   Who is correct?

   A. A only   C. Both A and B

   B. B only   D. Neither A nor B

JOB SHEET

Name _____   **Date** _____

DIAGNOSING ENGINE NOISES AND VIBRATIONS

Upon completion of this job sheet, you should be able to locate and recognize normal and abnormal engine noises and recommend repairs or further tests.

ASE Correlation

This job sheet is related to ASE Engine Repair Test content area: General Engine Diagnosis; tasks: Identify and interpret engine concern, and determine necessary action; research applicable vehicle and service information; diagnose engine noises and vibrations and determine necessary action.

Tools and Materials

Stethoscope

Describe the vehicle being worked on.

Year _____ Make _____ Model _____
VIN _____ Engine type and size _____

Procedure

Task Completed

Your instructor will provide you with a specific vehicle. Write down the information, and then perform the following tasks:

1. Check the oil level and correct as needed. ☐

2. Start the engine and allow it to idle. ☐

3. Are there any abnormal noises coming from any areas of the engine? ☐ Yes ☐ No

4. Use an available information system to locate any technical service bulletins related to abnormal engine noises.
   Describe your findings: _____
   _____
   _____
   _____

5. Use a stethoscope to listen to the bottom end of the block, the top end of the block, and to the valvetrain.
   Describe your results: _____
   _____
   _____
   _____

6. Raise the engine rpm to 2,500 and listen again for any unusual noises at the same locations.
Describe your results: _____
_____
_____
_____

7. Reduce the engine rpm to idle again and listen at the accessories for unusual noises such as squealing bearings or rough metallic sounds.
Describe your results: _____
_____
_____
_____

8. Based on your tests and results, what are your recommendations?
_____
_____
_____
_____

Instructor's Response _____
_____
_____
_____

JOB SHEET

Name _____ **Date** _____

ANALYZING SPARK PLUGS AND PERFORMING A POWER BALANCE TEST

Upon completion of this job sheet, you should be able to analyze spark plugs, perform a power balance test, and recommend repairs or further tests.

ASE Correlation

This job sheet is related to ASE Engine Repair Test content area: General Engine Diagnosis; tasks: Identify and interpret engine concern and determine necessary action; research applicable vehicle and service information; perform cylinder power balance test and determine necessary action.

Tools and Materials

Power balance tester
Insulated pliers

Describe the vehicle being worked on.

Year _____ Make _____ Model _____
VIN _____ Engine type and size _____

Procedure

Your instructor will provide you with a specific vehicle. Write down the information, and then perform the following tasks:

Task Completed

1. Check the oil level and correct as needed. ☐

2. Use an available information system to locate any technical service bulletins related to engine driveability concerns.
   Describe your findings: _____
   _____
   _____
   _____

3. Start the engine and allow it to idle. ☐

4. Record the engine idle speed and analyze the idle condition.
   Describe your results: _____
   _____
   _____
   _____

5. Prepare to perform a power balance test. If you are using an automated system, follow the instructions provided with the equipment. To perform a manual power balance test by disabling the spark to each cylinder in turn, follow these instructions: Using insulated pliers, remove the spark plug wire from cylinder number 1 and place the wire close to a good ground. Note the rpm drop and the idle condition. Describe your results: _____

_____

_____

_____

6. Wait one minute between tests, so you do not overheat the catalytic converter. Repeat the power balance test on each of the other cylinders and describe your results:

Cylinder number 2 _____

_____

Cylinder number 3 _____

_____

Cylinder number 4 _____

_____

Cylinder number 5 _____

_____

Cylinder number 6 _____

_____

Cylinder number 7 _____

_____

Cylinder number 8 _____

_____

7. Restore the engine to proper operating condition. Turn the engine off and allow it to cool before performing the spark plug analysis.

8. Analyze the results of the power balance test, and note any recommended repairs or tests.

_____

_____

_____

_____

9. Remove all spark plugs, and keep them in order so you can analyze the results of each plug. Describe your general results: _____

_____

_____

_____

Describe the condition of each of the spark plugs:

Cylinder number 1 _____

Cylinder number 2 _____

Cylinder number 3 _____

Cylinder number 4 _____

Cylinder number 5 _____

Cylinder number 6 _____

Cylinder number 7 _____

Cylinder number 8 _____

10. Based on your tests and results, what are your recommendations for further testing and repairs?

_____

_____

_____

_____

Instructor's Response _____

_____

_____

_____

Name _____ Date _____

OIL AND COOLANT CONSUMPTION DIAGNOSIS

Upon completion of this job sheet, you should be able to determine if the oil and coolant consumption on the vehicle you are working on is excessive.

ASE Correlation

This job sheet is related to ASE Engine Repair Test content area: General Engine Diagnosis; Removal and Reinstallation (R&R): Diagnose the cause of excessive oil consumption, coolant consumption, unusual engine exhaust color and odor; determine necessary action.

Tools and Materials

Coolant pressure tester
Service manual

Describe the vehicle being worked on:

Year _____ Make _____ Model _____
VIN _____ Engine type and size _____

Procedure

Task Completed

Your instructor will provide you with a specific vehicle. Write down the information, and then perform the following tasks:

1. Turn the engine off and let it rest for 5 minutes on a level surface. Check the oil and coolant levels. Describe your findings:

   _____
   _____
   _____
   _____

2. Use an available information system to locate any technical service bulletins related to oil and coolant consumption.
   Describe your findings: _____
   _____
   _____
   _____

3. Start the engine and allow it to idle.

   ☐

4. Check the exhaust tailpipe for blue, black, or white smoke. Describe your findings:

   _____
   _____
   _____

5. What are the API and SAE ratings for the oil that should be used in this vehicle?

_____

6. According to the service manual, what driving conditions must the driver have in order to consider them classified under the *severe maintenance* service schedule?

_____

_____

_____

7. If the vehicle you are working on was classified in the server service category, according to the service manual how often should the oil be changed? _____

8. Based on your tests and results, what are your recommendations for further testing and repairs?

_____

_____

_____

_____

Instructor's Response _____

_____

_____

_____

JOB SHEET

Name _____   Date _____

PERFORMING COMPRESSION TESTS

Upon completion of this job sheet, you should be able to perform a cranking compression test and recommend repairs or further tests.

ASE Correlation

This job sheet is related to ASE Engine Repair Test content area: General Engine Diagnosis; tasks: Identify and interpret engine concern, and determine necessary action; research applicable vehicle and service information; locate and interpret vehicle and major component identification numbers; perform cylinder compression tests; determine necessary action.

Tools and Materials

Compression tester
Oil can
Spark plug socket

Describe the vehicle being worked on.

Year _____ Make _____ Model _____
VIN _____ Engine type and size _____

Procedure

Task Completed

Your instructor will provide you with a specific vehicle. Write down the information, and then perform the following tasks:

1. Check the oil level, and correct as needed.   ☐

2. Locate the compression specifications for the engine you are working on, and record them below: _____
   _____

3. Start the engine, and allow it to idle until the engine is moderately warm. Allow it to cool slightly before removing the spark plugs.

4. Remove the spark plugs, and note their condition.
   Describe your results: _____
   _____
   _____
   _____

5. Prepare to perform a cylinder compression test. If you can, install a remote starter. Disable the fuel by locating the fuel pump fuse or relay and removing it. Disable the ignition system by removing the coil power wire or the ignition system fuse. Block open the throttle.
   _____

6. Install the compression tester adaptor into the number 1 spark plug hole, and crank the engine over five times.
   Record your results: _____
   Repeat for each cylinder, and describe your results.
   Cylinder number 2 _____
   Cylinder number 3 _____
   Cylinder number 4 _____
   Cylinder number 5 _____
   Cylinder number 6 _____
   Cylinder number 7 _____
   Cylinder number 8 _____

7. Analyze the results of the compression test, and note any recommended repairs or tests.

   _____
   _____
   _____
   _____
   _____
   _____

8. Identify the cylinder(s) with the lowest compression. Squirt 2 tablespoons of oil into the weak cylinder, and perform another compression test on that cylinder.
   Note the new results, and describe what problem this indicates:

   _____
   _____
   _____
   _____

9. Based on your tests and results, what are your recommendations for further testing and repairs?

   _____
   _____
   _____

Instructor's Response _____

_____
_____
_____

Name _____ Date _____

PERFORMING A RUNNING COMPRESSION TEST

Upon completion of this job sheet, you should be able to perform running compression tests and recommend repairs or further tests.

ASE Correlation

This job sheet is related to ASE Engine Repair Test content area: General Engine Diagnosis; tasks: identify and interpret engine concern, and determine necessary action; research applicable vehicle and service information; locate and interpret vehicle and major component identification numbers; perform cylinder compression tests, and determine necessary action.

Tools and Materials

Compression tester
Spark plug socket

Describe the vehicle being worked on.

Year _____ Make _____ Model _____
VIN _____ Engine type and size _____

Procedure

Task Completed

Your instructor will provide you with a specific vehicle. Write down the information, and then perform the following tasks:

1. Check the oil level and correct as needed. ☐

2. Record the general running compression specifications at idle and at 2,500 rpm:

   _____

   _____

3. Start the engine, and allow it to idle until the engine is moderately warm. Allow it to cool slightly before removing a spark plug.

4. Remove a spark plug from cylinder number 1, and note its condition. ☐
   Describe your results: _____

   _____

5. Install the compression tester adaptor into the number 1 spark plug hole, and run the engine. Release the pressure from the compression tester, and read the results at idle. Record your results: _____

6. Raise the engine speed to 2,500 rpm. Release the pressure from the compression tester, and record your results:

   _____

Repeat for each cylinder, and describe the spark plug condition and your idle and 2,500 rpm compression readings.

Cylinder number 2 _____

_____

Cylinder number 3 _____

_____

Cylinder number 4 _____

_____

Cylinder number 5 _____

_____

Cylinder number 6 _____

_____

Cylinder number 7 _____

_____

Cylinder number 8 _____

_____

7. Describe any significant differences between cylinders, and list the possible causes of low readings:

_____
_____
_____
_____
_____
_____

8. Based on your tests and results, what are your recommendations for further testing and repairs?

_____
_____
_____
_____

Instructor's Response _____

_____
_____
_____

Name _____ Date _____

PERFORMING A CYLINDER LEAKAGE TEST

Upon completion of this job sheet, you should be able to perform a cylinder leakage test and recommend repairs or further tests.

ASE Correlation

This job sheet is related to ASE Engine Repair Test content area: General Engine Diagnosis; tasks: Identify and interpret engine concern, and determine necessary action; Perform cylinder leakage tests; determine necessary action.

Tools and Materials

Cylinder leakage tester
Spark plug socket
Stethoscope

Describe the vehicle being worked on.

Year _____ Make _____ Model _____
VIN _____ Engine type and size _____

Procedure

_____
Task Completed

Your instructor will provide you with a specific vehicle. Write down the information, and then perform the following tasks:

1. Check the oil level, and correct as needed. ☐

2. Start the engine, and allow it to idle until the engine is moderately warm. Allow it to cool slightly before removing a spark plug. ☐

3. Remove a spark plug from a cylinder (preferably one with low compression), and note its condition.
   Describe your results: _____
   _____

4. Rotate the engine until it is at TDC on the compression stroke on the test cylinder. ☐

5. Apply shop air to the cylinder leakage tester and adjust the tester to read 0 percent. ☐

6. Install the cylinder leakage tester adaptor into the spark plug hole.

7. Connect the cylinder leakage tester adaptor in the spark plug hole to the leakage tester, and note the percent of leakage: ☐
   _____

   Is this an acceptable amount of leakage? _____

8. Use a stethoscope to listen at the oil dipstick. Describe your results:

_____

_____

9. Use a stethoscope to listen at the intake manifold. Describe your results:

_____

_____

10. Use a stethoscope to listen at the tailpipe. Describe your results:

_____

_____

11. Use a stethoscope to listen at the radiator fill. Describe your results:

_____

_____

12. Based on your tests and results, what are your recommendations for further testing and repairs?

_____

_____

_____

_____

_____

_____

_____

_____

Instructor's Response _____

_____

_____

_____

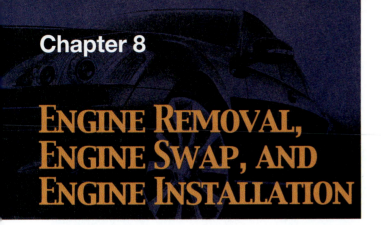

Chapter 8

ENGINE REMOVAL, ENGINE SWAP, AND ENGINE INSTALLATION

UPON COMPLETION AND REVIEW OF THIS CHAPTER, YOU SHOULD BE ABLE TO:

- Perform the steps required to prepare the engine for removal.

- Remove an engine from a front-wheel-drive vehicle.

- Remove an engine from a rear-wheel-drive vehicle.

- Determine the proper engine cleaning methods.

- Remove the engine accessories in preparation for engine repair or replacement.

- Remove and reinstall engine accessories for the installation of a rebuilt or remanufactured engine.

- Install an engine on a stand.

- Install an engine in a front-wheel-drive vehicle.

- Install an engine in a rear-wheel-drive vehicle.

Most engine rebuild operations require the removal of the engine from the vehicle. The procedure and tools used to remove the engine depend upon the vehicle's design. Most rear-wheel-drive vehicles require the engine to be removed from the hood opening, while many front-wheel-drive vehicles require engine removal from the bottom of the vehicle.

This chapter discusses the basic steps required to prepare an engine for removal, and for removing it from the vehicle. The typical procedures for removing an engine from a front-wheel-drive and rear-wheel-drive vehicle are discussed. In addition, general engine disassembly procedures are discussed.

> **Classroom Manual**
> Chapter 8

PREPARING THE ENGINE FOR REMOVAL

WARNING: Cleanliness is important when removing an engine. To prevent injury, keep your work area clean. If fluids are spilled or grease falls to the floor, clean it up immediately.

Before beginning the job of removing an engine from the vehicle, be sure to place fender covers on both sides and in the front of the vehicle (Figure 8-1). A big scratch on the fender would cost you or your shop money, and the customer would be upset, understandably.

The work of removing an engine from the vehicle is much cleaner if you can wash the engine and the engine compartment first. Many shops will wash every engine and engine compartment during major engine service because the customer is always pleased to see the engine "look like new." Other shops choose not to clean engines before removal, due to concerns about component damage and/or environmental regulations or concerns. Always consult local regulations and the vehicle service information for precautions and warnings before cleaning. You will need to protect electrical components such as the starter and generator. There may be additional electronic components such as computers or sensors that also need

SPECIAL TOOLS

Steam cleaner or pressure washer

Fender covers

Battery terminal puller

Drain pans

Antifreeze recycler

A/C reclaimer/ recycler

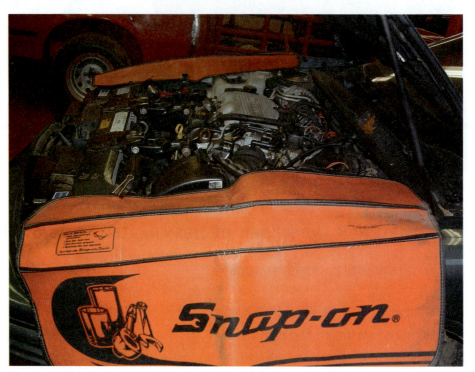

FIGURE 8-1 Don't let a careless mistake harm your reputation as an excellent engine repair technician.

A **steam cleaner** uses hot, pressurized water to clean an engine or other components.

to be removed or thoroughly covered. Follow the equipment manufacturer's instructions when using a **steam cleaner**, pressure washer, or cleaning chemicals to clean the engine.

Study the service information to determine what tools and equipment will be required to remove the engine, and to become familiar with the procedure for the vehicle you are working on. Until you have had a lot of practice, follow the service information's step-by-step procedure to be sure that you do not miss anything. You will often have to remove bolts that you can feel but not see. You could cause damage to components if you attempt to lift the engine with anything still attached.

With the vehicle centered in the service bay or over a frame contact hoist, perform the following steps to complete the preparation process:

WARNING: The steps described in this book are typical steps in preparing the engine for removal. Always refer to the service manual for all procedures and become familiar with all warnings and safety concerns.

CAUTION: Perform the engine cleaning process where a drain trap or waste water recycler is available.

1. Disconnect the battery negative terminal and isolate the cable. Remove the positive terminal next; then remove the battery.

> **CUSTOMER CARE:** Disconnecting the battery will erase radio station presets. Before disconnecting the battery, take a few minutes to write down the presets. Before returning the vehicle to the customer, reset the radio stations.

WARNING: Before disconnecting any electrical components or removing wire harnesses, always disconnect the battery first. Always disconnect the negative battery terminal first when removing the battery and connect it last when installing the battery. If the positive battery terminal is removed first, the wrench may slip and make contact from this terminal to ground. This action may overheat the wrench and burn the technician's hand.

WARNING: Never wear jewelry when working around the vehicle. Gold, silver, and copper are excellent conductors of electricity. In addition, your body is also a conductor of electricity. If the jewelry makes contact with a current-carrying wire or terminal, severe injury may result. Also, jewelry may become caught on objects or in fans and other rotating parts.

2. If the hood must be removed, mark the location of the hinge to the hood for reference during assembly, then remove the hood (Figure 8-2).
3. Lift the vehicle on the hoist and drain the engine oil.
4. Drain the engine coolant from the radiator and engine block. Removing the radiator cap will increase the flow of coolant through the drain.

WARNING: Do not open a radiator cap if the engine is warm. The release of hot coolant under pressure can cause serious burns.

5. If the transmission is being removed with the engine, drain its fluid. Disconnect shift linkages, transmission cooling lines, any vacuum hoses, clutch linkages, and drive shaft.
6. Disconnect the lower radiator hose and the transmission cooling lines (if needed).
7. If easily accessible while the vehicle is on the hoist, disconnect the cooling lines to the heater core.
8. Disconnect the exhaust system. Attempt to do this at the exhaust manifold if possible.
9. Disconnect any additional wiring harnesses, vacuum hoses, and cables that are easier to get to while the vehicle is on the hoist. Use some sort of identifying method to assure proper connections during engine replacement (Figure 8-3).
10. Lower the vehicle and remove the air intake ducts and air cleaner assembly (Figure 8-4).
11. Follow the service manual procedure to relieve fuel pressure. If a procedure is not listed in the service manual, attach a drain hose to the Schrader valve on the fuel rail.

SERVICE TIP:
When removing components from the engine, it is faster to remove assemblies rather than individual components. Also, it may not be necessary to remove all items from the engine compartment. The power steering pump, air conditioner compressor, cruise control servo, and so forth can be tied off to the side.

FIGURE 8-2 Marking the hood before removal so that it reinstalls correctly aligned.

FIGURE 8-3 Take the time to label electrical connectors and hoses during removal; it takes a lot less time than trying to figure out where everything should be reinstalled without clues. After a few engine removals, you will learn shortcuts that save time.

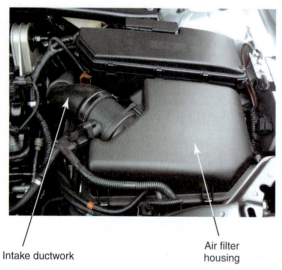

Intake ductwork

Air filter housing

FIGURE 8-4 Remove the intake air filter housing and ductwork.

FIGURE 8-5 These fuel lines have quick-disconnect fittings that come apart easily with the proper tool.

The other end of the drain hose must go into an approved container. When the pressure is completely relieved, disconnect the fuel feed line from the fuel rail. If the engine is equipped with a return fuel line from the pressure regulator, disconnect it also. To prevent fuel leakage, plug the fuel lines. Most modern vehicles use a quick-connect fitting on the fuel lines (Figure 8-5). These are separated by squeezing the retainer tabs together and pulling the quick-connect fitting assembly off of the fuel line nipple. The retainer will remain on the fuel line (Figure 8-6).

12. Disconnect the throttle cable to the throttle body (Figure 8-7).
13. Remove any accessory drive belts.
14. If the air-conditioning compressor needs to be disconnected from the pressure hoses, the system must be evacuated into a reclaimer unit. Follow all instructions and safety

FIGURE 8-6 Typical quick-disconnect fitting tools for fuel systems and other applications.

FIGURE 8-7 Disconnect the throttle cable.

procedures provided by the equipment manufacturer. In most cases, the compressor and bracket can be removed from the engine and wired off out of the way without disconnecting any lines.

15. Disconnect the air-conditioning compressor bracket, power steering unit, AIR pump, cruise control activators, and any other components that are attached to the engine block. Identify all wires or hoses that are disconnected, and cap any hoses that may leak fluid. If possible, lay the component to the side and secure it instead of disconnecting it and removing it from the engine compartment (Figure 8-8).

16. Disconnect the upper radiator hose.

17. Disconnect the electric cooling fan motor connector, then remove the radiator mounting. The radiator and cooling fans can usually be removed as a unit (Figure 8-9).

18. Disconnect any other wiring harnesses and vacuum hoses as required. Identify each connection. There may be individual connectors to each of the engine sensors and actuators; be sure to label them accordingly. Some engines use one or two large connectors to allow the engine harness to be disconnected from the chassis. If possible, disconnect the complete engine wiring harness at the bulkhead connector(s).

Additional steps may include removing the distributor cap and wires, the distributor, and the water pump. The service manual is always your best source of information concerning engine removal procedures.

CAUTION:
Never intentionally discharge the air-conditioning system into the atmosphere. The refrigerant used in these systems causes depletion of the ozone layer. As a professional employed within an industry traditionally associated with hazardous materials, it is your responsibility to conduct yourself in such a manner as to protect the environment and to improve the public perception of the industry. Also, severe penalties or fines can be issued for intentional discharge of R-12 or R-134a. Only tested and certified technicians can use the reclaimer/recycling equipment.

FIGURE 8-8 Whenever possible, remove components as assemblies and secure them in the engine compartment.

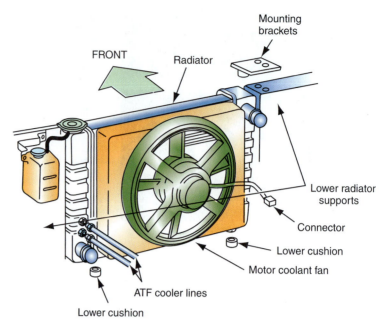

FIGURE 8-9 Remove the cooling fan and radiator assembly to prevent damage during engine removal.

REMOVING THE ENGINE

At this point, the procedure will vary depending on if the engine is removed from the bottom of the vehicle or through the hood opening. Many front-wheel-drive vehicles require removal of the engine through the bottom, while most rear-wheel-drive vehicles require the engine to come out the top. In addition, different steps are required between front- and rear-wheel-drive vehicles.

When removing the fasteners, pay close attention to their size and type. Many of the brackets used to secure the engine or accessories use several different size fasteners. Mark the fasteners so their proper location can easily be determined during reassembly. It is best to place fasteners in the same location to prevent mixing them.

Many manufacturers use a crankshaft position sensor attached above the flywheel or flex plate. This sensor must be removed before separating the engine from the bellhousing to prevent damage.

Front-Wheel-Drive Vehicles

In addition to the steps performed in preparing the engine for removal, the following steps may be required on front-wheel-drive vehicles before the actual process of removing the engine can be started (refer to Photo Sequence 14 for a picture by picture step of a typical FWD engine removal):

1. Remove the cross member (center beam) (Figure 8-10).
2. Disconnect the outer tie rod ends (Figure 8-11). Refer to the service manual for the proper procedure.
3. Disconnect the lower ball joint from the steering knuckle (Figure 8-12).
4. Remove the axles from the hubs and transaxle.

When the engine is removed through the bottom of the vehicle, use a special engine cradle and dolly to support the engine. If the vehicle manufacturer recommends engine removal through the hood opening, use an engine hoist (Figure 8-13). Regardless of the method of removal, the engine and transaxle are usually removed as a unit.

5. Adjust the pegs of the cradle to fit into the holes in the bottom of the engine block.

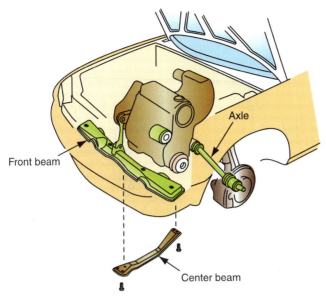

FIGURE 8-10 To remove the engine from a front-wheel-drive vehicle, it may be necessary to remove portions.

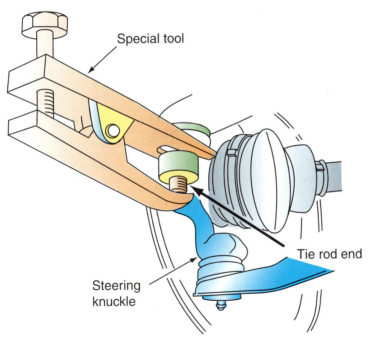

FIGURE 8-11 Use a tie rod end removal tool to disconnect the tie rod from the steering knuckle.

6. Remove the engine mount fasteners (Figure 8-14).
7. If required, remove the frame member from the vehicle (Figure 8-15). It may be necessary to disconnect the steering gear from the frame.
8. Double-check to assure all wires and hoses are disconnected from the engine.
9. Lift the vehicle using a frame contact hoist. As the vehicle is lifted, the engine remains on the cradle. During this process, continually check for interference with the engine and vehicle body. Also watch for any wires or hoses that may still be attached to the engine.

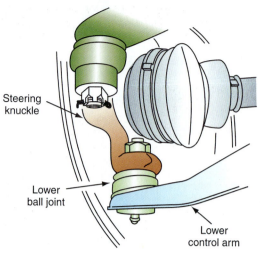

FIGURE 8-12 You may need to remove the lower ball joints in order to get the axles out.

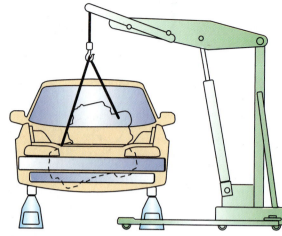

FIGURE 8-13 Some front-wheel-drive vehicles require the engine to be removed from the hood opening. Use an engine hoist to lift the engine.

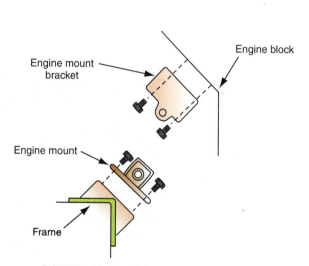

FIGURE 8-14 One of the last steps is to remove the engine mounts.

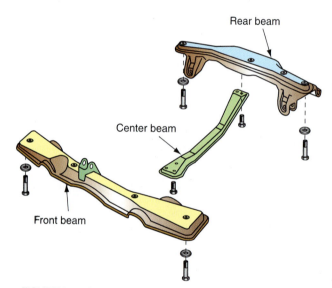

FIGURE 8-15 Some or all of the engine cradle's frame members may need to be removed to lower the engine through the bottom of the vehicle.

Rear-Wheel-Drive Vehicles

The engine is removed through the hood opening on most rear-wheel-drive vehicles, requiring the use of an **engine hoist**. Refer to the service manual to determine the proper engine lift points. If the transmission is being removed with the engine, it may be easier if you locate the hook of the engine hoist to the chain in such a manner that the engine tips a little toward the transmission.

Additional preparation steps depend on whether the engine and transmission are being removed together or if the engine is being removed alone. If the transmission is not being removed with the engine, support it with a transmission stand. Remove the bellhousing bolts and inspection plate. On vehicles equipped with automatic transmissions, remove the torque converter-to-flex plate bolts. Use a C-clamp or other brace to prevent the torque converter from falling. Also, mark the location of the torque converter in relation to the flex plate for reference during installation.

SPECIAL TOOLS

Fender covers
Engine hoist
Chain or sling
C-clamp
Transmission stand or jack

TYPICAL PROCEDURE FOR FWD ENGINE REMOVAL

P14-1 Disconnect negative cable first, the the positive cable and then remove the battery and battery holder.

P14-2 Remove the intake ducting and necessary vacuum lines and electrical connectors.

P14-3 Remove all throttle linkages.

P14-4 Relieve fuel pressure and disconnect the fuel supply (and return if applicable) line.

P14-5 Disconnect and label all necessary vacuum lines, electrical connectors and fluid lines.

P14-6 Remove the wheels.

P14-7 Disconnect the lower ball joints from the spindles.

P14-8 Remove the axle nuts.

P14-9 Remove the wishbone links.

TYPICAL PROCEDURE FOR FWD ENGINE REMOVAL

P14-10 Remove the axles from the hubs and pop them out of the transaxle.

P14-11 Remove the speedometer cable and other connections on the transaxle and engine.

P14-12 Remove the exhaust.

P14-13 Support the engine and remove the suspension and engine cradle.

P14-14 Use the engine hoist to lower the engine and transaxle onto the dolly underneath the vehicle. Make sure everything is disconnected, then raise the vehicle from the powerplant.

SERVICE TIP:
To lift the engine high enough to clear the vehicle, it may be necessary to adjust the length of the hoist boom and legs. Remember, this will also change the maximum amount of weight the hoist can lift.

If the transmission is being removed with the engine, remove the drive shaft after making an index mark with the companion flange at the differential. Disconnect the rear engine mount (Figure 8-16). Disconnect all shift and/or clutch levers.

After performing the preceding steps to prepare the engine for removal, follow the typical steps shown in Photo Sequence 15 to remove the engine through the hood opening on most rear-wheel-drive vehicles.

WARNING: Do not lift the engine by the intake manifold. This action may break the intake manifold or retaining bolts, causing the engine to drop suddenly and result in technician injury.

SPECIAL TOOLS
Engine stand
Engine hoist

MOUNTING THE ENGINE ON A STAND (PHOTO SEQUENCE 16)

To begin the disassembly process, attach the engine to an **engine stand** (Figure 8-17). If the transmission or transaxle was removed with the engine, it will be necessary to separate it from the engine first. Before removing the flywheel or flex plate, mark its reference to the crankshaft. Bolt the adjustable arms of the engine stand to the bellhousing attaching holes. Never work on the engine while it is hanging from the hoist.

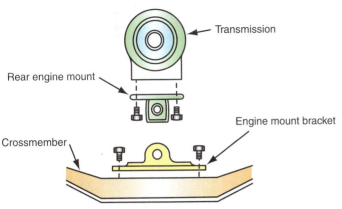

Transmission

Rear engine mount

Crossmember

Engine mount bracket

FIGURE 8-16 The rear engine mount will have to be disconnected if the transmission is coming out with the engine.

FIGURE 8-17 Proper installation of the engine to the stand is critical to your safety. Check the weight rating of the stand and thread the bolts in at least the distance equal to one and a half times the bolt diameter.

PHOTO SEQUENCE 15

TYPICAL PROCEDURE FOR RWD ENGINE REMOVAL

P15-1 Tools required to perform engine removal include fender covers, lifting strap or chain, basic mechanic's tools, and an engine hoist.

P15-2 Attach the engine lifting fixture to the engine.

P15-3 Disconnect the engine mounts.

P15-4 Double-check to assure all wires and hoses are disconnected from the engine.

P15-5 Lift the engine from the vehicle.

P15-6 During this lifting process, continually check for interference with the engine and vehicle body. Also watch for any wires and hoses that may still be attached to the engine.

TYPICAL PROCEDURE FOR MOUNTING AN ENGINE ON A STAND

P16-1 Locate the engine and crane near the engine stand.

P16-2 Adjust the mounting flange ears to fit the engine.

P16-3 Mount the flange onto the engine.

P16-4 Securely tighten the mounting bolts. They should extend into the engine at least as far as one and a half times the width of the bolt.

P16-5 Slide the mounting flange tube into the engine stand.

P16-6 Secure the engine with a pin through the stand and the flange.

P16-7 Remove the engine hoist and the chain.

P16-8 Now the engine is secure on the stand and ready for repair work.

WARNING: Check the load rating of the engine stand to assure it will be able to maintain the weight of the engine being installed.

INSTALLING A REMANUFACTURED ENGINE

In many cases you or your shop management will decide that it will be economically more profitable to install a remanufactured "crate" engine rather than to rebuild the existing engine in house. Remanufactured engines come with a warranty that relieves the shop of the liability that an error in your overhaul could cause the shop as well. With popular engines that are readily available as remanufactured units, this is definitely the trend in the automotive repair business.

A crate engine, as the name implies, comes in a crate fully assembled (Figure 8-18). You will need to swap over to the "new" engine all of the engine accessories, sensors, actuators, and manifolds, and often oil pans, valve covers, and other components. The best method of approaching the preparation of the new engine is to have it side by side with the old engine. Then you will know precisely what has to be removed from the old engine and installed on the new one. You will also be able to see exactly how the component mounts on the engine and to transfer it directly. It is always a good idea to check the manufacturer's technical service bulletins (TSBs) for information related to engine installation and removal. Organizations such as the Automotive Engine Rebuilders Association (AERA) will also offer advice in the form of TSBs that may be considered helpful when installing a crate engine.

You may start by disassembling the front end of the engine and removing the belt(s) hoses, lines and generator, power steering pump, air-conditioning compressor, and water pump (Figure 8-19). Some of these items may be hanging in the engine compartment, and you will reinstall them during the engine installation. Providing these components are in good condition, install them onto the new engine. If the crate engine does not come with a

FIGURE 8-18 Crate engines are available at aftermarket parts stores as well as from the manufacturer.

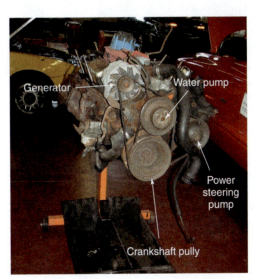

Generator

Water pump

Power steering pump

Crankshaft pully

FIGURE 8-19 Start removing components from the front end of the old engine and installing what is good onto the new engine.

SPECIAL TOOLS

Crankshaft
dampener puller
Crankshaft
dampener installer
Flywheel turning bar

new water pump, it is customary to fit a new one at this point unless it is has been replaced recently (Figure 8-20). Fit a new gasket using sealant only if it is recommended, and torque the water pump to the proper specification (Figure 8-21).

You may also need to remove the crankshaft pulley and harmonic balancer. Simply remove the bolts attaching the crankshaft pulley and pull it off (Figure 8-22). To remove the harmonic balancer, first remove the attaching bolt from the center of the pulley. Use a flywheel holder to keep the engine from turning; this bolt will be on tight (Figure 8-23). Once the bolt is removed, you will likely need to use a puller (Figure 8-24). Make sure the harmonic balancer is in good shape; the keyway should not be rounded, and the rubber must not be separating between the inner and outer sections. Use an installation tool to press the balancer onto the new engine (Figure 8-25). Reinstall the crankshaft pulley, and install a new accessory drive belt. You may need to follow the routing directions labeled under the hood. Release the tensioner and route the belt properly around all the pulleys (Figure 8-26). Allow the tensioner to tighten the belt if it is automatic or tighten the belt to specification using a

FIGURE 8-20 This pump has been seeping past the weep hole, it should definitely be replaced.

FIGURE 8-21 Install the new pump and attach the hoses. Be sure the clamps are in good condition or replace them.

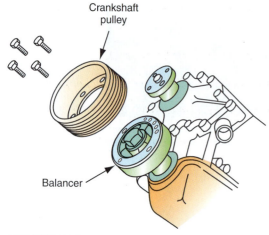

FIGURE 8-22 Remove the crank pulley for installation on the new engine.

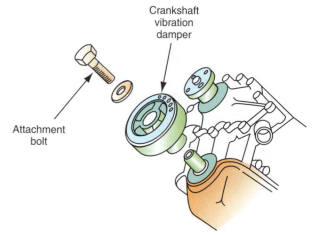

FIGURE 8-23 Use a flywheel holder to hold the engine and remove the harmonic balancer attaching bolt.

FIGURE 8-24 You will have to use a puller to remove most harmonic balancers. Be sure to place a round plug puller attachment at the end of the crankshaft to protect the threads.

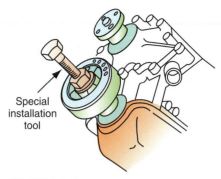

FIGURE 8-25 Use a special installation tool to protect the crankshaft when reinstalling the balancer.

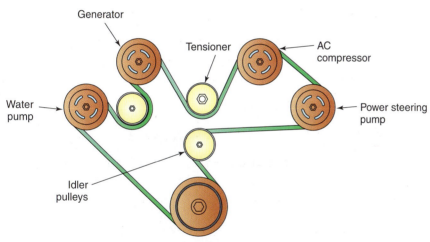

FIGURE 8-26 Route the belt carefully and then check the tension with a belt tension gauge.

manual tensioner. Confirm proper tension of the belt(s) using a belt tension gauge. Reinstall any other hoses and lines that were attached to the front of the engine. Check their condition and replace as needed. Remember that the customer will not want to have to bring his vehicle back for service in the near future after this big job. If a component is questionable, replace it.

Next you will need to remove the components from the top of the engine. This typically includes many brackets, sensors, and solenoids (Figure 8-27). Remove them one by one, and install them directly onto the new engine to avoid confusion about brackets and placement. The ignition system components must be swapped over whether the system uses a coil for each plug, an individual coil with a distributor, or a coil pack (Figure 8-28). Carefully inspect the wires for nicks or wear points or any signs of a gray powdery residue where the wire could have been arcing to ground. If the engine is equipped with an **exhaust gas recirculation (EGR) valve**, you will need to remove it from the engine (Figure 8-29). Clean the bottom of the EGR valve with a wire brush to remove carbon from the pintle. You will also need to

The **exhaust gas recirculation (EGR) valve** is an emission control device that cools the combustion chamber to reduce oxides of nitrogen emissions.

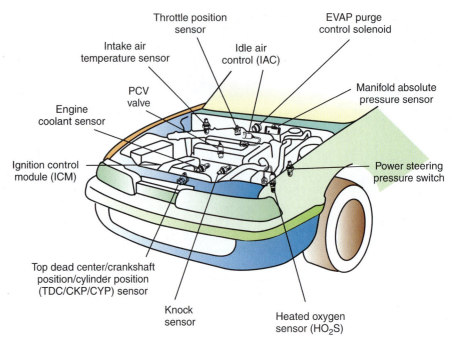

Throttle position
sensor

EVAP purge
control solenoid

Intake air
temperature sensor

Idle air
control (IAC)

PCV
valve

Manifold absolute
pressure sensor

Engine
coolant sensor

Ignition control
module (ICM)

Power steering
pressure switch

Top dead center/crankshaft
position/cylinder position
(TDC/CKP/CYP) sensor

Knock
sensor

Heated oxygen
sensor (HO$_2$S)

FIGURE 8-27 A modern engine has many sensors, switches, and solenoids that will have to be swapped over.

Corrosion

FIGURE 8-28 The corrosion on tower number one warrants replacement. That could easily cause an ignition misfire.

The **purge solenoid** is part of the evaporative emissions control system and is used to purge fuel vapors from the fuel tank into the intake manifold to be burned with the air-fuel mix.

remove and install solenoids such as the **purge solenoid** (Figure 8-30). There will be several sensors and vacuum lines that you will need to install on the new engine (Figure 8-31). It may be possible to remove the fuel rail, fuel pressure regulator, and injectors as an assembly and install it onto the new engine (Figure 8-32). Carefully check for any other components left on the old engine that are not installed on the new engine.

FIGURE 8-29 A typical EGR valve.

FIGURE 8-30 A purge control solenoid for the evaporative emissions control system.

FIGURE 8-31 Install all the sensors and vacuum lines onto the new engine.

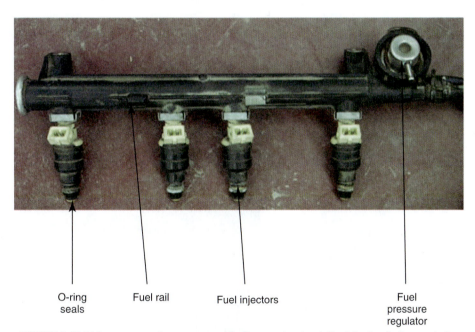

O-ring seals   Fuel rail   Fuel injectors   Fuel pressure regulator

FIGURE 8-32 This came apart as an assembly. Be sure to check the injector O-ring seals for damage and to lubricate them sparingly with petroleum jelly prior to installation.

Remove the intake manifold, and install it with a new gasket and the proper sealant if specified (Figure 8-33). Carefully lower the manifold into place without disturbing the gasket(s). Use the correct tightening sequence and torque specification (Figure 8-34). Swap the exhaust manifold to the new engine using a gasket if required and torque it properly (Figure 8-35). Be sure to use new studs and mounting hardware; the used hardware is usually damaged by rust.

You may need to swap the oil pan and valve covers onto the new engine. If the old engine was badly damaged, it may be quite a task to get the oil pan clean (Figure 8-36). Be sure to do an excellent job of cleaning it inside and out, to protect the new engine and to

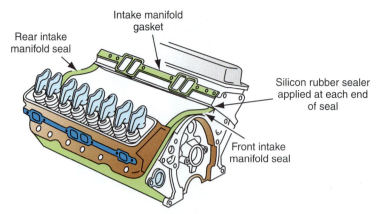

FIGURE 8-33 Install the new intake manifold with any sealants specified.

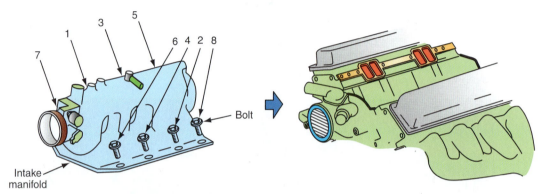

FIGURE 8-34 Torque the intake manifold into place using the specified (if available) or typical torque sequence.

FIGURE 8-35 Many technicians will clean and paint the old manifolds and covers to make the whole engine look like new.

FIGURE 8-36 This oil pan will require quite a bit of work before it will be suitable for installation on the new engine.

please the customer. Use a new gaskets and torque the oil pan into place using the appropriate sequence and proper torque specification. Similarly, clean and properly install the valve cover(s) with new gaskets and torque them properly (Figure 8-37).

Once you have the new engine assembled, you can bolt the flywheel or flexplate onto the back of the engine. Be sure to tighten these properly to avoid additional work and a frightening knock (Figure 8-38).

FIGURE 8-37 Use a high quality gasket and torque properly to avoid leaks that could bring the customer back dissatisfied.

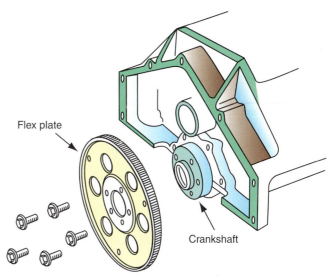

Flex plate

Crankshaft

FIGURE 8-38 The flexplate or flywheel attaches to the rear of the crankshaft.

On a manual transmission vehicle, this is an excellent opportunity to inspect the clutch disc and pressure plate (Figure 8-39). If it is significantly worn, now is the time to replace the clutch assembly. The labor to install a clutch later is quite costly; right now it is basically free. Use a clutch aligning tool to line up the clutch disc (Figure 8-40). Then, tighten the pressure plate in a diagonal or star fashion to the specified torque. Torquing the bolts around in a circle can warp a brand new pressure plate due to the spring tension.

Make a final check of the old engine to be sure that you have transferred all the required parts. Inspect the new engine to be sure you have tightened everything and installed it properly. Next you will install it back into the vehicle.

If the engine uses a mechanical fuel pump operated off an eccentric on the camshaft, install it. On some designs, the arm of the fuel pump must ride on top of the eccentric, while on other designs, it must ride on the bottom of the eccentric. Some engines use a rod between the camshaft eccentric and pump arm. The rod must be held up against the camshaft as the fuel pump is installed.

To complete the assembly, lubricate the oil filter gasket and fill the filter with oil. Install the oil filter. Do not overtighten the filter. Turn the filter three-quarter turn after the seal makes contact.

CAUTION:
Overtightening the oil filter may cause seal damage and result in oil leakage.

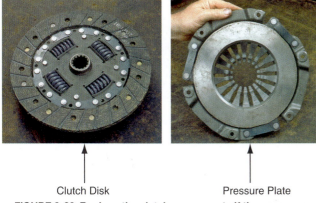

Clutch Disk   Pressure Plate

FIGURE 8-39 Replace the clutch components if they are worn. Doing so now will be a lot cheaper than doing so after the transmission and engine are reinstalled.

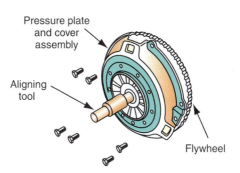

Pressure plate and cover assembly

Aligning tool

Flywheel

FIGURE 8-40 Use a clutch aligning tool to be sure the engine and transmission slide together.

CAUTION:
When setting the frame contact hoist, remember to consider the weight transfer with the engine installed.

CAUTION:
Never attempt to "suck" a transmission up to the engine. If the transmission does not seat, remove it and realign the clutch disc.

SPECIAL TOOLS
Engine cradle and dolly
Hoist

INSTALLING THE ENGINE

The process of engine installation is the next step. Before placing the engine into the vehicle, double-check all work thus far to be sure there are no loose fasteners or connections. When ready, locate the vehicle in the center of the service bay or over a frame contact hoist. Remove the engine from the engine stand and attach it to the engine hoist.

It may be easier to install the transmission on the engine before placing the engine into the vehicle. Regardless of when the transmission is installed, it must be properly centered before the bolts are torqued. When the transmission is attached to the engine, it should fit up to the block. If it does not fit up to the block, the clutch is not properly aligned.

Automatic transmissions use torque converters to make the link between the engine and the transmission. The torque converter should be installed onto the front pump of the transmission before the transmission is installed to the engine. Installing the torque converter to the flex plate first makes aligning the pump shafts into the converter difficult and may result in pump or converter damage. Align the scribe marks made between the flex plate and torque converter during disassembly.

The process of installing the engine into the vehicle is basically a reversal of the removal process. The same safety concerns applying to engine removal also apply to installing the engine. As with engine removal, the procedures vary between front-wheel and rear-wheel-drive vehicles. The procedure given here for front-wheel-drive vehicles assumes the engine is to be installed from under the vehicle. If the engine is installed from the hood opening, the procedure of placing the engine into the vehicle is similar to rear-wheel-drive vehicles.

Front-Wheel-Drive Vehicles

Many front-wheel-drive vehicles require the installation of the engine through the bottom of the vehicle. Prepare for engine installation by installing the engine into the cradle and dolly (if needed). Lift the vehicle on the hoist, and position the engine under the vehicle. Slowly lower the vehicle over the engine while guiding wires and hoses out of the way. As the vehicle and engine approach each other, align the engine and transmission mounts. This may require some shaking of the engine. Be careful not to damage any steering or suspension components. When the engine mounts are aligned, install the mounting bolts into the motor mounts (Figure 8-41). The vehicle can now be lifted to a comfortable working height.

Undercarriage installation includes the following procedures:

1. Installation of the axle shafts (Figure 8-42).
2. Installation of the transmission and engine braces.
3. Connection of the exhaust system to the exhaust manifold.
4. Installation of any splash shields and/or heat shields.
5. Connection of clutch and shift lever linkages, and adjustment as needed.

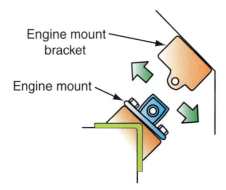

FIGURE 8-41 As you locate the engine in the engine compartment, line up the engine mounts before dropping the engine into place.

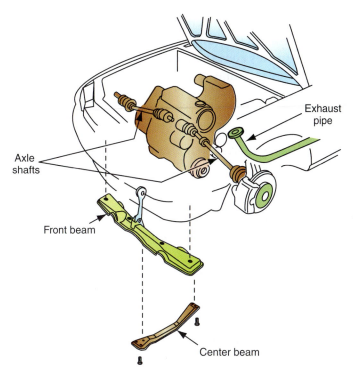

Axle shafts

Exhaust pipe

Front beam

Center beam

FIGURE 8-42 Lift the vehicle to reassemble the bottom components of the engine, suspension, and transmission.

6. Connection of any transmission cooling lines.
7. Connection of all wiring harnesses that are routed to the underside of the engine.
8. Installation of the tires and wheels, being sure to torque the lug nuts to the specified value.
9. Connection of the fuel delivery system.

After the undercarriage installation is completed, the vehicle can be lowered so upper engine connections can be made. Following are common procedures required at the upper engine:

1. Connection of the fuel lines to the fuel rail or carburetor.
2. Connection of the heater hoses.
3. Installation of engine ground straps.
4. Connection of all engine electrical connectors.
5. Connection of all vacuum and fluid hoses.
6. Connection of the throttle linkage and adjustment as needed.
7. Installation of the radiator and cooling fans. This may also include connection of the transmission cooling lines (if equipped).
8. Installation of the upper and lower radiator hoses.
9. Installation of the battery tray.
10. Installation of the air cleaner, air intake ducts, and vacuum hoses.
11. Installation and connection of any control modules.
12. Installation of the battery. Before making connections, make sure all wires are properly connected. Connect the battery, positive terminal first, then the negative terminal.
13. Fill the radiator.
14. Fill the crankcase with the proper oil type and quantity.
15. Fill the brake master cylinder or clutch cylinder as needed.
16. Fill the transmission with the proper fluid.
17. Follow the service manual procedures to fill the air-conditioning compressor with oil; then connect the hoses.

SERVICE TIP:
To assist lining up the engine mounts with the bushing try using a long aligning punch. The punch is strong, and can help to move the engine in the just the right direction.

SERVICE TIP:
Since water is easier to clean up than anti-freeze, fill the radiator with pure water until it is determined there are no leaks.

SERVICE TIP:
A technician must be EPA Section 609 certified for refrigerant recovery and recycling in order to recover and recharge the air conditioning system.

CAUTION
Make sure the engine hoist is capable of lifting the weight of the engine to a height sufficient to clear the engine compartment.

WARNING: Never wear jewelry when working around the vehicle. Gold, silver, and copper are excellent conductors of electricity. Your body is also a conductor of electricity. If the jewelry makes contact with a current-carrying wire or terminal, severe injury may result. Jewelry may also become caught on objects or in fans and other rotating parts.

Rear-Wheel-Drive Vehicles

On most rear-wheel-drive vehicles, it is possible to install the engine through the hood opening. Begin by attaching the engine lifting fixture to the engine. The following procedure is typical of most engine installations:

WARNING: Do not lift the engine by the intake manifold. This action may cause the intake manifold or retaining bolts to break, allowing the engine to drop suddenly, resulting in technician injury.

SPECIAL TOOLS

Engine hoist
Hoist

CAUTION:

Make sure to install the proper oil into the system. Oil used in R-12 and R-134a systems is not interchangeable.

1. Lift the engine high enough to clear the vehicle and center it over the motor mount towers. If the transmission is attached, it will have to be tipped to clear the bulkhead and lowered slowly in order to center it.
2. Slowly lower the engine until the engine settles into the front mounts.
3. Use a transmission jack to support the transmission. Locate the jack in an area that will not interfere with installation of the rear motor mount.
4. Secure the front motor mounts and cross members.
5. Install the rear motor mount and cross member.

Complete under vehicle installation by raising the vehicle on the hoist. Install transmission and engine braces and connect the exhaust system to the exhaust manifold. Make sure to install any splash shields and/or heat shields.

Connect the clutch and shift lever linkages, and adjust as needed. Many clutch assemblies use hydraulic cylinders to actuate the clutch. These require the air to be bled for proper operation. On automatic transmissions, connect the cooling lines. In addition, connect all wiring harnesses that are routed to the underside of the engine. Finally, connect the drive shaft using the marks made during removal to align it properly.

Lower the vehicle and complete the upper engine connections. These may include:

1. Connecting the fuel lines.
2. Connecting the heater hoses.
3. Installing engine ground straps.
4. Connecting all engine electrical connectors.
5. Connecting all vacuum and fluid hoses.
6. Connecting and adjusting the throttle linkage.
7. Installing the radiator and cooling fans. Connecting the transmission cooling lines (if equipped).
8. Installing the upper and lower radiator hoses.
9. Installing the battery tray.
10. Installing the air cleaner, air intake ducts, and vacuum hoses.
11. Installing and connecting any control modules.

Installation is complete after the following procedures are performed:

1. Install the battery. Before making connections, make sure all wires are properly connected. Connect the battery, positive terminal first, then the negative terminal.
2. Fill the radiator.
3. Fill the crankcase with the proper oil type and quantity.
4. Fill the brake master cylinder or clutch cylinder as needed.

5. Fill the transmission with the proper fluid.
6. Install and align the hood.
7. Follow the service manual procedure to fill the air-conditioning compressor with oil.

ENGINE START-UP AND BREAK-IN

The first 20 minutes of an engine's life is a critical time. Lubrication must be flawless, and coolant must be circulated efficiently. Worn parts have already lost their high spots and sharp edges, but new components must wear in; they need proper lubrication to avoid excessive scuffing. The increased friction of new parts creates higher engine temperatures that only a fully functional cooling system can control. The rings begin to break into the cylinder walls and develop a seal. The whole valvetrain must be properly adjusted and functional to allow the valves to wear into their seats. The bearings will be conforming to the shape of the journals. During the first couple of thousand miles of operation, the bearings will embed some fine metal particles into them as the sharp rings scrape the cylinder walls to seat. The new lifters and perhaps new camshaft will also experience some critical mating during the first half hour of operation.

If you are installing a crate engine, you may receive specific instructions to follow. Manufacturers may also include guidelines about engine start-up and break-in. Camshaft manufacturers may include additional instructions to follow for the break-in process. This section will give generalized advice that should apply to start-up and break-in for most new or freshly rebuilt engines. We will also discuss the important conversation you should have with the customer before handing over the keys.

> **CUSTOMER CARE:** The process you use to break in a new engine can affect engine performance and oil consumption during the first few thousand miles.

Start-Up and Initial Break-In

The initial start-up and the first 20 minutes of operation are critical to the health of the new or rebuilt engine. Before starting the engine, double check the fluids to be sure they are at the proper level; low coolant could cause engine overheating during the first 20 minutes of operation. A low oil level could be disastrous. The lubrication system should be primed before the engine actually runs, if possible, or the first minutes of operation could cause undue wear on new components. You should be prepared to evaluate the engine operation and sealing during its initial break-in period.

Priming the Lubrication System

If the oil pump is driven by the crankshaft, you will not be able to physically rotate the pump effectively to prime the system. It is best to use an oil priming tool that will deliver oil under pressure to the engine through an adaptor on the oil filter housing or in the oil pressure sending unit bore. Some parts stores carry pressurized aerosol priming oil cans that can spray oil throughout the engine via the oil sending unit port. If no priming tool is available, start the engine and immediately verify that the engine is generating adequate oil pressure. If it is not, turn it off quickly and locate the cause before damaging the engine.

On an older engine that drives the oil pump through the distributor shaft, you can physically rotate the shaft to prime the lubrication system. You should already have an oil pressure gauge installed on the engine. If not, remove the oil pressure switch and install a gauge now. What you want to do is rotate the oil pump until some oil pressure is seen on the gauge. You will not meet normal oil pressure specifications because you will not be able to spin the pump at a high enough rpm. You will know that oil has reached all areas of the engine. To rotate the pump, use a special drive tool to attach to the drive point of the pump. You may be able to use a 1/4-in. drive extension, a special tool with the correct flat drive on the end, or an old distributor drive shaft from another engine. Use a drill or a speed handle to rotate the pump until you can feel significant resistance. Continue rotating the pump for a minute or so to be sure that oil has reached the top of the engine.

Starting the Engine

Set wheel chocks before and after a wheel. If the shift linkage has been disturbed and is not properly adjusted, the engine could start in gear and lurch forward. Set the parking brake as an added precaution. Install appropriate exhaust extraction equipment; the engine will smoke while it runs for the first several minutes. Recheck the oil and coolant levels; you should be positive that the engine will have adequate lubrication and cooling. Crank the engine over until it starts or for 30 seconds maximum. If everything in and on the engine is properly connected and adjusted, it should start within two 30-second cycles of the starter. If the engine does not start, take another look at electrical connectors and vacuum lines. Check the engine to be sure it is getting proper spark and fuel.

Once the engine starts, it is important to hold the rpm to between 1,500 and 2,000 rpm for the first 20 minutes of operation. Block the throttle open or use an adjusting rod on the accelerator pedal, so that you can make other checks while the engine runs. The increased engine rpm ensures that good oil pressure will lubricate all areas of the engine adequately. It also ensures that the cylinder walls receive generous splash flow. Make sure that the oil pressure is within specifications.

The new rings will need good lubrication to prevent scuffing as they break in to the cylinders. A new camshaft and lifters will endure a lot of friction as they wear in together. Engine bearings will conform to their journals during the first minutes of operation. These areas of increased friction, and others, require the higher oil pressure and flow delivered at the increased engine rpm. Do not let the engine idle.

Monitor the oil pressure and be sure it stays within the specified psi. After a few minutes of operation, top off the coolant level once it has circulated throughout the engine. Replace the reservoir or radiator cap before the thermostat opens. Check the radiator inlet hose as the engine heats up; it should get hot and full as the engine warms up to temperature. If problems occur with oil pressure or cooling, or if you hear unusual noises, shut the engine down and investigate. Prevent extensive engine damage that could occur if you continue to run the engine while obvious problems exist.

While the engine is running at 1,500 to 2,000 rpm, check the engine for leaks. Raise the vehicle and be sure that no seals, gaskets, or plugs are leaking. If you find leakage, stop the engine and repair the leaks now before damage occurs. It may be frustrating to go backwards, but it is a lot better than having to redo major engine work. Listen for unusual noises all over the engine. Use a stethoscope, if needed, to isolate rattles or belt or bearing issues. Make sure that the engine is running on all its cylinders. The engine should look steady and sound smooth under the hood. The exhaust should have regular pulses of exhaust. Listen for vacuum or compression leaks and correct if needed.

Continually check the engine oil pressure and cooling system. Verify that the cooling fan comes on when the engine reaches the prescribed temperature. Use a pyrometer for the best accuracy and definitely if there is no coolant temperature gauge to monitor engine temperature. Check for proper charging system operation by placing the leads of your voltmeter across the positive and negative posts of the battery. If the generator, voltage regulator, and wiring are all good, you should read between 13.5 and 15.0 volts.

After 20 minutes of high rpm operation, return the engine to its normal idle speed. Set the ignition timing as specified if applicable. Shut the engine down, and remove the oil pressure gauge. Install the oil pressure switch, and verify proper operation. The light must come

CAUTION:
Cranking the engine for extended periods can permanently damage or destroy a starter motor.

on with the key in the RUN position; it should go out promptly after engine start-up. Make sure there is no oil leakage by the switch. Check all fluid levels before going on a road test.

Road Test

Prepare yourself for a 15- to 20-minute road test. Turn the radio off; you need to listen closely for rattles, knocks, or squealing. If any abnormal noises occur, make a mental note to repair the associated problem upon return. On modern engines, a knock sensor can pick up vibration from a rattle or knocking and the powertrain control module (PCM) will retard the timing and reduce power. If a rattle or knock is left alone, engine performance will suffer. If you can hear a noise, the customer will also likely pick it up. Figure out the cause, and repair it when you return to the shop. Pull over and stop the engine if a serious knocking noise occurs or if the oil light comes on. Attempting to limp the vehicle back to the shop could destroy your work.

During the road test, you need to begin the process of breaking in the rings. Some ring manufacturers say this is no longer necessary, but it is still a generally recommended procedure. Run the vehicle on the highway or open road. Accelerate to 50 mph (as the speed limit allows), and allow engine braking to bring the speed back down to 30 mph. Accelerate aggressively but not to wide-open throttle. You should stay below 75 percent of maximum engine load while still pushing the engine close to that threshold. Repeat this cycle a dozen times. The acceleration will help the rings seal with good pressure around them. The high vacuum during deceleration helps keep the cylinder walls and rings well lubed as the rings break in. Do not overrev the engine or drive at excessive speeds. The smoking from the exhaust should noticeably diminish after this process, although a little blue smoke is still normal.

Final Checks after Break-In

After your road test, repair any concerns noticed while you were driving. If you noticed a problem with the brakes, suspension, or exhaust system, mark your observations on the repair order so that you can communicate these to the customer when he picks up his vehicle. If the check engine or MIL light came on during your road test, extract the diagnostic trouble code (DTC) from the PCM. If the code relates to a component such as the throttle position sensor or manifold absolute pressure sensor, check the wiring connections and vacuum supply to the part as applicable. Next, diagnose the fault using the manufacturer's procedure.

Give a final check to all the fluid levels. Take the time to put the vehicle back up on the lift and make sure there are no leaks from the engine. Repair any faults now, before the customer can notice them and return angrily to you. A small oil leak could lead to serious engine damage once the customer is driving the vehicle. Take one last look under the hood at wires, hoses, vacuum lines, fuel lines, and belts.

Clean the steering wheel, seat, and fenders. Check the seat for any spots, and use an appropriate cleaner if needed. Put a fresh floor mat into the vehicle. Set the radio station presets to the channels you recorded earlier or to a variety of channels with good reception. Adjust the clock to the proper time. Set the trip odometer to zero, so the customer can keep an eye on the mileage.

Five-Hundred-Mile Service

After 500 miles of operation on a new or rebuilt engine, you should ask the customer to return for a service. Most shops will bill this into the engine overhaul charges so that it appears to be a "free" service to the customer. They are less likely to return if they have to lay out more money shortly after paying the large bill for major engine work. The service work should include an oil change. The oil becomes contaminated quickly with small metal particles as new engine components wear in. It is occasionally recommended to retorque the intake manifold. Cylinder head shifting can affect manifold bolt tightness. Check the adjustment of solid lifters after the valves have had time to wear into their freshly ground or new seats. Verify that the engine is running and idling properly. Finally, road test the vehicle, and be sure that it is performing as it should. Take any measures needed to avoid a customer comeback. Repair any concerns that could be related to your engine work. Make a note of other problems and discuss them with the customer.

> **CUSTOMER CARE:** It is best if you can speak directly with the customer when she comes to pick up her vehicle after major work. Be sure you are presentable and prepared to explain the work you have completed. Describe the problems you found and how you corrected them to ensure proper engine operation. Let the customer know that you have road tested the vehicle and that you know the engine is running properly. If you noticed any unrelated concerns, describe those to the customer so she is are not surprised. If an axle boot is cracked or the muffler is leaking, let her know before she drives the vehicle. The customer will be more aware of noises and faults after a major service. Be prepared with a rough estimate of what the additional work would cost.

Let the customer know that the battery had been disconnected. Explain that you set the clock and some radio stations but that those and seats presets or other memory functions may not be set exactly as they had been prior. Tell her that you set the trip odometer to zero so that she will be aware of when she should return for the 500-mile service. Explain the importance of returning for this oil change and checkup.

Before handing over the keys, explain how the customer should drive the vehicle for the first few hours and then for the next 2,000 miles. In the short term, the customer should avoid maximum engine load or extensive idling. The more she varies the engine speed, the better it is for the rings. Ask the customer to idle the engine for 30 seconds before shutting the engine down after a hard drive. During the first 2,000 miles of operation, the customer should avoid overrevving the engine or driving at excessive speeds. She can drive aggressively but not to the point of abuse.

Make sure that the customer is aware that the rings may not fully seat for a few thousand miles. This can cause increased oil consumption, which is normal. Ask her to check and refill the oil and other fluids at every fuel fill while the engine is still breaking in. If she keeps a note about the amount of oil she has had to add, you'll be able to gauge whether a problem is indicated.

A final step in excellent customer service is to call the customer after a week to make sure the engine is running properly and that she has no concerns. Remind her about the 500-mile service and its importance to the longevity of the engine. Thank her for her business and let her know you'll be looking forward to checking out her vehicle soon. This follow-up call can be an important one. It is hard to overcome the negative statements an unhappy customer might make to others in the community. A satisfied customer is the best advertisement you can get for your shop; take the time to be sure you've done the excellent work you have been trained to perform.

CASE STUDY

On my very first engine overhaul, I failed to properly tighten the flywheel bolts. When I started the engine up, I because the engine had a serious knock. Luckily, the flywheel bolts were accessible through a cover, and I checked them before disassembling anything on the engine. Once they were properly torqued, the engine ran well and lasted a good long time for the customer. I never made that mistake again!

TERMS TO KNOW

Engine hoist
Engine stand
Exhaust gas recirculation (EGR) valve
Purge solenoid
Steam cleaner

ASE-STYLE REVIEW QUESTIONS

1. Preparing the engine for removal is being discussed. *Technician A* says you should read through the service procedure for the specific vehicle before beginning.

   Technician B says that the air-conditioning refrigerant should be vented to the atmosphere.

   Who is correct?

   A. A only   C. Both A and B
   B. B only   D. Neither A nor B

2. *Technician A* says to disconnect the positive battery cable first, then the negative cable.

   Technician B says to disconnect the negative battery cable first.

   Who is correct?

   A. A only   C. Both A and B
   B. B only   D. Neither A nor B

3. *Technician A* says that when removing the engine from a front-wheel-drive vehicle, you should first remove differential.

   Technician B says that when removing an engine from a front-wheel-drive vehicle, it may be necessary to remove some steering and suspension components.

   Who is correct?

   A. A only   C. Both A and B
   B. B only   D. Neither A nor B

4. *Technician A* says that a front-wheel-drive engine may be removed from the top through the hood opening.

   Technician B says the transaxle may have to be removed with the engine on a front-wheel-drive vehicle.

   Who is correct?

   A. A only   C. Both A and B
   B. B only   D. Neither A nor B

5. *Technician A* says that you should use a special cradle to support the engine when it is removed from the bottom of the vehicle.

   Technician B says that you should remove the heavy accessories from the engine while it is on the hoist before mounting the engine on the stand.

   Who is correct?

   A. A only   C. Both A and B
   B. B only   D. Neither A nor B

6. *Technician A* says that you often need to remove the radiator before removing the engine.

   Technician B says to drain the coolant from the engine block as well as from the radiator.

   Who is correct?

   A. A only   C. Both A and B
   B. B only   D. Neither A nor B

7. *Technician A* says that when you are installing a crate engine you will usually install a new water pump if one does not come with the engine.

   Technician B says that you will need to swap components such as sensors and actuators.

   Who is correct?

   A. A only   C. Both A and B
   B. B only   D. Neither A nor B

8. *Technician A* says that a new or freshly rebuilt engine will run at higher engine temperatures during the break-in period.

   Technician B says that you should check oil pressure during the break-in period.

   Who is correct?

   A. A only   C. Both A and B
   B. B only   D. Neither A nor B

9. *Technician A* says that during the first road test you should accelerate and decelerate to full throttle several times to maximize oil pressure.

   Technician B says that after the road test you should put the vehicle back up on the lift and make a final check for fluid leaks.

   Who is correct?

   A.   A only   C.   Both A and B
   B.   B only   D.   Neither A nor B

10. *Technician A* says that any engine stand can hold engines that are eight cylinders or less.

    Technician B says that the attaching bolts must be threaded in at least as much as one and a half times the diameter of the bolts.

    Who is correct?

    A.   A only   C.   Both A and B
    B.   B only   D.   Neither A nor B

ASE CHALLENGE QUESTIONS

1. *Technician A* says that an ECR valve should be thoroughly cleaned of RTV and fuel before reinstallation.

   Technician B says that the purge solenoid admits fuel vapors from the fuel tank into the intake manifold.

   Who is correct?

   A.   A only   C.   Both A and B
   B.   B only   D.   Neither A nor B

2. *Technician A* says that you can remove the valve stem from the Schrader valve to drain the fuel from the fuel rail.

   Technician B says to install a pressure gauge to drain the fuel from the fuel rail.

   Who is correct?

   A.   A only   C.   Both A and B
   B.   B only   D.   Neither A nor B

3. *Technician A* says to remove all the engine accessories before mounting the engine on the stand.

   Technician B says to place the crate engine next to the old engine to help you transfer all the accessories to the proper location.

   Who is correct?

   A.   A only   C.   Both A and B
   B.   B only   D.   Neither A nor B

4. *Technician A* says that it is illegal to discharge R-12 or R-134a refrigerants to the atmosphere.

   Technician B says that you must be certified to operate air-conditioning reclaiming/recycling equipment.

   Who is correct?

   A.   A only   C.   Both A and B
   B.   B only   D.   Neither A nor B

5. *Technician A* says to let the engine idle for the first 15 minutes of operation after a rebuilt engine is installed.

   Technician B says that the oil should be changed after 500 miles of service on the rebuilt engine.

   Who is correct?

   A.   A only   C.   Both A and B
   B.   B only   D.   Neither A nor B

FWD ENGINE REMOVAL

Name _____ Date _____

Upon completion of this job sheet, you should be able to properly prepare and remove the engine from a vehicle.

ASE Correlation

This job sheet is related to ASE Engine Repair Test content area: General Engine Diagnosis; tasks: Remove an engine from a FWD vehicle.

Tools and Materials

Engine lift
Engine stand
Technician's tool set

Describe the vehicle being worked on:

Year _____ Make _____ Model _____
VIN _____ Engine type and size _____

Procedure

Task Completed

Your instructor will assign you (or your group) to a vehicle that needs to have its engine removed. Using the proper service manual, you will identify and perform the required steps to remove the engine.

1. Is the vehicle FWD? ☐ Yes ☐ No
   According to the service manual, does the engine come out from the top or the bottom? ☐ Top ☐ Bottom

2. Following all safety cautions and warnings, clean the engine and compartment. ☐

3. On a separate sheet of paper, make a list of any special tools and equipment that are required to remove the engine. After making the list, gather the tools and organize them in your area. ☐

4. Center the vehicle over the frame contact hoist. ☐

5. Disconnect and remove the battery. ☐

6. Lift the vehicle and drain all necessary fluids. ☐

7. Is the transmission coming out with the engine? ☐ Yes ☐ No

8. Make a list of all the components that can be removed easily while the vehicle is on the hoist.

   _____
   _____
   _____
   _____

9. Describe the method you will be using to assure correct electrical and vacuum line connections are made when the engine is reinstalled.

_____

_____

_____

_____

☐ _____

☐

10. Perform all steps necessary to prepare the bottom of the vehicle for engine removal.

11. Lower the vehicle and perform all steps needed to prepare upper engine for removal.

12. List the components that can be removed as units from the engine block.

_____

_____

_____

_____

_____

_____

13. Following the service manual procedure, remove the engine.

14. Once the engine is removed, attach it to an engine stand.

15. When attaching the engine to the stand, what is the minimum depth the bolts must thread into the block?

_____

Instructor's Response _____

_____

_____

Name _____   **Date** _____

RWD Engine Removal

Upon completion of this job sheet, you should be able to properly prepare and remove the engine from a vehicle.

ASE Correlation

This job sheet is related to ASE Engine Repair Test content area: General Engine Diagnosis; tasks: Remove an engine from a RWD vehicle.

Tools and Materials

Engine lift
Engine stand
Technician's tool set

Describe the vehicle being worked on:

Year _____ Make _____ Model _____
VIN _____ Engine type and size _____

Procedure

Task Completed

Your instructor will assign you (or your group) to a vehicle that needs to have its engine removed. Using the proper service manual, you will identify and perform the required steps to remove the engine.

1. Is the vehicle RWD?   ☐ Yes   ☐ No
   According to the service manual, does the engine come out from the top or the bottom?   ☐ Top   ☐ Bottom

2. Following all safety cautions and warnings, clean the engine and compartment.   ☐

3. On a separate sheet of paper, make a list of any special tools and equipment that are required to remove the engine. After making the list, gather the tools and organize them in your area.   ☐

4. Center the vehicle over the frame contact hoist.   ☐

5. Disconnect and remove the battery.   ☐

6. Lift the vehicle and drain all necessary fluids.   ☐

7. Is the transmission coming out with the engine?   Yes   ☐ No

8. Make a list of all the components that can be removed easily while the vehicle is on the hoist.

   _____
   _____
   _____
   _____

_____

9. Describe the method you will be using to assure correct electrical and vacuum line connections are made when the engine is reinstalled.

_____

_____

_____

_____

_____

☐   **10.** Perform all steps necessary to prepare the bottom of the vehicle for engine removal.

☐   **11.** Lower the vehicle and perform all steps needed to prepare upper engine for removal.

12. List the components that can be removed as units from the engine block.

_____

_____

_____

_____

_____

_____

_____

☐   **13.** Following the service manual procedure, remove the engine.

☐   **14.** Once the engine is removed, attach it to an engine stand.

15. When attaching the engine to the stand, what is the minimum depth the bolts must thread into the block?

_____

Instructor's Response _____

_____

_____

JOB SHEET

Name _____ **Date** _____

ENGINE INSTALLATION

Upon completion of this job sheet, you should be able to properly prepare and install the engine into a vehicle. This job sheet is related to ASE Engine Repair Test content area: General Engine Diagnosis; tasks.

Tools and Materials

Engine lift
Technician's tool set
Vehicle lift

Describe the vehicle being worked on:

Year _____ Make _____ Model _____
VIN _____ Engine type and size _____

Procedure

Task Completed

Your instructor will assign you (or your group) to a vehicle that needs to have its engine removed. Using the proper service manual, you will identify and perform the required steps to remove the engine.

1. Is the vehicle FWD or RWD? ☐ FWD ☐ RWD
   If FWD, according to the service manual, does the engine get installed from the top or the bottom? ☐ Top ☐ Bottom

2. Prepare the engine compartment for reinstallation by tying back any hoses, lines, or components that may be in the way. ☐

3. If necessary, be sure that the suspension components are not going to block engine installation. ☐

4. Make sure that the transmission is properly mounted and fastened to the engine if it was removed with the engine. ☐

5. Center the vehicle over the frame hoist.

6. If the engine is being installed from the bottom, raise the vehicle on the lift. Center the engine under the vehicle and *slowly* lower the engine into alignment with the mounts. Reattach the engine mounts and secure the components required from underneath the vehicle. Make a list of components that need to be reinstalled from underneath the vehicle.

   _____
   _____
   _____
   _____

7. If the engine is being installed from the top, center the engine crane over the engine compartment. Slowly lower the engine into alignment with the engine mounts and secure the engine mounts. Raise the vehicle and describe the components that need to be reinstalled from underneath the vehicle.

_____
_____
_____
_____
_____

8. Lower the vehicle and attach all the hoses, lines, electrical connectors, and components that need to be reattached. Describe the process and components.

_____
_____
_____
_____
_____
_____

9. Double check your work so far. Then, fill the engine and other components with fluids. Describe what fluids need to be filled.

_____
_____
_____
_____
_____

10. Install an oil pressure gauge into the system so you can monitor oil pressure at start-up.

11. Start the engine and follow the break-in procedures. Describe the procedures and your results here.

_____
_____
_____
_____
_____

Instructor's Response _____
_____
_____

Chapter 9

CYLINDER HEAD DISASSEMBLY, INSPECTION, AND SERVICE

BASIC TOOLS

Basic mechanic's tool set
Service manual

UPON COMPLETION AND REVIEW OF THIS CHAPTER, YOU SHOULD BE ABLE TO:

- Remove a cylinder head from an engine.

- Properly disassemble a cylinder head and prepare it for inspection and cleaning.

- Perform a complete visual inspection of the valve and determine any failures or damage.

- Properly determine wear of the valves by use of measuring instruments and by comparing results with specifications.

- Inspect the cylinder head casting for cracks.

- Determine the amount of mating surface warpage and recommend needed repairs.

- Measure valve guide wear, clearance, and taper, and determine needed repairs.

- Perform a complete visual inspection of the valve seats and properly measure valve seat width and runout.

- Straighten aluminum cylinder heads.

- Correct camshaft bearing bore alignment in an overhead camshaft (OHC) cylinder head.

- Repair or replace worn valve guides.

- Replace integral and insert valve seats.

- Grind valves and obtain correct contact pattern.

- Recondition valve faces and stem tips.

- Identify proper valve face-to-seat contact patterns.

The cylinder head can be removed with the engine installed in the vehicle or as part of engine disassembly on an engine stand. With the cylinder head removed from the engine block, it is ready to be disassembled, cleaned, and completely inspected. The cylinder head should be checked for cracks after it is cleaned. Work performed on a cracked cylinder head would be wasted. Disassembly generally includes removal of the rocker arms, valve springs, valves, **core plugs**, and oil **gallery plugs**. It is important for today's technician to properly inspect the cylinder head and its components to determine the presence of defects. Just as important, the cause of the failure should be identified. This chapter covers typical cylinder head disassembly procedures and inspection. Keep track of the measurements obtained and the results of the inspections on the work order or in a notebook for future references.

Once the technician has determined the type of machine work required to restore the cylinder head, the actual work is the next step. Using the information gathered during the inspection process, the technician should have a good idea of how much machining will be required for each operation. It is important to perform the different machining and reconditioning operations in a logical order. This will prevent having to repeat operations.

The valve faces and seats are usually reconditioned as a matter of common practice whenever the cylinder heads are disassembled. This chapter provides general procedures for reconditioning the cylinder head. In addition, this chapter discusses other major cylinder

> **Core plugs** are used to cap the holes used to create hollow passages for coolant during the foundering process.

> **Gallery plugs** are used to cap the drilled oil passages in the cylinder head or engine block.

head reconditioning operations, including: straightening aluminum cylinder heads, valve guide repair, reconditioning valve seats, grinding valves, and reconditioning cast rocker arms. The diversity of the different types of equipment available to perform cylinder head reconditioning operations makes it impossible to cover each procedure with each machine. The general procedures are provided in this text. Always refer to the equipment manufacturer's instructions. If you are not familiar with the equipment, seek instruction before attempting to operate it.

Cylinder Head Removal

Cylinder head removal can be performed as an in-vehicle service if a total engine rebuild is not necessary. Follow the manufacturer's service information for cylinder head removal. It will offer the most accurate instructions for proper service. Our discussions are general and are to give you a sense of the procedure. They are not a replacement for specific service instructions. The removal procedure will vary depending on whether the engine is an OHC or OHV engine. Regardless of the design, some typical precautions should be observed. These include:

1. Do not loosen the intake manifold or head bolts when the engine is still warm. These components and the block should cool together to prevent warpage.
2. When removing the intake manifold bolts, reverse the tightening sequence as described in the service manual.
3. When removing the cylinder heads, loosen the head bolts in the reverse of the tightening sequence listed in the service manual. Loosen all the bolts first, and then go back and remove them.

OHV Engine Head Removal

To remove the cylinder heads from an OHV engine, begin by removing the valve covers (Figure 9-1). If the engine is in the vehicle, you may need to remove electrical connectors, vacuum lines, the throttle cable, and various brackets and components to gain access to the

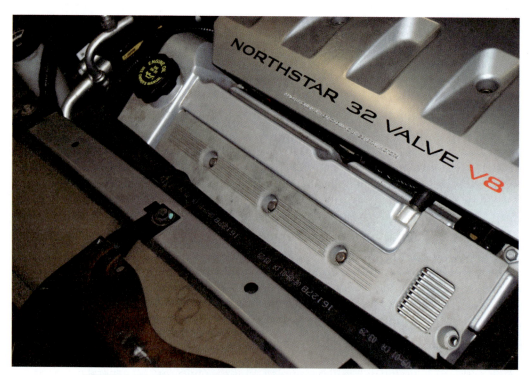

FIGURE 9-1 In this case, you'll have to remove the ignition cassette in order to gain access to the valve cover bolts.

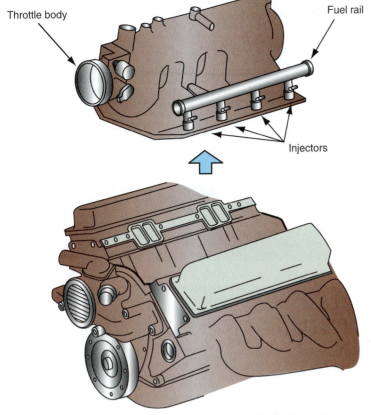

FIGURE 9-2 Remove the intake manifold with the fuel rail and injectors attached.

covers. Mark these lines, connectors, and brackets for reference during reassembly. Use special care when removing plastic, aluminum, and magnesium valve covers. They will not bend; if you crack them, they must be replaced. Tin and stamped steel valve covers are a little more forgiving.

If the engine is equipped with a distributor, it may need to be removed. Prior to removing it, make an index mark to aid in proper installation. Remove the intake duct work and other components that will hinder the removal of the intake manifold.

When removing the intake manifold, remove it in as much of an assembly as possible (Figure 9-2). The intake manifold can usually be removed with the injectors, fuel rail, and throttle body still attached.

Next, remove the exhaust manifold from the cylinder head (Figure 9-3). In some areas of North America the exhaust studs and nuts may be quite rusty. You will usually replace the studs, but it is easier to remove them if they are not broken. It is good practice to spray these fasteners with penetrating oil and to allow them to soak a few minutes before removing them. In some cases, you will need to use some heat on the nuts before attempting to loosen them.

Remove the rocker arm assemblies and pushrods. The method of removing the rocker arms depends on the mounting design. If the rocker arms are located on a single shaft, remove the fasteners that attach the shaft to the cylinder head. If the rocker arms are stud mounted or pedestal mounted, remove each one individually by removing the adjusting nut. If the rocker arms and push rods are going to be reused, it is a good practice to mark or store them in their original positions to assure they are installed in the proper positions during reassembly.

Next, remove the cylinder head bolts in reverse order of the tightening sequence. Break each bolt loose first; then go back and remove them. Pay attention to the sizes of the bolts as you remove them. Mark the bolts to identify their locations for reassembly (Figure 9-4).

FIGURE 9-3 Soak the rusty exhaust manifold nuts with penetrating oil or use heat to remove them.

FIGURE 9-4 Whether this is your first time rebuilding an engine or not, keep the bolts organized.

SERVICE TIP:
Never attempt to "slide" the head off the cylinder block. You may damage the head or block by moving the dowel pins in the block (Figure 9-5).

The cylinder head should now be ready for removal. It may be necessary to pry the cylinder head to break it loose. Make sure to pry in a location that will not damage the sealing surfaces. Do not use excessive force to pry the cylinder head loose. If it does not come off, make sure all of the bolts have been removed. On engines that use wet liners, leave the center bolt loose in the head and place a 2 × 4 inch block of wood against the cylinder head. Hit the wood sharply with a hammer to rotate the cylinder head. This will break the sleeves loose from the head and prevent them from being lifted out of the block as the head is lifted.

OHC Engine Head Removal

Begin cylinder head removal by removing the valve cover. This may require the removal of several accessory items first. Next, remove the intake manifold (Figure 9-6). Then remove the exhaust manifold and gasket from the cylinder head (Figure 9-7).

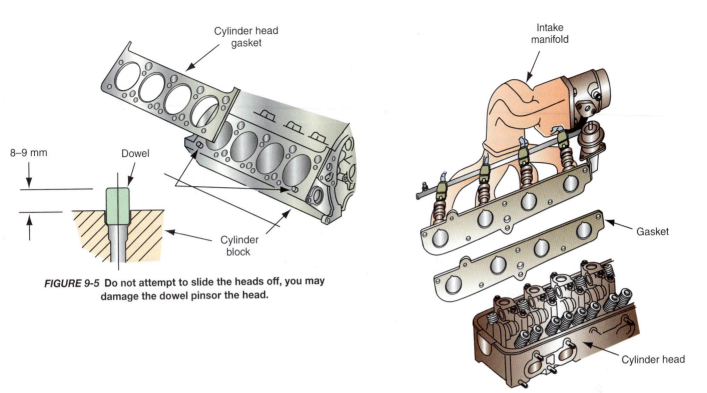

FIGURE 9-5 **Do not attempt to slide the heads off, you may damage the dowel pins or the head.**

FIGURE 9-6 **Removing the intake manifold from an OHC engine.**

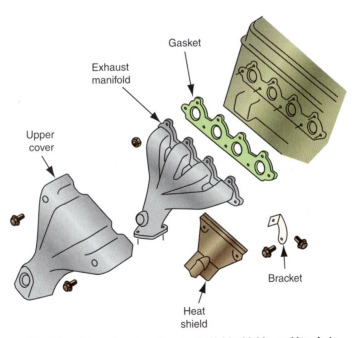

FIGURE 9-7 **Removing the exhaust manifold, shields, and brackets.**

On an overhead camshaft engine, you will have to disturb the timing mechanism, either a timing chain or a timing belt. In many cases, there is a special tool that bolts to the camshaft sprockets, or in between the chain on a dual overhead camshaft engine, to hold the timing chain in place for cylinder head service. This tool prevents you from having to disassemble the front cover and access the tensioner. It will hold the chain in place while you remove the cylinder head. Follow the service manual instructions to install the tool before removing the camshaft or sprockets. If you do need to remove the belt or chain, be

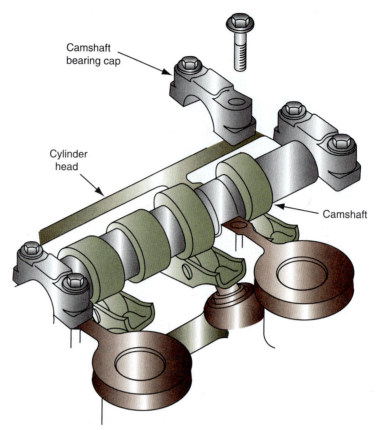

FIGURE 9-8 Some OHC engines use split camshaft bearings. Note the markings and remove the caps to remove the camshaft from the engine.

Camshaft bearing cap

Cylinder head

Camshaft

sure to rotate the engine to TDC number 1 and note the timing marks on the camshaft sprocket. If the engine is equipped with a variable valvetrain, you will have to use special tools to remove the servo and camshaft. These are explained in the manufacturer's service manual.

In some cases, you will need to remove the rocker arm shaft assemblies to gain access to the cylinder head bolts. Be sure to mark the rocker arms so that they can be returned to their original positions on the camshaft. Other engine designs necessitate the removal of the camshaft. Some camshafts slide out of the bearing supports; others have split caps that allow the camshaft to be lifted from the cylinder head (Figure 9-8). Note the markings on the caps before disassembly. It is critical that the caps be returned to the same location and installed in the same direction when reassembling. When removing the caps, follow the sequence in the service manual to prevent warping or breaking of the camshaft or damage to the threads in the cylinder head. As the camshaft is removed, mark or organize the followers to identify their original positions. There may be adjusting shims with each follower; it is important to keep them with their respective valve and follower assemblies.

DISASSEMBLING THE CYLINDER HEAD

The cylinder head is attached to the block above the cylinder bores and is usually constructed of cast iron or aluminum. Disassembling the cylinder head usually begins with cleaning it with mild solution in a parts washer or using a steam cleaner. Sludge can build up around the rocker arms and valve spring areas. Excessive sludge may indicate the engine was neglected. This may include improper oil change intervals or not allowing the engine to warm properly before driving short distances. Removing this sludge will make the task of cylinder head disassembly easier and safer. However, before cleaning the head, inspect the combustion chamber. The type, color, and quantity of carbon buildup can provide hints to identify problem areas.

> **CUSTOMER CARE:** If excessive sludge is found throughout the cylinder head and crankcase, tactfully explain to the customer the need to change the engine oil at the mileage interval recommended by the vehicle manufacturer.

Normal carbon buildup is a light layer evenly covering the combustion chamber. If the carbon is thick and black in color, this indicates excessive oil induction into the combustion chamber. Oil induction is also identifiable by an oily residue on the carbon. This could be the result of worn valve stem seals, valve guides, or rings. Dry, black, ash-type deposits indicate a rich air-fuel mixture or ignition misfire. Some common causes include a ruptured fuel pressure regulator, excessive fuel pressure, leaking injectors, or faulty spark plug cables, coils, or spark plugs.

If the combustion chamber is clean and the metal is shiny, engine coolant may have leaked into it (Figure 9-9). This can be caused by a damaged head gasket, cracked cylinder head, cracked engine block, or blown intake gasket.

After the initial inspection of the cylinder head is complete, the head should be cleaned. Before cleaning the cylinder head, remove the core and galley plugs. Remove carbon deposits from the combustion chambers with a scraper or wire wheel, then clean the cylinder head in a parts cleaner or use a steam cleaner. This cleaning will make cylinder head inspection easier and disassembly safer. It will be necessary to clean the cylinder head again after the valves and valvetrain components are removed. Once the initial cleaning is complete, inspect the cylinder head for cracks and other obvious damage. If the engine has been overheated, the cylinder head should be checked for cracks using magnaflux if it is cast iron or through pressure testing or fluorescent dye if it is aluminum. Aluminum heads should always be checked for cracks. If the cylinder head assembly has excessive damage, it may be more cost efficient to replace it instead of repairing it. If the cylinder head passes this preliminary inspection, it is ready for disassembly.

SPECIAL TOOLS
Parts cleaner

See Chapter 4 for information on removing core plugs.

A core plug may be called a frost plug. If antifreeze protection is inadequate and coolant freezes in the cylinder head, the core plugs are usually pushed out of the head.

FIGURE 9-9 The number 1 piston is clean from a blown headgasket, allowing coolant into the cylinder.

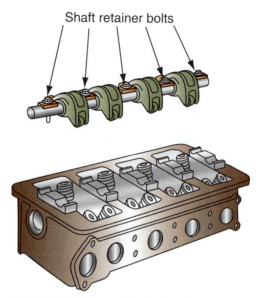

Shaft retainer bolts

FIGURE 9-10 **Remove the shaft retainer bolts to remove the rocker arm shaft.**

The rocker arms usually must be removed before the valves can be accessed for removal. Some overhead cam (OHC) cylinder head designs have the camshaft running a follower directly on the valve, while others use rocker arms. Rocker shafts are removed as an assembly by removing the shaft retainer bolts (Figure 9-10). If the rocker arms are fitted under the camshaft, it may be necessary to remove them by prying against the spring retainer and sliding the rocker arm out.

Valve Removal

SPECIAL TOOLS

Valve spring
compressor
Soft-faced hammer
Pencil magnet
Organizing board

Before beginning the task of removing the valves from the cylinder head, review these safety concerns:

1. Check the condition of the valve spring compressor before using it. If the tool is worn or damaged, do not use it.
2. Always wear safety glasses when removing the valve springs. When removing the keepers, the valve springs must be compressed. If the spring comes loose, both the keepers and the spring may be propelled at a high rate of speed and force.
3. When removing the valve springs, do not stand in front of the springs. Turn the heads so you are facing the combustion chamber and the springs are facing away from you.
4. Do not compress the valve spring more than needed to remove the keepers. Overcompressing the spring may result in damage to the compressor and cause the spring to come loose (Figure 9-11).

Before removing the valves from the head, measure valve stem height. After the springs are removed, measure the height of the valve stem tip from the head casting (Figure 9-12). This can be done using calipers or by special tools designed for this function (Figure 9-13). Record the measurements for reference when performing valve reconditioning.

When disassembling the cylinder head, consider the following:

1. Organize the components so their original positions in the head are maintained. This can be done using a board with holes drilled in it to hold the valves, valve springs, and rocker arms.
2. When handling OHC cylinder heads, remember some of the valves will be held open by the camshaft. Do not set the head down on the combustion chamber side, or damage to the valves may result. Mounting the head on head stands will prevent this damage, and make the disassembly procedure easier (Figure 9-14).

FIGURE 9-11 Use a valve spring compressor to compress the spring and remove the keepers.

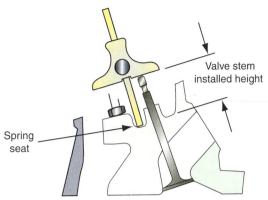

FIGURE 9-12 Measuring valve stem tip height.

FIGURE 9-13 Valve stem height gauge.

Head stands

FIGURE 9-14 Use head stands to hold the head up off the bench.

3. The valve retainers may be stuck to the keepers. Break the varnish loose before attempting to compress the spring by gently striking the retainer with a soft-faced hammer with an old piston pin (or small piece of pipe).

4. Measure stem tip height prior to removing the valve. After the valve springs are removed, use vernier calipers or a special valve stem tip height gauge to measure the distance from the cylinder head casting to the tip of the stem. The measurements should be within specifications and consistent between valves. Stem tip height greater than specifications indicates a recessed valve seat or a necked valve stem. Correct valve stem height is important for proper rocker arm geometry and lifter operation. If the seats and valves are reconditioned, valve stem tip height must be restored. Valve stem tip height must be within specifications after assembly or the valves may be held open.

5. The tip of the valve stem may become "mushroomed" as the rocker arm pounds against it (Figure 9-15). Use a file or cup-type grinding wheel to debur the stem before removing the valve. Do not drive a valve stem through the guide. This can result in guide damage.

Follow Photo Sequence 17 on page 347 to disassemble a typical OHC cylinder head. The procedure for disassembling the (OHV) cylinder head is similar to that for an (OHC) cylinder head, except there are no camshafts.

SERVICE TIP:
Always refer to the service manual for the engine you are working on. Photo Sequence 17 is a typical example that may assist in the event a service manual is not available.

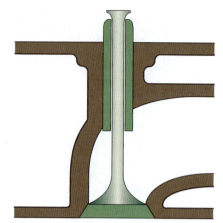

FIGURE 9-15 Remove any tip "mushrooming" before removing the valve from the guide.

WARNING: Wear approved eye protection when disassembling the cylinder head. Parts may come loose suddenly and become airborne.

Classroom Manual

Chapter 9

Valve stem seals are used to control the amount of oil traveling down the valve stem and to provide lubrication. These seals are constructed of synthetic rubber or plastic and are designed to control the flow of oil, not prevent it. There is no reason to reuse valve stem seals; they should always be replaced when reconditioning the cylinder head. However, inspection of the seals may confirm your diagnosis or indicate that additional inspection is required (Figure 9-16).

Remove the O-ring and umbrella seals by simply lifting them off. Positive-lock seals can be removed by prying them off with a small screwdriver, pliers, or a special removal tool (Figure 9-17).

As the cylinder head is disassembled and inspected, keep good notes concerning worn or damaged parts. In your notes, indicate if machining or replacement parts are needed. After inspecting the components of the cylinder head and referring to your notes, you will be able to provide an accurate evaluation to the customer. You will be able to help the customer determine the most cost-effective way of correcting any faulty conditions. Depending on the shop, sources for parts include vehicle manufacturer's dealerships, chain retail and jobber stores, independent parts suppliers, and recycled parts distributors. It may be less

Positive
lock seal

Umbrella
oil shedder

FIGURE 9-16 Inspect the old valve stem seals to confirm your diagnosis.

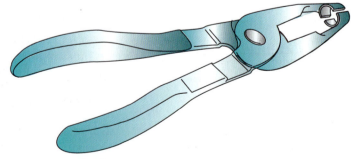

FIGURE 9-17 Special valve stem seal pullers.

TYPICAL PROCEDURE FOR DISASSEMBLING AN OHC CYLINDER HEAD

P17-1 Tools required to perform this task include a cylinder head holding fixture, a valve spring compressor, a dial indicator, dial calipers, a soft-faced hammer, an old piston pin, a pencil magnet, and a service manual.

P17-2 Mount the cylinder head holding fixture. Then reverse the tightening sequence to loosen the rocker arm shaft attaching bolts, and remove the rocker arm shafts from the cylinder head. Be sure to maintain the original order of the rocker arms.

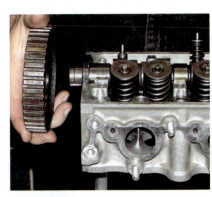

P17-3 Remove the camshaft timing belt sprocket.

P17-4 Measure camshaft end play by attaching the dial indicator so the contact point is against the front of the camshaft. Zero the dial while the camshaft is pried rearward.

P17-5 Pry the camshaft forward and note the indicator reading.

P17-6 Remove the camshaft from the cylinder head. If the head is equipped with cam followers, lash adjusters, or lash adjusting shims, remove them at this time.

P17-7 Use a soft-faced hammer and an old piston pin to loosen the keepers from the retainers. Do not use heavy blows.

P17-8 Use a valve spring compressor to relieve valve spring tension. Compress the spring only enough to remove the keepers.

P17-9 Remove the keepers from the top of the valve stem. A pencil magnet can be used to lift off the keepers.

expensive to purchase a used part and recondition it than to attempt to recondition the part that was originally on the engine; for example, if a cylinder head with integral valve seats is found to have four worn seats requiring replacement, and two combustion chambers have cracks between the valve seats, it may be more beneficial to purchase a used cylinder head requiring only the seats to be cut.

It may be necessary to remove manifold studs and rocker arm studs before the head can be machined. The studs can be either threaded in or press fit. Use a stud remover to remove threaded-in studs (Figure 9-18). Special stud pullers are available to remove press-fit studs (Figure 9-19). An alternative method for pulling press-fit studs is to place a deep socket or steel tube over the stud; then install a washer and nut onto the threads (Figure 9-20). As the nut is tightened against the washer, the stud is pulled from the cylinder head.

> **CUSTOMER CARE:** If metal erosion is evident and the apparent cause is not coolant leakage, advise the customer to avoid the use of chemicals not recommended by the vehicle's manufacturer. Use of these types of chemicals may void the vehicle warranty.

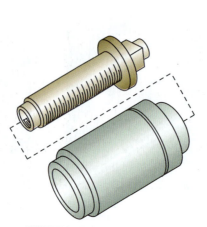

FIGURE 9-18 Use a stud remover to remove threaded studs from the head.

FIGURE 9-19 Stud puller used to remove press-fit studs.

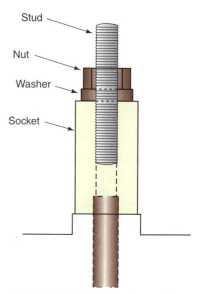

FIGURE 9-20 Using a socket and nut to draw out press-fit studs.

FIGURE 9-21 Use a guide brush to clean carbon from the valve guides.

Complete this segment of cylinder head disassembly by inspecting the cylinder head coolant and oil passages for dents, scratches, and corrosion. With the cylinder head completely disassembled, it is ready for cleaning. Some areas of the head and valves may require the use of a wire brush to remove stubborn carbon (Figure 9-21). Make sure all oil passages are open and clean.

VALVE INSPECTION

Modern automotive engines use poppet valves, which are opened by applying a force against their tips. The valve is closed by spring pressure. This opening and closing action results in wear of the valve face at the contact point with the seat (Figure 9-22). If this wear is excessive, the valve will not seal properly, resulting in poor engine performance and possible burned valves. Begin inspection of each valve using a close visual inspection. A normally worn valve will need to be cleaned of carbon. You will be able to see the seating contact area on the valve face (Figure 9-23). The valve in Figure 9-21 shows normal wear and will be refinished and put back into service. Look for indications of burning, galling, warpage, and bends. If the valve fails visual inspection, it must be replaced. Any valves passing the

Classroom Manual
Chapter 9

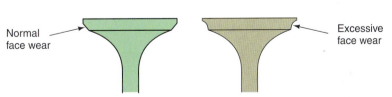

Normal face wear

Excessive face wear

FIGURE 9-22 Comparison between a normal valve face and a badly worn valve face.

FIGURE 9-23 This valve shows little carbon buildup and normal wear on the valve face; it should clean up nicely.

visual inspection must be measured for wear before reusing. Areas of the valve requiring measuring include the face, margin, head, and stem (Figure 9-24). Always compare the measurements against specifications.

Inspecting the Valve Head

All valves must be inspected for damage and wear. The valve must be completely clean in order to properly inspect it. The fillet area may become coated with hard carbon deposits (Figure 9-25). Before placing the valve into a cleaner, remove as much of the carbon as possible. Using an old valve to knock off the deposits works well (Figure 9-26). Since the fillet shape affects the flow of the air-fuel mixture around the valve, leaving any carbon in this area will disrupt the flow of the air-fuel mixture. A fine wire brush, buffing wheel, or parts tumbler can be used to remove stubborn carbon from the valve.

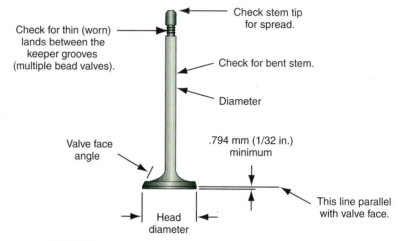

FIGURE 9-24 High wear areas of the valve that must be checked.

FIGURE 9-25 This valve has heavy carbon buildup that must be thoroughly cleaned.

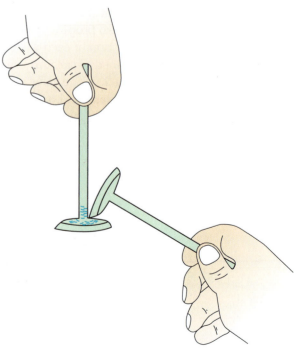

FIGURE 9-26 Use an old valve to knock carbon off the fillet area.

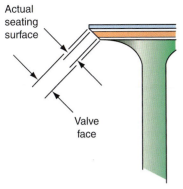

FIGURE 9-27 Proper seat contact should be centered on the valve face.

FIGURE 9-28 If the margin is worn to less than 1/32 inch, the valve must be replaced.

Note the contact location of the valve face and seat. Proper seat contact is in the middle of the face (Figure 9-27). Incorrect contact areas can usually be corrected. Minor pits, grooves, and scoring can be removed during the refacing operation. Excessively worn valve faces require valve replacement.

Visually inspect the margin for excessive wear. The margin allows for machining of the valve face and for dissipation of heat away from the valve head. As the face wears, the margin may become thinner. If the margin is worn away, the valve may not dissipate the heat sufficiently, resulting in a burned valve, cupped head, or warped stem. If the margin is worn away to the point where the face and head form a sharp edge, the valve must be replaced (Figure 9-28).

Use an outside micrometer or calipers to measure the diameter of the valve heads. Compare the results with the manufacturer's specifications. When measuring the valve head, check in at least two directions to inspect for out-of-round.

Use vernier calipers or a machinist's rule to measure the width of the margin. The size of the valve margin must meet or exceed the manufacturer's specifications for the valve face to be reconditioned. If specifications are not available, most manufacturers require at least 1/32 in. (0.8 mm) margin width. Also, the margin must be measured after refacing the valve to assure the margin is still at least 1/32 in. (0.8 mm) wide.

Valve Stem Inspection

The stem guides the valve through its linear movement. The stem also has valve keeper grooves machined into it close to the top (Figure 9-29). These grooves are used to retain the valve springs. The stem rides in a valve guide located in the cylinder head. The continual

Classroom Manual
Chapter 9

SERVICE TIP:
Burned valves are often caused by improper seat contact. This can be caused by worn valve stems or valve guides. Inspect all components before replacing the damaged valve.

FIGURE 9-29 Inspect the keeper groove(s) near the top of the valve stem.

SPECIAL TOOLS

Straightedge
Feeler gauge set
Bore gauge
Inside micrometer
Outside micrometer

movement of the valve may result in wear on the portion of the stem traveling in the guide. As the stem wears, the oil clearance increases. This results in excessive oil being digested into the combustion chamber and the formation of deposits on the valve.

When visually inspecting the valve stem, look for galling and scoring. Either of these can be caused by worn valve guides, carbon buildup, or off-square seating. If the valve stem is scored, the valve is usually replaced. Also look for indications of stress. A stress crack near the fillet of the valve indicates excessive valve spring pressures or uneven seat pressure. Stress cracks near the top of the stem can be caused by excessive valve lash, worn valve guides, or weak valve springs. If a stress crack is found, replace the valve.

Inspect the tip of the valve stem for wear and flattening. These conditions can be caused by improper rocker arm alignment, worn rocker arms, or excessive valve lash. Separation of the tip from the stem can be caused by these same conditions. Light wear can be corrected when the valve is reconditioned. If the stem is flattened excessively, causing the overall length of the valve to be below specifications, replace the valve.

Another condition to look for during the visual inspection is necking. Necking is identified by a thinning of the valve stem just above the fillet (Figure 9-30). The head pulls away from the stem, causing the stem to stretch and thin. This condition is caused by overheating or excessive valve spring pressures. In addition, necking can be caused by exhaust gases circling around the valve stem.

If the valve stem passes visual inspection, it must be measured for taper and wear. Measure the stem diameter in three locations using an outside micrometer (Figure 9-31). Measure the stem at the top, in the middle, and near the fillet. The diameter of the stem must meet specifications. As a general rule, taper should not be more than 0.001 in. (0.025 mm). Always refer to the manufacturer's specifications when measuring valve stems. Many modern engines use tapered stems to provide additional clearance between the stem and guide close to the head of the valve. This additional clearance allows for expansion due to heat and reduces scuffing, galling, and sticking valves. A tapered stem usually has a diameter 0.001 in. (0.025 mm) smaller at the head than at the tip. If the valve stem diameter is smaller than specifications or excessively tapered, it must be replaced.

Bent valve stems are not always identified during a visual inspection, yet a bent valve cannot be reused. The easiest method of detecting a bent stem is during the valve reface operation. With the stem secured in the chuck of the valve machine, observe the face as it approaches the grinding wheel. If the valve head wobbles, the valve stem is bent.

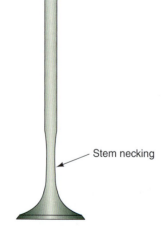

— Stem necking

FIGURE 9-30 A valve with a necked stem must be replaced.

FIGURE 9-31 Visually inspect the valve stem for scoring and measure it for wear.

CYLINDER HEAD INSPECTION

With the valves and valvetrain components removed from the cylinder head and the casting thoroughly cleaned, the head must be inspected to determine its serviceability. The cylinder head casting, valve guides, and valve seats are visually inspected and measured.

See Chapter 5 in the Classroom Manual for crack detection and repair methods.

Visually inspect the cylinder head for cracks and other damage. Remember, not all cracks may be visible to the eye. Therefore, it may be necessary to perform additional tests if a crack is suspected. Common locations for cracks are between the valve seats, from the spark plug hole and valve seats, or any location where the metal is thinned. If a crack is detected, the head must be replaced or the crack repaired. If the crack is to be repaired, perform this operation before performing any machining on the head. Some manufacturers do not recommend repairing cracks and suggest the cylinder head be replaced. Always follow the manufacturer's recommendations.

Inspect the mating surface to the block and the old gasket for indications of leakage. The leakage can be from the coolant passages into the combustion chamber, from the coolant passages to the outside of the head, or compression leakage between cylinders. Compression leaks are identifiable by trails of carbon adhering to the cylinder head. Also, aluminum heads with corrosion around the water passages must be resurfaced. Also look for gasket etching. If the etching is deep enough to catch your fingernail as you move across it, the head must be resurfaced.

Next check the cylinder head's mating surface to the block for warpage, using a straightedge and feeler gauge (Figure 9-32). Warpage can occur in any direction on the head surface. Measure for warpage along edges and three ways across the center (Figure 9-33). The head surface must be thoroughly clean to get accurate measurements. When checking for warpage on the ends of the head, be sure to rock the straightedge to the opposite side of the head. If this is not done, only half of the warpage will be measured. Compare the resulting measurements with specifications. In the absence of specifications, a general rule of thumb is 0.003 in. (0.08 mm) for any 6-in. (153 mm) length or 0.006 in. (0.18 mm) overall. Excessive warpage may indicate the engine was overheated. If this is suspected, closely inspect the cooling system components.

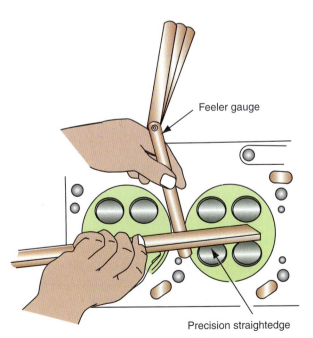

Feeler gauge

Precision straightedge

FIGURE 9-32 Use a straightedge and feeler gauges to determine warpage of the cylinder head.

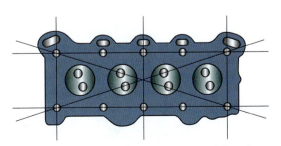

FIGURE 9-33 Measure for warpage around the edges and across the middle of the cylinder head.

The feeler gauge can be used as a "go, no-go" gauge by using a feeler gauge the thickness of the maximum allowable warpage. If the feeler gauge fits under the straightedge, the warpage is over the maximum limit. If the feeler gauge fits under the straightedge, increase the thickness to determine the amount of warpage. Continue to increase the thickness until the feeler gauge has a noticeable drag as it is pulled from under the straightedge. This information is necessary to determine if the cylinder head can be resurfaced. If too much material must be removed from this surface, the compression ratio will change and deck height will be lowered. These changes may result in valve-to-piston contact.

Use the straightedge and feeler gauge method to check for warpage of the intake and exhaust manifold mating surfaces. As a general rule, the maximum amount of warpage allowed in these areas is 0.004 in. (0.1 mm). In addition, check the valve cover rail for warpage. The spark plug side of an OHC cylinder head is usually warped more than the other side.

If the cylinder head warpage is excessive, the cylinder head must be resurfaced. This operation is usually completed in an automotive machine shop rather than an automotive repair shop.

On overhead camshaft engines, the camshaft bores or saddles must be inspected for warpage. The most common causes of camshaft bearing wear in an OHC engine are misalignment of the bearing bores or saddles and lack of lubrication. As with most components, begin with a thorough visual inspection. Scoring or galling of the bearing surfaces may indicate one of the preceding problems.

To check the camshaft bores or saddles of an overhead camshaft cylinder head for proper alignment, use a straightedge and feeler gauge (Figure 9-34). Alignment should not be out more than 0.003 in. (0.08 mm). It may be possible to correct camshaft bore alignment if the manufacturer allows it. This can be done by align boring and using oversized bearings or by using heat and special clamping fixtures to straighten the cylinder head. If the manufacturer does not recommend straightening or boring, the head must be replaced.

Next measure the diameter of the bearing bores. If the cylinder head uses bearing caps, they must be installed and properly torqued before taking any measurements. Check for taper and out-of-round by measuring several locations in the bore. Oil clearance can be determined by measuring the journal of the camshaft and subtracting it from the measurement of the bore, then dividing the difference by two. If the cylinder head uses camshaft bearing caps, oil clearance can be checked using plastigage (Figure 9-35).

Excessive oil clearance may be corrected on some cylinder heads by align boring the camshaft bores and installing oversize bearings. Another method is to remove stock from the mating surfaces of the camshaft bearing caps. If the manufacturer does not provide for these corrective actions, the head must be replaced. Excessive clearance can be the result of lack of lubrication or oil contamination.

SERVICE TIP:
If the camshaft bearing bores are out of alignment, do not correct the problem at this time. Other operations such as resurfacing and straightening will affect bore alignment. Because of this, do bearing bore alignment after all other operations are completed.

FIGURE 9-34 Use a straightedge and feeler gauges to check camshaft bore alignment.

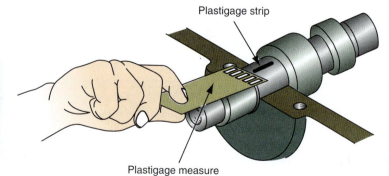

Plastigage strip

Plastigage measure

FIGURE 9-35 On cylinder heads equipped with camshaft bearing caps, the oil clearance can be checked with plastigage.

Valve Guide Inspection

Valve guides are designed with very tight tolerances to the valve stem. Worn valve guides are common causes for excessive oil consumption. Also, unusual valve seat wear patterns are an indication of valve guide wear. Before measuring the valve guide for wear, use a guide cleaner or bore brush to clean the guide. This assures any contamination is removed and accurate readings will be obtained. Refer to the appropriate service manual for the recommended procedure to measure valve guide clearance. One of the following methods is generally used:

1. *Dial bore gauge.* Zero the dial bore gauge using a special fixture. This is usually done by placing two identical new valves in the fixture and locating the dial bore gauge stem between them. This sets the correct size to zero the gauge. Insert the bore gauge into the valve guide and rotate it. Also move the gauge through the length of the guide to determine taper. The measurement indicated on the gauge is the valve guide clearance.

2. *Small hole gauge.* With the tool inserted into the guide, turn the thumb screw until the fingers of the tool contact the guide walls (Figure 9-36). After locking the tool, remove it from the guide. Then measure the expanded fingers with an outside micrometer to determine the guide bore size (Figure 9-37). Measure the guide in three locations to determine wear and taper. This provides a measurement for the valve guide bore. Subtract the diameter of the valve stem from this reading to determine valve guide clearance.

3. *Valve rock method.* Install a new valve with a special sleeve tool into the valve guide (Figure 9-38). The special tool maintains the correct height of the valve during the check. Set up a dial indicator so the contact point is against the head of the valve at a right angle (Figure 9-39). The check is made in the direction of rocker arm movement. Note the reading on the dial indicator as the valve is rocked back and forth. The amount of wear is half that indicated by the dial.

4. *Special valve stem clearance tool* (Figure 9-40). Place the tool over the valve stem until it is fully seated, and then tighten the set screw. The tool maintains the specific height of the valve by allowing the valve to drop off of its seat until the tool rests on the upper surface of the guide. A dial indicator is set up to measure movement of the tool as the valve is rocked back and forth. The clearance measurement is the reading obtained divided by two, the division factor of the tool.

SPECIAL TOOLS
Bore brush
Dial bore gauge
Small hole gauge
Outside micrometer
Dial indicator
Valve stem
clearance tool
Machinist's rule

The valve guide clearance is the difference between the valve stem diameter and the guide bore.

SERVICE TIP:
If the special fixture is not available, use an outside micrometer set to the size of the valve stem, then zero the bore gauge to the micrometer.

SPECIAL TOOLS
Machinist's rule
Seat width scale
Valve seat
runout gauge

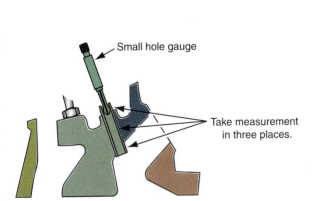

FIGURE 9-36 A small hole gauge used to measure guide diameter. Measure the valve guide in three locations to determine the amount of taper and bellmouthing.

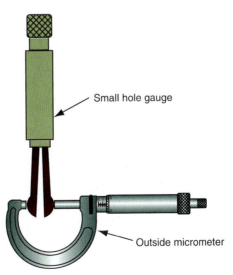

FIGURE 9-37 Use a micrometer to measure the size of the small hole gauge.

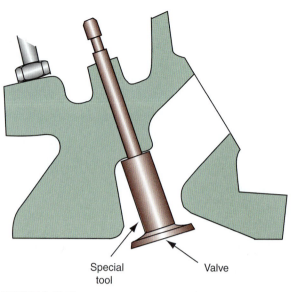

Special   Valve
tool

FIGURE 9-38 The special tool maintains the proper height to
measure valve movement.

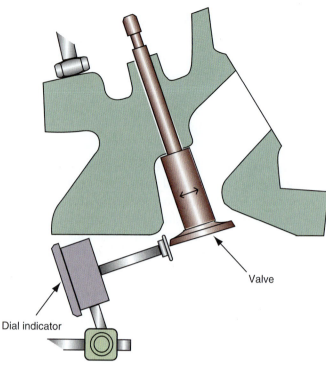

Valve

Dial indicator

FIGURE 9-39 Rocking the valve will indicate guide clearance on the
dial indicator.

Valve stem
clearance tool

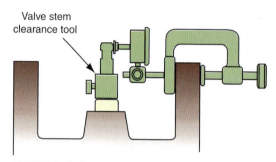

FIGURE 9-40 Some manufacturers have special tools
to measure guide clearance.

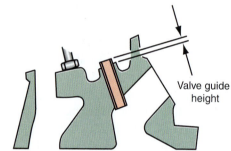

Valve guide
height

FIGURE 9-41 Some manufacturers require a
valve guide height measurement.

The desired stem-to-guide clearance is between 0.001 and 0.003 in. (0.025 and 0.080 mm) for intake valves and 0.0015 to 0.0035 in. (0.04 to 0.09 mm) for exhaust valves. As the valve travels back and forth through the guide, wear can occur. This wear is accelerated if there is a lack of lubrication, if the valve seat is worn, or if the valve stem is bent.

Some manufacturers require a valve guide height measurement. This measurement is the measured distance between the spring seat and the top of the guide (Figure 9-41).

Integral valve guides with excessive wear can be reconditioned. If the cylinder head is equipped with guide inserts, they can be removed and replaced with new ones.

Unequal forces created by valvetrain components can result in normal bellmouthing of the valve guide (Figure 9-42). This type of wear is due to the angle at which most valves are installed into the combustion chamber. The angle causes the valve stem to create wear in one direction at the top of the guide and in the opposite direction at the bottom of the guide. When measuring the diameter of the valve guide bore, measure it in three locations to check for excessive bellmouthing and taper. Worn valve guides can cause excessive oil consumption, exhaust smoke, and burned valves.

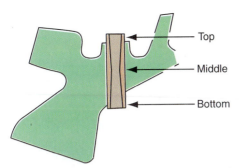

FIGURE 9-42 Bellmouth wear is normal, but it must be corrected before servicing the seats and fitting the valves.

Excessive valve guide wear is often found on exhaust valves closest to the exhaust gas recirculation (EGR) valve. If a severely worn exhaust valve guide is found, it is recommended that the valve's spring be closely inspected. Hot exhaust gases escaping past the guide will heat the spring, causing it to lose its strength.

Valve Seat Inspection

When the valve closes, it must form a seal to prevent loss of compression. This seal is provided by the contact of the valve face with the valve seat (Figure 9-43). In addition to sealing compression pressures, the seat and face provide a path to dissipate heat from the valve head to the cylinder head.

There are two basic types of valve seats: integral and inserts. Most cast-iron cylinder heads have integral valve seats machined into the cylinder head.

Valve seats must be inspected for cracks and excessive wear (Figure 9-44). Some heads with integral seats may be reconditioned even if the seats are damaged. This is done by boring out the original seat to accept an insert. Cylinder heads equipped with inserts must be inspected for loose seats. Pry on the seat to load it while checking for looseness.

Use a machinist's rule or a special seat width scale to measure the width of the valve seat (Figure 9-45 and Figure 9-72). If the valve seat width is not within specifications, the valve seat can be reconditioned.

Due to the assortment of problems that can arise from improper seating, **seat concentricity** should be checked before and after cylinder head reconditioning. Use a valve seat runout gauge with the arbor installed into the valve guide bore (Figure 9-46). Slowly rotate the gauge around the valve seat while observing the gauge readings. Compare the

Seat runout or concentricity is a measure of how circular the valve seat is in relation to the valve guide.

If the valve guide is worn, the valve seat runout reading will not be accurate.

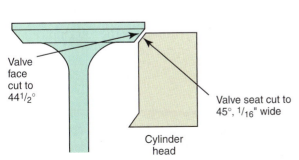

FIGURE 9-43 Proper valve face-to-seat contact is critical to proper engine performance.

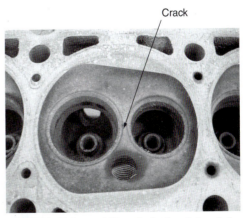

FIGURE 9-44 Inspect the cylinder head closely for cracks. This head is cracked between the seats.

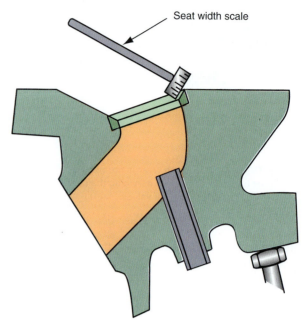

FIGURE 9-45 Measuring valve seat width.

FIGURE 9-46 Using a valve seat runout gauge to measure seat concentricity.

SPECIAL TOOLS

Spring compressor
Air hose adapters

CAUTION:
An accurate runout reading will not be obtained if the guides are worn. Always check valve seat runout after resurfacing the valve seat or the valve guides.

test results with specifications, generally 0.002 in. (0.050 mm). Runout is the total indicator reading. If runout exceeds specifications, the valve seat will require refacing.

MACHINE SHOP SERVICES

The following section describes cylinder head straightening, valve guide repair, and valve seat replacement. These services are generally performed at a machine shop or engine repair specialty shop. If you work in a general service repair shop or new car dealership, it is unlikely that you will perform this work. You will still perform the inspections of the cylinder head and instruct the machine shop on what needs repair. It is also important to review the work that was performed before reassembling the cylinder head.

Straightening Aluminum Cylinder Heads

Warped aluminum heads may be straightened, provided the manufacturer allows this operation. If the manufacturer does not recommend straightening, replace the head. Straighten an aluminum cylinder head after the valve guides are installed, but before the valve seats are reconditioned. Also, straighten the cylinder head before any crack repair operations are done.

Remove all screws, bearings, rocker arm studs, core plugs, oil gallery plugs, and any other hardware. The cylinder head is attached to a special fixture (Figure 9-47). A fixture can be made from a steel plate about 2 in. (50 mm) thick, surfaced on both sides to remove any warpage. Holes are drilled through the plate to attach the head. The attaching bolts align only with the center bolt holes of the cylinder head. This allows for expansion over the full length of the cylinder head. Coat the attaching bolt threads with an antiseize compound to prevent thread damage.

Use the inspection report for the cylinder head to determine the size of shims required. Place brass shim stock one-half the total amount of warpage under each end of the cylinder head (Figure 9-48). The shims must extend the full width of the cylinder head gasket surface. Tighten the attaching bolts to a torque of no more than 25 pounds-feet (18 Nm).

Place the cylinder head and fixture into a thermal oven heated between 450° and 500°F (232° and 260°C) for about 5 hours. Then shut off the oven and allow the heads to cool inside the oven. Recheck the warpage to determine if the process requires repeating.

SPECIAL TOOLS
Brass shim stock
Straightening fixture
Thermal oven

WARNING: **The cylinder head will retain heat for a long time. Wear a good pair of insulated gloves when removing the cylinder head from the oven.**

OHC Bearing Bore Alignment

The straightening process may correct any camshaft bore out-of-alignment; however, always recheck the bore alignment using a straightedge and feeler gauges. Also, recheck bore taper and out-of-round at this time. If it is determined the bore must be aligned, this can be done by line boring or align honing.

If the cylinder head uses bearing caps to retain the camshaft, it is possible to restore the bearing bores to their original size without using oversize bearings. The inside diameter of the bearing bore is decreased by removing material from the cap at the parting line.

SPECIAL TOOLS
Bore bar
Dial bore gauge

WARNING: Always wear approved eye protection when grinding or cutting.

Head straightening fixture
FIGURE 9-47 The head is placed on a thick steel fixture and bolted down to straighten it.

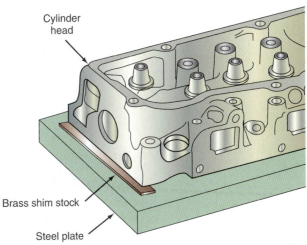

Cylinder head

Brass shim stock

Steel plate

FIGURE 9-48 Preparing the cylinder head for straightening. The shims must fit the entire width of the head.

VALVE GUIDE REPAIR

Valve guide repair is usually one of the first operations performed when reconditioning the cylinder head. The valve guides must be repaired or replaced before any valve seat work is performed. The guide is used as a pilot during seat replacement and seat refinishing; if the guide is worn, the seat may not be properly centered. There are several options available to the technician if it is determined the valve guides are excessively worn. These options include:

- Knurling the guide
- Reaming the valve to accept oversize valve stems
- Replacing insert guides
- Installing guide liners
- Replacing integral guides with false guides
- Installing coil inserts

Regardless of the method used, it is vital that the valve guide is centered so the valve face will be concentric in its seat. Also, after the guide is repaired, always clean it before assembling the valves. Most operations will leave metal chips and debris that will result in premature failure if not removed.

Knurling

Knurling can be done using a thread-type arbor (Figure 9-49) or a wheel-type tool (Figure 9-50). Knurling arbors are generally available in 5/16, 11/32, and 3/8 in. (7.937, 8.731, and 9.52 mm) sizes. In addition, knurling arbors are available in 0.005 in. (0.13 mm) over standard to accommodate oversize valve stems. Following is a typical procedure using the thread-type knurler:

1. Clean the valve guides.
2. Select a knurling arbor the normal size of the valve guide.
3. Lubricate the guide and arbor.
4. Install the arbor into the guide from the combustion chamber side.

Knurling is a procedure in which a special bit is forced through a bore to swell the metal and decrease the bore size. The bore is then reamed to the specified diameter. Knurling is effective on valve guides with wear less than 0.005 in. (0.13 mm) over the specified clearance.

FIGURE 9-49 This cylinder head bench has an assortment of knurling bits.

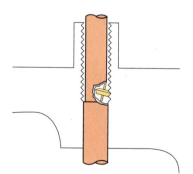

FIGURE 9-50 A wheel-type knurler.

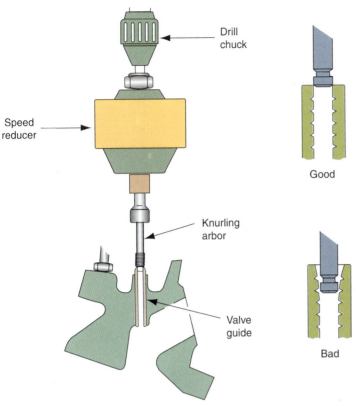

Drill chuck

Speed reducer

Knurling arbor

Valve guide

Good

Bad

FIGURE 9-51 Driving the knurling arbor into the valve guide.

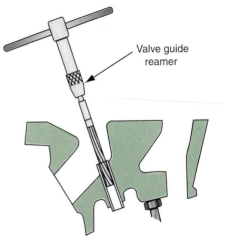

Valve guide reamer

FIGURE 9-52 After the guide is knurled, a reamer is run through to obtain the desired clearance.

5. Use a tap handle to hand start the arbor into the guide. Be careful not to rock the arbor while starting it.
6. Using an electric drill equipped with a speed reducer (1,200 to 1,600 rpm), drive the arbor into the guide (Figure 9-51). The arbor will feed itself down the guide. Do not back the arbor through the guide. Once the arbor is through the guide, release it from the drill chuck. If the motor stalls, remove the motor and finish the guide by hand.
7. Clean the guide with solvents and a bore brush.
8. Select the correct-size reamer to obtain one-half the specified clearance. The reamer size selected is usually 0.001 to 0.002 in. (0.025 to 0.050 mm) larger than the valve stem diameter.
9. Insert the pilot end of the reamer into the guide and use a tap handle to turn the reamer clockwise (Figure 9-52). If a cast-iron guide is used, do not lubricate the guide. If the guide is bronze, it must be lubricated with high pressure lubricant.
10. If a closer clearance is desired, use a guide hone until the desired clearance is obtained.
11. Use a deburring tool to remove irregularities in the guide.

WARNING: **Always wear approved eye protection whenever performing this task.**

Oversize Valve Stems

If the valves are going to be replaced anyway, it may be more cost efficient to install valves with oversize stems. The only operation involved is to ream the valve guide to the next oversize, then replace the valve to obtain the correct clearance. Valves are available in these standard stem oversizes: 0.003, 0.005, 0.006, 0.010, 0.013, and 0.015 inch.

Classroom Manual
Chapter 9

SPECIAL TOOLS
Knurling bit
Electric driver
Speed reducer
Reamer
Guide hone
Deburring tool

SPECIAL TOOLS
Reamer

Guide Replacement

Guides that are worn too excessively to be knurled can be replaced using special inserts. If the original valve guide is insert designed, the old insert can be driven out and a new one can be installed. On integral guides, the technician may choose to use coil inserts, **liners,** or false guides.

Insert Guides. The old insert guide must be pressed out using a special guide replacement tool (Figure 9-53). The tool fits into the guide and has a shoulder slightly smaller than the outside diameter of the guide. The tool can be used with a press or a heavy hammer. Using a press is best since it applies a constant force. Special drivers that attach to an air hammer can be used.

The insert usually has a 0.001 to 0.002 in. (0.025 to 0.050 mm) interference fit into the cylinder head. The direction in which the guide is removed varies between manufacturers. For this reason, always refer to the appropriate service manual for the recommended procedure. Some manufacturers use guides with an angle cut on one end. Never attempt to remove or install these guides by pressing against the angle cut.

WARNING: **Always wear approved eye protection when driving or pressing guides out or installing them.**

Integral Guides. A variety of machining operations makes it possible to recondition cylinder heads with worn integral valve guides. Depending on the amount of wear, the guide can be repaired using a liner, false guide, or coil inserts. If valve guide wear is within 0.030 in. (0.80 mm) of standard, a thin-wall liner can be used to restore original clearances.

If the wear is greater than 0.020 in. (0.50 mm), the guide can be drilled out to accept a **false guide** (Figure 9-54).

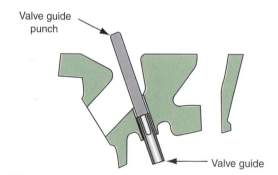

FIGURE 9-53 Removing the old guide with a guide driver.

FIGURE 9-54 False guides can be installed to correct excessive valve guide wear.

Another accepted method of guide repair involves the installation of a bronze coil insert. This method involves threading the inside of the original guide bore to accept a coil. The coil insert reduces the inside diameter of the guide. The coil is finished to obtain the desired clearance. Like knurling, the clearance can be halved since the threads trap sufficient oil.

REPLACING VALVE SEATS

If the cylinder head uses insert valve seats, excessively worn seats should be replaced and ground to assure proper sealing. Lightly worn insert seats and integral seats can usually be cleaned up by resurfacing the seat. If the seat is damaged too extensively to be restored by surfacing, the seat will have to be replaced. Always resurface new valve seats to assure proper concentricity.

Insert Valve Seat Replacement

The procedure for replacing insert valve seats is as follows:

1. Remove the original insert by prying it out of the head or using a special seat remover (Figure 9-55). Remove all inserts that are to be replaced.
2. Thoroughly clean the insert counterbore. Check this area closely for cracks. If any cracks are found, they will have to be repaired before proceeding.
3. If the counterbore does not require recutting to accept oversize inserts, go to step 8. Recut the counterbore using a pilot installed in the guide and mount the base and ball shaft assembly to the sealing surface of the cylinder head (Figure 9-56).
4. Set an outside micrometer to the recommended counterbore size. Use the micrometer to set the cutter head to this size. The recommended size is usually set to provide a 0.005 to 0.008 in. (0.13 to 0.20 mm) interference fit.
5. Place the counterbore cutter over the pilot and ball shaft.
6. Set the depth of the insert at the feed screw.
7. Turn the stop collar to cut the counterbore. Apply cutting oil to the cutter at regular intervals. Continue until the stop collar reaches the preset depth.

SPECIAL TOOLS

Seat remover
Counterbore cutter
Pilot and ball shaft
Driver motor
Thermal oven
Insert driver

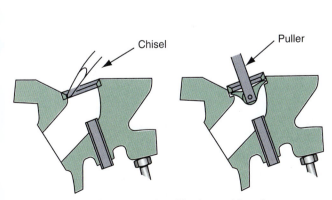

FIGURE 9-55 Removing the old valve seat inserts.

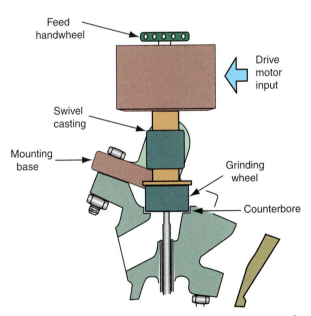

FIGURE 9-56 Cutting the counterbore to accept a new seat insert.

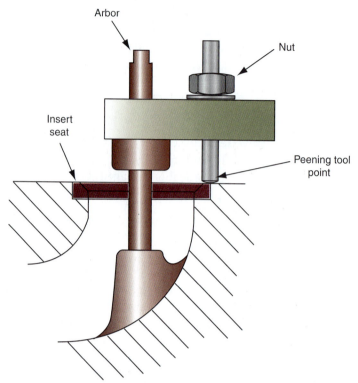

Arbor

Nut

Insert
seat

Peening tool
point

FIGURE 9-57 Peening the valve seat insert assures it will not come
loose during operation.

8. Place the cylinder head in a thermal oven set at 350° to 400°F (177° to 204°C). While the cylinder heads are heating, place the inserts in a freezer.
9. Use an insert driver to press the new insert into the counterbore. Make sure the insert is flush and fully seated.
10. After all inserts are installed, stake or peen the edge around the outer edge of the insert (Figure 9-57).

Integral Valve Seat Replacement

If the cylinder head is equipped with integral valve seats, a counterbore is cut into the original seat to accept a new seat insert. The counterbore must provide a 0.005 to 0.008 in. (0.13 to 0.20 mm) interference fit with the seat insert. In order for the counterbore to be centered, the valve guides must be reconditioned before doing this operation.

The procedure for counterboring and installing insert seats into a cylinder head with integral seats is similar to the insert replacement discussed previously. Removal and installation of integral-type valve seats is as follows:

1. Thoroughly clean the area around the seats.
2. Install a pilot into the guide and mount the base and ball shaft assembly to the sealing surface of the cylinder head.
3. Set an outside micrometer to the recommended counterbore size. Use the micrometer to set the cutter head to this size.
4. Place the counterbore cutter over the pilot and ball shaft.
5. Set the depth of the insert at the feed screw.
6. Turn the stop collar to cut the counterbore. Apply cutting oil to the cutter at regular intervals. Continue until the stop collar reaches the preset depth.
7. Place the cylinder head in a thermal oven set at 350° to 400°F (177° to 204°C), while freezing the inserts.

8. Use an insert driver to press the new insert into the counterbore. Make sure the insert is flush and fully seated.
9. After all inserts are installed, stake or peen the outer edge of the insert.

WARNING: Always wear approved eye protection when performing this task.

REFINISHING VALVES AND VALVE SEATS

Most valve machining work is sent out to a local machine shop. There the technicians perform this work daily and can efficiently produce refinished cylinder heads. Many general repair shops cannot afford to purchase the high tech equipment required to refinish late-model cylinder heads. The general repair shop will often make more money on the job by subletting the machine work out to a local and trusted machine shop.

In some cases, the job of refinishing the valves and valve seats may still be performed in the general repair shop. Many older facilities still have the equipment and will use it to perform the valve work if the cylinder head is in good condition otherwise. First the valves are all refinished; then the seats are refinished and the valves fitted to the seat to assure that the contact area is correct on the valve face.

Valve Reconditioning

A valve grinding machine is required to reface the valves (Figure 9-58). You want to cut as little metal off the valve face as possible while still removing all cupping and pitting. Make sure the valve face and stem are clean. This assures that the cutting stone will not get loaded with carbon and that the valve will sit squarely in the mounting chuck. Make sure you have enough margin left to cut the valve face. There are several different types of valve grinders available; you will have to become familiar with the one used in your shop. These guidelines apply to most valve grinding equipment:

1. Refer to the manufacturer's specification to determine at what angle the valve face should be cut. Sometimes an **interference angle** is used. This is a difference in angle between the valve face and the valve seat of 1/2 to 1 degree. It promotes better sealing by increasing the valve seating pressure.
2. Adjust the valve holding bench to the specified angle.
3. Use the diamond tip stone dresser to put a good clean finish on the stone. Use lubrication while cutting. Take as little material off as necessary. Make several fast

SERVICE TIP:
To remove difficult inserts, weld a small bead around the inside of the seat. This will shrink the seat, allowing easier removal.

CAUTION:
Always wear approved eye protection when performing this task.

SPECIAL TOOLS
Counterbore cutter
Pilot and ball shaft
Driver motor
Thermal oven
Insert driver

SERVICE TIP:
For cylinder heads with the valve guides set at an angle in relation to the cylinder head surface, it will be necessary to set the base in the direction of the centerline of the valves.

FIGURE 9-58 A typical shop valve grinding machine.

FIGURE 9-59 Pass the stone across the diamond cutting tip to clean the face of the stone.

FIGURE 9-60 Adjust the stops on the bench so that you cannot nick the valve neck and so that you use the whole stone.

SPECIAL TOOLS

Valve machine

An **interference angle** may or may not be specified by the manufacturer. It means that the valve face and valve seat are cut at angles that are 1/2 to 1 degree different from each other to improve sealing.

passes across the diamond while you are removing material; then finish the stone by moving it slowly across the diamond several times to smooth the surface finish (Figure 9-59).

4. Place the valve in the holding chuck as close to the neck as possible. Be sure the balls of the chuck are contacting the machined part of the stem. Spin the valve and check for runout of the valve head. You can set up a dial indicator to measure runout as you spin it in the chuck. Replace a valve that has over 0.002 in. (0.050 mm) runout as shown by the total indicator reading (TIR).

5. If possible, set the mechanical stops on the bench (Figure 9-60). Move the valve in close to the stone. You want the valve to move all the way across the stone so as not to develop a groove in your freshly dressed stone. Adjust the stops so the valve does not travel beyond the end of the stone; this can nick the valve as it runs back up onto the stone. You also want to set the stops so that you prevent the stone from contacting the neck of the valve. If you nick the neck of the valve, the valve must be replaced.

6. Turn the chuck and the stone on, and adjust the lubrication so it hits the valve face.

7. Slowly move the valve into contact with the stone, while simultaneously moving the stone back and forth across the valve (Figure 9-61). If equipped, set the depth indicator to zero. Move the stone quickly, and adjust the valve into the stone slowly, until you can see no more pitting. Then put a nice finish on the face by making slow passes across the valve. Move the valve away from the stone and turn off the machine. Look at the depth indicator to see how much material you removed from the valve.

8. Carefully inspect the valve, and continue grinding if pits or cupping are still seen.

9. Measure the margin, and be sure you have enough left to reuse the valve. If not, discard the valve; if the valve is okay, proceed to the next step.

10. Next you need to dress the valve tip to make a clean smooth surface to contact the rocker arm. You also want to remove the same amount of material from the tip as you did from the face so that you don't disturb the rocker arm geometry. If you have depth indicators, simply match the amount of material you take off the tip with what you removed from the face. On machines without this measuring equipment, you will have to gauge this by feel and then measure valve installed height.

11. Dress the tip cutting stone, usually on the end of the valve grinder (Figure 9-62).

12. Place the valve in the holder, and adjust the lubrication to hit the tip. Move the stone into the valve tip slowly while rocking the valve back and forth across the stone. Proceed until the tip is fully freshened.

FIGURE 9-61 Draw the valve into the stone while both are rotating. Set the oil so that it is running over the valve face.

FIGURE 9-62 Run the valve tip back and forth across the stone to dress the tip.

13. Finally, place the valve in the chamfering holder. Very gently push the valve against the stone to produce a slight chamfer. All you are trying to do is to eliminate the sharp edge of the tip; a deep chamfer is not desired. (Figure 9-63).

14. Repeat the process for each of the valves. Many technicians will refinish the face of each valve and then refinish the tips to avoid resetting the stops on the stone bench.

The end result should be a perfectly cleaned-up valve face and tip with a margin at least as thick as specified.

Valve Seat Refinishing

The valve seats must be fully refinished to provide a good sealing surface for the fresh valves. You will have to achieve a good finish on the seat and the proper seat width. Remember, typical specifications are 1/16 in. on the intakes and 3/32 in. on the exhausts, but check your particular specifications. It is best to clean the valve seats and port area with a light wire brush if the head has not been professionally cleaned. This saves the stones and cutters from getting loaded up with carbon. There are several types of valve seat refinishing

SERVICE TIP:
Be careful not to remove too much material from the grinding stone in one pass. Grinding stones and diamond tip dressers are expensive, but can last a long time with good maintenance and if used as directed.

FIGURE 9-63 Rotate the valve in the holder and put a light chamfer on the edge of the valve tip.

FIGURE 9-64 These diamond seat cutters are long lasting and do an excellent job of refinishing the seats.

FIGURE 9-65 The machine shop uses this bench for much of the seat work.

SPECIAL TOOLS

Pilots
Stone assortment
Stone holder
Power head
Diamond dressing tool
Lapping stick
Prussian blue
Machinist's rule
Valve seat concentricity gauge
Dressing bit

equipment. A diamond cutter type is shown in Figure 9-64. This bench includes valve guide pilots, knurling tools, diamond seat cutters, and an overhead drill for cutting the seat (Figure 9-65). Another type of seat cutter is an adjustable carbide cutter that can cut three angles on the seat in one shot. Other seat cutting equipment uses stones of different sizes and angles. This has been the traditional method for years, but equipment is often being replaced by the other styles. Any type of equipment is designed to achieve the same effects on the valve seats.

Carbide Seat Cutters

To use the carbide cutter, adjust the cutters to fit the seat (Figure 9-66). The cutter shown has a guide pilot in it. With any cutting procedure, you must select a proper fitting pilot that fits securely in the guide, to assure that you cut the seat squarely in relation to the guide. Some cutters are designed to be used with a drill; others you can turn by hand. Rotate the

FIGURE 9-66 This adjustable carbide cutter is an efficient seat cutting method; it cuts all three angles at once and provides the correct seat width.

tool clockwise. Follow the equipment instructions. Remember, these cutters cut the seat and finish it to the correct width. It performs a three-angle valve job. The seat is typically cut at 45 degrees, and there is a 15-degree or 30-degree angle on the top of the seat and a 60-degree or 75-degree angle cut on the lower part of the seat, in the throat.

Stone Seat Cutters

The stone-type refinishing equipment, whether diamond or traditional, uses very similar processes. We'll discuss the stone seat cutting procedure here. When using a stone, it must be dressed like the stones on the valve grinding machine. Diamond cutters do not require refinishing; they should theoretically last forever unless the diamonds break free from the tool.

To refinish a seat using a stone cutter:

1. Select a 45-degree stone (or 30-degree if that is the seat angle) that fits the seat without scraping on any portion of the combustion chamber. It must be large enough to cut the whole width of the seat (Figure 9-67).
2. Fit the stone onto the drilling tool (Figure 9-68).
3. Place a drop of oil on the stone dressing shaft, and place the stone and tool onto the shaft. Select the proper angle at which the diamond cutter will cut the stone.
4. Apply the power head (special drill) on top of the stone holding fixture, and spin the stone. Support the weight of the power head, and hold it squarely on the stone. Slowly move the diamond cutter into the stone while moving it quickly up and down across the stone (Figure 9-69). Finish the stone by taking several slow passes across the stone.
5. Make sure the guide is clean, and select the proper pilot (Figure 9-70). The pilots are tapered. They should fit snugly and not quite bottom out on the guide. They are sized as 9/32 (as an example), 9/32 +1 (0.001 in.), +2, +3, −1, and so on for each different size guide.
6. Place a drop of oil on the pilot, and place the stone and its holder over the pilot.
7. Spin the stone on the seat while supporting the weight of the power head and holding it squarely above the stone. Spin the stone for just a few seconds at a time; check the seat between cuts. You want to take little material off but remove all pitting from the seat (Figure 9-71).

FIGURE 9-67 Stones are still a fine way to refinish seats; you just need some patience.

FIGURE 9-68 Fit the stone on its holder to get it ready for use with the power head.

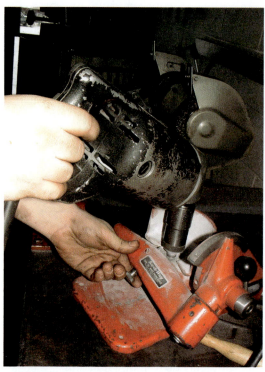

FIGURE 9-69 Pass the diamond across the stone to freshen up the stone.

FIGURE 9-70 Use this tool to secure the pilot into the guide. It is also very helpful when removing the pilot.

FIGURE 9-71 Make short cuts on the seat with the stone. Check the seat often; you want to take as little material off as possible.

Fitting the Valve and Seat

Once the valve and seat are clean, you must narrow the seat (usually) and check to see where the valve face is contacting the seat. Check the manufacturer's specifications and compare the seat width to the specification (Figure 9-72). You want the area of contact to

FIGURE 9-72 Measure the width of the seat using a machinist's rule. Intake seats are typically 1/16", and exhausts are often 3/32".

FIGURE 9-73 Rotate the valve in the head with a lapping stick to see where the seat is contacting the valve.

be near the center of the face, with some area above and below the contact patch still left on the face. To check where the seat is contacting, dab a little **Prussian blue** on the valve face and use a lapping stick to spin the valve on the seat (Figure 9-73).

The Prussian blue will be rubbed away from the contact patch. If the contact is high toward the head, you will cut a 15-degree or 30-degree angle on the top of the seat to narrow the seat and adjust the valve contact down (Figure 9-74). This is called **topping**, and it is the most common requirement when using refinished valves because they tend to sink down in the head as they wear. If the seat is wide and the area of contact is low on the valve face, toward the tip, cut the seat with a 60-degree or 75-degree stone in the throat. **Throating** will narrow the seat and move the area of contact up toward the head. Continue adjusting until you have the correct seat width and contact area on the valve face. Always finish the seat with a short burst with the 45-degree stone. This removes any burrs that could be left on the seating surface by the other cuts. Make a final check of the contact area using Prussian blue. Next you should check the concentricity or runout of the seat. If it is concentric, it means that it is centered to the valve guide. Use a dial indicator seated in the guide to check that the seat runout is less than 0.001 in.

Prussian blue is a thick ink used to show the area of seat contact on the valve face.

Topping a seat with a 15-degree or 30-degree stone narrows the valve seat and lowers the area of contact toward the tip.

Throating a seat with a 60-degree or 75-degree stone narrows the valve seat and raises the area of contact toward the head.

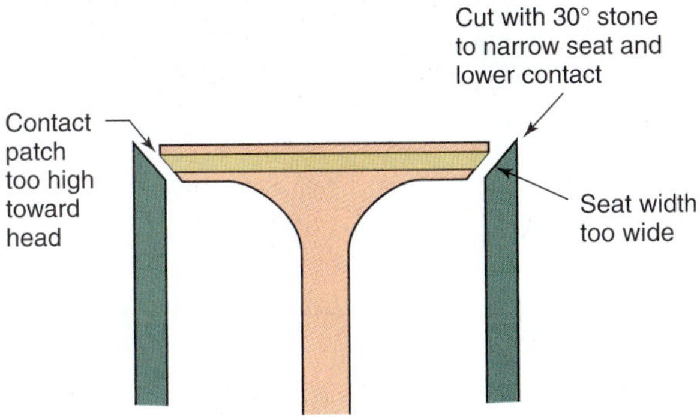

Cut with 30° stone to narrow seat and lower contact

Contact patch too high toward head

Seat width too wide

FIGURE 9-74 Cut a 30-degree angle on the top of the seat to narrow the 45-degree seat and move the contact down toward the tip.

Some technicians will perform a three-angle valve job on every job. This means they will cut the 45-degree seating angle, a 15-degree or 30-degree topping cut, and a 60-degree or 75-degree cut in the throat. In production work on a stock engine, this is not required.

Finally, you need to check the integrity of the valve face to the seat. You can apply a vacuum to the valve seal stand or the top of the guide, and make sure each valve can hold a vacuum. Alternately, you can pour clean parts cleaner or mineral spirits on top of the valves, and make sure none leaks down within 1 minute. Photo Sequence 18 on page 373 shows step by step directions for refinishing valve seats.

Valve Measurement

Valve stem height is the height of the valve tip as measured from the base of the head to the top of the valve tip.

Valve stem height should be measured to determine if more material needs to be ground off the valve stem tip. After the seat and face refinishing, the valve stem will sit higher above the head. This measurement is particularly important on engines with nonadjustable hydraulic lifters. If you install the lifter and the tip is extending too far, the lifter may not allow the valve to close. Measure stem height (if specified), using a machinist's rule, from the base of the head to the top of the tip (Figure 9-75). Compare the measurement to specifications. If the stem is too tall, grind the required amount off of the tip.

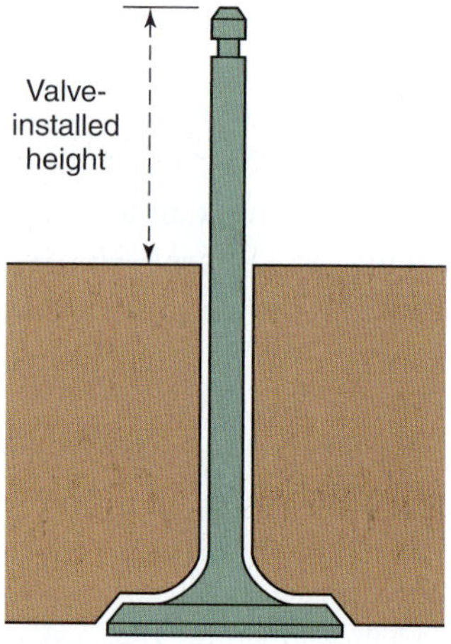

FIGURE 9-75 **Measure the height of the valve stem from the cylinder head to the valve tip.**

Terms to Know

- Core plugs
- False guide
- Gallery plugs
- Interference angle
- Liners
- Prussian blue
- Seat concentricity
- Throating
- Topping
- Valve stem height

CASE STUDY

The technician had diagnosed an oil consumption concern as worn valve guides. With the cylinder head removed from the engine block, the technician cleaned, disassembled, and carefully inspected it. The guides were excessively worn, but upon closer examination it was found that some of the seats were not concentric. After consulting with and explaining the condition to the vehicle owner, the technician performed the repair. Simply replacing or reconditioning the valve guides would have resulted in a comeback, as the valves would not have sealed properly. Careful inspection saves time and money, and earns you respect from your customers.

TYPICAL PROCEDURE FOR GRINDING VALVE SEATS

P18-1 Select the correct size and grit of stones required. Select the stones required for the seat, for topping, and for throating.

P18-2 Thread each stone onto a stone holder. Using a holder for each stone will reduce the amount of time required to recondition the seats and will keep stone location consistent.

P18-3 Dress each stone by attaching the stone holder to the grinder motor and using a diamond dressing tool. The stones should be dressed before use on each seat. If the surface finish is rough or shows indications of chatter, redress the stone.

P18-4 Select the proper size pilot and pass it through a cloth saturated with oil. Do not squirt oil into the valve guide.

P18-5 Insert locating pilots into each of the valve guides by pushing the pilot down while twisting it until it firmly wedges in place. The pilot should extend at least 2 1/2 inches from the seat to fit inside the stone holder.

P18-6 Locate the lifting spring over the pilot. Lubricate the pilot with a few drops of motor oil.

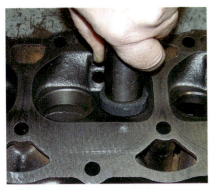

P18-7 Place the stone holder, with the proper stone for the seat, over the pilot. Check its fit onto the seat.

P18-8 Attach the driver motor to the holder. Hold the driver with the pistol grip in one hand, supporting the weight. Cradle the front of the driver with the other hand.

P18-9 Turn on the motor and work the stone into the seat. When working the grinding stone, lift it on and off the seat at a rate of 120 times per minute to prevent excessive pressures on the stone and seat.

TYPICAL PROCEDURE FOR GRINDING VALVE SEATS

P18-10 Remove only enough material to provide a new surface all around the seat. Grind for only a few seconds and stop to check the work.

P18-11 Check the width of the seat with a scale.

P18-12 Top and throat the valve seat to obtain the proper seat width.

P18-13 Check valve seat concentricity by placing the fixture over the guide pilot.

P18-14 Check the fit of the valve into the seat by coating the valve face with a light layer of Prussian blue, then placing the valve into the seat.

P18-15 Rotate the valve back and forth one-quarter turn while applying pressure to the center of the valve.

P18-16 Remove the valve and inspect the contact pattern on the valve face and seat.

P18-17 If the valve seat-to-face contact requires correction, this is done by topping and throating the seat. If the seat contact is high, topping the seat will lower the contact area. When the seat is finished, the surface must be free of any chatter marks.

ASE-STYLE REVIEW QUESTIONS

1. Reconditioning valves is being discussed.

   Technician A says the valve is stroked across the stone at the same time it is fed in.

   Technician B says when the last pass is complete, back the valve out of the stone. Who is correct?

   A. A only
   B. B only
   C. Both A and B
   D. Neither A nor B

2. *Technician A* says valve head runout can be checked with the valve in the valve machine.

   Technician B says a total indicated reading (TIR) greater than 0.002 inch (0.05 mm) on the valve head requires valve replacement.

   Who is correct?

   A. A only
   B. B only
   C. Both A and B
   D. Neither A nor B

3. Guide replacement is being discussed.

   Technician A says, when selecting valve guide inserts or liners, match guide materials to valve stem materials.

   Technician B says all guides are removed toward the combustion chamber.

   Who is correct?

   A. A only
   B. B only
   C. Both A and B
   D. Neither A nor B

4. Installing guide inserts is being discussed.

   Technician A says to measure the amount the guide protrudes from the head casting so the new guides can be installed to the same height.

   Technician B says to cut a chamfer on the insert if the insert does not have one.

   Who is correct?

   A. A only
   B. B only
   C. Both A and B
   D. Neither A nor B

5. Reconditioning OHC cylinder heads is being discussed.

   Technician A says cylinder head warpage does not affect camshaft bearing bore alignment.

   Technician B says if the head is to be straightened by heat, brass shim stock the thickness of the total amount of warpage is placed under each end of the cylinder heads.

   Who is correct?

   A. A only
   B. B only
   C. Both A and B
   D. Neither A nor B

6. Reconditioning valve seats is being discussed.

   Technician A says if the seat is too narrow, it is corrected by using the same degree stone as the valve seat angle.

   Technician B says if the valve seat is too wide, throat the seat. Who is correct?

   A. A only
   B. B only
   C. Both A and B
   D. Neither A nor B

7. *Technician A* says that valve guides can be pearlized to restore them to their original condition.

   Technician B says that worn valve guide inserts can be pressed or driven out and new ones installed.

   Who is correct?

   A. A only
   B. B only
   C. Both A and B
   D. Neither A nor B

8. *Technician A* says to try to take as much material off the valve stem tip as you did off the face.

   Technician B says you should put a light chamfer on the tip. Who is correct?

   A. A only
   B. B only
   C. Both A and B
   D. Neither A nor B

9. *Technician A* says that the valve guide must be clean and not worn excessively before repairing valve seats.

   Technician B says to pick a guide pilot that slides down to the bottom of the guide easily.

   Who is correct?

   A. A only
   B. B only
   C. Both A and B
   D. Neither A nor B

10. A valve seat is too wide. The area of contact is high toward the head of the valve. The seat should be cut with a _____ stone.

    A. 30-degree
    B. 44-degree
    C. 45-degree
    D. 60-degree

ASE Challenge Questions

1. What preparation must be done to the valve guides before measuring valve seat runout?
   A. Use 30W engine oil to lubricate the guide.
   B. Clean the valve guide with a bore brush.
   C. Ream the valve guide to its original diameter.
   D. Use a pilot to hold the guide in place.

2. *Technician A* says that a worn-out valve guide will provide an inaccurate valve seat runout measurement.
   Technician B says that the valve seat-to-face contact area provides a path for heat from the valve head to dissipate.
   Who is correct?
   A. A only
   B. B only
   C. Both A and B
   D. Neither A nor B

3. An acceptable valve seat width is:
   A. 1/16 in. (1.58 mm).
   B. 3/16 in. (4.76 mm).
   C. 7/32 in. (5.56 mm).
   D. 1/4 in. (6.35 mm).

4. All of the following are considerations when grinding valve seats EXCEPT:
   A. Dress the stone before grinding any seat.
   B. Remove only enough material to provide a new surface.
   C. Use transmission fluid to lubricate only the stone.
   D. Do not apply pressure to the grinding stone.

5. *Technician A* says that if the valve seat is too narrow, the valve may burn.
   Technician B says that if the valve seat is too wide, the valve may not seal properly.
   Who is correct?
   A. A only
   B. B only
   C. Both A and B
   D. Neither A nor B

JOB SHEET

Name _____ Date _____

REMOVING THE HARMONIC BALANCER AND PULLEY

Upon completion of this job sheet, you should be able to properly remove the harmonic balancer or crankshaft pulley.

ASE Correlation

This job sheet is related to ASE Engine Repair Test content area: General Engine Diagnosis; Engine Block Assembly, Diagnosis, and Repair: Remove, inspect or replace crankshaft vibration damper (harmonic balancer).

Tools and Materials

Technician's tool set
Harmonic balancer or gear puller

Describe the vehicle being worked on:

Year _____ Make _____ Model _____
VIN _____ Engine type and size _____

Procedure

Task Completed

Your instructor will provide you with a specific engine. Write down the information, then perform the following tasks:

1. Locate the correct procedure for removing the crankshaft vibration (harmonic) balancer for your specific vehicle. According to the service manual, what special tools are necessary?

    _____

2. Remove any belts and other components that must be removed prior to installing the special pulling tool. ☐

3. Install the crankshaft or gear puller on the crankshaft balancer. ☐

4. Grease the threads of the puller and start to remove the balancer. ☐

5. After removing the balancer, inspect it for damage. If a woodruff key was used, make sure to keep it in a safe location. ☐

Instructor's Response _____

_____

_____

JOB SHEET

Name _____ **Date** _____

CYLINDER HEAD REMOVAL

Upon completion of this job sheet, you should be able to properly remove a cylinder head from an engine.

ASE Correlation

This job sheet is related to ASE Engine Repair Test's content area: Cylinder Head and Valvetrain Diagnosis and Repair; task: Remove cylinder head, disassemble, clean, prepare for inspection according to manufacturer's procedures.

Tools and Materials

Basic hand tool set

Describe the vehicle being worked on:

Year _____ Make _____ Model _____

VIN _____ Engine type and size _____

Procedure

_____

Task Completed

Use an engine in a vehicle or on a stand to perform the following steps to remove the cylinder head.

1. Locate the service information procedure for removing the cylinder head from your engine. Read through the procedure, and list in order the components that must be removed from the head before removal from the engine.

   _____

   _____

   _____

   _____

2. Install safety glasses and fender covers. Disconnect the negative battery cable.   ☐

3. Remove all wiring connectors, vacuum lines, and hoses as needed. Mark them carefully for reinstallation.   ☐

4. Follow the manufacturer's procedure to remove the valve cover(s). What type of valve cover is used on your engine? _____

5. Follow the manufacturer's procedure to remove the intake manifold. Be sure to follow the loosening sequence or reverse the tightening sequence. Describe the pattern in which you loosened the attaching fasteners.

   _____

   _____

☐

6. Spray the exhaust manifold nuts with penetrating oil and allow them to soak for a minute. Remove the exhaust manifold while following any special instructions offered in the service information.

Instructor's Response _____

_____

_____

Name _____ Date _____

CYLINDER HEAD DISASSEMBLY

Upon completion of this job sheet, you should be able to disassemble a cylinder head properly.

ASE Correlation

This job sheet is related to ASE Engine Repair Test's content area: Cylinder Head and Valvetrain Diagnosis and Repair; task: Remove cylinder head, disassemble, clean, prepare for inspection according to manufacturer's procedures.

Tools and Materials

Parts washer
Valve spring compressor

Describe the vehicle being worked on:

Year _____ Make _____ Model _____
VIN _____ Engine type and size _____

Procedure

Task Completed

Using the cylinder head(s) that you removed from the engine and the proper service manual, you will identify and perform the required steps to disassemble the cylinder head.

1. On what page of the service manual are the cylinder head disassembly procedures found?

   _____

2. What components of the cylinder head must be removed prior to removing the valves?

   _____
   _____
   _____

3. List any special tools and equipment required to disassemble the cylinder head. After making the list, gather the tools and organize them in your area.

   _____
   _____
   _____
   _____
   _____
   _____

4. Clean the cylinder head and mount the head to the holding fixture. ☐

5. If working on an OHV cylinder head, go to step 12. On OHC cylinder heads, reverse the tightening sequence to loosen the rocker arm shaft attaching bolts, and remove the bolts. ☐

☐ 6. Remove the rocker arm shafts from the cylinder head. Be careful to maintain the original order of the rocker arms.

☐ 7. Remove the camshaft timing belt sprocket.

☐ 8. Measure the camshaft end play and record your reading.

_____

☐ 9. Remove the camshaft from the cylinder head.

☐ 10. If the head is equipped with cam followers, lash adjusters, or lash adjusting shims, remove them at this time. Be careful to maintain proper order of all components.

☐ 11. Use a soft-faced hammer and an old piston pin to loosen the keepers from the retainers. Do not use heavy blows. Be sure the head is mounted on head stands to avoid bending valves.

☐ 12. Use a valve spring compressor to compress the valve springs. Compress the springs only enough to remove the keepers.

☐ 13. Remove the keepers from the top of the valve stem.

☐ 14. Slowly release the valve spring compressor until it can be removed. Remove the valve spring retainer, valve spring, valve spring shims, and stem seal. Maintain proper order of the valve spring components to the cylinder head.

15. Measure the thickness of any spring shims used and record the results.

_____

_____

_____

16. What type of stem seal is used on this cylinder head?

_____

17. Measure the valve stem tip height and record the results.

_____

_____

_____

☐ 18. Use a file to remove any mushrooming at the top of the valve stem.

☐ 19. Remove the valve from the cylinder head.

☐ 20. Repeat this procedure for all valves. Organize the components so their original positions in the cylinder head are maintained.

☐ 21. Are there any studs that must be removed from the cylinder head?   ☐ Yes   ☐ No
If yes, are they pressed in or threaded?   ☐ Pressed   ☐ Threaded
Use the proper procedure to remove the studs.

☐ 22. Remove any core plugs that may be installed into the head.

Instructor's Response _____

_____

_____

JOB SHEET

Name _____ Date _____

CYLINDER HEAD CRACK DETECTION

Upon completion of this job sheet, you should be able to properly check a cylinder head for cracks.

ASE Correlation

This job sheet is related to ASE Engine Repair Test content area: Cylinder Head and Valvetrain Diagnosis and Repair; tasks: Clean and visually inspect a cylinder head for cracks; check gasket surface areas for warpage and surface finish; check passage condition.

Tools and Materials

Service manual
Technician's tool set
Magnetic particle inspection equipment
Penetrant dye crack detection equipment

Describe the vehicle being worked on:

Year _____ Make _____ Model _____
VIN _____ Engine type and size _____

Procedure for Iron Cylinder Heads

Task Completed

Using an assigned engine and the proper service manual, you will identify and perform the required steps to detect cracks on an iron cylinder head.

1. Check for technical service bulletins for articles related to engine crack detection. Describe what you found.

   _____

2. After cleaning the cylinder head, place the electromagnet in one area of the cylinder head gasket surface. ☐

3. Turn the electromagnet on and dust the iron powder in between the magnets. ☐

4. If a surface or near surface crack is present, the powder will collect near the crack. Did this occur?

   _____

5. Continue to perform this testing method on the remainder of the cylinder head, paying special attention to hot spots such as in between valve seats and around coolant passages. Describe what the results of this test were.

   _____
   _____

Procedure for Aluminum Cylinder Heads

Using an assigned engine and the proper service manual, you will identify and perform the required steps to detect cracks on an iron cylinder head.

1. Check for technical service bulletins for articles related to engine crack detection. Describe what you found.

   _____

☐ 2. After cleaning the cylinder head using conventional methods, spray the suspected area with the supplied spray bottle of special cleaner and wipe it dry.

☐ 3. After the cleaner dries, spray the area with the penetrant and wait 5 minutes.

☐ 4. After the penetrant dries, spray the area with the developer and wait 5 minutes.

5. If a surface or near-surface crack is present, the developer will outline the crack in a different color. Use of a black light will aid in seeing any cracks. Did this occur?

   _____

6. Continue to perform this testing method on the remainder of the cylinder head, paying special attention to hot spots such as in between valve seats and around coolant passages. Describe what the results of this test were.

   _____

   _____

Instructor's Response _____

_____

_____

JOB SHEET

Name _____ Date _____

MEASURING CYLINDER HEAD WARPAGE

Upon completion of this job sheet, you should be able to properly measure a cylinder head for warpage.

ASE Correlation

This job sheet is related to ASE Engine Repair Test content area: Cylinder Head and Valvetrain Diagnosis and Repair; tasks: Clean and visually inspect a cylinder head for cracks; check gasket surface areas for warpage and surface finish; check passage condition.

Tools and Materials

Service manual
Technician's tool set
Straightedge
Feeler gauge set
Cylinder head stand (set)

Describe the vehicle being worked on:

Year _____ Make _____ Model _____
VIN _____ Engine type and size _____

Procedure

Task Completed

Using the cylinder head(s) that you removed from the engine and the proper service manual, you will identify and perform the required steps to measure for gasket surface warpage. The cylinder head gasket surfaces that are to be checked must be cleaned of any old gasket material before measuring.

1. Using the service manual, check for any technical service bulletins concerning gasket surface warpage measurement. ☐

2. Place the cylinder head on a set of stands so that they can be held steady during the test. ☐

3. Lay the straightedge lengthwise across the center of the cylinder head gasket surface and use the feeler gauge set to check for the maximum gap between the gauge and the straightedge. What was the maximum measurement found?

   _____

4. Perform this step a minimum of three times lengthwise and two times diagonally (or as directed to do so by the manufacturer's service manual). What is the maximum amount of clearance (warpage) that was found?

   _____

5. What is the manufacturer's maximum specification for warpage?

   _____

6. Continue to perform this method for the intake and exhaust gasket mounting surfaces. What did you find?

_____

7. Based on your inspection and measurements, describe what procedures must be done or what components must be replaced?

_____

_____

Instructor's Response _____

_____

_____

JOB SHEET

Name _____ Date _____

MEASURING VALVE STEM-TO-GUIDE CLEARANCE

Upon completion of this job sheet, you should be able to properly inspect and measure the valves and guides.

ASE Correlation

This job sheet is related to ASE Engine Repair Test content area: Cylinder Head and Valvetrain Diagnosis and Repair; tasks: Inspect valve guides for wear; check valve stem-to-guide clearance; determine necessary action.

Tools and Materials

Service manual
Technician's tool set
Dial indicator
Small hole gauge set
Outside micrometer

Describe the vehicle being worked on:

Year _____ Make _____ Model _____

VIN _____ Engine type and size _____

Procedure

Using the cylinder head(s) that you removed from the engine and the proper service manual, you will identify and perform the required steps to measure for valve stem-to-guide clearance. The cylinder head and valve guides must be thoroughly cleaned first.

1. Use a small hole gauge and an outside micrometer to measure each valve guide's I.D. and then use the outside micrometer to measure each valve stem's O.D. and record differences below.

Cylinder #	Intake	Intake	Exhaust	Exhaust
1 (measured)				
1 (specification)				
2 (measured)				
2 (specification)				
3 (measured)				
3 (specification)				
4 (measured)				
4 (specification)				

Task Completed

5 (measured)				
5 (specification)				
6 (measured)				
6 (specification)				
7 (measured)				
7 (specification)				
8 (measured)				
8 (specification)				

☐   2. Install one valve at a time back into the valve guide with a small piece of rubber vacuum hose (about ½ inch in length) around the valve stem to hold it above the valve seat.

3. Attach a dial indicator to the cylinder head and place the tip of the indicator on the head of the valve so you can measure the sideways movement of the valve (valve stem-to-guide clearance). Measure and record each valve's clearance and compare it to specification.

Cylinder #	Intake	Intake	Exhaust	Exhaust
1 (measured)				
1 (specification)				
2 (measured)				
2 (specification)				
3 (measured)				
3 (specification)				
4 (measured)				
4 (specification)				
5 (measured)				
5 (specification)				
6 (measured)				
6 (specification)				
7 (measured)				
7 (specification)				
8 (measured)				
8 (specification)				

☐   4. Compare the measurements from #3 and #1. Are any of the measurements out of the manufacturer's specifications? _____

5. Using a flashlight, visually check for cracks in and around the valve guide.

6. Using the outside micrometer, measure each valve stem's O.D. in two areas. First measure the O.D. of the valve stem above where the valve normally rides in the valve guide (this is the shiny, worn part). Then measure the O.D. of the valve stem in the worn portion (shiny part where the valve sits in the guide). The difference between the two is how much the valve is worn. What is the maximum amount of valve wear? _____

7. Describe the service procedures that would be applicable if the cylinder head had excessive valve stem-to-guide clearance (include valve and guide service options).

_____

_____

8. Based on your inspections and measurements, what service procedures would you recommend?

_____

_____

Instructor's Response _____

_____

_____

Name _____ Date _____

INSERT VALVE SEAT REPLACEMENT

Upon completion of this job sheet, you should be able to replace insert valve seats in a cylinder head properly.

ASE Correlation

This job sheet is related to ASE Engine Repair Test's content area: Cylinder Head and Valvetrain Diagnosis and Repair; task: Inspect and resurface valve seats according to manufacturer's procedures.

Tools and Materials

Cylinder head
Valve seat remover
Valve seat bore cutter
Warming oven
Valve seat insert drivers

Describe the vehicle being worked on:

Year _____ Make _____ Model _____

VIN _____ Engine type and size _____

Describe the type of valvetrain used in the engine and the metal used in the cylinder head.

_____

_____

Procedure

Using the cylinder head(s) that you removed from the engine (or one provided to you by your instructor) and the proper service manual, you will identify and perform the required steps to replace valve seat inserts.

Task Completed

1. Clean the cylinder valve seat area. ☐

2. Using a pry tool or special remover, pry out the old insert. ☐

3. Clean the insert bore. ☐

4. Inspect for any cracks. Report any cracks to your instructor. ☐

5. Will the insert bore require oversizing?
   ☐ Yes   ☐ No
   If yes, go to step 6. If no, go to step 12.

6. What size bore is needed to provide the proper interference? _____

7. Set the cutter head to the proper size. ☐

☐   **8.** Set up the pilot and ball shaft.

☐   **9.** Place the cutter over the pilot and ball shaft.

☐   **10.** Set the depth of the insert at the feed screw.

☐   **11.** Cut the bore to the proper depth.

         12. Place the cylinder head(s) into a thermal oven. What temperature should the oven be set at? _____

☐   **13.** If available, place the new inserts into a freezer.

☐   **14.** Remove the heads from the oven and install the insert using the proper driver.

☐   **15.** Stake or peen the edge of the insert.

Instructor's Response _____

_____

_____

JOB SHEET

Name _____ Date _____

VALVE SEAT GRINDING

Upon completion of this job sheet, you should be able to resurface the valve seat properly.

ASE Correlation

This job sheet is related to ASE Engine Repair Test's content area: Cylinder Head and Valvetrain Diagnosis and Repair; task: Inspect and resurface valve seats according to manufacturer's procedures.

Tools and Materials

Valve seat grinding stones
Valve seat grinder
Valve seat concentricity tool

Describe the vehicle being worked on:

Year _____ Make _____ Model _____

VIN _____ Engine type and size _____

Describe the type of valvetrain in the engine.

_____

Procedure

Task Completed

Using the cylinder head(s) that you removed from the engine during Job Sheet 19 and the proper service manual, you will identify and perform the required steps to resurface the valve seat.

1. What are the valve set angles for your cylinder head?
   Intake _____ Exhaust _____

2. What is the angle needed to throat the seat?
   Intake _____ Exhaust _____

3. What is the angle needed to top the seat?
   Intake _____ Exhaust _____

4. Select the correct size and grit of stones required. Select the stones required for the seat, for topping, and for throating. ☐

5. Thread each stone onto a stone holder. ☐

6. Dress each stone. ☐

7. Select the proper size pilot and pass it through a cloth saturated with oil. ☐

8. Insert a pilot into each of the valve guides by pushing it down while twisting it until it is firmly wedged in place. The pilot should extend at least 2 ½ inches (63.5 mm) from the seat to fit inside the stone holder. ☐

☐

9. Locate the lifting spring over the pilot. Lubricate the pilot with a few drops of motor oil.

☐

10. Place the stone holder, with the proper stone for the seat, over the pilot. Check its fit onto the seat.

11. Attach the driver motor to the holder.

☐

12. With the stone off of the seat, turn on the motor and work the stone into the seat. When working the grinding stone, lift it on and off the seat at a rate of 120 times per minute to prevent excessive pressures on the stone and seat.

☐

13. Remove only enough material to provide a new surface all around the seat.

☐

14. Use a scale and measure and record the width of the seat. _____

☐

15. Top and throat the seat as needed to obtain the proper seat width.

☐

16. Check valve seat concentricity by placing the fixture over the guide pilot.
Specification _____ Actual _____

17. Check the fit of the valve into the seat by coating the face with a light layer of Prussian blue, then placing the valve into the seat.

☐

18. Rotate the valve back and forth 1/4 turn while applying pressure to the center of the valve. A small suction cup can be used.

☐

19. Remove the valve and inspect the contact pattern on the valve face and seat. Where is the contact located? _____

20. What action is required to correct the contact? _____

21. Repeat for all valve seats.

☐

Instructor's Response _____
_____
_____

Name _____ Date _____

RECONDITIONING VALVES

Upon completion of this job sheet, you should be able to recondition the valves properly.

ASE Correlation

This job sheet is related to ASE Engine Repair Test's content area: Cylinder Head and Valvetrain Diagnosis and Repair; task: Inspect and resurface valves according to manufacturer's procedures.

Tools and Materials

Valve refacing equipment

Describe the vehicle being worked on:

Year _____ Make _____ Model _____

VIN _____ Engine type and size _____

Procedure

Task Completed

Using the cylinder head(s) that you removed from the engine during Job Sheet 19 and the proper service manual, you will identify and perform the required steps to recondition the valves.

1. What is the specified face angle for the valves?

   Intake _____ Exhaust _____

2. Does the manufacturer recommend an interference angle?

   ☐ Yes   ☐ No

3. Set the spindle head to the specified angle. Lock into place.   ☐

4. Dress the stone with a diamond tool.   ☐

5. Clamp the valve into the spindle chuck. What special concerns must you be aware of when setting the stem depth into the chuck?

   _____

   _____

6. Set the stroke length. What special concerns are associated with this step?

7. Correctly position the valve face and the stone.   ☐

8. Advance the stone into the valve slowly until light contact is noted. Begin to stroke the valve across the stone.   ☐

9. Note the micrometer reading. _____

10. Continue to dress the face, feeding it 0.001 inch (0.25 mm) per pass until the face is clean.   ☐

11. Note the micrometer reading. _____

12. What was the total amount of face material removed? _____

13. Measure and note the valve margin. _____
Is the margin still within specifications? ☐ Yes ☐ No
If no, what must be done?

_____

☐   **14.** Use a piece of emery cloth to break the margin.

15. Measure and note valve head runout.
Specification _____ Actual _____

☐   **16.** Inspect the valve face. Report any defects to your instructor.

☐   **17.** Remove the valve from the spindle chuck, and set it into the valve stem holding fixture.

☐   **18.** Dress the stone used to grind the valve stem.

19. How much material needs to be removed from the tip of the stem? _____

20. Why is it important to stem the valve?

_____

_____

_____

☐   **21.** Stem the valve, followed by chamfering the tip.

Instructor's Response _____

_____

_____

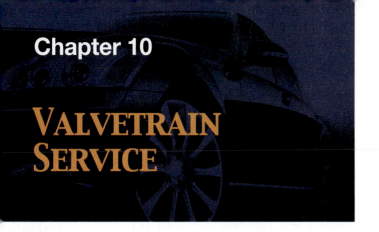

VALVETRAIN SERVICE

UPON COMPLETION AND REVIEW OF THIS CHAPTER, YOU SHOULD BE ABLE TO:

- Inspect the camshaft for straightness.
- Measure the camshaft lobes and journals and determine needed repairs.
- Inspect solid and hydraulic lifters and determine needed repairs.
- Perform the leak-down test on hydraulic lifters and accurately interpret the results.
- Inspect the pushrods and determine needed repairs.
- Describe the methods used to correct rocker arm geometry.

- Recondition rocker arms and replace studs.
- Evaluate and measure valve springs.
- Adjust the valvetrain during installation of hydraulic lifters.
- Adjust valve clearances on engines using mechanical lifters.
- Properly reassemble the cylinder head.
- Install a cylinder head.
- Replace valve seals with the cylinder head installed on the engine.

The valvetrain works to open and close the valves at the proper time. As the engine is run, the components of the valvetrain wear and stretch, causing valve opening to be altered.

This chapter discusses the methods used to inspect and repair the valvetrain, reassemble the cylinder head, adjust the valves, diagnose a failed head gasket, and replace worn valve stem seals on the car. Remember, before deciding to rebuild components such as camshafts and lifters, the cost of rebuilding components must be compared to the cost of replacing them. These components can usually be purchased new at less expense than rebuilding; however, there may be instances when rebuilding is a viable option.

INSPECTING AND SERVICING THE VALVETRAIN

The camshaft, lifters, rocker arms, and pushrods work together to open and close the valves. If any of these components are worn or damaged, valve operation is adversely affected.

Removing and Replacing the Camshaft, OHV Engines

Before removing the camshaft on an OHV engine, the timing chain and camshaft sprocket must be removed as we'll discuss in the next chapter. The rocker arms, pushrods, and valve lifters must be removed so the valve lifters do not interfere with the camshaft removal. Other components driven by the camshaft, such as the oil pump and/or distributor, must be removed prior to camshaft removal. On a rear-wheel-drive vehicle with the engine in the vehicle, the radiator usually has to be removed to allow camshaft removal. After these components are removed, the camshaft may be removed from the front of the engine.

All the camshaft journals and lobes must be thoroughly lubricated with the engine manufacturer's specified oil or an approved pre-lubing compound before installing the camshaft. The camshaft bearings must also be well lubricated.

Install the camshaft in the camshaft bearing openings. Hold the camshaft in a horizontal position during installation, and avoid scraping the cam lobes across the bearings, which results in bearing scratches. After the camshaft is installed, the valve lifters, pushrods, and rocker arms may be installed, followed by the oil pump and distributor. Install the camshaft sprocket and timing chain, and be sure the camshaft is properly timed to the crankshaft. Install the other components that were removed to gain access to the timing chain and sprocket. Be sure all fasteners are tightened to the specified torque.

Removing and Replacing the Camshaft, OHC Engines

Read through the service information for the vehicle you are working on before attempting to remove the camshaft from an overhead camshaft engine. Often there are special tools or procedures that can make your work both safe and efficient. The following procedure outlines the steps required to remove the camshafts on a late-model Asian overhead camshaft V6 engine:

1. Remove the valve covers.
2. Rotate the engine to top dead center firing on cylinder number 1.
3. Remove the camshaft sprocket bolts, but do not disturb the sprockets.
4. Install the sprocket holding fixtures in between the camshaft sprockets, and remove the sprockets from the camshafts.
5. Loosen the camshaft cap retaining caps in stages, following the proper sequence. Failure to do so can break the camshaft.
6. Remove the camshafts and inspect them thoroughly.
7. Remove and label the rocker arms, if needed, to be able to reassemble them in their original positions.

Removing Camshafts and Rocker Arms, DOHC Engine with Variable Timing Electronic Control (VTEC)

Before removing the camshafts on a DOHC engine, the timing belt must be removed, as we'll discuss. Follow this procedure to remove the camshafts:

1. Remove the pulley to camshaft retaining bolts on each camshaft and remove the camshaft pulleys (Figure 10-1). Identify each pulley so it is reinstalled in the original location.

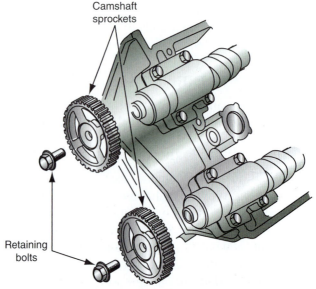

FIGURE 10-1 Removing the camshaft pulleys.

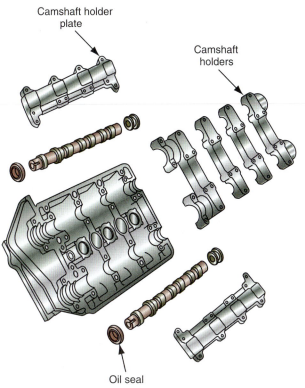

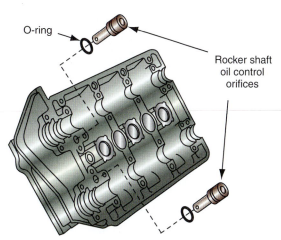

FIGURE 10-3 Removing rocker shaft oil control orifices.

FIGURE 10-2 Removing camshaft holder and holder plates.

2. Identify all camshaft holder plates and camshaft holders so they will be reinstalled in their original locations. Loosen the rocker arm locknuts and adjusting screws, and then remove the retaining bolts in the camshaft holder plates and camshaft holders (Figure 10-2).
3. Remove the camshafts from the cylinder heads.
4. Remove the retaining bolt on each rocker shaft oil control orifice, and pull out the oil control orifices (Figure 10-3).
5. Remove the VTEC solenoid from the cylinder head, and then remove the sealing bolts from the cylinder head (Figure 10-4).
6. Place an elastic around each triple set of rocker arms to keep them together. Identify all rocker arm sets or lay them out in order after removal, because they must be reinstalled in the original positions.
7. Screw a 12 × 1.25 mm bolt into each rocker arm shaft, and pull these shafts from the cylinder heads toward the transmission end of the engine (Figure 10-5).
8. Remove the rocker arm sets from the cylinder head.

Measuring Camshaft End Play and Camshaft Bearing Clearance, DOHC Engine with Variable Timing Electronic Control (VTEC)

Follow these steps to measure the camshaft end play:

1. With the rocker arms removed, place the camshafts in their original positions on the cylinder head. Be sure the camshaft journals and bearing surfaces are clean.
2. Place a light coating of the engine manufacturer's specified oil on the camshaft holder and holder plate bolt threads. Install the camshaft holders, holder plates, and retaining bolts. Tighten these bolts in the proper sequence to the specified torque.
3. Mount a dial indicator so the indicator stem is contacting the front of the camshaft (Figure 10-6). With the camshaft pushed rearward, zero the dial indicator.

SPECIAL TOOLS
Outside micrometer
V-blocks
Dial indicator
Pushrod adapter

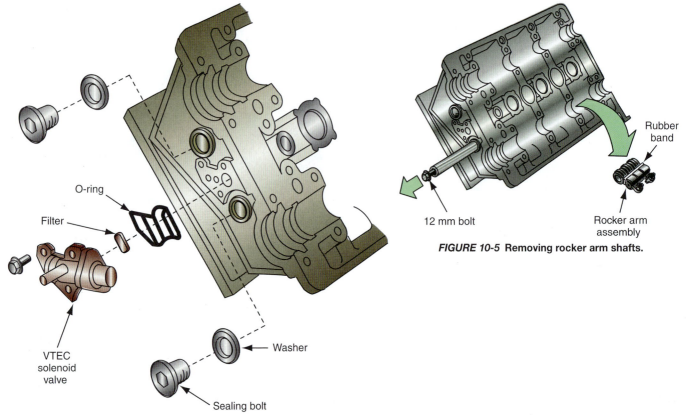

FIGURE 10-4 Removing VTEC solenoid and sealing bolts.

FIGURE 10-5 Removing rocker arm shafts.

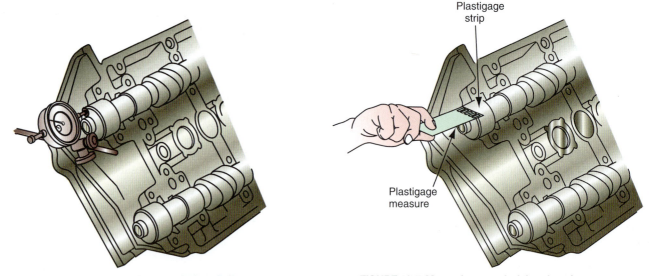

FIGURE 10-6 Measuring camshaft end play.

FIGURE 10-7 Measuring camshaft bearing clearance.

The **camshaft** contains a group of lobes that open and close the intake and exhaust valves at the correct time.

4. Pull the camshaft forward, and read the end play on the dial indicator.
5. Remove the camshaft holder plates and holders, and place a Plastigage strip across each camshaft journal.
6. Install the camshaft holders and holder plates, and tighten the retaining bolts in the proper sequence to the specified torque.
7. Remove the camshaft holders and holder plates, and measure the Plastigage width to determine the camshaft bearing clearance (Figure 10-7).

8. If the camshaft bearing clearance is more than specified, and the camshaft has been replaced, the cylinder head must be replaced. If the camshaft has not been replaced, remove the camshaft, and measure the camshaft runout. If the runout is within specifications, replace the cylinder head. When the runout is not within specifications, replace the camshaft, and recheck the bearing clearance.

Inspecting the Camshaft

Thorough inspection of the **camshaft** is important due to the many factors that apply to valve operation, and ultimately engine performance. These factors include lift, **duration,** and **overlap**. The proper relationship among these three aspects of the camshaft can only be maintained if the lobes are not worn (Figure 10-8).

Begin camshaft inspection with a visual inspection of the lobes and journals. Both of these must be free of scoring and galling. Normal lobe wear is a pattern slightly off center, with a wider wear pattern at the **nose** than the **heel** (Figure 10-9). The off-center wear is a result of the slight taper (0.0007 to 0.002 inch) of the lobe used in conjunction with the convex shape of the lifter to rotate the lifter. If the wear pattern extends to the edges of the lobe, the lifter will not rotate. If there is any scoring or galling, the camshaft will have to be replaced or reconditioned. Also inspect the fuel pump eccentric and distributor drive gear (if equipped).

The camshaft lobe lift can be measured using an outside micrometer. Lift is determined by measuring the lobe from heel to nose and subtracting the **base circle** measurement from the initial measurement (Figure 10-10). Another method is to install the camshaft onto V-blocks and use a dial indicator to measure the amount of lift. Locate the plunger of the indicator on the heel of the lobe, and zero the indicator. Rotate the camshaft until the highest reading is observed. This is the amount of lobe lift. Check camshaft straightness before checking lift in this manner.

Camshafts with high durations cannot be measured using a micrometer since a true base circle measurement will not be obtainable. This is due to the ramps beginning more toward the heel. In these instances, use a dial indicator to measure the lift of the pushrod.

Duration is the number of degrees that a cam lobe holds a valve open.

Overlap is the number of degrees that the crankshaft rotates when both intake and exhaust valves are open near TDC on the exhaust stroke.

Classroom Manual
Chapter 10

The **nose** is the area just before and after the peak of the lobe. The camshaft **heel** is 180° from the lobe high point. This is where the valve will be closed.

The camshaft **base circle** is the diameter across the sides of the camshaft lobe with the lobe facing upwards.

SERVICE TIP:
Worn camshaft lobes may cause a heavy clicking noise at 1,500 to 2,000 rpm.

Lobe-separation angle
(lobe centers)

Overlap

TDC

Exhaust closed

Intake opens

Lift

Lift

Intake duration

Exhaust duration

Intake lobe

Exhaust lobe

Intake closed

Exhaust opened

FIGURE 10-8 Wear on the camshaft lobe can affect several factors of valve timing and duration.

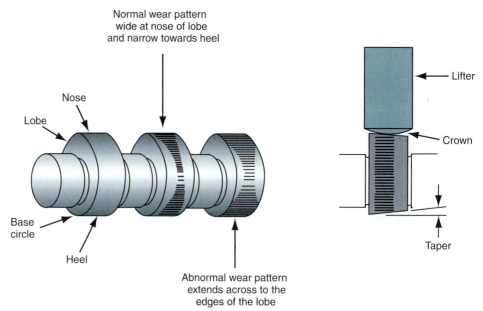

FIGURE 10-9 Normal wear patterns are slightly off center on the lobe.

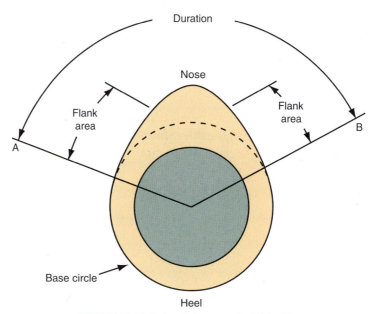

FIGURE 10-10 Determining camshaft lobe lift.

Lobe lift is the distance that a cam lobe moves the contacting valvetrain component as the cam lobe rotates.

Another method of measuring **lobe lift** involves installing the camshaft into the engine block with the lifters and pushrods. A special cup-shaped adapter is installed onto the dial indicator to accept the pushrod (Figure 10-11). With the pushrod at its lowest point of travel, zero the dial indicator. Continue to rotate the camshaft until the maximum amount of deflection of the dial is observed. This represents the total lift of the camshaft.

Measure the camshaft journals for size, taper, and out-of-round using an outside micrometer (Figure 10-12). Most overhead valve engines have different size journals, with the largest being toward the front of the engine. This is done to make camshaft removal and installation easier. To determine the amount of out-of-round, measure each journal in two different directions, and compare to specifications. Also check for journal taper by measuring at each end of the journal (Figure 10-13).

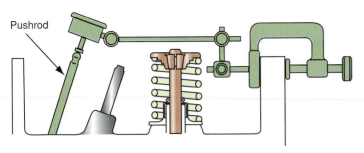

Pushrod

FIGURE 10-11 Using a special adapter to measure camshaft lobe lift with the camshaft installed in the engine.

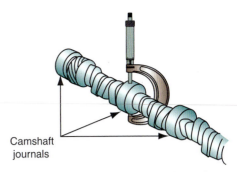

Camshaft journals

FIGURE 10-12 Worn camshaft journals can cause valve timing to change as the engine is running.

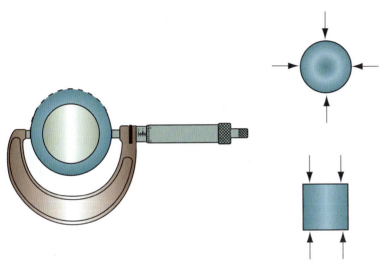

FIGURE 10-13 Compare each journal to specifications and check for taper.

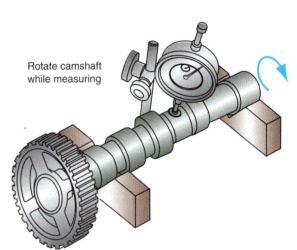

Rotate camshaft while measuring

FIGURE 10-14 Measuring camshaft warpage.

Once the new camshaft bearings are installed into the engine block, the clearance can be determined. The bearings may need to be honed to obtain the required clearance. This will not be possible if the journals are worn beyond limits.

Camshaft straightness can be checked by placing it in V-blocks. The camshaft is supported in the blocks by the journals on both ends. Install a dial indicator onto the central journal (Figure 10-14). Observe the dial indicator as the camshaft is rotated. A total indicated reading (TIR) of 0.002 in. (0.050 mm) indicates excessive warpage.

With the camshaft set up to measure straightness, also check base circle runout. Locate the dial indicator plunger on the heel of the lobe, and zero the gauge. Slowly rotate the camshaft until the lowest gauge reading is obtained. This is the point where the camshaft lobe ramp starts. Note this reading, then rotate the camshaft in the opposite direction until the gauge indicates the start of the other ramp. As a rule, total runout should not exceed 0.001 in. (0.025 mm). If the runout is excessive, the lifters may pump up when the valve is closed.

INSPECTING AND SERVICING VALVETRAIN COMPONENTS, OHV ENGINE
Inspecting the Lifters

Inspect the camshaft contact surface face of the **lifter.** It must be smooth, with a centered circular wear pattern. To reduce wear, the lifter rotates in its bore. Part of this rotating action is accomplished by a convex contour machined in the face. If the wear extends to the edge of the face, it indicates the lifter is flat or concave. If the wear pattern runs across the

CAUTION:
If it is determined the camshaft requires replacement, always replace the lifters. The camshaft and lifters become wear mated within a short time, and used lifters will rapidly wear a new camshaft. It is possible to replace a single lifter without replacing the camshaft.

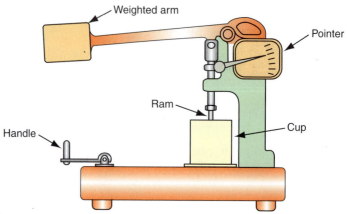

FIGURE 10-15 Using a special lifter leak-down tester to determine lifter condition.

SPECIAL TOOLS

Leak-down tester
Special tester fluid

Valve **lifters** are sometimes called tappets because of the sound they make during operation, especially if they are loose.

Leak-down is the relative movement of the plunger in respect to the lifter body.

SERVICE TIP:
Worn or sticking valve lifters may cause a clicking noise at low speed, especially when the engine is warming up.

lifter face, it indicates the lifter was stuck in its bore. Lifter crown can be checked by laying a straightedge across the face. If the straightedge does not rock, the lifter must be replaced.

The lifter body should be polished and absent of any signs of scoring or scuffing. The pushrod seat should be polished and smooth. If the seat shows signs of a ridge, the lifter must be replaced.

Hydraulic lifters passing visual inspection should be tested for **leak-down.** A special leak-down tester is used for this operation (Figure 10-15). Test results are determined by the length of time, in seconds, required before the tester overcomes the hydraulic pressures in the lifter. Normal leak-down range is between 20 and 90 seconds. Always refer to the manufacturer's specifications. A lifter that fails the first leak-down test should be retested, since air may be trapped in it. If the test results indicate a failure or marginal results, replace the lifter.

To perform a leak-down test:

1. Fill the cup with enough fluid to cover the lifter, and place the lifter into the fluid.
2. Position the tester ram into the pushrod seat of the lifter, making sure the ram is centered.
3. Adjust the ram length until the pointer aligns with the set mark on the lever, and tighten the jam nut.
4. Pump the weight arm to move the lifter plunger through its entire range of travel. This is done to purge air from the lifter. Continue to pump the lifter until a resistance is felt.
5. Raise the arm to return the lifter to its full height, then lower the ram onto the pushrod seat. Time how long it takes for the weight to compress the lifter. During this test, turn the hand crank on the tester so the cup rotates at a rate of one turn every 2 seconds.

Roller lifters are inspected in the same manner. The roller must be smooth and cannot be twisted to the body. In addition, inspect the mechanism used to prevent lifter rotation. The lifter must be parallel to the camshaft lobe at all times. To prevent rotation, some manufacturers use a pivot link between two lifters, while others use some type of lock. These mechanisms cannot have any twist in them.

If the lifters are to be reused, it is important to maintain their original positions on the camshaft. When a lifter is determined unserviceable, replace the set. Also, replace the camshaft whenever the lifters are replaced. Make sure the replacement lifters have the same body diameter, operating height, oil groove position, and oil groove width as the original equipment lifters.

It may occasionally be necessary to disassemble a hydraulic lifter to clean it or to inspect it. If this is necessary, disassemble one lifter at a time and pay attention to the order of the check valve assembly.

Lifter disassembly is done by pressing down on the pushrod seat with your thumb and using a screwdriver to remove the snap ring (Figure 10-16). Remove the plunger from the

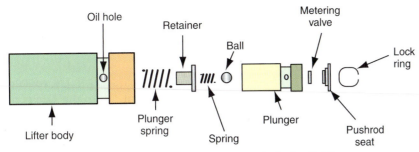

FIGURE 10-16 Internal components of a hydraulic lifter.

body; this may require tapping the body. Next, remove the check valve assembly. After thoroughly cleaning the lifter components in solvent, visually inspect them for signs of wear. If the internal components pass inspection, the lifter can be reassembled. During reassembly, cleanliness of the parts is of utmost importance.

Inspecting the Pushrods

Inspect **pushrods** for wear and bending. Any pushrod failing inspection must be replaced. Visually inspect the two tip surfaces for wear. Normal wear results in a smooth and polished finish. Wear on the side of the pushrod indicates the pushrod guide is defective and it too must be replaced. Finally, check the pushrod for warpage by rolling it on a sheet of glass or using a special fixture (Figure 10-17). Total pushrod runout should be less than 0.003 in. (0.08 mm).

On hollow pushrods, clean the oil passage with a piece of wire and solvent. The oil passage opening should be circular and have a well-defined edge. If the hole is oval shaped or the edge is chamfered, replace the pushrod.

Inspecting and Servicing Rocker Arms and Shafts

When rocker arms are removed, always set them out in order so they can be installed in their original positions. Inspect all rocker arms for wear in the pushrod contact area, the tip that contacts the valve stem, and the bore in the center of the rocker arm. A slight amount of wear on the end of the rocker arm that contacts the valve stem may be removed with a valve grinding machine. A special attachment on these machines is available for rocker arm resurfacing. Do not remove excessive material from the rocker arm, because this action changes the rocker arm geometry. Rocker arm wear in the pushrod contact area or in the bore requires rocker arm replacement.

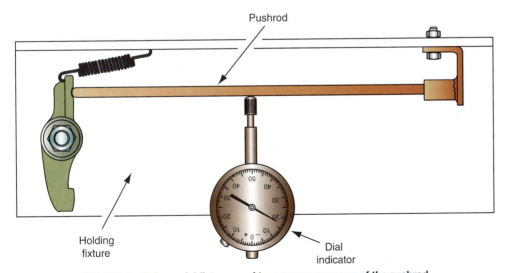

FIGURE 10-17 A special fixture used to measure warpage of the pushrod.

On individual rocker arms, inspect the rocker arm for wear in the pivot contact area. Wear in this area requires rocker arm replacement. Worn pivots must also be replaced. Inspect all rocker arm shafts for wear, scoring, cracks, blocked oil passages, and a bent condition. If these conditions are present, replace these shafts.

Use shop air pressure and an air gun to blow out all oil passages in the rocker arms and shafts. Be sure the oil passages in the cylinder head that supply oil to the rocker arms and shafts are not restricted. If the oil for the rocker arms is supplied through the pushrods, be sure all the pushrod oil passages are not restricted.

REPLACING ROCKER ARM STUDS

If any of the rocker arm studs are damaged, worn, or loose in the cylinder head, they must be replaced. Broken or damaged press-in studs can be replaced by pulling the original stud and replacing it with a standard press-in stud. If the original press-in stud is loose, the technician can oversize the hole and install oversize studs. Press-in oversize studs are available in 0.006, 0.010, and 0.015 inch oversize.

If desired, it is possible to replace press-in studs with threaded studs. Begin by removing the original studs using a stud puller (Figure 10-18). If the new stud has a jam nut, the stud boss height must be reduced the same amount as the height of the nut. Use a milling tool to remove the required amount of material. The stud boss is ready to be tapped to accept the new threaded stud.

The rocker arm is constructed of cast iron, stamped steel, or aluminum. The contact areas with the pushrods and valve stems are usually hardened. Regardless of the type of mounting used by the manufacturer, all rocker arms must be inspected for wear at the three high stress areas. These areas are the pivot, pushrod contact, and valve stem contact surfaces (Figure 10-19).

The pushrod contact area of the rocker arm should be round and the wear even. Normal wear will result in a shiny surface. A ridge buildup around the pushrod contact area indicates the rocker was hammering or too much valve lash. Pits, scores, and galling generally indicate lack of lubrication or excessive heat. Valve stem wear should be centered in the arm. If the wear is not centered, the rocker arm pivot may be worn or the stud may be bent. If these are good, then carefully inspect the valve guide and seat for wear. In addition to these checks, inspect the side of the rocker arm for any indications of cracks. Pay close attention around the pivot contact area. Stamped steel rocker arms are not resurfaceable and should be replaced if found to be worn.

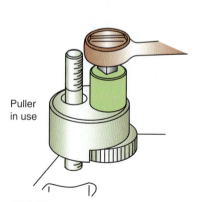

FIGURE 10-18 Using a stud remover to pull a rocker stud.

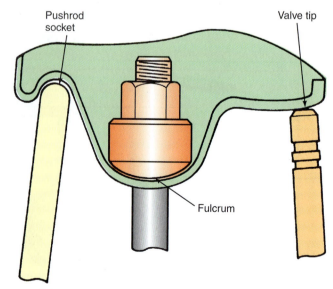

FIGURE 10-19 The three wear areas of the rocker arm.

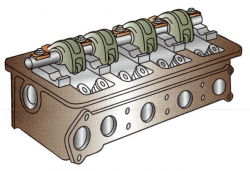

FIGURE 10-20 Shaft mounted rocker arms.

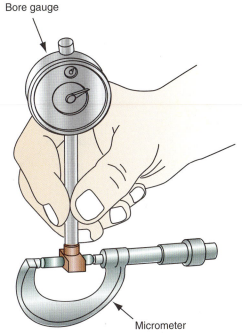

Bore gauge

Micrometer

FIGURE 10-21 Shaft-to-bore clearance and taper can be checked by using an outside micrometer set to the size of the rocker arm shaft, then zeroing the small bore gauge to the micrometer.

SPECIAL TOOLS
Sheet of glass
Outside micrometer
Bore gauge

The method the manufacturer uses to mount the rocker arm will require unique inspection procedures. Following are some of the additional inspections that should be made based on the type of mounting:

1. Shaft-mounted rocker arms are located on a heavy shaft running the length of the cylinder head (Figure 10-20). The shaft is supported by stands. Inspect it for wear on the shaft where the rocker arm pivots. The shaft should be free of scoring or galling. Normal wear is indicated by a slight polishing but no ridges. Wear in this location is usually due to lack of lubrication or excessive heat. Also check the shaft for warpage by simply rolling it on a flat surface (such as a sheet of glass). Shaft-to-bore clearance can be checked by using an outside micrometer to measure the diameter of the shaft where the rocker rides. Use a bore gauge and zero it to the micrometer (Figure 10-21). Insert the bore gauge into the rocker arm and note the dial reading. Out-of-round can be checked at the same time as clearance by measuring in at least two directions. Compare results to specifications. If wear is excessive, the rocker arm must be replaced. Typical oil clearance specification is a maximum of 0.005 in. (0.13 mm). Visually inspect the springs and spacers used to separate the rocker arms for warpage, breakage, scoring, and other types of unusual wear patterns. Replace any that are found to be defective or worn.

2. Stud-mounted rocker arms use a split ball for the pivot point of the rocker arm (Figure 10-22). Pedestal-mounted rocker arms are similar to the stud-mounted assembly, except the rocker pivots on a split shaft (Figure 10-23). With both of these styles, the pivot fulcrum must be inspected for wear. Look for nicks, scoring, or scuffs. If the pivot or rocker arm is worn in these areas, replace both components. Scoring or scuffing indicates a lack of lubrication. Stud-mounted rocker arms can be lubricated through a hollow pushrod or by small holes drilled through the mounting stud. Pedestal-mounted rocker arms are usually lubricated by an oil passing through a hollow pushrod. If poor lubrication is suspected, the stud should be replaced. On positive stop rocker arm stud nuts, check the shoulder for damage or fractures (Figure 10-24). Also

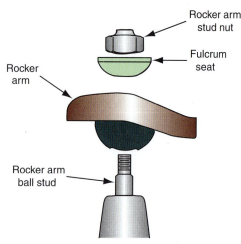

Rocker arm stud nut

Fulcrum seat

Rocker arm

Rocker arm ball stud

FIGURE 10-22 Stud-mounted rocker arm.

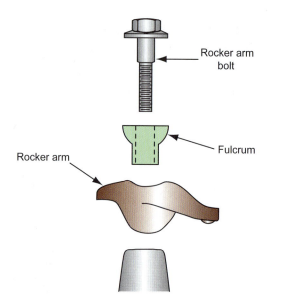

Rocker arm bolt

Fulcrum

Rocker arm

FIGURE 10-23 Inspect the pivot fulcrum and contact surface for nicks, scoring or scuffing.

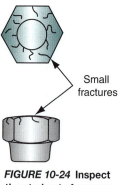

Small fractures

FIGURE 10-24 Inspect the stud nuts for wear or damage.

SERVICE TIP:
If the engine is equipped with stamped steel rocker arms, it is a good practice to replace them during a rebuild. The cost of these components is very low and can be considered inexpensive insurance.

check the rocker arm's contact surface with the pivot fulcrum. If the pivot or rocker arm is worn in this area, replace both components.

The follower may have a flat pad, sliding pad, or rollers where it rides against the camshaft lobe. This area must be visually inspected for wear. Check the rest of the follower much like the rocker arm.

Rocker arm studs may pull out of the head or wear from contact with the rocker arm. Inspect the stud very closely. If it is loose, install a new stud.

Correcting Rocker Arm Geometry

If the **rocker arm** is not contacting the valve stem tip in its center at 50 percent valve opening, the valve guide will be exposed to excessive side thrust. To correct rocker arm geometry, one of the following methods can be used (Figure 10-25):

■ Changing the length of the pushrod
■ Machining or shimmying the rocker arm stand
■ Correcting valve stem height

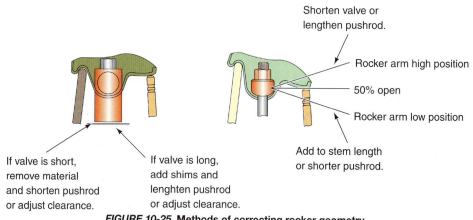

Shorten valve or lengthen pushrod.

Rocker arm high position

50% open

Rocker arm low position

Add to stem length or shorter pushrod.

If valve is short, remove material and shorten pushrod or adjust clearance.

If valve is long, add shims and lenghten pushrod or adjust clearance.

FIGURE 10-25 Methods of correcting rocker geometry.

Inspecting and Servicing Rocker Arms and Shafts, DOHC Engine with Variable Valve Timing Electronic Control (VTEC)

Inspect the contact pad on all rocker arms for scoring and wear.

Inspect the rocker arm pistons for scoring and free movement in the rocker arms (Figure 10-26). Scored or sticking pistons require rocker arm and piston replacement. Inspect the rocker arm shafts for wear and scoring. Shafts that are visibly worn and/or scored must be replaced. Measure the rocker arm shaft diameter with a micrometer in the first rocker arm position on one end of the shaft. Install a bore gauge in the micrometer with the micrometer set at the rocker arm shaft diameter, and zero the bore gauge (Figure 10-27). Install the bore gauge in the matching rocker arm opening, and determine the rocker arm to shaft clearance by reading the amount the gauge pointer moved from the zero position (Figure 10-28). Repeat this procedure for all the rocker arms. Replace any rocker arms and shafts if this clearance is not within specifications.

Inspect each lost motion assembly for scoring or wear. Replace all scored or worn assemblies. If the lost motion assembly bores in the cylinder head are scored, replace the cylinder head. When the lost motion assembly protrusion is pressed gently with finger pressure, it should sink slightly (Figure 10-29). Increasing the force on the protrusion should result in further sinking. If any lost motion assembly does not move smoothly when finger pressure is applied to the protrusion, replace the assembly.

Valve Springs

Begin valve spring inspection with a visual inspection. If a spring is broken or obviously shorter than the others, replace it (Figure 10-30). In addition, look for cracks, excessive corrosion, pitting, and nicks. If a valve spring has any of these defects, replace it.

If the spring passes visual inspection, other checks are required before reusing it. These checks include **spring squareness**, free length, valve close pressure, and valve open pressure.

A spring that is not square can cause excessive side pressure on the valve stems. This pressure can cause wear or stress on the stem, guide, and valve seat. The **free length** of the

Classroom Manual
Chapter 10

The **rocker arm** is a pivoting lever used to transfer the motion of the pushrod to the valve stem.

SERVICE TIP: If correction is done by milling or shimming the rocker arm stand, alter the stand height one-half the total amount needed at the stem tip.

SERVICE TIP: Rocker arms must be replaced in sets of three.

Classroom Manual
Chapter 10

SPECIAL TOOLS
Square
Machinist's rule
Spring tension tester
Torque wrench

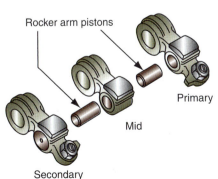

Rocker arm pistons

Primary

Mid

Secondary

FIGURE 10-26 Inspecting rocker arms.

Bore gauge

Micrometer

FIGURE 10-27 Transferring rocker arm shaft diameter reading from a micrometer to a small bore gauage.

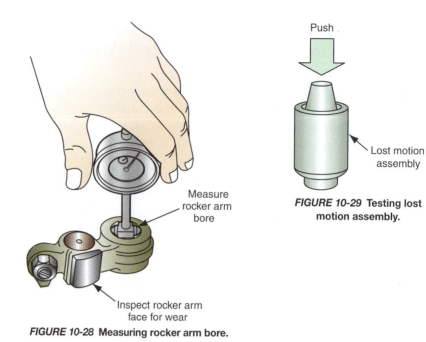

Measure
rocker arm
bore

Inspect rocker arm
face for wear

FIGURE 10-28 Measuring rocker arm bore.

Push

Lost motion
assembly

FIGURE 10-29 Testing lost
motion assembly.

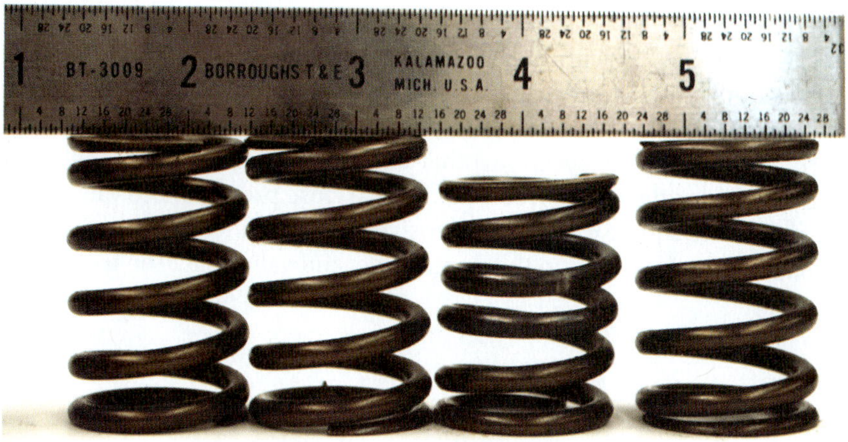

FIGURE 10-30 Any valve springs that are below free height
specifications must be replaced.

spring must be checked and compared to specifications. Valve springs below the minimum length indicate weak springs. Valve springs under the specified height may be the result of excessive heating. Check the cylinder head and valves for other indications of overheating. Springs that are below the specified height can cause burned or floating valves. Also, the valve spring free length should not vary more than 1/32 in. (0.8 mm) between springs. It may be possible to use **spring shims** to correct for spring height instead of replacing the spring.

To inspect the valve spring for squareness and free length, place the spring next to a square and rotate it (Figure 10-31). Watch for any warpage of the spring as evidenced by a gap between the top of the spring and the square. A general rule is the spring should not have more than a 2.3-degree angle. The gap between the spring and square will be different based on the length of the spring. For this reason, general specifications for all springs should be avoided. The basic formula for determining gap is:

Spring length × 0.039 = Maximum gap

where 0.039 is the product of 2.3 times 0.017 (the maximum amount of gap allowed per degree per inch).

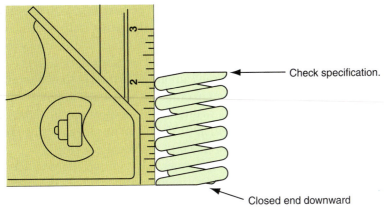

Check specification.

Closed end downward

FIGURE 10-31 Spring squareness can be checked by placing the spring next to a square and rotating the spring.

For example, if the spring free length is 2 inches, the maximum gap would be 2 × 0.039 = 0.078 inch. To convert to metric measurements, use spring length × 0.43 = max. gap in mm.

While checking squareness, also note the length of the spring as it is rotated. The length should not vary by more than 1/32 in. (0.8 mm). If the spring is not square or free length is not within specifications, replace the spring.

Valve closed and open pressures are tested using a special spring tension tester (Figure 10-32). It is possible for a spring to pass the free height check but fail the tension tests. This is due to changes in the metal structure as a result of temperature changes. Springs not within specifications must be replaced.

Spring pressure is measured at the two lengths at which the spring operates: valve open and valve closed. To test the spring, begin by checking that the gauge reads zero. Place the spring onto the tester's table. Pull down on the lever until the spring is compressed to the specified installed height as indicated on the scale. Read the pressure indicated on the dial

Spring shims are used to correct the installed height of the valve spring. Details concerning shim use and selection are provided later in this chapter.

SERVICE TIP: Check new springs for squareness and warpage before installing them.

Seat pressure indicates spring tension with the spring at installed height with the valve closed.

Open pressure indicates spring tension with the spring compressed and the valve fully open.

FIGURE 10-32 Testing the valve spring for proper seat and open pressures.

required to achieve the specified height. Compare the dial reading with the manufacturer's specifications. This reading provides the spring's **seat pressure** (valve closed pressure). Next, compress the spring to the valve open height specification. This **open pressure** specification is determined by the maximum camshaft lift; for example, if the installed height specification is 1 13/16 inches and the cam lift is 7/16 inch, then compress the spring until the height of the spring is 1 3/8 inches. Compare the tension reading obtained at the valve open height to specifications. While maintaining the valve open height, check for coil bind by trying to insert a 0.010- to 0.012-inch (0.25 to 0.30 mm) feeler gauge between each coil. If the feeler gauge cannot fit between the coils, replace the spring.

Repeat the pressure tests for all outer and inner valve springs. Remember that the inner spring rides on a stepped portion of the spring retainer, and you will have to allow for this; for example, if the step measures 1/8 inch, the inner spring must be compressed 1/8 inch more than the outer spring.

Spring tension should be within 10 percent of the manufacturer's specification, with no more than a 10-pound difference between springs. If the spring tensions are outside of specifications, the springs must be replaced.

Another style of spring tension tester uses a torque wrench instead of a gauge (Figure 10-33). To adjust the table of this tool, turn it until the surface is in line with the specified height marked on the threaded stud (Figure 10-34). Make sure the zero mark is facing toward the front. Locate the valve spring over the stud on the table, and lift the compressing lever to reset the tone device. Install a needle- or dial-type torque wrench onto the tension tester. Compress the spring with the torque wrench until the ping is heard. Note the reading on the torque wrench. Spring tension is determined by multiplying the torque wrench reading by two. Valve open tension tests are performed in the same manner.

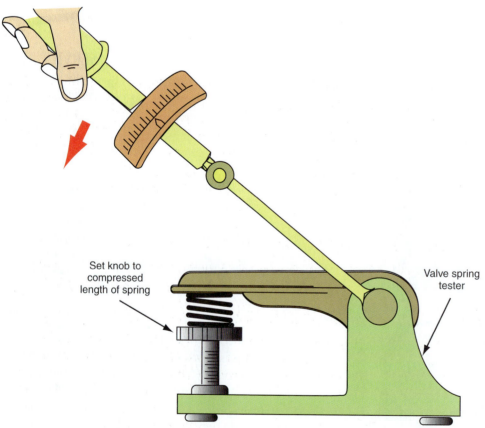

Set knob to compressed length of spring

Valve spring tester

FIGURE 10-33 Another style of valve spring tension tool uses a torque wrench instead of a dial to indicate pressures.

FIGURE 10-34 Adjust the table height to the installed height of the spring.

CYLINDER HEAD REASSEMBLY

The cylinder head is now ready for reassembly. Begin this procedure by replacing any studs removed during the reconditioning operation. Next, clean all oil galleys and coolant passages, and install the plugs. If the cylinder head uses a core plug to seal the back camshaft bore, do not install this plug until the camshaft is fitted. Coat threads with liquid pipe thread sealer or Teflon tape. Core and galley plugs are driven into the cylinder head using a special driver. Make sure to drive the plug in the correct direction. Coat the sides of the plug with a good water- and oil-resistant sealer. Be careful to keep any chemical sealers off the end of the plug where they can be washed into the engine oil. Start all plugs by hand, then torque them to specifications. Most plugs use a taper pipe thread. These plugs will not fully seat until they are tightened.

Installed Spring Height

Before installing the valves, polish the valve stems using a fine crocus cloth and solvent, and then thoroughly clean the valves again. Lubricate the valve stems with a generous amount of engine oil or assembly lube, and install the valves into their correct ports. After the valves are installed in the head, check spring height. This can be done without installing the valve spring. Install the valve and the spring retainer; then measure the distance between the cylinder head and valve spring retainer (Figure 10-35). Stem height increases due to resurfacing of the valve face and seat. Stem height is corrected by removing material from the stem tip. This operation restores the proper rocker arm geometry. However, the valve spring keeper grooves are higher from the valve spring seat now than they were originally. A telescoping gauge or vernier calipers can be used to check spring height (Figure 10-36). The valve retainer and keeper are installed without the valve spring to make this measurement. If the installed height is greater than specifications, select the correct shim to return the spring height to specifications. If the spring height measurement is greater than 0.060 in. (1.5 mm) over specifications, the valve seat will need to be replaced.

To determine the shim size required, measure the installed height of the valve spring. Compare the measurement with specifications. If the measured distance is greater than specifications, subtract the measurement from the specifications to determine the difference. The difference is the correct size shim required.

Installing Valve Stem Seals

The point at which the valve stem seals are installed during reassembly of the head depends upon the type of seal. If the cylinder heads use positive lock valve stem seals, they may need to be installed prior to installing the valves. A seal installation tool is used for this purpose (Figure 10-37). Other types of positive seals are installed over the valve stems.

CAUTION:
Be sure all sealers used are oxygen-sensor safe. Some sealers will cause the sensor to fail.

SPECIAL TOOLS
T-gauge
Micrometer

CAUTION:
Use of shims in excess of recommendations will overstress the valve springs and overload the camshaft lobes. Shims should never be used to correct for a weak spring.

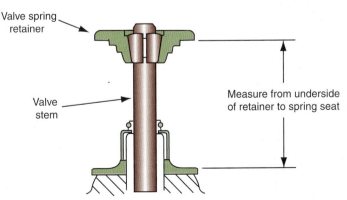

FIGURE 10-35 Measuring valve spring installed height without the valve spring installed.

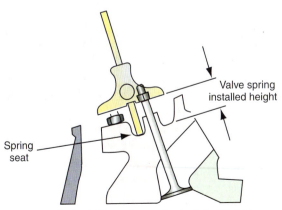

FIGURE 10-36 Measuring spring installed height.

413

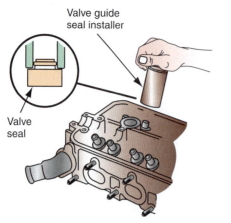

FIGURE 10-37 Using a special tool to install the valve stem seal.

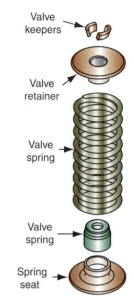

FIGURE 10-38 Many aluminum cylinder heads use a steel spring seat. Install the seat before installing the valve stem seal.

Most manufacturers use different size seals for the intake valves than for the exhaust valves. The seals are usually color coded or identified in some manner. Always refer to the appropriate service manual to determine proper seal location.

Installing Positive Stem Seals.

Many aluminum cylinder heads use a steel spring seat between the head and the valve spring (Figure 10-38). The seal must fit onto the seat properly. Fit the seat first, then install the seal. After all the seals are installed, place the cylinder head into a work stand for preparation for valve installation.

Following is a general procedure for the seals if the valve stem seal is to be installed over the valve stems:

1. If it has not already been done, polish the valve stem using a fine crocus cloth and solvent. Be sure to thoroughly clean the valve again before installing it.
2. Lubricate the valve stem with a generous amount of engine oil or assembly lube.
3. Install the valve into its correct port.
4. Place the plastic sleeve over the end of the valve stem to protect the seal as it is installed. The sleeve is included in the valve stem seal kit.
5. Place the seal over the stem and slide it down until it contacts the top of the valve guide.
6. Use the installation tool to press the seal over the guide until it is flush with the top of the guide.

Installing Umbrella-Type Seals.

Umbrella-type seals are installed after the valves are located into the cylinder head. Before installing the valve, polish the valve stem using a fine crocus cloth and solvent; then thoroughly clean the valve again. Lubricate the valve stem with a generous amount of engine oil or assembly lube, and install the valve into its correct port. Place the stem seal over the stem and push it down until it touches the valve guide boss. There is no need for special tools or press fitting this type of seal.

Installing O-Ring Stem Seals.

The O-ring seal is installed after the valve springs have been located over the valve stem and are compressed. With the spring compressed, install the seal into the bottom groove. It is a good practice to lubricate the seal with engine oil before installing it. Check the seal to make sure it did not twist during installation.

Installing the Valves, Springs, and Keepers

If the cylinder head uses positive stem seals requiring installation prior to valve installation, press them onto the valve guide bosses at this time. Start at one end of the cylinder head and work toward the other, working with one valve at a time. The following general procedure can be used to install the valves into the cylinder head (Figure 10-39):

WARNING: Wear approved eye protection when assembling the cylinder head. In addition, face the springs away from your body as they are being compressed.

1. Polish the valve stems using a fine crocus cloth and solvent. Be sure to thoroughly clean the valve again before installing it.
2. Lubricate the valve stems with a generous amount of engine oil or assembly lube.
3. Locate the valve into the correct port. Do not mix the valves. The valve should install easily into the guide.
4. Install the correct size spring shim with the serrated side resting against the cylinder head.
5. Install umbrella-type or positive-type valve stem seals (if equipped) over the valve stem.
6. Place the valve spring over the stem. When installing the valve spring, make sure it is facing in the correct direction. Variable rate springs are installed with the tightly coiled end of the spring face toward the head. If the valve spring has a taper, the larger end fits against the spring seat.
7. Locate the spring retainer over the valve spring.
8. Use an approved spring compressor to compress the valve spring (Figure 10-40). Compress the spring only enough to install the keepers.
9. If the cylinder head uses an O-ring-type valve seal, install it onto the bottom groove while the spring is compressed.
10. With the spring compressed, install the keepers (Figure 10-41). Make sure they are seated into their grooves before releasing the valve spring compressor.
11. Tap the top of the valve stem with a plastic hammer (Figure 10-42). This is to assure the keepers are properly seated.
12. Attempt to turn the valve spring by hand. If it is properly installed, the valve spring should not rotate.

CAUTION:
Be sure to install the correct length of umbrella seal. If a seal that is too short is installed, the up-and-down motion will create a pumping action. This will result in excess oil entering the combustion chamber.

SPECIAL TOOLS
Valve spring compressor

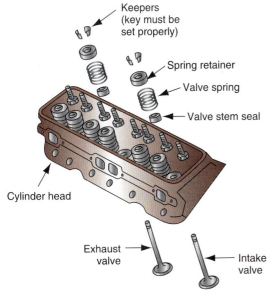

FIGURE 10-39 Typical cylinder head assembly.

FIGURE 10-40 Compress the spring and use a dab of grease to hold the keepers in place.

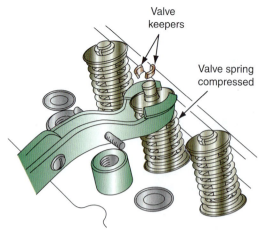

FIGURE 10-41 The spring must be compressed to install the keepers.

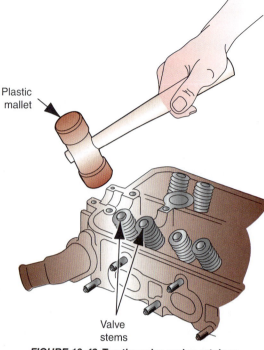

FIGURE 10-42 Tap the valve spring retainer with a plastic hammer to be sure that the keepers are fully seated.

13. Repeat for all valves.
14. Lubricate the stem tips with a multipurpose grease.
15. Install the rocker arms, fulcrums, or shafts. Do not tighten the rocker arm bolts at this time.

   In addition, it is a good practice to test the valve seat sealing after the valves are installed. To do this, install the spark plugs, then placing the cylinder head with the combustion chambers facing up. Fill the chambers with solvent, and let the heads sit for a few minutes. If the seats leak, solvent will leak down the stem and into the ports.

OHC Cylinder Head Assembly

Following is a general procedure for assembling an OHC engine cylinder head with camshaft bearing caps. Follow the preceding procedure to install the valves; then return here to continue with the assembling process:

1. Assemble the rocker arm assembly following the manufacturer's procedures (Figure 10-43).
2. Clean the camshaft and cylinder head bores.
3. Lubricate the bores and journals with engine oil or assembly lube.
4. Install the camshaft into the lower half of the bores.
5. Rotate the camshaft so the keyway is facing up at the 12:00 position. This is top dead center (TDC) for number 1.
6. Insert correct size plastigage strips on each journal.
7. Position the rocker arm assembly (or bearing caps) in place and loosely install the bolts. Check the bolts before reusing them; the threads must be clean and undamaged.
8. Tighten each rocker arm two turns at a time following the recommended tightening sequence.
9. Finish torquing the rocker arm assembly bolts to the final torque value.

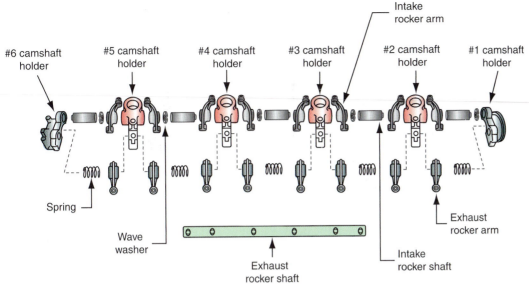

Intake
rocker arm

#6 camshaft holder | #5 camshaft holder | #4 camshaft holder | #3 camshaft holder | #2 camshaft holder | #1 camshaft holder

Spring

Wave washer

Exhaust rocker shaft

Intake rocker shaft

Exhaust rocker arm

FIGURE 10-43 Rocker arm assembly.

10. Remove the rocker arm assembly and measure the plastigage at its widest point. This is the amount of oil clearance. Check the results against the specifications.
11. Remove the plastigage from the journals.
12. Slide the camshaft seal onto the camshaft.
13. Some manufacturers require liquid sealers to be used in specified locations. Use the specified sealer type.
14. Position the rocker arm assembly in place and loosely install the bolts.
15. Tighten each rocker arm two turns at a time following the recommended tightening sequence.
16. Finish torquing the rocker arm assembly bolts to the final torque value.
17. Seat the camshaft by pushing it all the way to the rear of the cylinder head.
18. Install a dial indicator against the front of the camshaft and zero the gauge (Figure 10-44).
19. Push the camshaft back and forth to measure end play. Compare the results with specifications.
20. Remove the dial indicator.
21. Install back plates (if equipped) and the camshaft pulley. Torque all bolts to proper specifications.

For other styles of camshaft mounting, follow the manufacturer's recommended torque values and sequences. Installing the camshaft and followers may require the use of special tools. Verify bearing oil clearances before considering the procedure completed. Coat all components with engine oil before installing them.

CAUTION:

Make sure the rocker arms are in the proper position on the valve stems. In addition, loosen and back off the valve adjustment screws before installing the rocker arm assembly. If these screws are not backed off, the valves may be forced against the pistons, resulting in valve damage.

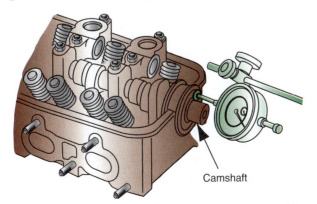

Camshaft

FIGURE 10-44 Checking camshaft endplay after reassembly.

VALVE ADJUSTMENT

Valve adjustment can be performed before the engine is installed after an overhaul or replacement. To perform a periodic adjustment, you will remove the valve cover(s) and follow the same procedures. The adjustment must be correct to provide long valve life and excellent engine performance. If there is too much clearance (lash) in the valvetrain, it will clatter and the valves won't open as much as they should. Too little valve lash can cause valve burning and poor engine performance. Some hydraulic lifters are adjustable and are set after replacement or reinstallation. Other hydraulic lifters are not adjustable. Mechanical lifters require periodic adjustment. One method of adjustment involves loosening or tightening an adjustment screw on the rocker arm (Figure 10-45). Another process involves measuring the lash and adjusting the clearance by changing the size of a shim installed between the camshaft and the follower. We'll look at the general methods of adjusting each type of valvetrain.

Adjustment Intervals

Adjustable hydraulic lifters should be adjusted during installation of new or used lifters or whenever the valvetrain has been disassembled. An example would be after an on-the-car valve seal replacement. After reinstalling the rockers, the lifters must be adjusted. This initial adjustment should last as long as the valvetrain components or until they are disassembled again. If during normal service the valvetrain becomes noisy, check the lifter adjustment if no visible damage is seen before condemning the lifters.

Mechanical or solid lifters require periodic adjustment. They should be adjusted at the specified interval. This may be as little as every 15,000 miles or as long as every 90,000 miles. Adjustment every 30,000 or 60,000 miles is a common recommendation. If the valves clatter, they should be adjusted regardless of the specified mileage.

Symptoms of Improper Valve Adjustment

When the valvetrain has too much lash, it cannot open the valve as far as it is designed to do. You can identify the problem by listening under the rocker cover for the telltale valvetrain clatter. It is a light, fast, tapping or ticking noise from under the valve cover(s). Loose valve adjustment will also affect engine performance. When the valves don't open fully or

FIGURE 10-45 These rocker arms have adjusting screws held in place with a locking nut. Use a feeler gauge to measure the clearance between the rocker arm and the valve tip. Adjust the screw to the correct clearance and tighten the locknut.

for as long as they should, volumetric efficiency decreases. As you know, this will cause a loss of power, particularly at higher rpm. The lack of fresh air will also affect emissions as the engine may run too rich.

If there is less than zero clearance, it means the valves are tight. They are held open more than they should be and may not close fully when on the camshaft's base circle. This is a very dangerous situation; the valves can burn quickly. The valves need to fully seat in order to dissipate their heat. If the exhaust valves open too early in the combustion stroke, they are subject to too much heat. In severe cases, the engine will run poorly as the valves stay open too long, and overlap will be too great. The engine may backfire through the intake and exhaust. The engine might also turn over irregularly, as though the valve timing is off. It will also significantly increase hydrocarbon and carbon monoxide emissions.

Adjusting Hydraulic Lifters

Some manufacturers use nonadjustable hydraulic lifters. In these cases, the proper "adjustment" method is to torque the rockers on their stands to the proper specification. If serious changes (valve and seat grinding) have been made that could affect the rocker geometry, pushrods or rocker stand shims of different length can be fitted. Follow the manufacturers' procedures for determining when you must perform these "adjustments." It is not common on modern passenger vehicle or light-duty truck engines to have to make any changes to the existing setup. The valves are smaller than they used to be, and you usually can't take enough material off the valves to alter the valvetrain geometry significantly before the valve needs replacement.

Commonly, hydraulic lifters are adjustable. The goal in adjusting hydraulic lifters is to center the plunger in the lifter bore (Figure 10-46). If the plunger is centered, it can adjust the valvetrain tighter or looser as needed without bottoming out at one end. Centering it gives the plunger equal travel in both directions. If it were adjusted too tight, the plunger would sit near the bottom of the bore and have little adjustment left as the valves wear.

The procedure for adjusting hydraulic lifters is similar on most vehicles, but it is critical to look at the manufacturers' service information to determine how much to tighten the rocker. When reassembling an engine after a rebuild, you should perform a preliminary adjustment during engine assembly and then make a final adjustment after the engine has run. You can perform the adjustment procedure while the engine is off or when it's running. It's messier with the engine running because oil is splashed around. Oil deflectors fit on top of the pushrod end of the rockers to help minimize the oil splash (Figure 10-47).

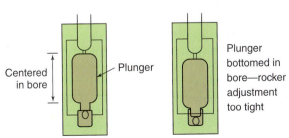

FIGURE 10-46 Follow the correct procedure to center the plunger in the bore. This will allow the lifter to adjust the valvetrain both looser and tighter.

FIGURE 10-47 These oil deflectors clip over the rocker arms and around the pushrod, blocking the oil splash hole on top of the rocker.

In either case, you'll need to remove the valve covers. To adjust the valves with the engine running:

1. Loosen a rocker bolt until you can hear the valve clatter; this raises the plunger toward the top of its bore.
2. Tighten the nut until the clatter goes away. Now the valve is at zero lash with the plunger still near to the top.
3. Tighten the rocker nut however many turns specified by the manufacturer, typically a half to one and a half turns. This centers the plunger in the bore.
4. Repeat for each valve.

To adjust the valves with the engine off:

1. Rotate the engine to TDC number 1. One half of the rockers should be loose with the cam on the base circle; these can be adjusted. You should retrieve this service information and have it with you. For example, one engine may specify to adjust intake and exhaust number 1, intake on number 5, and exhaust on number 2. Then you rotate the crankshaft 360° and adjust intake and exhaust number 6, intake on number 2, and exhaust on number 4 and number 5. As an alternate method, you can rotate the engine until you reach TDC number 1 compression and adjust those valves. Then rotate the engine, and watch the rockers for the next cylinder in the firing order to come up to TDC on compression. You will watch the intake rocker come up and then become loose as the piston approaches TDC on the compression stroke. This puts the camshaft on the base circle for all the valves in that cylinder. Adjust them, and continue to rotate the engine through the firing order, adjusting as you go.
2. To make the adjustment, tighten (if necessary) the rocker just until the lash is removed between the rocker arm and the valve and the pushrod does not spin freely or there is no lash between the rocker and the valve (Figure 10-48).
3. Tighten the rocker arm nut the specified number of turns.
4. Repeat for each valve.

When the valves are adjusted, replace the valve cover gasket and properly torque the cover. Wipe up all excess oil that may have spilled on the exhaust manifold or other

FIGURE 10-48 Tighten the rocker nut on its stud until the pushrod resists turning. Rotate the nut the specified number of degrees further to achieve the proper adjustment.

components. Start the engine. Make sure that it cranks over and starts and runs normally. Listen for any valvetrain clatter. Readjust if necessary.

ADJUSTING MECHANICAL FOLLOWERS WITH SHIMS

Engines that use mechanical followers with the camshaft riding above them may use shims to adjust the valves (Figure 10-49). To check the clearances, locate the clearance specifications in the service information. The clearances are also often listed in the engine information on the decal on the underside of the hood. The specifications will be given for when the engine is either hot or cold. An example could be 0.008 in. to 0.012 in. intake, 0.016 in. to 0.020 in. exhaust when hot. Be sure to adjust the valves with the engine properly heated or cooled. First you will check the clearance, and then you will tighten or loosen the clearance by fitting a thicker or thinner shim, respectively.

On some engines, you can remove the shim by depressing the follower and using a magnet or special tool to pull the shim out. There are several different tools available for this job, depending on the vehicle (Figure 10-50). On other engines, you may have to remove the camshaft to access the shims; they may be placed under the follower. On engines with removable shims, you can adjust one valve at a time. When you have to remove the cam to replace shims, you'll need to carefully record all your clearances first. Then you can remove the cam and bring all the clearances within specification. On either type, always double check your adjustment after the new shim is installed.

To begin adjustment, rotate the engine, using the crank bolt, to TDC number 1. You will be able to check half of the valves. Run a feeler gauge under the camshaft on top of the follower. When you fit a gauge that has a little drag as you pass it under the cam, you have found the clearance (Figure 10-51). Record the clearance. You can see which clearances can be checked by looking for the ones where the base circle of the camshaft is on the follower. Rotate the engine another 360° and you can check the remaining valves. You can look for the specified valve sequence in the service information if you have any difficulty determining which valve should be adjusted. Other technicians will use a remote starter and "bump" the engine over until each cylinder in turn is at TDC, or until each camshaft lobe is facing straight up.

To adjust the clearance, you'll change the thickness of the shim. Use a micrometer to measure the thickness of the current shim. Calculate the needed change in clearance for the valve, and select a shim that is much thicker or thinner. Let's look at an example:

1. Cylinder number l exhaust clearance is 0.009 in. The specification is 0.016 in. to 0.020 in.
2. The valve clearance should be 0.007 in. to 0.011 in. greater.
3. The existing shim size is 0.156 in.

CAUTION:
When using a remote starter to turn the engine over, be sure the transmission is in neutral, the parking brake is set, and the ignition key is off.

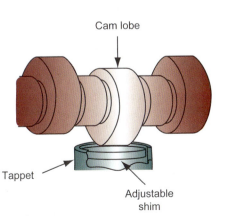

FIGURE 10-49 On this engine the shim sits on top of the tappet or follower.

Cam lobe

Tappet

Adjustable shim

FIGURE 10-50 This tool can be used on some engines to depress the follower so a magnet can be used to extract the shim.

FIGURE 10-51 Carefully check the clearance between the shim and the camshaft while the camshaft is on its base circle.

4. The clearance needs to be 0.007 in. to 0.011 in. greater, so the shim must be that much thinner.
5. Select a shim: 0.156 in. to 0.010 in. = 0.146 in.
6. Install the new shim and recheck the clearance.

After you have adjusted and checked each of the valves, crank the engine over for a few seconds and recheck the clearances. You need to be sure that all the shims are properly seated in the followers. Readjust as needed.

ADJUSTING VALVES USING ADJUSTABLE ROCKER ARMS

Another common method of adjusting mechanical valvetrains is through the rocker arm. One side of the rocker may ride on the camshaft or a solid lifter. The other side can work on the valve. One end of the rocker arm has a locknut and adjusting screw. To adjust the valve, you can loosen the locking nut and turn the screw in or out to tighten or loosen the valve adjustment respectively (Figure 10-52). The clearance specifications will be similar to those with shim-adjusted valve lash. To adjust the valves:

1. Obtain the lash specifications, and be sure the engine is at the desired temperature.
2. Rotate the engine to TDC number 1, and note which valves can be adjusted using the service information.
3. Using feeler gauges, determine the clearance on the valve you are adjusting. The feeler gauge should fit in the clearance under the adjustment screw with light drag (Figure 10-53).
4. If the valve needs adjustment, loosen the locknut just enough that you can turn the screw. Insert the desired thickness (clearance) feeler gauge under the lash adjuster and *lightly* tighten the adjuster screw while alternately feeling the clearance. When there is just the right amount of drag, tighten the locknut (Figure 10-54).
5. Recheck your adjustment. This is essential as the adjustment can change as the locknut pulls the adjuster screw up. It may be necessary to adjust the valve a little snugly and then have it loosen up as you tighten the locknut. It is essential to recheck the adjustment after tightening the locknut.
6. Repeat for all the other valves adjustable at TDC number 1. Rotate the crank 360°, and complete the adjustment of the valves.

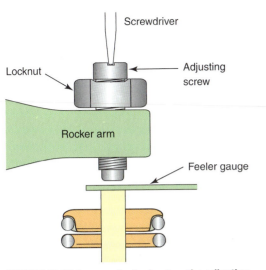

FIGURE 10-52 Loosen the locknut on the adjusting screw, fit the appropriate feeler gauge between the screw and the valve tip, and adjust the screw until some drag is felt on the feeler gauge.

FIGURE 10-53 Check the clearance first to be sure adjustment is needed.

FIGURE 10-54 Three hands to hold the wrench, the screwdriver, and the feeler gauges would make this job easier.

7. Properly install the valve cover and start the engine. It should turn over smoothly.
8. Allow the engine to warm up, and listen for any excessive valvetrain clatter and feel for rough running.

Installing the Cylinder Head

If the engine is an OHC type, the cylinder heads must be installed before the timing components. On OHV engines, the timing components can be installed before the heads. The procedures for cylinder head installation are similar. On some OHC engines, the camshaft and valvetrain components can be installed into the head before the head is placed onto the engine block. Use the following checklist along with the appropriate service manual as the cylinder heads are installed.

1. Any bolts that enter into coolant or oil passages must have approved sealer applied to the threads.
2. Be sure that the block and cylinder head surfaces are perfectly clean. As described earlier, the head bolt holes must also be cleaned.
3. Install any dowels that were originally equipped on the deck (Figure 10-55).

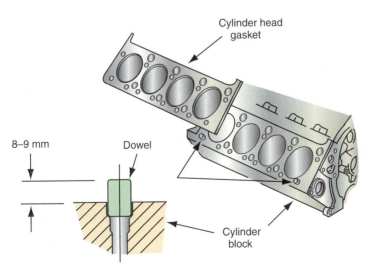

FIGURE 10-55 Alignment dowel installation.

SPECIAL TOOLS
Torque wrench
Alignment dowels

SERVICE TIP:
If locating dowels are not used, alignment pins can be made to assist in head installation. Thread a couple of locating pins in the corners of the deck surface. Slide the cylinder head gasket over the pins and locate it on the deck surface. Next, slide the cylinder head over the pins and onto the gasket. Install the head bolts fingertight; then remove the alignment pins and install the correct bolts.

423

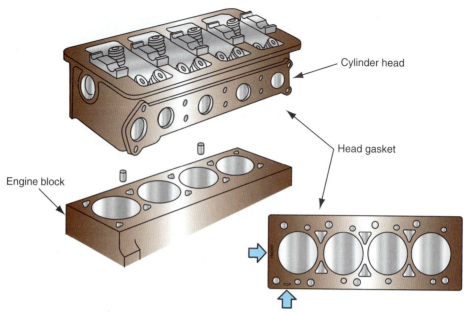

Cylinder head

Head gasket

Engine block

FIGURE 10-56 Most cylinder head gaskets have markings to indicate proper installation direction.

4. Locate any markings on the cylinder head gasket that identify the proper direction for installation (Figure 10-56).
5. Position the new head gasket over the dowels and check for proper fit.
6. Locate the crankshaft a few degrees before TDC so all pistons are below deck height.
7. Position the cylinder head over the dowels. On V-type engines, make sure the correct head is installed onto the correct bank.
8. Install and torque the cylinder head bolts. Most manufacturers specify multiple steps to torquing the head bolts. Follow these instructions to prevent warpage (Figure 10-57). Also follow the correct torque sequence. Torque-to-yield cylinder head bolts are used in many engines. These bolts must be tightened to the specified torque and then rotated a specific number of degrees (Figure 10-58). Torque-to-yield bolts must be replaced each time the cylinder head is removed and replaced.

ON-CAR SERVICE

You can replace worn valve seals or springs without removing the cylinder head from the engine. Worn valve seals will cause oil consumption. Valve seals often wear as a result of worn valve guides (Figure 10-59). The customer will likely notice a big cloud of blue smoke

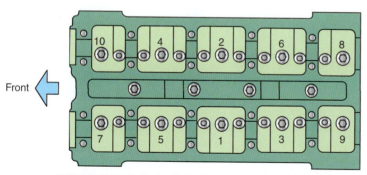

Front

FIGURE 10-57 Typical cylinder head torque sequence.

FIGURE 10-58 A torque angle gauge allows you to accurately measure the degrees of turning.

just after starting the vehicle. The oil on the top of the head leaks down past the valve seals as the vehicle sits; when it starts, the oil in the combustion chamber burns. Blue smoke is also often seen after a period of idling or deceleration. High engine vacuum pulls oil past the weak intake valve seals into the combustion chamber. The resulting blue smoke is a telltale sign of worn seals. This pattern of smoking is different than oil consumption caused by worn rings. Ring troubles generally lead to more consistent smoking that is worse when the engine is under a load.

Before replacing the valve seals, it is a good idea to check for valve guide wear. Often the seals are damaged because the guides are allowing the valves to wobble back and forth in the seals. Check the engine vacuum at idle and at 2,500 rpm. If the vacuum fluctuates at idle but smoothes out at higher rpm, suspect worn valve guides. The valve is not being guided correctly onto its seat at lower rpms. You can also check running compression. The compression would be lower at idle when the valve has more time to rock back and forth in the guide. Running compression would return to a nearly normal value at 2,500 rpm.

To replace a valve seal with the head installed, you must work one cylinder at a time. Remove the valve cover(s) first. The critical issue during this service is to prevent the valve from dropping into the engine when the keepers are removed. That mistake could require that you remove the cylinder head. Place the first cylinder at TDC on the compression stroke. Remove the camshaft, rocker arm, and lifter or follower as needed to gain access to the valve retainer. Use a cylinder leakage tester or regulated shop air to apply 100 psi of air pressure into the cylinder with the valves closed. This will keep the valves up in the head while you release the valve retainer and keepers. Place a socket over the valve retainer, and give it a light rap to break any varnish loose; this may make it much easier to compress the spring.

Use a valve spring compressor to compress the spring and remove the keepers (Figure 10-60). Set the spring, retainer, and keepers safely aside. Note which end of the spring faces up; often tighter coils are placed at the bottom toward the head. Use a screw driver or seal puller to remove the old valve seal. Check for damage or excessive scoring on the valve stem that could ruin a new valve seal. Lubricate the inner lip of the new valve seal with oil and install it onto the valve. Many positive lock seals must be driven down onto the head to seat properly. Use a valve seal installation tool to fit the seal into its proper position. If the correct tool is not available, you can use a deep socket that matches the diameter of the seal but will easily slide over the valve stem. Be sure the socket is deep enough to tap the seal down without hitting the valve tip. Knocking the tip could release the air pressure in the cylinder and cause you to drop a valve into the cylinder, necessitating head removal.

CAUTION:
Do not rotate the camshaft or crankshaft without the timing belt properly installed unless instructed to do so in the service manual.

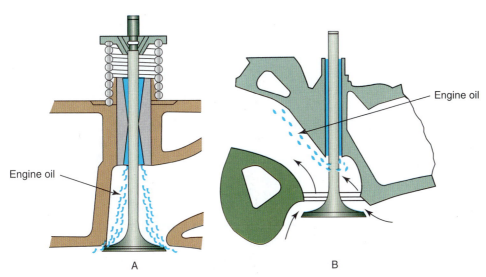

A   B

FIGURE 10-59 Worn valve seals allow oil to be pulled into the combustion chamber through the guides. It burns and exits the exhaust port and tailpipe as blue smoke.

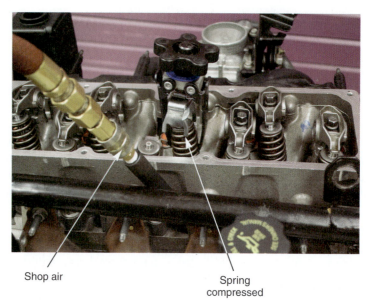

Shop air

Spring
compressed

FIGURE 10-60 Compressed air holds the valve up while you compress
the spring to remove the keepers.

PHOTO SEQUENCE 19

TYPICAL PROCEDURE FOR ADJUSTING VALVES

P19-1 Prepare the vehicle and assess
the task.

P19-2 Read through and follow the
service information procedures.

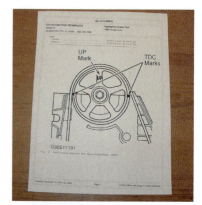

P19-3 This service information
provides a diagram of how to line up
the camshaft to adjust the valves.

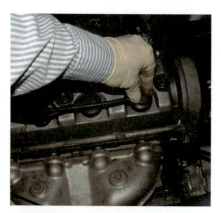

P19-4 Remove the necessary wiring and
linkage to remove the valve cover.

P19-5 Remove the valve cover following
any sequence specified.

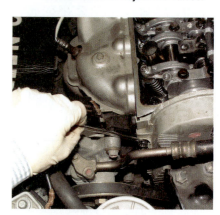

P19-6 Remove the upper timing belt
cover to access the timing marks on the
camshaft sprocket.

TYPICAL PROCEDURE FOR ADJUSTING VALVES

P19-7 With the valve cover and timing cover off you have plenty of room to work.

P19-8 Rotate the engine through the access port in the wheel well to rotate the engine to line up the marks on the camshaft. Do not rotate the engine over with the camshaft sprocket bolt.

P19-9 Check the adjustment of the valves using feeler gauges and compare to specifications.

P19-10 Loosen the locknut and turn the adjusting screw to achieve the proper adjustment. Double check your work after you retighten the locknut.

P19-11 Properly torque the new valve cover gasket and valve cover into place.

P19-12 Be sure to reinstall all hoses, connectors and linkages you disconnected to remove the valve cover.

P19-13 Make sure your work is clean and that the engine runs smoothly and quietly before you call the job complete.

REPLACING VALVE SEALS ON THE VEHICLE

P20-1 Remove all necessary wiring and vacuum lines from the intake ducting.

P20-2 Remove the intake ducting to access the valve cover.

P20-3 Move wiring and remove the bolts as needed to remove the valve cover.

P20-4 Remove the valve cover.

P20-5 Rotate the engine to TDC firing on the cylinder you are going to work on first. You can watch the intake valve close and go a bit further until the piston is at TDC.

P20-6 Apply 100 psi of air pressure to the cylinder to hold the valves up.

P20-7 Remove the rocker arms and compress the valve spring. Remove the keepers and the spring assembly. Set it aside for reinstallation.

P20-8 Pry the old seal off or use a valve seal removing tool if available.

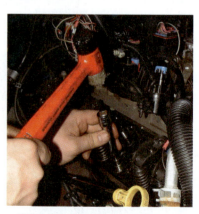

P20-9 It is best to use a valve seal installation tool made for the job, but if you are careful you can find just the right size deep socket to drive the new seal on without damaging the seal or popping the valve open.

REPLACING VALVE SEALS ON THE VEHICLE

P20-10 Replace the valve spring and keepers. Give a light rap on the valve retainer to be sure the keepers are fully seated.

P20-11 Adjust the rocker arm by removing the lash and turning the adjusting bolt one turn further. Check the specifications, this will vary with different vehicles.

P20-12 Be sure to torque the valve cover properly to prevent a comeback.

P20-13 Fully reassemble the engine and be sure it runs quietly, smoothly and without leaks.

Refit the valve spring, retainer, and keepers. When the valve is back in place, take the socket and lightly rap again on the retainer to check that the keepers are fully seated. Repeat the procedure for the other valve(s) on this cylinder. Then move to the next cylinder and repeat the process to replace the whole set of valve seals.

VALVE SPRING REPLACEMENT

Weak or broken valve springs can cause valve float and valve burning. When the valve hangs open like this, you can notice misfiring and power loss at higher rpms. If the valve spring is broken, the valve may never close. This can cause a steady misfire and popping out of the intake or exhaust, depending on which spring is at fault. A broken valve spring must be replaced immediately to prevent the valve from burning.

The procedure to replace valve springs is the same as for replacing valve seals. As a matter of fact, technicians should replace the valve seals while they have the valves disassembled. Once the valve spring is removed, carefully inspect it for cracks, proper free length, squareness, and opening tension. You want to be sure that you have correctly diagnosed the problem and that spring replacement will cure the symptom.

TERMS TO KNOW

Base circle

Camshaft

Duration

Heel

Leak-down

Lifter

Lobe lift

Nose

Open pressure

Overlap

Pushrods

Rocker arm

Seat pressure

Spring free length

Spring shims

Spring squareness

CUSTOMER CARE: Always concentrate on thorough and quality work to achieve customer satisfaction. Most customers do not mind paying for vehicle repairs if they believe that the work is necessary and performed properly. This is evidenced in the following case study. A follow-up phone call to assure that the customer is satisfied a few days after the service work demonstrates that you consider quality work and customer satisfaction a priority.

CASE STUDY

A customer came in very upset about his vehicle. He said the engine was blowing blue smoke out the tailpipe and he didn't have enough money to rebuild the engine. The technician drove the vehicle and noticed that the engine did not smoke during acceleration as it would if the rings were worn. Instead, he noticed smoking after a period of idle. To confirm his suspicion that the rings were in reasonable condition, he performed a cylinder leak-down test. The results showed less than 15 percent leakage on all four cylinders. He also performed a vacuum test to be sure that the valve guides didn't indicate excessive wear. The technician went back to the customer and reported his findings. He asked the customer if he had noticed that it smoked heavily during the first start-up of the day. When the customer told him that was indeed the case, the technician let him know that the cure, new valve seals, would be less than $200. The diagnosis and repair resulted in a very happy customer.

ASE-STYLE REVIEW QUESTIONS

1. *Technician A* says that cylinder head and block deck resurfacing may affect rocker arm geometry.
   Technician B says that intake manifold replacement will change rocker arm geometry.
   Who is correct?
   A. A only
   B. B only
   C. Both A and B
   D. Neither A nor B

2. *Technician A* says if face wear on the lifter runs across the face, it indicates the lifter was stuck in its bore.
   Technician B says that a worn lifter can cause valvetrain clatter.
   Who is correct?
   A. A only
   B. B only
   C. Both A and B
   D. Neither A nor B

3. *Technician A* says that installing a modified camshaft may affect rocker geometry.
   Technician B says that using different pushrods can correct rocker geometry.
   Who is correct?
   A. A only
   B. B only
   C. Both A and B
   D. Neither A nor B

4. *Technician A* says to check the pushrod for warpage and tip wear.
   Technician B says that wear on the side indicates that the pushrod guide is defective.
   Who is correct?
   A. A only
   B. B only
   C. Both A and B
   D. Neither A nor B

5. *Technician A* says that a worn camshaft lobe can cause reduced engine power.
   Technician B says that a bent camshaft should be straightened.
   Who is correct?
   A. A only
   B. B only
   C. Both A and B
   D. Neither A nor B

6. *Technician A* says that if you replace the camshaft, you should replace the lifters.
   Technician B says that lifters should be slightly concave on their face.
   Who is correct?
   A. A only
   B. B only
   C. Both A and B
   D. Neither A nor B

7. *Technician A* says to measure valve installed height when reassembling the cylinder head.
   Technician B says to shim the valve springs if the valve installed height is below specifications.
   Who is correct?
   A. A only
   B. B only
   C. Both A and B
   D. Neither A nor B

8. Valve spring free length should be measured:
   A. With the valve and spring installed
   B. With a machinist's rule
   C. With the valve in its fully opened position
   D. With the spring on the bench

9. A shim-type valvetrain is being adjusted. The exhaust clearance is specified at 0.016 in. to 0.020 in. (0.41 mm to 0.51 mm). The existing clearance is 0.027 in. (0.69 mm). To correct the clearance, install a shim _____ thick.
   A. 0.146 in. (3.7 mm)
   B. 0.142 in. (3.6 mm)
   C. 0.130 in. (3.3 mm)
   D. 0.126 in. (3.2 mm)

10. *Technician A* says that you should replace the valve seals when replacing the valve springs.
    Technician B says that you should check the condition of the valve guides before replacing the valve seals on the car.
    Who is correct?
    A. A only
    B. B only
    C. Both A and B
    D. Neither A nor B

ASE CHALLENGE QUESTIONS

1. *Technician A* says that on a RWD vehicle, the radiator usually needs to be remove in order to replace the camshaft.
   Technician B says that you can crack an overhead camshaft it you do not follow a bolt loosening sequence.
   Who is correct?
   A. A only
   B. B only
   C. Both A and B
   D. Neither A nor B

2. *Technician A* says that rocker arms should be replaced in their original positions if the camshaft will be reused.
   Technician B says that you typically resurface a rocker arm by about 0.002 in. to 0.004 in. to restore a good mating finish with the camshaft.
   Who is correct?
   A. A only
   B. B only
   C. Both A and B
   D. Neither A nor B

3. *Technician A* says that a VTEC engine should have zero camshaft endplay.
   Technician B says that you should check the camshaft bearing clearance on a VTECH engine with Plastigage.
   Who is correct?
   A. A only
   B. B only
   C. Both A and B
   D. Neither A nor B

4. *Technician A* say that lobe life is determined by subtracting the base circle diameter from the distance from the bottom of the base circle to the end of duration.
   Technician B says that worn camshaft lobes may cause a heavy clicking noise at 1,500 to 2,000 rpm.
   Who is correct?
   A. A only
   B. B only
   C. Both A and B
   D. Neither A nor B

5. *Technician A* says that lifters are typically replaced during an engine overhaul.
   Technician B says that too little valve clearance can cause valvetrain clatter.
   Who is correct?
   A. A only
   B. B only
   C. Both A and B
   D. Neither A nor B

Name _____ Date _____

VALVE STEM SEAL REPLACEMENT ON AN ASSEMBLED ENGINE

Upon completion of this job sheet, you should be able to replace the valve stem seals while the engine is still in the vehicle or on an assembled engine.

ASE Correlation

This job sheet is related to ASE Engine Repair Test content area: Cylinder Head and ValveTrain Diagnosis and Repair: Replace valve stem seals on an assembled engine; inspect valve spring retainers, locks/keepers, and valve lock/keeper grooves; determine necessary action.

Tools and Materials

Valve refacing equipment

Describe the vehicle being worked on:

Year _____ Make _____ Model _____

VIN _____ Engine type and size _____

Procedure

Task Completed

Using the cylinder head(s) that you removed from the engine during Job Sheet 19 and the proper service manual, you will identify and perform the required steps to recondition the valves.

1. What is the specified face angle for the valves?
   Intake _____ Exhaust _____

2. Does the manufacturer recommend an interference angle?

   ☐ Yes   ☐ No

3. Set the spindle head to the specified angle. Lock into place.　　　　　☐

4. Dress the stone with a diamond tool.　　　　　☐

5. Clamp the valve into the spindle chuck. What special concerns must you be aware of when setting the stem depth into the chuck?
   _____
   _____

6. Set the stroke length. What special concerns are associated with this step?

7. Correctly position the valve face and the stone.　　　　　☐

8. Advance the stone into the valve slowly until light contact is noted.
   Begin to stroke the ☐ valve across the stone.　　　　　☐

9. Note the micrometer reading. _____

☐ 10. Continue to dress the face, feeding it 0.001 inch (0.25 mm) per pass until the face is clean.

RECONDITIONING VALUES (CONTINUED)

11. Note the micrometer reading. _____

12. What was the total amount of face material removed? _____

13. Measure and note the valve margin. _____
Is the margin still within specifications?   ☐ Yes   ☐ No
If no, what must be done?

_____

☐ 14. Use a piece of emery cloth to break the margin.

15. Measure and note valve head runout.
Specification _____   Actual _____

☐ 16. Inspect the valve face. Report any defects to your instructor.

17. Remove the valve from the spindle chuck, and set it into
☐   the valve stem holding fixture.

☐ 18. Dress the stone used to grind the valve stem.

19. How much material needs to be removed from the tip of the stem? _____

20. Why is it important to stem the valve?

_____
_____
_____

☐ 21. Stem the valve, followed by chamfering the tip.

Instructor's Response _____
_____
_____

Name _____ **Date** _____

VALVE SPRING INSPECTION

Upon completion of this job sheet, you should be able to properly reassemble the cylinder head.

ASE Correlation

This job sheet is related to ASE Engine Repair Test content area: Cylinder Head and Valvetrain Diagnosis and Repair: Inspect valve springs for squareness and free height comparison; determine necessary action.

Tools and Materials

Valve spring height measuring equipment
Valve spring compressor and scale
Dial caliper

Describe the vehicle being worked on:

Year _____ Make _____ Model _____

VIN _____ Engine type and size _____

Describe the type of valvetrain in the engine:

_____

Procedure

Use the valve springs that you removed from the engine. Write down the information, and then perform the following tasks:

1. What is the specification for the spring free length (sometimes called free height)?
   Intake _____ Exhaust _____

2. Using the dial calipers, measure the free length (free height) of each spring:

   Number 1   Intake _____/_____   Exhaust _____/_____
   Number 2   Intake _____/_____   Exhaust _____/_____
   Number 3   Intake _____/_____   Exhaust _____/_____
   Number 4   Intake _____/_____   Exhaust _____/_____
   Number 5   Intake _____/_____   Exhaust _____/_____
   Number 6   Intake _____/_____   Exhaust _____/_____
   Number 7   Intake _____/_____   Exhaust _____/_____
   Number 8   Intake _____/_____   Exhaust _____/_____

   Place an * in the above chart for any springs that are out-of-specifications.

3. Inspect each spring for cracks and signs of rotation and wear. Describe what you found.

_____

_____

4. Place each spring in the spring compressor scale and measure the force at the specified height. Record your findings below:

Number 1   Intake _____/_____   Exhaust _____/_____

Number 2   Intake _____/_____   Exhaust _____/_____

Number 3   Intake _____/_____   Exhaust _____/_____

Number 4   Intake _____/_____   Exhaust _____/_____

Number 5   Intake _____/_____   Exhaust _____/_____

Number 6   Intake _____/_____   Exhaust _____/_____

Number 7   Intake _____/_____   Exhaust _____/_____

Number 8   Intake _____/_____   Exhaust _____/_____

Place an * in the above chart for any springs that are out-of-specifications.

5. Based on your tests and results, what are your recommendations?

_____

_____

_____

Instructor's Response _____

_____

_____

JOB SHEET

Name _____ Date _____

INSPECTING THE CAMSHAFT

Upon completion of this job sheet, you should be able to properly inspect the camshaft for wear and damage.

ASE Correlation

This job sheet is related to ASE Engine Repair Test's content area: Cylinder Head and Valvetrain Diagnosis and Repair; tasks: Inspect and measure camshaft journals and lobes.

Tools and Materials

Service manual

Micrometer

Camshaft

Describe the vehicle being worked on:

Year _____ Make _____ Model _____

VIN _____ Engine type and size _____

Procedure

Using the camshaft from the engine that you have disassembled and the proper service manual, you will identify and perform the required steps to inspect the camshaft.

_____
Task Completed

1. What page of the service manual covers the procedure for inspecting the camshaft?

   _____

2. Perform a visual inspection of the camshaft, and note your results below.

   _____

   _____

3. Describe the wear pattern on the lobes.

	Intake Lobes	Exhaust Lobes
Cylinder 1	_____	_____
Cylinder 2	_____	_____
Cylinder 3	_____	_____
Cylinder 4	_____	_____
Cylinder 5	_____	_____
Cylinder 6	_____	_____
Cylinder 7	_____	_____
Cylinder 8	_____	_____

4. Put an * in the previous chart for any lobes that show abnormal wear.   ☐

5. What is the lobe lift specification?

   Intake lobes _____ Exhaust lobes _____

6. Measure the lift for each lobe, and record your results.

	Intake Lobes	Exhaust Lobes
Cylinder 1	_____	_____
Cylinder 2	_____	_____
Cylinder 3	_____	_____
Cylinder 4	_____	_____
Cylinder 5	_____	_____
Cylinder 6	_____	_____
Cylinder 7	_____	_____
Cylinder 8	_____	_____

7. Put an * in the previous chart for any lobes that are out-of-specification.

8. List the journal measurements below:

	Size	Taper	Out-of-Round
Journal 1	_____	_____	_____
Journal 2	_____	_____	_____
Journal 3	_____	_____	_____
Journal 4	_____	_____	_____
Journal 5	_____	_____	_____

9. Put an * in the previous chart to note any journals that are out-of-specification.

10. Measure the camshaft for straightness, and record your results.

   _____

11. Measure each lobe for base circle runout, and record your results.

	Intake Lobes	Exhaust Lobes
Cylinder 1	_____	_____
Cylinder 2	_____	_____
Cylinder 3	_____	_____
Cylinder 4	_____	_____
Cylinder 5	_____	_____
Cylinder 6	_____	_____
Cylinder 7	_____	_____
Cylinder 8	_____	_____

12. Put an * in the previous chart for any lobes that are out-of-specification.

13. Based on your evaluation of the camshaft, what is your recommendation?

_____
_____
_____

Instructor's Response _____
_____
_____

JOB SHEET

Name _____ Date _____

INSPECTING THE LIFTERS (LASH ADJUSTERS)

Upon completion of this job sheet, you should be able to properly inspect the lifters for wear and damage.

ASE Correlation

This job sheet is related to ASE Engine Repair Test's content area: Cylinder Head and Valvetrain Diagnosis and Repair; tasks: Inspect and replace hydraulic and mechanical lifters/lash adjusters.

Tools and Materials

Service manual

Leak-down tester

Valve lifters

Describe the vehicle being worked on:

Year _____ Make _____ Model _____

VIN _____ Engine type and size _____

Procedure

Task Completed

Using the lifters removed from the engine that you have disassembled and the proper service manual, you will identify and perform the required steps to inspect the lifters.

1. What page of the service manual covers the procedure for inspecting the lifters?

_____

2. Perform a visual inspection of each lifter, and note your results below.

	Intake Lifters	Exhaust Lifters
Cylinder 1	_____	_____
Cylinder 2	_____	_____
Cylinder 3	_____	_____
Cylinder 4	_____	_____
Cylinder 5	_____	_____
Cylinder 6	_____	_____
Cylinder 7	_____	_____
Cylinder 8	_____	_____

3. Put an * in the previous chart for any lifters that show abnormal wear or damage. ☐

4. What is the specification for lifter leak-down testing?

_____

_____

5. Perform a leak-down test of each lifter, and record the time below.

	Intake Lifters	Exhaust Lifters
Cylinder 1	_____	_____
Cylinder 2	_____	_____
Cylinder 3	_____	_____
Cylinder 4	_____	_____
Cylinder 5	_____	_____
Cylinder 6	_____	_____
Cylinder 7	_____	_____
Cylinder 8	_____	_____

6. Put an * in the previous chart for any lifters that are out-of-specification.

7. Based on your test results, what is your recommendation?
_____
_____
_____
_____
_____

Instructor's Response _____
_____
_____

Name _____ Date _____

INSPECTING THE PUSHRODS

Upon completion of this job sheet, you should be able to properly inspect the pushrods for wear and damage.

ASE Correlation

This job sheet is related to ASE Engine Repair Test's content area: Cylinder Head and Valvetrain Diagnosis and Repair; tasks: Inspect pushrods, rocker arms, rocker arm pivots, and shafts for wear, bending, cracks, looseness, and blocked oil passages; repair or replace as required.

Tools and Materials

Service manual
Pushrods

Describe the vehicle being worked on:

Year _____ Make _____ Model _____

VIN _____ Engine type and size _____

Procedure

Task Completed

Using the pushrods removed from the engine that you have disassembled and the proper service manual, you will identify and perform the required steps to inspect the pushrods.

1. What page of the service manual covers the procedure for inspecting the pushrods?

_____

2. Perform a visual inspection of each pushrod, and note your results below.

	Intake Pushrods	Exhaust Pushrods
Cylinder 1	_____	_____
Cylinder 2	_____	_____
Cylinder 3	_____	_____
Cylinder 4	_____	_____
Cylinder 5	_____	_____
Cylinder 6	_____	_____
Cylinder 7	_____	_____
Cylinder 8	_____	_____

3. Put an * in the previous chart for any lifters that show abnormal wear or damage. ☐

4. Inspect each pushrod for warpage, and record your results.

	Intake Pushrods	Exhaust Pushrods
Cylinder 1	_____	_____
Cylinder 2	_____	_____
Cylinder 3	_____	_____
Cylinder 4	_____	_____
Cylinder 5	_____	_____
Cylinder 6	_____	_____
Cylinder 7	_____	_____
Cylinder 8	_____	_____

5. Put an * in the previous chart for any pushrods that are bent.

☐ 6. Clean the oil passage through each pushrod.

☐ 7. Inspect the oil passages for oval shape or edge chamfering.

☐ 8. Based on your test results, what is your recommendation?

_____

_____

_____

_____

_____

Instructor's Response _____

_____

_____

Name _____ Date _____

CHECKING AND ADJUSTING VALVE CLEARANCE

Upon completion of this job sheet, you should be able to properly check and adjust the valve clearance.

ASE Correlation

This job sheet is related to ASE Engine Repair Test's content area: Cylinder Head and Valvetrain Diagnosis and Repair; tasks: Adjust valves on engines with mechanical or hydraulic lifters.

Tools and Materials

Service manual
Engine with adjustable valvetrain
Feeler gauge

Describe the vehicle being worked on:

Year _____ Make _____ Model _____

VIN _____ Engine type and size _____

Procedure

Using a provided engine and the proper service manual, you will identify and perform the required steps to check and adjust valve clearance.

1. What page of the service manual covers the procedure for adjusting the valve clearance?
   _____

2. What is the specification for valve clearance?
   Intake valves _____   Exhaust valves _____

3. Is the specification for: ☐ HOT   ☐ COLD

4. What valves are adjusted with the engine at TDC number 1 compression stroke?
   _____
   _____

5. What location must the crankshaft be in to adjust the remaining valves?
   _____

6. What method is used to adjust the valve clearance?
   ☐ ADJUSTING NUT   ☐ SHIMS   ☐ SCREW   ☐ PUSHROD LENGTH

7. With the engine at TDC number 1 compression stroke, measure the valve clearance, and record your results below.

	Intake Valves	Exhaust Valves
Cylinder 1	_____	_____
Cylinder 2	_____	_____
Cylinder 3	_____	_____
Cylinder 4	_____	_____
Cylinder 5	_____	_____
Cylinder 6	_____	_____
Cylinder 7	_____	_____
Cylinder 8	_____	_____

☐ 8. Put an * in the previous chart for any valve clearances that are out-of-specification.

☐ 9. Correct any valves that are out-of-specification.

☐ 10. Rotate the crankshaft to the position determined in step 6.

11. Measure the clearance of the remaining valves, and record your results below.

	Intake Valves	Exhaust Valves
Cylinder 1	_____	_____
Cylinder 2	_____	_____
Cylinder 3	_____	_____
Cylinder 4	_____	_____
Cylinder 5	_____	_____
Cylinder 6	_____	_____
Cylinder 7	_____	_____
Cylinder 8	_____	_____

☐ 12. Put an * in the previous chart for any valve clearances that are out-of-specification.

☐ 13. Correct any valves that are out-of-specification.

Instructor's Response _____

_____

_____

Name _____ Date _____

REASSEMBLING THE CYLINDER HEAD

Upon completion of this job sheet, you should be able to properly reassemble the cylinder head.

ASE Correlation

This job sheet is related to ASE Engine Repair Tests's content area: Cylinder Head and Valvetrain Diagnosis and Repair; task: Reassemble and install cylinder heads and gaskets; replace and tighten fasteners according to manufacturer's procedures.

Tools and Materials

Torque wrench
Valve spring height measuring equipment
Valve stem height measuring equipment
Valve spring compressor

Describe the vehicle being worked on:

Year _____ Make _____ Model _____

VIN _____ Engine type and size _____

Describe the type of valvetrain in the engine.

_____

Procedure

Task Completed

Using the cylinder head(s) that you removed from the engine and the proper service manual, you will identify and perform the required steps to reassemble the cylinder head(s):

1. Replace any studs that were removed. ☐

2. Clean all oil and coolant passages. ☐

3. Install oil and coolant passage core plugs and cups. ☐

4. Polish the valve stems with a fine crocus cloth and solvent. Make sure all valves are clean. ☐

5. Lubricate the valve stem and install the valve into the cylinder head. Check spring height and record results:

   Number 1   Intake _____   Exhaust _____
   Number 2   Intake _____   Exhaust _____
   Number 3   Intake _____   Exhaust _____
   Number 4   Intake _____   Exhaust _____
   Number 5   Intake _____   Exhaust _____

Number 6   Intake _____   Exhaust _____

Number 7   Intake _____   Exhaust _____

Number 8   Intake _____   Exhaust _____

6. What is the specification for the spring height?
   Intake _____ Exhaust _____
   Place an * in the above chart for any valves that are out-of-specifications.

7. Indicate below what size shim will be required to correct installed spring height:

   Number 1   Intake _____   Exhaust _____
   Number 2   Intake _____   Exhaust _____
   Number 3   Intake _____   Exhaust _____
   Number 4   Intake _____   Exhaust _____
   Number 5   Intake _____   Exhaust _____
   Number 6   Intake _____   Exhaust _____
   Number 7   Intake _____   Exhaust _____
   Number 8   Intake _____   Exhaust _____

8. Install the valve stem seals. To install most positive lock seals, the valve will need to be removed before the seal is installed. Some seals are installed after the springs are installed.
   What type of seals are used? _____
   If different size seals are used for the intake and exhaust valves:
   What color are the intake valve seals? _____
   What color are the exhaust valve seals? _____

9. Install the valves (one at a time) into the cylinder head.

10. Install the correct size spring shim as determined above.

☐ 11. Depending on seal type, install the stem seal.

☐ 12. Place the valve spring over the stem. Is the valve spring directional?   ☐ Yes   ☐ No

☐   If yes, which direction do they face? _____

13. Locate the retainer over the valve spring.

☐ 14. Use an approved spring compressor to compress the spring just far enough to install the keepers. If O-ring seals are used, install them now.

☐ 15. Install the keepers, making sure they are locked in place.

16. Slowly release the valve spring compressor and remove it.

☐ 17. Use a soft-faced hammer and tap the top of the valve stem to assure the keepers are
☐   properly installed.

☐ 18. Attempt to rotate the valve spring by hand. Note any that rotate.

   Number 1   Intake _____   Exhaust _____
   Number 2   Intake _____   Exhaust _____

Number 3   Intake _____   Exhaust _____

Number 4   Intake _____   Exhaust _____

Number 5   Intake _____   Exhaust _____

Number 6   Intake _____   Exhaust _____

Number 7   Intake _____   Exhaust _____

Number 8   Intake _____   Exhaust _____

19. If the valve spring rotates, what must be done?

20. Repeat for all valves. ☐

21. Lubricate the stem tips with a multipurpose grease. ☐

22. On OHV cylinder heads, install the rocker arms, fulcrums, or shafts. Do not tighten the rocker arm bolts at this time. ☐

23. Turn the cylinder head over and test for leakage. Indicate your results below:

Number 1   Intake _____   Exhaust _____

Number 2   Intake _____   Exhaust _____

Number 3   Intake _____   Exhaust _____

Number 4   Intake _____   Exhaust _____

Number 5   Intake _____   Exhaust _____

Number 6   Intake _____   Exhaust _____

Number 7   Intake _____   Exhaust _____

Number 8   Intake _____   Exhaust _____

24. If leakage is detected, consult your instructor.

IF YOU ARE WORKING ON AN OHC ENGINE, COMPLETE THE FOLLOWING STEPS: ☐

25. Using the service manual procedures, assemble the rocker arm assembly.

26. Clean the camshaft and the camshaft bores. ☐

27. Lubricate the camshaft bores with engine oil or assembly lube. ☐

28. Install the camshaft into the cylinder head. ☐

29. Rotate the camshaft so it is in position for TDC number one cylinder. How is this identified? _____ ☐

30. If the camshaft uses journal caps, place a piece of plastigage on each camshaft journal. ☐

31. Make sure the cylinder head is lifted so the valves can open if needed. ☐

32. If all one assembly, install the rocker arm shaft and caps (or the journal caps if separate assemblies). ☐

33. Following the service manual procedure, torque the cap bolts (or arm assembly) to specifications. Specified torque _____ ☐

34. Remove the caps by reversing the torque sequence and measure the plastigage.

Bore 1 _____ Bore 2 _____
Bore 3 _____ Bore 4 _____

35. Compare measurements to clearance specifications. Specifications _____
Place an * in the previous chart to identify any bores that are out-of-specification.
If any are out-of-specification, consult your instructor and determine needed repairs.

_____

_____

☐ **36.** Remove the plastigage and install the camshaft seal.

☐ **37.** Reassemble the caps (rocker arm assembly) and tighten in proper sequence and to proper torque.

38. Measure camshaft end play.
Actual _____ Specifications _____
If not within specifications, consult your instructor and determine needed repairs.

_____

_____

39. Install back plates (if used) and the camshaft timing sprocket.
Is the sprocket directional? ☐ Yes ☐ No
If yes, how is direction determined?

Instructor's Response _____

_____

_____

Chapter 11

TIMING MECHANISM SERVICE

UPON COMPLETION AND REVIEW OF THIS CHAPTER, YOU SHOULD BE ABLE TO:

- Recognize and diagnose the symptoms of a jumped or broken timing mechanism.

- Perform a chain/belt deflection test and interpret the results.

- Inspect the timing belt or chain tensioner.

- Inspect the sprockets, idler pulleys, and guides.

- Discuss the possible extent of damage on an interference engine with the customer before beginning the job or offering an estimate.

- Replace and properly time the timing mechanism on an OHC engine using a timing belt.

- Replace and properly time the timing mechanism of an OHV engine using a timing chain.

- Replace and properly time the timing mechanism of an OHC engine using a timing chain.

- Replace and properly time the timing mechanism of an engine with variable valve timing.

SERVICE TIP:
"Crrraaaaaaaaaank, crank, crank, crrraaaaaaaaaank, crank, crank, pop, pop." That's an engine cranking over at irregular speeds and backfiring; it's a tell-tale sign of a valve timing mechanism that's jumped out of time. If the engine is old enough to have adjustable ignition timing, check that out too; it can cause the same symptoms.

INTRODUCTION

In order to prepare for valve timing mechanism service, you must carefully inspect the components and decide which need replacement. Even when the timing mechanism is being replaced during a major engine overhaul, you should carefully inspect for damage of components to determine if there was a particular cause for the wear. Belts, chains, and gears must be inspected for wear. Tensioners, sprockets, and guides must also be thoroughly evaluated to make a good decision about which components should be replaced. You want to provide repairs in a cost-effective manner but without the risk of premature failure of your work. In many cases, you will replace multiple components in the timing mechanism. You should use the manufacturers' specific procedures when that information is available or if you have any doubt about your diagnosis. There are also industry standard procedures for determining the condition of belts, chains, and gears, which we'll discuss here. These inspections will well serve to give you an overview of the inspection process.

WARNING: It is important to know whether the engine with a suspected timing mechanism problem is an interference or free-wheeling type. If it is an interference engine, major damage to the engine may already have occurred. You could also incur damage if you rotate the engine to test compression or rotate the engine by hand when the valves are hitting the top of the pistons. Check the service information for the vehicle you are working on to be sure you provide professional service for your customers.

Backlash is the small amount of clearance between the gear teeth that allows room for expansion as the gears heat up.

SYMPTOMS OF A WORN TIMING MECHANISM

We mentioned that a worn timing chain may slap against the cover or against itself on a quick deceleration and make a rattling noise. Use a stethoscope in the area of the chain to confirm your preliminary diagnosis, and then proceed with the checks described in the next section to confirm the extent of the damage (Figure 11-1).

A customer may also report that the engine seems to have poor acceleration from starts but that it runs well, perhaps even better, at higher rpms. When there is slack in the chain, the camshaft timing is behind the crank. This is called retarded valve timing. This improves high end performance while sacrificing low end responsiveness.

Timing gears may clatter on acceleration and deceleration. The engine does not have to be under a load; just snap the throttle open and closed while listening under the cover for gear clatter. The customer is unlikely to notice the reduced low-end performance before the gears break from too much **backlash**.

Customers with worn or cracked timing belts will notice nothing until their engines stop running or begin running very poorly. What you and they should be paying close attention to is the mileage on the engine and the recommended belt service interval. Skipping a recommended replacement is at the very least an invitation to meet a tow truck driver.

SYMPTOMS OF A JUMPED OR BROKEN TIMING MECHANISM

When a timing chain, belt, or gear breaks and no longer drives the camshaft(s), you can often tell by the sound of how the engine turns over. It cranks over *very* quickly and no sounds even resembling firing occur (Figure 11-2). The engine has no compression. Hopefully it has been towed to your shop; that's the only way it'll run again.

When a timing chain, belt, or gear skips one or more teeth, jumps time, it sets the valve timing off significantly. Some engines will run when the timing is off one or even two teeth,

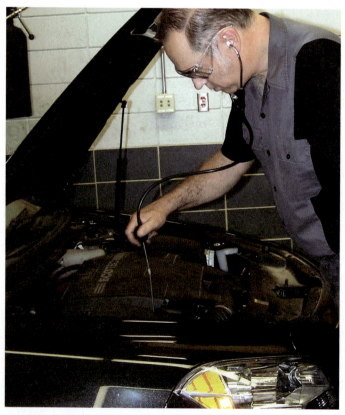

FIGURE 11-1 Listen for a slapping or rattling noise on deceleration.

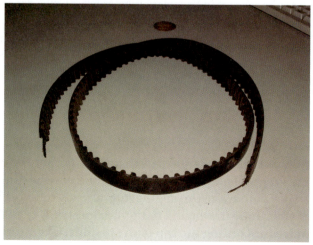

FIGURE 11-2 This snapped timing belt left the customer stranded on the highway.

but not well. Usually you can hear the problem as the engine cranks over. It turns over unevenly; the engine speeds up and then slows down as you crank. Idle quality, acceleration, and emissions will be affected. If the timing is off significantly, it will pop out of the intake or exhaust or backfire under acceleration. Often the engine will not even start.

Diagnosing a Jumped or Broken Timing Mechanism

If the engine just spins over fast and does not run at all, you may be able to crank the engine over and look for valvetrain movement through the oil filler cap. If there is no movement, the mechanism is broken. When you cannot see the valvetrain components through the oil fill cap, it may be easiest to pull the upper timing cover off. When a customer said that it was running fine yesterday and today it stopped running and cranks over quickly, a failed timing mechanism is a common cause. One good way to verify a jumped or broken timing mechanism or to confirm your suspicions is to perform a cranking compression test. If the engine is running poorly because of an engine that jumped time, your diagnosis may require a compression test. Crank the engine over and check each cylinder. When the timing has jumped, the compression will be low on all the cylinders. Frequently it'll be as low as 50 or 75 psi on every cylinder (Figure 11-3).

Another quick check for a broken timing mechanism is to check cranking vacuum. Instead of the normal 3–6 in. Hg vacuum, you will see no vacuum if the mechanism is broken. The vacuum will be low if the belt or chain has jumped time. If the engine runs, engine vacuum will be lower than the usual 18–22 in. Hg vacuum found at idle on a healthy engine.

Sometimes when the engine runs fairly well but not correctly, the compression may not be low enough to confirm jumped timing. If you still suspect that the valve timing is off, rotate the engine to TDC number 1 compression using the crankshaft or camshaft indicator. If you can locate TDC on the cam sprocket pretty easily, you may be able to detect TDC of the number 1 piston by a mark on the crankshaft pulley. If there is no mark on the crankshaft pulley, you may still be able to pick out jumped time without removing the crankshaft pulley and accessories. Remove the upper timing cover. Rotate the engine by hand until the TDC number 1 marking is lined up on the camshaft (Figure 11-4). Remove the spark plug from cylinder number 1. Place a straw or a suitable plastic piece on top of the piston. Rotate the engine to assure that the straw starts declining within a couple of degrees of crankshaft rotation. Rotate the engine two full revolutions and carefully observe the straw as you approach TDC number 1 on the camshaft; if the timing is on, the piston must reach the top of its travel and stop briefly right when the camshaft timing mark lines up.

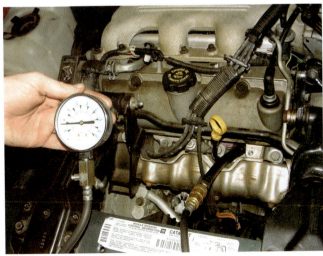

FIGURE 11-3 Low compression on most or all cylinders can point to a timing belt/chain that jumped a gear and changed the valve timing.

FIGURE 11-4 The arrow on the camshaft sprocket and the dot on the cylinder head designate TDC number 1 when they are aligned.

TIMING CHAIN INSPECTION

To inspect a timing chain for wear, you can measure the amount of movement that the crankshaft makes before the camshaft(s) begins to move. Generally this should be less than 8°, but check the manufacturer's specifications. Use the crankshaft timing marks when available to watch the degrees of crankshaft movement. Many newer engines do not have timing marks on the crank pulley and front cover because ignition timing is rarely adjustable now. If that is the case, use a torque angle gauge on a ratchet as you turn the crankshaft. On an older engine that has a distributor, remove the distributor cap and watch for rotor movement as a sign that the chain is rotating the cam or an auxiliary sprocket. When an engine does not use a distributor, it may be possible to detect movement of the camshaft through the oil fill cap. Remove the cap and see if you can clearly view the camshaft or a rocker. If you can, rotate the crankshaft counterclockwise then clockwise to measure the amount of slack before movement of the cam or valvetrain. Sometimes it will be necessary to remove a valve cover to get a good view of the movement of the cam or rockers. If the crankshaft rotates more than 8° before the camshaft moves, the timing chain and/or tensioner is worn beyond usable limits (Figure 11-5).

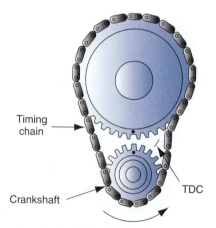

Timing chain

Crankshaft

TDC

FIGURE 11-5 Rotate the crankshaft backwards to get all the slack on one side. Next, measure the rotation before the camshaft moves as you rotate the slack out of the chain.

If you have the timing cover off of a cam in the block engine, you can try to deflect the chain. If you can deflect it less than one quarter of an inch, the chain is generally satisfactory for continued service as long as the chain itself and the sprockets are in good shape. Deflection of one half of an inch requires immediate attention. Be sure to consult the service information for the deflection specifications.

You can use this same procedure on an overhead cam timing chain by removing the valve cover. Replacement is recommended when there is more than one half of an inch of deflection or as specified by the manufacturer.

Inspect the chain itself for shiny spots on the edges of the chains; sometimes they slap against the metal cover. Look for worn or flat rollers on a roller-type chain. On a flat-link silent chain, look for wear at the attachment pins and joints; shiny spots indicate friction points.

TIMING BELT INSPECTION

To inspect a timing belt, you really need to pull the top timing cover off of the engine. This is usually a plastic cover with an upper and lower section. For inspection purposes, it will only be necessary to remove the upper cover (Figure 11-6). You will be looking for cracks similar to those in serpentine belts. Rotate the engine around to be able to view the whole belt. Look on the underside of the belt at the cogs or teeth that hold the belt in the sprocket. Any broken or missing teeth dictate immediate replacement. Figure 11-7 shows some typical timing belt failures. If the belt shows signs of oil or coolant contamination, the belt should be replaced and the leak must also be fixed as part of the repair. Be sure to diagnose the cause of the leak and include that procedure and part(s) in your estimate to the customer. As we've discussed, there is no reason to try to eek out another few months on a damaged belt; it could result in a very expensive repair.

Clip

FIGURE 11-6 Remove the upper timing cover to get a look at the belt. This cover is easily removed with accessible clips.

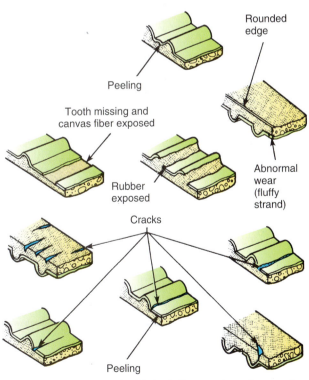

Rounded edge

Peeling

Tooth missing and canvas fiber exposed

Rubber exposed

Abnormal wear (fluffy strand)

Cracks

Peeling

FIGURE 11-7 Any fault with the belt warrants replacement.

TIMING GEAR INSPECTION

To check timing gears for excessive wear, remove the timing cover. You should listen before you take the cover off, so you'll begin to learn what good and bad gears sound like. Put a ratchet on the crankshaft bolt and rock it lightly back and forth between the backlash of the gears. You don't want to turn it so much that the camshaft actually turns. Just listen for the clacking of the gears. It may be so bad you'll know it the very first time, but take the cover off to confirm your diagnosis.

Worn teeth will be sharp, shiny, and pointed (Figure 11-8). If you are not sure by visual inspection, you can measure the gear backlash. This is the small clearance left between any gear set that allows room for expansion as the gears heat up. You can measure the clearance between any two teeth. Typical gear backlash is 0.004 in. to 0.008 in.; you may be able to visually see that this set of gears has more than that. If you are unsure, take feeler gauges and rock the crank so the teeth are touching and then go backwards as far as you can without rotating the camshaft. Measure the clearance between the gears and compare your reading to the actual specification. Excessive backlash or visual damage to the gears warrants replacement.

SPROCKET INSPECTION

Recommended practice is to replace the sprockets used when a timing chain is replaced. The job is not usually a simple one, and the sprockets have typically been in service for over 100,000 miles. The teeth of the sprockets wear with the chain. If you had a low-mileage failure, you would carefully inspect the sprocket teeth. Worn teeth will be shiny and sharper; good sprockets will have an unworn flat spot at the top of the tooth (Figure 11-9). The teeth may actually be rounded on the face from wear with the rollers.

Rounded edge shows minor wear.

FIGURE 11-9 This timing chain sprocket was replaced with the chain. Although the wear was not severe, the timing mechanism had a lot of miles on it.

Good teeth

Worn teeth

FIGURE 11-8 Worn teeth develop a sharp pointed edge.

Sprockets are not normally replaced when the timing mechanism uses a belt. The belt is very soft compared to the steel sprockets and do not wear rapidly. You should always carefully inspect them, however, because they can fail. A sharp edge on a timing belt sprocket will cause premature failure of a new belt. Compare each of the cogs to the others and make sure they are in fine condition. Always check the manufacturer's TSBs for information about replacing the sprockets. If you have any doubt, replace them.

TENSIONER INSPECTION

The timing chain hydraulic-type tensioner is commonly replaced with a high mileage timing chain job. Again, it makes good sense to take care of all the components in the system. Most tensioners have a minimum length that the plunger should stick out from the body. Refer to the manufacturer's specifications, and if the plunger length is not within specifications, replace it. You should also check for oil leakage past the front seal from which the plunger protrudes. If you see wetness around the seal, you should replace the tensioner.

The pulley type tensioner used with belts is also commonly replaced when the service interval is long, every 90,000 miles, for example. If the timing belt is being replaced for the second time and it was not replaced the first time, you should definitely recommend replacement (Figure 11-10). An engine that requires belt replacement every 60,000 miles and is being serviced at 120,000 miles deserves a new tensioner to prevent premature failure. To inspect the tensioner for faults, check for scoring on the pulley surface. Scratches, sharp edges, or grooves can damage a new belt rapidly. Also check the pulley's bearing. Rotate it by hand and feel for any roughness. If you can hear it scratching or feel coarseness, replace it. Also check the pulley for side play; grasp it on its ends and check for looseness. If it wobbles at all, replace it.

A timing belt automatic tensioner is also typically replaced during belt replacement. If there are any signs of oil leakage, replacement is mandatory. If the belt replacement interval is relatively frequent, it may be possible to use the automatic tensioner through the life of two timing belts. The best advice is to follow the recommendation of the manufacturer and carry out any inspections or measurements indicated.

TIMING CHAIN GUIDE INSPECTION

You should replace the guides on a high mileage timing chain replacement. They are synthetic rubber or Teflon˚ or some other material that is much softer than the chain. It is recommended practice to replace the guides. To inspect a guide for wear, look at the surface

Belt tensioner pulley
FIGURE 11-10 Replace the belt tensioner pulley if it has excessive mileage or if you can feel any roughness or wobbling in the bearing.

where the chain rides. If the grooves are deeper or wider or you can see any signs of wear, be safe and replace the guide.

REUSE VERSUS REPLACEMENT DECISIONS

When in doubt, replace a worn valve timing mechanism component. A premature failure can cause extensive damage, and you could lose a customer. When you choose to reuse components, make sure you are confident that they will last an appropriate amount of time for the customer and the engine. Your repair work should result in reliable and long-term service.

TIMING MECHANISM REPLACEMENT

The actual repair procedures for service of timing chains, belts, and gears vary immensely. Some OHC timing chains can be replaced in a few hours with the engine in place; others require that the engine be removed from the vehicle. It is common to have to remove the cylinder head and/or timing cover. Timing belts can generally be replaced with the engine in the vehicle. They can be simple and straightforward or time-consuming and a little tricky. A timing belt on a transversely mounted V6 engine in a minivan, for example, may leave you little room to access the front end of the engine. Replacing the timing mechanism on cam-in-the-block engines using either a chain or a gear set is usually uncomplicated. In every case, it is critical that you set all the shafts into perfect time with the crankshaft. As we've discussed, setting the valve timing off by one tooth will impact engine drivability and emissions. Similarly, an engine with balance shafts not timed properly will vibrate noticeably. Often it is too easy to miss the timing of a shaft by one tooth if you aren't paying close attention and double checking your work (Figure 11-11). This is important and time-consuming service work; you only want to do it once! Many technicians will replace the camshaft and crankshaft seals while they have access to them during the job. If the water pump is driven by the timing belt, you will usually replace that as well. Always check idle quality and engine performance after a replacement job to be sure the engine is operating as designed. This text cannot offer a replacement for the manufacturer's service procedures. Instead we will offer a couple of specific examples of the procedures for replacing different timing mechanisms. You'll still need manufacturers' specific information to achieve proper timing mark alignment and to perform the job in the most efficient manner. You may also need special tools to hold the camshaft sprockets in position while replacing the belt or chain(s).

> **CUSTOMER CARE:** When repairing an engine with a broken timing mechanism, it is important to know whether the engine is an interference engine or not. If it is, advise the customer of the potentially very high added cost of removing the cylinder head and replacing valves. Also make sure he understands that even more serious damage, such as a valve stuck into a piston, could have occurred. Even on freewheeling engines, it's wise to inform the customer that valve damage is possible. Discuss these issues with the customer or service advisor before offering an estimate for the work.

TIMING CHAIN REPLACEMENT ON OHC ENGINES

Replace the timing chain(s) on an engine during a thorough overhaul if noise or symptoms indicate excessive wear, or if the chain has broken. Timing chain replacement procedures vary significantly in the specifics of what must be removed in order to access the chain

FIGURE 11-11 **This engine has balance shafts.**

and sprockets. You will usually have to remove the valve cover(s), the front engine or timing cover, the engine drive belt(s), any accessories blocking access to the front cover, and the harmonic balancer. Sometimes you will need to remove the front engine mount as well. You must follow the manufacturer's specific instructions to perform the job most efficiently.

Below is a general outline of procedures for replacing the timing chain on a popular DOHC engine:

1. Disconnect the negative battery cable.
2. Drain the coolant and remove the reservoir.
3. Remove the serpentine engine drive belt (Figure 11-12).
4. Loosen the generator and swing it back out of the way.
5. Remove the upper timing cover fasteners.
6. Remove the engine mount assembly. These bracket bolts may be torque to yield bolts and must be replaced with new special bolts.
7. Raise the vehicle and remove the right wheel.
8. Remove the inner fender splash guard.
9. Use a suitable puller to remove the harmonic balancer.
10. Remove the lower front cover fasteners.
11. Lower the vehicle and remove the front cover.
12. Rotate the engine clockwise until the camshafts' keyways are at 12:00 (Figure 11-13). The holes in the camshaft sprocket should line up with the holes in the timing housing.
13. The crankshaft sprocket keyway should be at 12:00 if the valve timing is correct.
14. Remove the timing chain tensioner and guides.

SERVICE TIP:
When rotating an engine, you should always use the crankshaft bolt rather than the camshaft(s) bolt. You could break a camshaft trying to rotate the engine with it. Also, turn the engine in its normal direction of rotation unless specified otherwise. Engines rotate clockwise when looking at them from the front.

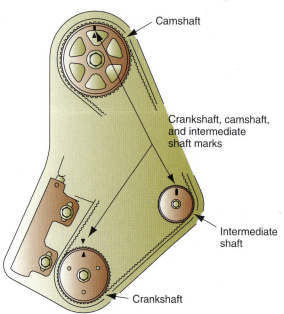

Camshaft

Crankshaft, camshaft, and intermediate shaft marks

Intermediate shaft

Crankshaft

FIGURE 11-12 Remove the serpentine belt and accessories as needed to access the timing cover.

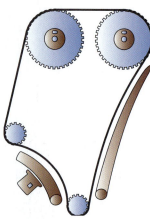

FIGURE 11-13 Line up the timing marks before disassembling the timing chain.

15. Make a front marking on the cam and crank sprockets using a paint stick or correction fluid in case they will be reused.
16. Remove the timing chain.
17. Carefully inspect all the components and determine which, if any, you will reuse (Figures 11-14 –11-16).
18. Clean the front cover and mating surface thoroughly using a plastic scraper on aluminum surfaces.
19. Install the camshaft and crankshaft sprockets with the front markings out. Torque them to the specified value.
20. Make sure the camshaft sprocket holes are still lined up with the holes in the housing. If not, rotate the crankshaft 90 degrees off TDC to provide valve clearance, and turn the cams to line up the holes. Install a dowel through the sprocket into the hole in the timing housing to retain the sprockets (Figure 11-17).

FIGURE 11-14 This chain shows wear on the shiny spots from the friction between the pins and the links.

FIGURE 11-15 This timing chain guide has a deep groove worn in it from the chain.

FIGURE 11-16 The tensioner plunger did not protrude the specified length; it was replaced.

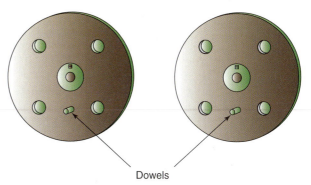

Dowels

FIGURE 11-17 Insert the dowels through the sprockets and into the head. This holds the camshafts in perfect alignment while you install the chain.

21. Turn the crankshaft backwards (counterclockwise) to TDC until the keyway is in the 12:00 position and the marks on the sprocket and the engine block align.

22. Install the chain around the crank sprocket, idler sprocket, and exhaust cam sprocket. Be sure all slack is on the tensioner side of the chain. Remove the intake cam dowel, and rotate the sprocket just a little to install the chain with no slack between the two cam sprockets. Relax the cam, and refit the dowel pin. If it doesn't slide in, retime the sprocket and chain. All the chain slack must be on the tensioner side.

23. Raise the vehicle, and recheck the crank sprocket alignment marks; adjust as needed.

24. Use a small screwdriver to release the ratcheting mechanism in the tensioner (Figure 11-18). Fully depress the tensioner and install a retaining pin through the hole in the plunger and housing (Figure 11-19). Install the tensioner and torque to specification.

25. Install the guides and torque properly. Remove the retaining pin in the tensioner.

26. Remove the camshaft dowels and rotate the engine two complete revolutions. Recheck the alignment marks to be sure they are *perfect*; adjust as needed (Figure 11-20).

27. Replace the front seal, and lubricate its lip with chassis grease or oil.

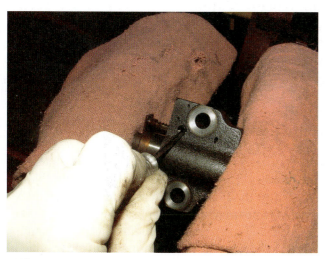

FIGURE 11-18 Use a tiny screwdriver to release the ratcheting mechanism while you retract the tensioner.

FIGURE 11-19 Slide a small Allen head or pin through the hole in the tensioner body to retain the tensioner in its retracted position.

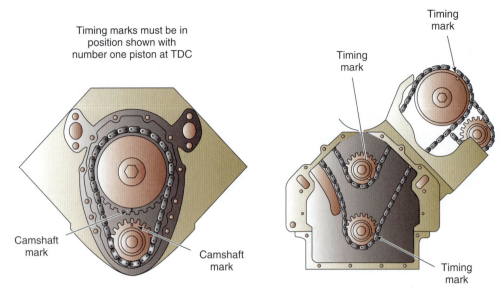

Timing marks must be in position shown with number one piston at TDC

Timing mark

Timing mark

Timing mark

Camshaft mark

Camshaft mark

Timing mark

FIGURE 11-20 Line up the timing marks exactly as shown in the service information.

28. Reassemble in reverse order using appropriate torque specifications and procedures.
29. Idle the engine, and road test the vehicle to confirm proper performance.

As you can see from these instructions, the work of replacing a timing chain requires time, concentration, and attention to the details of valve timing alignment. Follow the manufacturer's procedures for reinstallation, to ensure all bolts are torqued properly and components installed correctly.

TIMING BELT REPLACEMENT

Check the maintenance manual for the vehicle whenever a vehicle comes to your shop for service. Routine maintenance is often neglected by customers because many expect technicians to inform them when a procedure is due. This is critically important when it comes to timing belt replacement. Many consumers have no idea what a timing belt is, much less that it needs to be replaced periodically. Explain the importance of timing belt maintenance to your customers. Remember that when the timing belt also drives the water pump, it is common practice to install a new water pump when replacing the belt. The safest way to determine if a belt needs replacement is through a visual inspection and by mileage intervals.

> **CUSTOMER CARE:** As a technician, it is important that you read technical service bulletins when performing service work. One manufacturer decided to change the timing belt replacement interval to every 15,000 miles because the belts were snapping so often! Other manufacturers have lengthened the service interval to over 100,000 miles due to improvements in timing belt materials.

The interval on timing belt replacement varies widely. On modern vehicles the interval typically ranges between 60,000 and 100,000 miles. There are some new timing belts that do not have scheduled replacement. Inspect them at the requested intervals and replace them when signs of wear are evident. Some older vehicles may require replacement every 30,000 or 45,000 miles. All these examples must make it clear that you'll need to check the maintenance information to be sure. Timing belts are replaced during a thorough engine overhaul as well.

To replace the timing belt, you'll need to remove the upper and lower timing covers, the harmonic balancer, and any accessories that are in the way of those covers. The procedure will be similar to the timing chain replacement, but the plastic timing belt covers do not seal oil. If the timing belt drives the water pump, this is an excellent opportunity to replace it. Sometimes a timing belt replacement is a pretty quick and simple job. Read through the whole service procedure before beginning, so that you'll be familiar with all the steps required (Figure 11-21). Let us look at a typical example of the procedure to replace a timing belt:

1. Remove the negative battery cable.
2. Remove the accessory drive belts.
3. Lift the vehicle and remove the right wheel and inner fender splash guard.
4. Remove the engine drive belt(s) and the harmonic balancer using a suitable puller.
5. Lower the vehicle and support the engine with a jack.
6. Remove the right engine mount and bracket from the engine.
7. Remove the timing belt cover.
8. Rotate the engine to TDC number 1 firing, and note that the marks on the cam and rear timing cover line up. Check the crankshaft sprocket to be sure its notch lines up with the notch on the rear timing cover. (If the engine is equipped with an auxiliary or balance shaft, make sure that its marks are also properly aligned (Figure 11-22).)
9. Rotate the tensioner pulley counterclockwise to loosen it until you can slide a small Allen wrench or pin through the locking hole in the pulley and into the hole in the tensioner bracket. This relieves the tension on the belt.
10. Remove the timing belt. Do not rotate the camshaft or crankshaft.
11. Scrutinize the sprockets, tensioner pulley, and water pump; replace components as needed.
12. Install the new components and new timing belt, ensuring that the belt fits tightly around all the components. The slack should be on the tensioner pulley side.
13. Rotate the tensioner pulley just enough to pull out the locking pin. Allow the automatic tensioner to properly tension the belt. Note: Some tensioners are not automatic.

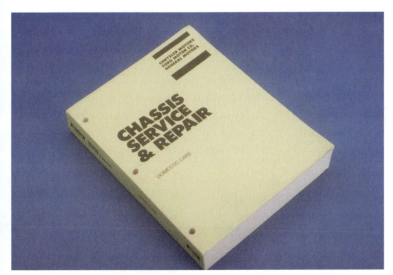

FIGURE 11-21 Read through the service information first so you understand the procedure.

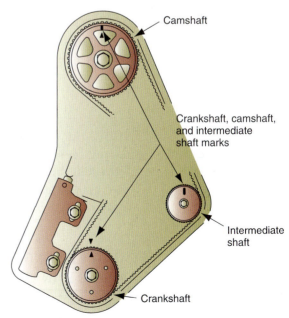

FIGURE 11-22 Check all the alignment marks before disassembly so you know how they should line up during reassembly.

FIGURE 11-23 Use a belt tension gauge when the tensioner is manual to be sure you make the correct adjustment.

FIGURE 11-24 The timing marks should line up perfectly after two revolutions of the engine.

On these you will have to loosen the locking nut, rotate the pulley on its eccentric bolt until the proper tension is achieved, and tighten the locking bolt. Use a belt tension gauge to confirm proper tension (Figure 11-23). The belt tension gauge fits over the belt and deflects the needle further the tighter the belt is. Read the tension on the gauge, compare to specifications and adjust as needed.

14. Check the camshaft and crankshaft alignment marks. Rotate the engine two complete revolutions, and check the marks again. Adjust as needed (Figure 11-24).
15. Reassemble in reverse order following the recommended procedures.
16. Idle the engine, and road test the vehicle to confirm proper performance.

CUSTOMER CARE: Do not omit what seem like little details on a job. It is critical that you replace the plastic timing covers securely on the engine. They keep debris off the belt, but they also prevent water or ice from leaking in. If ice or even a lot of water gets onto the belt, it can skip teeth even when it is relatively new and properly tensioned. This has happened many times to careless technicians.

TIMING CHAIN OR GEAR REPLACEMENT ON CAMSHAFT-IN-THE-BLOCK ENGINES

Replace an in-the-block timing chain or gear set when an engine is overhauled, when excessive wear is detected, or after the mechanism fails. Rotate the engine to TDC number 1 using the mark on the harmonic balancer. To access the timing mechanism, you will generally have to disconnect the negative battery clamp and remove the cooling fan if it is mounted on the

TYPICAL PROCEDURE FOR REPLACING A TIMING BELT ON AN OHC ENGINE

P21-1 Disconnect the negative cable from the battery prior to removing and replacing the timing belt.

P21-2 Carefully remove the timing cover. Be careful not to distort or damage it while pulling it up. With the cover removed, check the immediate area around the belt for wires and other obstacles. If some are found, move them out of the way.

P21-3 Align the timing marks on the camshaft's sprocket with the mark on the cylinder head. If the marks are not obvious, use a paint stick or chalk to mark them clearly.

P21-4 Carefully remove the crankshaft timing sensor and probe holder.

P21-5 Loosen the adjustment bolt on the belt tensioner pulley. It is normally not necessary to remove the tensioner assembly.

P21-6 Slide the belt off the camshaft sprocket. Do not allow the camshaft pulley to rotate while doing this.

P21-7 To remove the belt from the engine, the crankshaft pulley must be removed. Then the belt can be slipped off the crankshaft sprocket.

P21-8 After the belt has been removed, inspect it for cracks and other damage. Cracks will become more obvious if the belt is twisted slightly. Any defects in the belt indicate it must be replaced. Never twist the belt more than 90 degrees.

P21-9 To begin reassembly, place the belt around the crankshaft sprocket. Then reinstall the crankshaft pulley.

TYPICAL PROCEDURE FOR REPLACING A TIMING BELT ON AN OHC ENGINE

P21-10 Make sure the timing marks on the crankshaft pulley are lined up with the marks on the engine block. If they are not, carefully rock the crankshaft until the marks are lined up.

P21-11 With the timing belt fitted onto the crankshaft sprocket and the crankshaft pulley tightened in place, the crankshaft timing sensor and probe can be reinstalled.

P21-12 Align the camshaft sprocket with the timing marks on the cylinder head. Then wrap the timing belt around the camshaft sprocket and allow the belt tensioner to put a slight amount of pressure on the belt.

P21-13 Adjust the tension as described in the service manual. Then rotate the engine two complete turns. Recheck the tension.

P21-14 Rotate the engine through two complete turns again, then check the alignment marks on the camshaft and the crankshaft. Any deviation needs to be corrected before the timing cover is reinstalled.

water pump snout. You may have to drain the coolant and remove the water pump; if you do, look over the pump closely and replace it if it has defects or extended mileage on it. Remove the engine drive belts. Use a puller (generally) to remove the crankshaft pulley and any other accessories or brackets that may be blocking access to the timing cover bolts. Remove the timing cover bolts and timing cover and note the markings—usually small dots, notches, or lines—on the gears or sprockets.

Inspect the chain or gears for damage, so you are confident that your diagnosis will repair the original concern. Examine the sprockets for the timing chain; they are typically replaced with the chain. Scrape the timing cover and mating surface clean. Replace the gears or chain and sprockets (Figure 11-25). When replacing a timing chain, place the chain around the sprockets, and adjust the sprocket until the timing marks line up; then reinstall the sprockets onto the crankshaft and camshaft. With gears, mate them together so that the timing marks are aligned, and then slide them over the shafts (Figure 11-26). In both cases,

FIGURE 11-25 Fit the chain over the sprockets so the alignment dots point toward each other.

FIGURE 11-26 Slide the sprockets over the shafts with the marks aligned. Recheck the marks after two engine revolutions.

the shafts will have to be properly set up so the gears or sprockets will fit over the keys on the cam- and crankshafts. Rotate the engine two full revolutions, and be sure that the marks line up perfectly.

Reinstall the timing cover with a new gasket and any recommended sealant. Torque the bolts in the designated sequence; this is typically in a diagonal fashion around the cover. Reassemble the rest of the components, and verify proper engine performance.

Timing Chain Replacement on Engines with OHC and Balance Shafts

There is no room for error in replacing the timing chains on an engine with overhead camshafts and balance shafts. The valve timing must be perfect in order for the engine to perform properly and achieve good fuel economy and proper emissions. The balance shafts must be timed precisely to prevent engine vibration. There is no substitute for the manufacturer's service information. Make sure you have any special tools required to perform a professional job. In some cases you will have to remove the engine from the vehicle to perform the job. In other cases, you can perform the work with the engine in the vehicle. To make the job simplest, place the engine at TDC number 1 firing, remove the timing cover, and make a careful note of the timing marks. This will help you during the reassembly process. Carefully inspect for wear of all the timing chain components: guides, tensioners, sprockets, and chains. Because the task is usually a time consuming and expensive proposition, most technicians will replace those components just listed. Never be satisfied with close enough when it comes to lining up the timing marks; this is a job you only want to do once!

TYPICAL PROCEDURE FOR REPLACING TIMING CHAINS ON A DOHC ENGINE WITH BALANCE SHAFTS

P22-1 Set the engine at TDC number 1 compression and remove the valve cover and crank pulley.

P22-2 Remove the timing cover.

P22-3 Remove the tensioner, do not rotate the engine and loosen the camshaft sprockets.

P22-4 Remove the camshaft sprockets and chain.

P22-5 You may need to remove the cylinder head to access all the guides. Remove the guides and primary chain.

P22-6 The secondary chain runs the balance shafts and water pump. Remove, inspect, and replace any worn guides. It is common practice to replace guides, sprockets, chains, and tensioners when replacing chains, on a higher mileage engine.

P22-7 Align the balance shafts and crankshaft sprockets exactly as shown in the service manual. To set the chain up, you compress the tensioner and install a pin to relieve pressure. When all is lined up, remove the pin to tension the chain.

P22-8 Rotate the engine two complete revolutions and be sure the timing marks line up perfectly.

P22-9 Install the crankshaft sprocket for the primary chain.

TYPICAL PROCEDURE FOR REPLACING TIMING CHAINS ON A DOHC ENGINE WITH BALANCE SHAFTS

P22-10 Line the chain up on the camshaft sprocket in its correct position and torque the sprocket to specification.

P22-11 With the primary chain installed, check the alignment marks on the camshaft and crankshaft sprockets.

P22-12 Torque the tensioner and rotate the engine another two revolutions to recheck your timing marks. You only want to do this job once.

P22-13 Reassemble the engine using the proper torque specifications, gaskets, and sealants if required.

TIMING CHAIN REPLACEMENT ON ENGINES WITH VVT SYSTEMS

In order to successfully replace the timing chain(s) on a VVT engine, you will need to closely follow the manufacturer's procedure and have access to any special tools required. The Cadillac VVT system uses three timing chains to drive the four overhead camshafts. The primary chain drives each intermediate camshaft sprocket. A chain is used, one on each head, from the intermediate sprocket to drive the top camshaft sprockets with the **camshaft position actuators**. The actual procedure is not dramatically different from other timing chain replacements, but there are more steps because of the multiple chains. What follows is a simplified outline of the steps to replace the timing chain on the Cadillac VVT engine.

1. Place the engine at TDC number 1 on the compression stroke.
2. Remove the camshaft covers. You will need to remove the brake booster vacuum hose, the fuel injector sight shield, and the positive crankcase ventilation (PCV) tube from the left cover. Also remove the ignition module, the camshaft sensors, the ground strap on the left cover, and the dipstick tube securing bolt (Figure 11-27).
3. Remove the front timing cover and oil pump.
4. Install the special camshaft holding tools to the backs of the camshafts to prevent them from rotating.

> A **camshaft position actuator** is a hydraulically actuated device that changes the camshaft in relation to the crankshaft.

5. Use a paint stick to mark the location of the black links of the chains in relation to the timing marks on the camshaft actuators and intermediate sprockets (Figure 11-28).
6. Remove the bolts attaching the timing chain tensioners. They will expand as you remove them (Figure 11-29).
7. Remove the timing chains, camshaft actuators, and sprockets.
8. Clean and inspect all components, and replace any parts that show signs of wear, such as scoring; rounded edges; or shiny, sharp teeth.
9. Reinstall the camshaft actuators and sprockets, with the timing marks positioned according to the manufacturer's diagrams.
10. To reassemble with a new chain, locate the black links and reposition them in the same locations on the sprockets (Figure 11-30).
11. Align the timing marks as indicated by the manufacturer's diagrams (Figure 11-31).
12. Compress and lock the timing chain tensioners, and reinstall them using the proper torque specification (Figure 11-32).
13. Release the tensioners and recheck the timing marks. Adjust as needed.
14. Rotate the engine two complete revolutions, and confirm that the timing marks are properly aligned.
15. Reassemble the engine and check for oil leaks, proper idle, and quality engine performance.

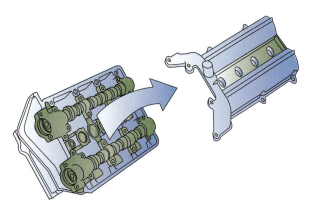

FIGURE 11-27 Remove the camshaft cover.

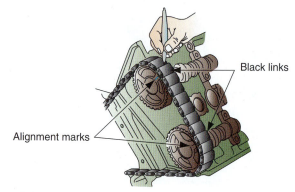

Black links

Alignment marks

FIGURE 11-28 Use a paint stick to mark the black links and the alignment marks on the sprockets.

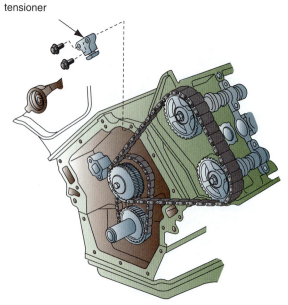

Chain tensioner

FIGURE 11-29 Remove the timing chain tensioners.

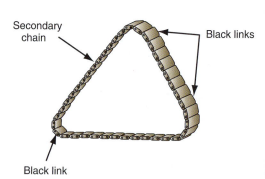

Secondary chain

Black links

Black link

FIGURE 11-30 Use the black links on the new chain to align with the marks on the sprockets.

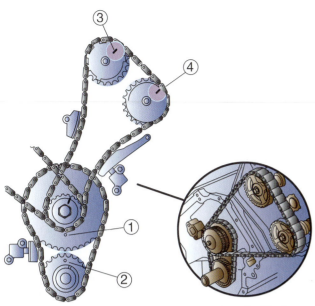

FIGURE 11-31 The upper numbers 1-4 show the black links aligned with the marks on the camshaft sprockets. The 1 and 2 on the intermediate and crankshaft sprockets show those alignment marks.

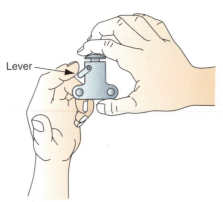

Lever

FIGURE 11-32 Collapse the lever to retract the tensioner then hold the lever down to retain the tensioner until it is installed. Sometimes a pin can be installed to hold the ratcheting mechanism back.

TERMS TO KNOW

Backlash

Camshaft position actuator

ASE-STYLE REVIEW QUESTIONS

1. *Technician A* says that a timing chain is generally replaced every 60,000 or 90,000 miles.
   Technician B says the timing chain is typically replaced during a major engine overhaul.
   Who is correct?
   A. A only   C. Both A and B
   B. B only   D. Neither A nor B

2. The best way to properly determine if a timing belt needs replacement is to:
   A. Ask the customer when it was last done.
   B. Look at the maintenance information to determine the recommended interval.
   C. Remove the top timing cover to inspect the condition of the belt.
   D. Replace it at every tune up to be safe.

3. *Technician A* says the timing chain is removed before the tensioner.
   Technician B says to install the tensioner, you have to lock the plunger into the body.
   Who is correct?
   A. A only   C. Both A and B
   B. B only   D. Neither A nor B

4. *Technician A* says that some tensioner pulleys automatically adjust belt tension.
   Technician B says that on manually adjusted belts, you set the belt with one inch of deflection per foot of open belt.
   Who is correct?
   A. A only   C. Both A and B
   B. B only   D. Neither A nor B

5. *Technician A* says that the timing chain cover seals oil in.

   Technician B says the plastic timing belt covers can be eliminated if they crack during removal.

   Who is correct?

   A. A only
   B. B only
   C. Both A and B
   D. Neither A nor B

6. *Technician A* says you need to rotate the engine two complete revolutions to make a final check of the timing marks.

   Technician B says you should always road test a vehicle after timing mechanism replacement.

   Who is correct?

   A. A only
   B. B only
   C. Both A and B
   D. Neither A nor B

7. *Technician A* says that worn timing belts will make the engine run better until they break.

   Technician B says that low-end acceleration may deteriorate with a worn timing chain.

   Who is correct?

   A. A only
   B. B only
   C. Both A and B
   D. Neither A nor B

8. *Technician A* says that worn timing gears may clatter on acceleration or deceleration.

   Technician B says to verify worn timing gears by checking for excessive end play.

   Who is correct?

   A. A only
   B. B only
   C. Both A and B
   D. Neither A nor B

9. *Technician A* says that timing chain tensioners are typically replaced with the chain.

   Technician B says you should measure the spring tension to determine if they can be reused.

   Who is correct?

   A. A only
   B. B only
   C. Both A and B
   D. Neither A nor B

10. *Technician A* says that timing chain guides should last the life of the engine.

    Technician B says that worn timing belt pulleys will show scoring and/or roughness in the bearing.

    Who is correct?

    A. A only
    B. B only
    C. Both A and B
    D. Neither A nor B

ASE CHALLENGE QUESTIONS

1. *Technician A* says that being off on valve timing as little as three teeth can result in valve-to-piston contact.

   Technician B says that advancing the camshaft one tooth will improve high rpm breathing.

   Who is correct?

   A. A only
   B. B only
   C. Both A and B
   D. Neither A nor B

2. *Technician A* says that most engines are timed at BDC of the compression stroke for cylinder number 1.

   Technician B says that valve timing should be adjusted every 60,000 miles.

   Who is correct?

   A. A only
   B. B only
   C. Both A and B
   D. Neither A nor B

3. *Technician A* says that you may need to use a block of wood to hold the camshafts in place when replacing the timing chains.

   Technician B says that if the crankshaft rotates less than 20° before the camshaft moves, the chain is acceptable.

   Who is correct?

   A. A only
   B. B only
   C. Both A and B
   D. Neither A nor B

4. Each of the following is a likely symptom of a valve timing problem EXCEPT:

   A. Detonation
   B. Backfiring
   C. Rough idle
   D. Poor fuel economy

5. *Technician A* says that you have to remove the engine to replace the timing mechanism on a variable valve timing system.

   Technician B says that the camshaft actuators must be properly timed during reassembly.

   Who is correct?

   A. A only
   B. B only
   C. Both A and B
   D. Neither A nor B

Name _____ **Date** _____

INSPECTING THE TIMING CHAIN OR BELT

Upon completion of this job sheet, you should be able to inspect the timing chain or belt properly.

ASE Correlation

This job sheet is related to ASE Engine Repair Test's content area: Cylinder Head and Valvetrain Diagnosis and Repair; tasks: Inspect and replace camshaft drives (includes checking gear wear and backlash, sprocket chain wear, overhead cam drive sprockets, drive belts, belt tension, tensioners, and cam sensor components).

Tools and Materials

Service manual

Describe the vehicle being worked on:

Year _____ Make _____ Model _____
VIN _____ Engine type and size _____

Procedure

Using an assigned engine and the proper service manual, you will identify and perform the required steps to inspect the valvetrain timing chain or belt.

1. Does your assigned engine use a timing belt or a timing chain? _____

2. What page of the service manual covers the procedure for checking timing belt or chain wear? _____

3. What is the specification for chain or belt deflection? _____

4. What is the measured deflection? _____

5. If a chain and sprocket system is used on your engine, inspect the camshaft and crankshaft sprockets for wear and damage. Record your findings below.
   _____
   _____

6. If a timing belt is used on your engine, inspect it for cracks and separations. Note any defects below.
   _____
   _____

7. If your engine uses a timing belt, what type of belt tensioner is used?
   Hydraulic ☐   Mechanical ☐

8. Inspect the tensioner, and note any defects below.

_____

_____

9. Based on your inspection, what is your recommendation?

_____

_____

_____

Instructor's Response _____

_____

_____

JOB SHEET

Name _____ Date _____

OHV VALVETRAIN TIMING

Upon completion of this job sheet, you should be able to time the valvetrain on an OHV engine properly.

ASE Correlation

This job sheet is related to ASE Engine Repair Test's content area: Cylinder Head and Valvetrain Diagnosis and Repair; tasks: Time camshaft(s) to crankshaft.

Tools and Materials

Service manual
Torque wrench
OHV engine

Describe the vehicle being worked on:

Year _____ Make _____ Model _____
VIN _____ Engine type and size _____

Procedure

Task Completed

Using an assigned engine and the proper service manual, you will identify and perform the required steps to time the valvetrain on an OHV engine.

1. What page of the service manual covers the procedure for setting valve timing?
   _____

2. Align the timing marks so the engine is at TDC. How do you know when TDC is obtained?
   _____
   _____

3. Following the service manual procedure, remove the timing chain and sprockets. ☐

4. Install the new chain over both the camshaft sprocket and crankshaft sprocket. ☐

5. Align the pip marks to their proper location. ☐

6. Following the service manual procedure; install the chain and sprockets. ☐

7. What is the torque specification for the camshaft sprocket bolt? _____

Instructor's Response _____
_____
_____

Name _____ Date _____

SOHC Valvetrain Timing

Upon completion of this job sheet, you should be able to time the valvetrain on an SOHC engine properly.

ASE Correlation

This job sheet is related to ASE Engine Repair Test's content area: Cylinder Head and Valvetrain Diagnosis and Repair; tasks: Time camshaft(s) to crankshaft.

Tools and Materials

Service manual
Torque wrench
SOHC engine

Describe the vehicle being worked on:

Year _____ Make _____ Model _____
VIN _____ Engine type and size _____

Procedure

Task Completed

Using an assigned engine and the proper service manual, you will identify and perform the required steps to time the valvetrain on an SOHC engine.

1. What page of the service manual covers the procedure for setting valve timing?
   _____

2. Align the timing marks so the engine is at TDC. How do you know when TDC is obtained?
   _____
   _____

3. Following the service manual procedure, release the belt tension. ☐

4. Following the service manual procedure, remove the timing belt. ☐

5. While preventing the camshaft from rotating, remove the camshaft sprocket. ☐

6. While preventing the crankshaft from rotating, remove the crankshaft sprocket. List any special tools required to remove this sprocket. ☐
   _____
   _____

7. Install the camshaft sprocket. What is the torque specification for the sprocket bolt?
   _____
   _____

8. Install the crankshaft sprocket. List any special tools required to install the sprocket.

_____

_____

☐

9. Following the service manual procedure, install the timing belt.

10. Briefly describe the procedure for adjusting belt tension.

_____

_____

_____

_____

_____

Instructor's Response _____

_____

_____

Name _____ Date _____

VARIABLE VALVE TIMING

Upon completion of this job sheet, you should be able to understand the procedure for service of a timing mechanism on a DOHC engine with variable valve timing.

Tools and Materials

DOHC VVT engine
Service information

Describe the vehicle being worked on:

Year _____ Make _____ Model _____

VIN _____ Engine type and size _____

Procedure

1. Using a suitable form of service information, locate a late-model engine that uses a DOHC engine with variable valve timing.

2. In a few sentences, describe the basic operation of the system.

_____
_____
_____
_____
_____
_____
_____
_____

3. Read through the service procedure. List any special tools required to perform a replacement of the timing chain.

_____
_____

4. Briefly outline the steps to remove the timing chains(s).

1. _____
2. _____
3. _____
4. _____
5. _____
6. _____
7. _____
8. _____
9. _____

10. _____

11. _____

12. _____

13. _____

14. _____

5. Describe the procedure to time the camshaft sprockets, intermediate shafts, and crankshaft sprockets. Use a diagram if needed.

_____

_____

_____

_____

_____

_____

_____

_____

_____

6. What is the torque specification for the crankshaft sprocket? _____

Instructor's Response _____

_____

_____

Chapter 12

INSPECTING, DISASSEMBLING, AND SERVICING THE CYLINDER BLOCK ASSEMBLY

UPON COMPLETION AND REVIEW OF THIS CHAPTER, YOU SHOULD BE ABLE TO:

- Properly disassemble the engine block.
- Clean the engine assembly and individual components.
- Perform a complete visual inspection of the cylinder block and determine needed repairs.
- Properly measure the cylinder block for deck warpage and other damage.
- Measure and evaluate the cylinder bores for taper and out-of-round.
- Properly refinish cylinder bores.
- Measure the main bore for wear or misalignment; determine needed repairs.

- Perform a thorough visual inspection of the crankshaft and determine needed repairs.
- Accurately measure crankshaft warpage.
- Measure the journals, seal diameters, flange, and vibration dampener mating surface of the crankshaft for wear, and determine needed repairs.
- Remove core plugs and gallery plugs for engine cleaning.
- Properly remove and inspect the harmonic balancer.
- Inspect the flywheel and determine needed service.
- Evaluate the engine block to determine if repairs or replacement would be more economical.

ENGINE DISASSEMBLY

The engine disassembly procedure is an important part of your engine repairs. Work like a detective, carefully inspecting each component as it is removed. You need to pay attention to all the clues to ensure that you find every fault and each potential problem during your service. You should only have to remove an engine once to make all the necessary repairs. Speed is not the most important factor when disassembling the engine. It will come apart easily, but your organization is what will allow you to put it back together efficiently as well. Separate and label bolts and mark brackets. Write down problems as you see them so you will remember to repair them and so you can refer to them as you evaluate the extent of the damage. Save the old parts and gaskets until the overhaul is complete. You may need to compare new parts to the old parts, and the customer may want to look at damaged components.

When you are replacing an engine, it is easiest to have the replacement engine next to the old one. That will make it clear which components need to be transferred to the new engine and will show you how they are mounted.

ENGINE PREPARATION

Make sure the engine is well secured on the engine stand (Figure 12-1). Before beginning the engine disassembly, clean the outside of the engine well enough that you will not be getting dirt or grime into it. Make a clean work area for yourself with enough room to lay parts

SERVICE TIP:
The best engine repair technician I have met took some ribbing for how much time he spent disassembling the engine. He carefully marked each set of bolts or threaded them lightly back into place; everything was very well organized. Not once in the years I worked with him did he have an engine come back for as much as an oil leak or a rattle.

FIGURE 12-1 This high-mileage, late-model, turbocharged engine is already in need of an overhaul.

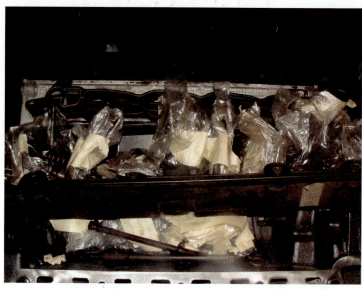

FIGURE 12-2 Separate and label the bolts to help yourself with reassembly. Many parts such as lifters and rocker arms should be reinstalled in their original locations if they are to be reused.

out for examination and proper storage. Many parts if reused will have to be installed back into their original location. Have coffee cans, plastic containers with lids, or plastic bags available for storing the many sets of bolts you will remove. Use masking tape and a permanent marker for labeling (Figure 12-2). Do not mix bolts from the oil pan, valve cover, timing cover, rockers, and main and rod caps into one big bin. Digging through a pile of bolts trying to find the correct ones is not an efficient or reliable way to reassemble an engine. Spray the exhaust manifold studs, the crankshaft snout to **harmonic balancer** seam, and any other rusty bolts with penetrating oil and allow it to soak while you proceed. Follow the manufacturer's procedure for complete disassembly instructions. It will provide torque sequences for removing major components such as the head, main caps, and intake manifold.

The **harmonic balancer** attaches to the front of the crankshaft and reduces torsional vibrations.

CYLINDER HEAD REMOVAL

You can remove the cylinder head with the engine either in or out of the vehicle. The procedure is similar with the engine installed. You may have to remove some of the hoses, connections, and cables that we discussed in engine removal to gain access to the valve cover(s) and head(s). Be sure the engine is cool when loosening intake or head bolts to prevent warpage. The following is a general outline of the procedure to remove a cylinder head:

1. Remove the intake manifold using the reverse of the tightening sequence (unless a loosening sequence is provided). You can usually leave the injectors, fuel rail, and pressure regulator attached to the manifold. On some V engines, you will have to remove the pushrods before you can lift the manifold out (see step number 5 below).
2. Remove the exhaust manifold(s). In some areas of North America, the exhaust studs and nuts may be quite rusty (Figure 12-3). You will usually replace the studs, but it is easier to remove them if they are not broken. Many times it is safest to use some heat on the nuts before attempting to loosen them.
3. Remove the valve cover(s).

FIGURE 12-3 These exhaust manifold nuts will require heat to loosen them. Replace the studs while the cylinder head is off.

FIGURE 12-4 This lifter removing tool can help remove sticky lifters without damaging the lifter or the bore.

4. If the engine has an overhead cam, you will have to remove the timing belt or chain. Rotate the engine to TDC number 1 and locate the timing marks on the camshaft(s) pulley(s). Make notes about the TDC markings to facilitate installation. Some manufacturers call for removal of the timing cover to access the tensioner. On other vehicles, it is possible to remove the tensioner from the outside of the cover. Release the tension on the chain or belt. Remove the camshaft sprocket from the camshaft and inspect it for wear.

5. On an engine with the camshaft in the block, loosen the rocker arms and remove the pushrods. It is a good idea to mark their position so they can be installed in the same location. Inspect the pushrods for bends or wear.

6. Remove the lifters from their bores. If you might reuse them, mark their correct location with a marker or masking tape; they should be returned to their original bore and contact the same lobe on the camshaft. The lifters can be sticky in their bores; a special tool can help remove them (Figure 12-4).

7. If you are working on a V engine, mark the cylinder heads left and right. Left and right are determined by looking at the engine from the flywheel end.

8. Loosen the head bolts using the loosening sequence. If one is not specified, use the reverse of the tightening sequence (Figure 12-5). Check for any differences in the bolts, and if they differ, mark their proper locations. If the bolts are to be reused, clean them and inspect the threads for damage. On many newer engines, the manufacturers use torque-to-yield bolts and specify replacement after one use.

9. Knock the cylinder head(s) with a plastic dead blow hammer to try to loosen it. You may have to use a pry bar to release the head. Be careful not to damage the sealing surface of the head. If the head does not pry off easily, check to be sure all the bolts are removed. Use the service information to locate all the cylinder head bolts to prevent damaging the head.

10. Remove the cylinder head and gasket. Carefully inspect the gasket for signs of leakage or detonation. Detonation will often distort the steel fire ring on the gasket that surrounds the cylinder bore.

SERVICE TIP:
One time a new technician had a V6 engine almost fully put together after an overhaul. He started putting the brackets and accessories back on, but two of the brackets would not bolt up. On one head, there were no mounting holes showing at all. After a period of frustration, he realized the heads were installed on the wrong sides.

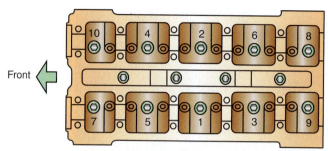

FIGURE 12-5 Reverse this tightening sequence to loosen the head bolts.

TIMING MECHANISM DISASSEMBLY

To uncover the timing mechanism, you will have to disassemble the front end of the engine, including the timing cover, any remaining accessories or brackets, and the harmonic balancer. Remember to mark the position and bolts for brackets and accessories.

Removal of the harmonic balancer can be a challenging project. This straightforward operation has turned many simple repairs into a daylong exercise. Improper removal procedures can lead to crankshaft damage and the need for expensive replacement. In most cases, a front pulley is attached to the harmonic balancer to drive the accessory belt(s) (Figure 12-6). Remove the bolt and washer that retains the harmonic balancer (Figure 12-7). The harmonic balancer can be designed as a slip fit or an **interference fit.** Slip-fit balancers are simply removed by sliding (or gently prying) the balancer off. A key or flat area on the crankshaft prevents rotation of the pulley on the crankshaft. Most harmonic balancers are interference or press fit onto the end of the crankshaft. A bolt may also be used, and it may be necessary to use a flywheel holding tool to keep the engine from spinning. Use a puller to remove the harmonic balancer (Figure 12-8). It is important to place a spacer between the threaded end of the puller and the crankshaft. If you fail to do this, the puller threads can thread into the crankshaft and seriously damage it. Usually there are holes threaded in the end of the balancer, so you can mount the puller to the balancer and then push on the crankshaft. Sometimes you will have to use jaws on the outside of the balancer (Figure 12-9). In either case, be sure to protect the internal threads of the crankshaft. Install a spacer button on the end of the crankshaft so the puller cannot thread into the crank.

Interference fit components are slightly smaller than the shaft they fit onto or the bore they fit into. This makes a puller or a press necessary for disassembly and reassembly.

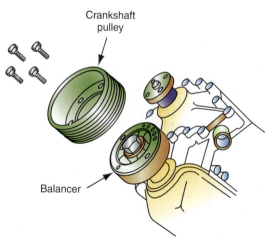

FIGURE 12-6 Removing the crankshaft pulley.

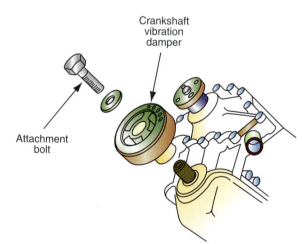

FIGURE 12-7 Remove the attaching bolt before pulling the balancer off.

FIGURE 12-8 The right puller makes short work of pulling the balancer off safely.

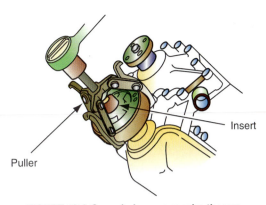

FIGURE 12-9 Some balancers require the use of a jaw-type puller.

With the balancer off, remove the timing cover. Be sure to mark where bolts go when they are different lengths; you may want to insert them into the proper hole of the cover and tape them into place. If the oil pump is mounted on the front of the crankshaft, remove it and set it aside for evaluation later (Figure 12-10).

You may have to remove brackets, accessories, or the water pump to access the timing cover (Figures 12-11 and 12-12). Before disassembling the timing mechanism, rotate the engine to TDC number 1 compression so that the timing marks line up (Figure 12-13). It is helpful to see how the marks should line up when the timing is correct. Check the chain or gears for excess slack or backlash. Remove the tensioner and guides if applicable. Remove the camshaft or crankshaft sprocket to release the chain or belt (Figure 12-14).

SERVICE TIP:
Many late-model overhead camshaft engines have complex timing chain mechanisms and variable cam timing components that require special service tools and procedures. Refer to the manufacturer's specific procedures before disassembling the timing mechanism.

FIGURE 12-10 This crank-driven oil pump must be removed before the bottom end can be disassembled.

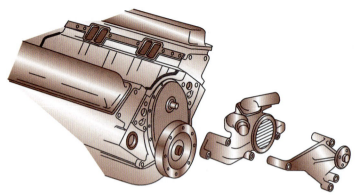

FIGURE 12-11 Removing the water pump to access the timing cover.

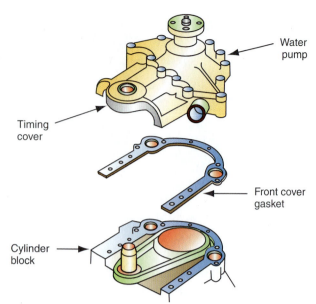

FIGURE 12-12 In some cases, it is faster to remove the water pump and timing cover together. If the job is a thorough overhaul, it is customary to replace the water pump anyway.

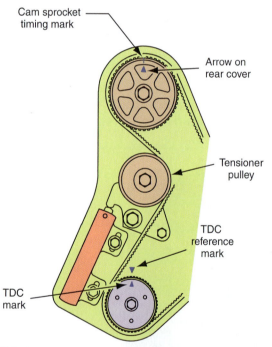

FIGURE 12-13 Align the timing marks before removing the timing belt so you'll know how they align during reassembly.

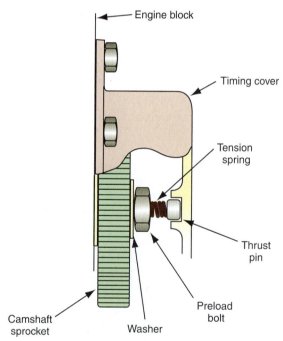

FIGURE 12-14 A thrust pin may be used to keep the camshaft from walking in the block.

You may need to use a puller to remove the crankshaft sprocket (Figure 12-15). You may be able to pry it off, but do not use a hammer; you will chip the teeth. If the engine has timing gears, remove them now. Carefully inspect the timing mechanism and sprockets to determine whether they should be reused. When performing a thorough overhaul on a high-mileage engine, you should generally replace the timing components. Record your recommendation.

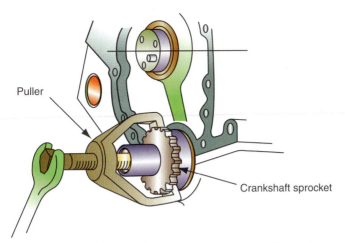

Puller

Crankshaft sprocket

FIGURE 12-15 Using a puller to remove the crankshaft sprocket.

ENGINE BLOCK DISASSEMBLY

Proceed methodically through the block disassembly. There are several special procedures and critical inspections to perform. Keep your eyes open and your mind on the job; pay attention to details. Be careful not to cause any damage as you disassemble the engine.

The engine should still be upright. Remove the oil pan using the reverse of the tightening sequence or a diagonal pattern around the pan. Many engine stands have drain pans that rest on their frame so the engine oil will be contained. If not, place some rags underneath the engine to keep your work area safe.

Removing the Camshaft

If you are working on an OHV engine, remove the valve lifters and pushrods if this has not already been done. The lifters will normally be replaced during a thorough overhaul, but you should organize the lifters and pushrods so they could be replaced in their original positions. Remove the camshaft retainer or thrust plate before removing the camshaft carefully from the block (Figure 12-16). You may need to use an **impact screwdriver** on countersunk Phillips head screws. Use care to remove the camshaft; banging it against the block can chip the lobes and require replacement. Set the camshaft safely aside for later examination.

> An **impact screwdriver** twists the screw as you hit the end of the driver with a hammer. It works very well to remove tightly fastened, rusty, or recessed screws.

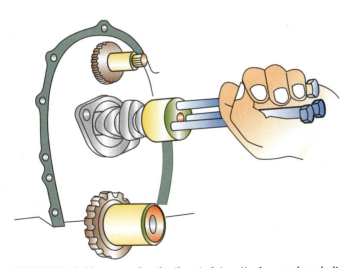

FIGURE 12-16 After removing the thrust plate, attach some long bolts into the camshaft to give yourself leverage to pull it straight out.

Ring Ridge Removal

When the engine is running, the top ring does not reach the top of the cylinder bore. It is constantly cleaning and scraping most of the cylinder wall, but at the very top it is pushing carbon up onto a ridge. Remove this **ring ridge** before attempting to get the pistons out. If the pistons are driven up past the ring ridge, they can easily break. Also, if new rings and bearings are installed without removing the ridge, the top ring can hit the bottom of the ring ridge and snap the land or break the ring. Either result would be disastrous after a fresh rebuild. If you can feel the ridge with your fingernail, remove it before extracting the pistons.

Move the piston down toward bottom dead center and place the ring ridge remover on top of the piston (Figure 12-17). Tighten the top bolt down to center and tighten the tool in the cylinder. Adjust the cutting bit so that it touches the cylinder wall just below the ridge. Then rotate the nut and cutting tip up and out through the top of the cylinder. Do not take a second pass; you could remove too much material from the cylinder wall. Repeat for each of the cylinders.

Piston Removal

Before removing the pistons, check the connecting rod caps for markings. They may be numbered or have punch marks on both halves of the cap. If they are not marked, use a paint stick or correction fluid to mark the rods on the parting faces of the cap halves. Number one is at the front (harmonic balancer) end of the engine (Figure 12-18). If the rod

> A **ring ridge** develops at the top of the cylinder that the rings scrape and wear down. Carbon also builds up in this area to create a smaller diameter region of the cylinder bore.

FIGURE 12-17 Turn the ridge reamer in a clockwise direction to thread the cutters up the cylinder to remove the ridge.

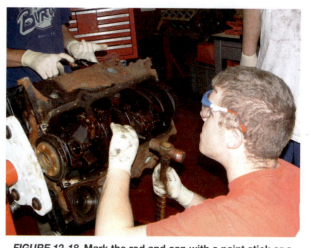

FIGURE 12-18 Mark the rod and cap with a paint stick or a number punch. The caps must be returned to the same rod and bolted in the same direction.

FIGURE 12-19 Protect the crankshaft and cylinder walls by using rubber hose on the ends of the connecting rod bolts.

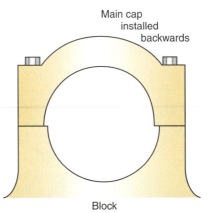

Main cap installed backwards

Block

FIGURE 12-20 With the main caps on backwards, the main bore is no longer round. This can cause premature bearing wear or can seize the crankshaft.

caps are installed on the wrong connecting rods or even backwards on the same rod, the bore will not be perfectly round. This will cause crankshaft and bearing damage as the crank is forced to turn through a distorted bore.

Remove one connecting rod cap. You may need to tap the rod bolts and side of the cap lightly with a brass hammer to loosen the caps. Place a piece of vacuum line over the rod bolts so that they don't scrape the crankshaft or cylinder wall as you tap the piston out (Figure 12-19). Use a plastic mallet to drive the piston out. You will be replacing the bearings, so it is appropriate to tap on the bearing half in the rod. Don't let the piston drop; pull it out the rest of the way once its head is out of the top of the block. Leave the bearing halves in so you keep them with the correct connecting rod for inspection. Lightly thread the cap back on the rod and store in a safe place where the pistons will not get scratched. Remove all the pistons.

Crankshaft Removal

Once the connecting rods and pistons have been removed, the crankshaft can come out. The main bore in which the crankshaft lies was drilled out with one long boring bar. This is called line boring. As a result of this process, the caps and block will form round bores only if the caps are installed in the same spot and in the same direction. If a main cap is put on backwards, it will cause the crank to seize (Figure 12-20). Check the main caps for markings that indicate number and direction. If there are none, use a center punch or number punches to mark the block and caps.

Loosen the main caps in two stages in the reverse of the tightening sequence. In the first stage, turn each main cap bolt about one turn to loosen it. Then follow the proper sequence again and fully loosen the bolts (Figure 12-21). Remove the main caps, leaving the bearings installed for a thorough inspection. If the bearings are in their proper spots, you will be able to see if there is any pattern to the wear. You may be able to pick out something as serious as a twisted main bore or a bent crankshaft by looking at the bearings. Remove the crankshaft and store it on end to prevent it from bending or sagging.

Main bearing cap bolt loosening sequence

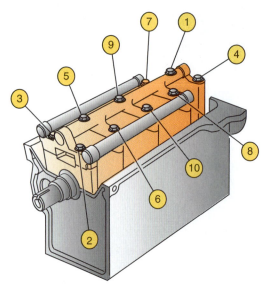

FIGURE 12-21 Follow the recommended sequence for removing the main bearing cap bolts; this will prevent crankshaft warpage.

Core and Oil Gallery Plugs

Many times you will send the engine block out to a machine shop to have it thoroughly cleaned and perhaps have the cylinders bored. Technicians in the machine shop will then remove the oil gallery plugs and clean the passages. They will also remove and replace the freeze plugs on any old or high-mileage engine. Your shop may also choose to perform these operations in house if properly equipped.

To remove core plugs, you can use a punch or chisel to hit on one side of the plugs. Be careful not to drive the plug into the block. With the plug turned out of the block, grab it with pliers or vise grips and pull it out of the block (Figure 12-22). Alternately, thread a sheet metal screw into the center of the plug. Put vise grips on the screw and tap or pull it out with the plug attached. Be sure to lightly sand the corrosion out of the sealing surface of the bore with emery cloth before installing new core plugs.

Oil gallery plugs have tapered threads that seal very tightly. Though you would think they'd be constantly lubricated with oil, the reality is that they can become quite rusty and difficult to remove. Use the proper tool to remove them; this may be a square plug, an Allen

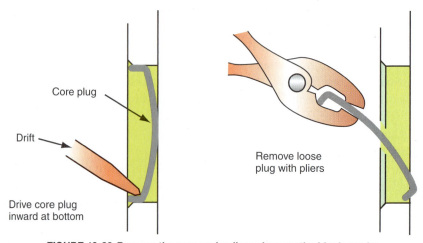

Core plug

Drift

Remove loose plug with pliers

Drive core plug inward at bottom

FIGURE 12-22 Remove the core and gallery plugs so the block can be thoroughly cleaned.

head, or an eight-point socket. If they are particularly difficult, heat them with a torch and drop a little wax on them and try again. It may help to use an impact gun; just be sure to apply plenty of force against the plug to prevent stripping the head. Use stiff bristle, long gallery brushes and hot soapy water to remove all sludge and debris from the oil galleries. Blow the passages out with pressurized air to get any residual particles out and to dry the galleries.

A thorough cleaning of the block and all components is an essential step in engine repair. Any residual contamination left in the engine can lead to premature failure. Proper cleaning is required during the rebuilding process (Figure 12-23).

With the engine fully disassembled and the parts well organized, you will be able to inspect each of the components thoroughly. With this information, you will be able to make appropriate repair decisions. It is common to see significant damage to the cylinder bores and crankshaft, for example, and to determine during disassembly that engine replacement would offer a more efficient and cost-effective resolution (Figure 12-24).

With the engine block disassembled, it should be thoroughly cleaned and inspected. Before cleaning the block, remove the core and gallery plugs. After the block is cleaned, coat it with a water-repellant solution. This is important in the cylinder and journal bores to prevent the formation of surface rust.

This chapter discusses the process of inspecting the block casting, main bearing bores, camshaft bores, cylinder bores, piston assemblies, and crankshaft. Keep a record of the needed parts and machining procedures required before reassembling the engine. Also, this chapter covers the procedures for reconditioning the components of the cylinder block assembly.

FIGURE 12-23 The block is thoroughly cleaned in a hot tank before inspection and machining.

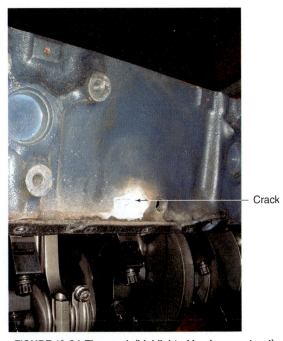

FIGURE 12-24 The crack (highlighted by dye penetrant) in this racing engine will be pinned to save the block.

SPECIAL TOOLS

Precision
straightedge
Feeler gauges

Magnafluxing is
a crack detection
method that can be
used on any ferrous
metal component.
Magnets are placed
at either end of the
component, and
liquid with metal
filings are poured
over the component.
If there is a crack,
the filings line up
across the crack
and are highlighted
under a black light.

The **deck** is the top
of the engine block
where the cylinder
head is attached.

INSPECTING THE CYLINDER BLOCK

Begin block inspection by first giving it a thorough visual inspection. Inspect the bores and journals for wear patterns and the entire casting for cracks. Look for cracks especially in the cylinder walls and in the lifter valley. If an engine had a coolant consumption problem or if it was overheated, use a special procedure to check for cracks that cannot be detected visually. Cast iron blocks can be checked by **magnafluxing.** To magnaflux the block you apply magnets at either end of the casting and spread a solution on the block that contains metal filings. If there is a crack the filings will line up on either side of it, attracted by the opposite charges across the crack. To check for a crack on an aluminum block pour a special dye penetrant over the block. If there is a crack the dye will settle into it and highlight the crack under a black light so you can see it. You may have to send the block out to the machine shop to have these procedures performed (Figure 12-24). Also, use this time to clean and inspect all oil passages. Some of these passages can be quite small and easily plugged (Figure 12-25). Use a small bore brush or a piece of wire to clean all oil passages.

If the block passes visual inspection, it must be checked for deck warpage, cylinder wall wear, crankshaft saddle alignment, camshaft bore wear, and lifter bore wear.

Checking for Deck Warpage

Visually inspect the **deck** for scoring, corrosion, cracks, and nicks. If a scratch in the deck is deep enough to catch your fingernail as you run it across the surface, the deck needs to be resurfaced.

Measure deck warpage using a precision straightedge and feeler gauge. To obtain correct results, the deck must be completely clean. Check for warpage across the four edges and across the center in three directions (Figure 12-26). The amount of warpage is determined by the thickest feeler gauge that will fit between the deck and the straightedge (Figure 12-27). Compare the results with specifications. The deck can be resurfaced if the cylinder block dimensions will still be within specifications after machining (Figure 12-28). Even if the deck

FIGURE 12-25 Make sure all oil passages are
thoroughly cleaned.

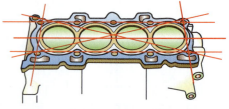

FIGURE 12-26 Measure the deck in several
directions to determine the amount of
warpage.

FIGURE 12-27 Clean the deck thoroughly and then use a straightedge and feeler gauges to measure for warpage. If the deck is not flat, the head gasket will not seal.

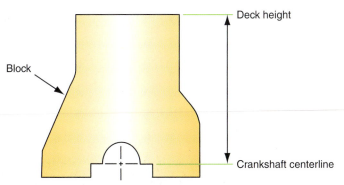

FIGURE 12-28 The deck may be resurfaced if the overall block height will still be within specifications.

is within specifications, it is a good practice to put a new surface finish on the deck so the gasket will seal properly. This can be done by using a light cut on the milling machine or using a special disc to clean the deck.

If the deck is warped, it is important to determine how much material the manufacturer will allow to be removed. If more material will have to be removed than the manufacturer allows, the block will have to be replaced or a thicker head gasket will have to be used (if available). Removing more material than allowed may result in piston-to-valve contact. In addition, removing stock from the deck surfaces may affect intake manifold bolt alignment. V-type engines require both decks to be machined the same amount to keep the compression ratio equal on both sides.

Inspecting and Measuring Cylinder Wall Wear

After visually inspecting the cylinder bores, use a dial bore gauge, an inside micrometer, or a telescoping gauge to measure the bore diameter. First, check to see if the cylinders have been bored to an oversize on a previous rebuild. Oversize pistons usually have a stamped number on the piston head to indicate the size. If the piston has no numbers, bore oversize can be checked by measuring the cylinder diameter near the bottom of the bore. Since this is a nonwear area, if the bore is larger than original specifications, the cylinder has been oversized. This information is important when ordering new pistons and rings. Common piston oversizes are 0.020, 0.030, 0.040, and 0.060 inch. Metric oversizes are in 0.50 mm increments.

Piston movement in the cylinder bore produces uneven wear throughout the cylinder. The cylinder wears the most at 90 degrees to the piston pin on the thrust sides of the cylinder walls. Wear also occurs most at the top of ring travel, where combustion pressures and temperatures are hottest and there is the least lubrication. Wear decreases toward the bottom of the cylinder, resulting in taper.

Taper in the cylinder bore causes the piston ring gaps to change as the piston travels in the bore. Out-of-round wear is caused by the thrust forces exerted by the piston. Another cause of out-of-round wear is gasoline washing away the oil film from the cylinder walls during cold engine operation or other high-fuel conditions.

Normally, the most cylinder wear occurs at the top of the ring travel area. Pressure on the top ring is at a peak and lubrication at a minimum when the piston is at the top of its stroke. A ridge of unworn material will remain above the upper limit of ring travel.

SPECIAL TOOLS
Dial bore gauge
Micrometer
Telescoping gauge

Taper is the difference in diameter at different locations in a bore or on a journal.

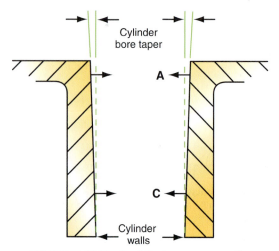

FIGURE 12-29 To check for taper, measure the cylinder diameter at A and C. The difference between the two readings is the amount of taper.

FIGURE 12-30 A dial bore gauge makes measuring taper quick work.

Below the ring travel area, wear is negligible because only the piston skirt contacts the cylinder wall.

A properly reconditioned cylinder must have the correct diameter, have no taper or out-of-roundness, and the surface finish must be such that the piston rings will seat to form a seal that will control oil and minimize blowby.

Taper is the difference in diameter between the bottom of the cylinder bore and the top of the bore just below the ridge (Figure 12-29). Subtracting the smaller diameter from the larger one gives the cylinder taper. Using a dial bore gauge is the easiest way to measure for taper (Figure 12-30). Set the gauge to zero near the bottom of ring travel and drag the gauge up to the top of ring travel. The reading on the dial is the amount of taper. Repeat the procedure a few times to be sure your readings are consistent. Often, some taper is permissible. Later model, 1990s and newer engines may specify as little as 0.002 in. (0.0508 mm) while many older engines allow up to 0.006 in. (0.1524 mm). The only way to be certain is to check the manufacturer's specifications for the engine you are working on. If an engine has too much taper, it must be bored. Some taper is permissible, but normally not more than 0.006 in. (0.1524 mm). If the taper is less than that, reboring the cylinder is not necessary.

Cylinder **out-of-roundness** is the difference between the cylinder's diameter when measured parallel with the crank and then perpendicular to the crank (Figure 12-31). Often, a cylinder bore is checked for out-of-roundness with a dial bore gauge; however, a telescoping gauge can also be used (Figure 12-32).

Measure the cylinder near the top of ring travel on the thrust sides and then on the nonthrust sides. Compare the readings on the dial of a bore gauge; the difference is out-of-round.

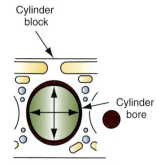

FIGURE 12-31 Measure the bore diameters on the thrust and nonthrust sides.

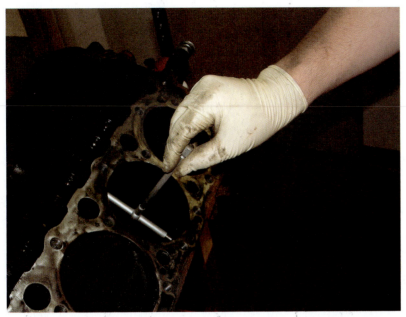

FIGURE 12-32 Measure the bore diameters just below the ring ridge. The difference between the two readings is the amount of cylinder out-of-round.

When using a telescoping gauge and micrometer, subtract the diameter of the nonthrust side from the diameter of the thrust side. This gives you the measurement of out-of-round. Look at the example below:

	Cylinder #1	Cylinder #2	Cylinder #3	Cylinder #4
Thrust Diameter	3.656"	3.656"	3.656"	3.657"
Nonthrust Diameter	−3.651"	−3.652"	−3.650"	−3.651"
Out-of-Round	0.005"	0.004"	0.006"	0.006"

The original diameter of the cylinders is 3.650 in. Compare your measurements to the specified diameter; the engine could have been bored oversize. You would need to know this when ordering parts. This engine has cylinder out-of-round between 0.004 in. and 0.006 in. On an older engine, this may be just barely acceptable. Many newer engines specify a maximum of 0.001 in. (0.0254 mm) to 0.002 in. (0.0508 mm) of out-of-round. Some newer engines allow for absolutely no out-of-round or taper, while some older engines may allow up to 0.005 in. (0.127 mm). It is important to check the specifications for the engine you are working on to make a good repair decision. If the out-of-round is beyond the allowable limit, the engine should be bored to produce a powerful and long-lasting engine. When new rings are installed in a cylinder with too much out-of-round blowby, compression loss and oil consumption can result. When an engine requires boring, new oversized pistons will have to be fitted. Given today's trend toward replacement, these findings may dictate engine replacement rather than overhaul.

Checking Main Bearing Bore Alignment

Over the life of the engine, the main bearing bores can become misaligned (Figure 12-33). The main bearings will usually compensate for this by wearing unevenly; however, if new main bearings are installed into an engine with the bores misaligned, the crankshaft will have increased resistance to turning. If excessive misalignment is not corrected, the new main bearings will fail prematurely. Upon inspection of the old main bearings, it may be

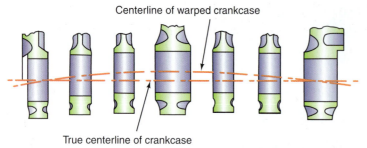

Centerline of warped crankcase

True centerline of crankcase

FIGURE 12-33 The crankcase can warp, causing saddle misalignment.

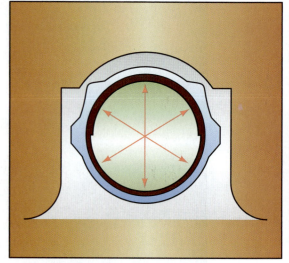

FIGURE 12-34 Measure the main bore in different locations to check for out-of-round.

possible to determine if the bores are misaligned. A warped crankcase will result in bearing wear on one side of the insert.

Warpage of the main bearing bores can be checked with a precision straightedge and a feeler gauge. The straightedge must be longer than the length of the engine block. With the main bearing inserts removed, install the bearing caps and torque the bolts to specifications. This stresses the crankcase to provide accurate measurements. Place the straightedge into the bore. Then select a feeler gauge half the thickness of the maximum oil clearance. If the feeler gauge fits under the straightedge on any of the saddles, the block is warped. Follow Photo Sequence 23 for the procedures on performing this measurement. Repeat this procedure at two other parallel positions in the bores.

A second method is to use an arbor ground to 0.001 in. (0.025 mm) less diameter than the minimum saddle bore specification. The arbor must be long enough to fit into all bores simultaneously. Place the arbor into the saddles with the bearings removed and install the bearing caps. Torque the bolts to specifications. Attempt to rotate the arbor using a foot-long bar. If it will not turn, the crankcase is warped.

Another method is to coat the crankshaft main bearing journals with Prussian blue and install the crankshaft into the saddles. New bearings are installed for this method. With the bearing caps installed and torqued, rotate the crankshaft two revolutions. Then turn the engine upside-down and rotate the crankshaft through two additional rotations. This ensures that the weight of the crankshaft will fall on both bearing halves. Remove the crankshaft from the block, being careful to lift it straight up. The areas of contact on the bearing will be blue. Acceptable alignment is indicated when 75 percent or more of the bearing is blue.

If the amount of warpage is not excessive, it can be corrected. This is done by **line boring** the saddles and installing the appropriate oversize bearings.

Engine Block Bore Measurements

SPECIAL TOOLS

Dial bore gauge
Micrometer
Telescoping gauge
Small-bore gauge

All bores of the engine block should be measured to determine size, taper, and out-of-round. This includes the main bearing, camshaft, and lifter bores. In particular, the main bearing bores are susceptible to wear. When a main bearing becomes excessively hot, the bearing bore size decreases. Bearing bores can be checked with a telescoping gauge, an inside micrometer, or a dial bore gauge. When measuring main bearing bores, the caps must be installed and the bolts properly torqued. Check bore diameter in three directions (Figure 12-34). The vertical measurement should not be larger than any of the others. A larger reading in the vertical direction indicates the bore is stretched. Out-of-round measurements of less than 0.001 in.

TYPICAL PROCEDURE FOR CHECKING MAIN BORE ALIGNMENT

P23-1 This engine uses a single piece main cap or "bedplate" to hold the crankshaft. Loosen the bedplate or the main caps to begin the check of the main bore alignment.

P23-2 Lift the bedplate or main caps off and remove the crankshaft.

P23-3 The block mating surface must be clean and the main bearings are removed.

P23-4 Use a plastic mallet if the bearings come out hard. Keep them in order so you can look for patterns of wear that can indicate problems.

P23-5 Torque the bedplate or main caps back on in sequence and to specification.

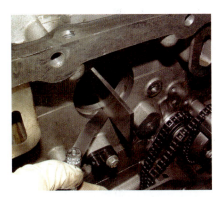

P23-6 Lay a machined straightedge the length of the bore and check for any warpage. Generally no more than 0.001 in. is allowed but as always check the specifications.

P23-7 Check in the center of the main bore as well.

P23-8 Flip the engine over to check the lower half of the bores. A misaligned bore can cause premature bearing wear or even a seized crankshaft.

(0.025 mm) are acceptable, provided the vertical reading is not the largest. If the bore measurements are out of specifications, the block will require line boring before assembly. An engine that requires line boring is a good candidate for replacement.

Check the camshaft bores in the same manner. If the wear is out of limits, the bores can be honed to accept oversized bearings, if available. Visually inspect the lifter bores for scoring or wear. Clean the bores with a stiff brush to remove any varnish. Measure the diameter of the bores and compare to specifications. If necessary, the lifter bores can be honed to a standard oversize to fit oversize lifters if available.

INSPECTING THE CRANKSHAFT

As the pistons are forced downward on the power stroke, pressure is applied to a crankshaft, causing it to rotate. The crankshaft transmits this torque to the drivetrain, and ultimately to the drive wheels. These pressures and rotational forces eventually cause wear and stress on the crankshaft. The lack of proper lubrication or the addition of abrasives greatly accelerates this wear. Before reusing the crankshaft, a thorough inspection of suspect areas is required. These areas include main bearing journals, connecting rod journals, fillets, thrust surfaces, oil passages, counterweights, seal surfaces, flange, and vibration dampener journal. Generally, crankshaft inspection begins with a thorough visual inspection, followed by checking for warpage and measuring the journals for excessive wear. An additional check is to inspect for stress cracks using magnafluxing (Figure 12-35). Provided the wear is not outside specified limits, the crankshaft can be reconditioned.

First, visually inspect the crankshaft for obvious wear and damage. This includes inspecting the threads at the front of the crankshaft, the keyways, and the **pilot bushing** bore. When inspecting the journals, run your fingernail across their surfaces to feel for nicks and scratches. If a journal is scored, it will have to be **polished** before accurate micrometer readings can be obtained. Remember to inspect the area around the fillet very closely. Stress cracks can develop in this area (Figure 12-36). If the crankshaft is going to be used under extreme conditions, or if there are any doubts concerning cracks, magnaflux the crankshaft.

Clean the oil passages by running a length of wire or a small bore brush through them, followed by spray cleaner (such as spray carburetor cleaner). Inspect the passage openings. There cannot be any nicks in these areas. If there is a rise in the metal around the hole, it will have to be removed by polishing or grinding the journal.

The **pilot bushings** or pilot bearings are pressed into the centerline of the crankshaft at the back. They are used to support the front of the transmission's input shaft.

Polishing is the process of removing light roughness from the journals by using a fine emery cloth. Polishing can be used to remove minor scoring of the journals that does not require grinding.

FIGURE 12-35 Despite severe damage to the journals under the hoop, this $1,400 racing crank will be repaired if it is not cracked.

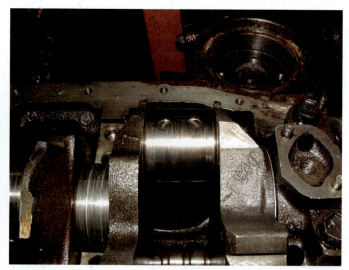

FIGURE 12-36 This scored crank should be polished, if not ground or replaced. If reused, it should be checked carefully for cracks in the fillets and probably magnafluxed.

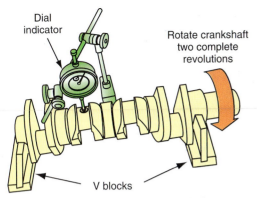

Dial indicator

Rotate crankshaft two complete revolutions

V blocks

FIGURE 12-37 A setup to check crankshaft warpage.

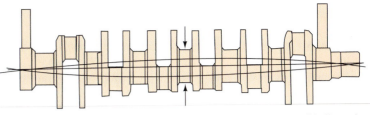

FIGURE 12-38 The actual amount of warpage is half of the total indicated reading.

Checking Crankshaft Warpage

The loads placed on the crankshaft can cause it to warp into a bow shape. If a warped crankshaft is installed with new bearings, the bearings will fail prematurely. Always check crankshaft straightness before returning it to service. If the crankshaft is out of specifications, it should be replaced; however, some machine shops use a special press to straighten the crankshaft.

To check crankshaft warpage, place the crankshaft into two V-blocks or between lathe pins so they support the end main bearing journals (Figure 12-37). Install a dial indicator on the middle main bearing journal with the plunger positioned in the three o'clock position on the journal. Rotate the crankshaft one revolution while observing the dial indicator. The amount of crankshaft warpage is 50 percent of the **total indicator reading (TIR)** (Figure 12-38). Remember, any journal out-of-round will affect the TIR. If specifications are not available, a general rule is that TIR should not exceed half of the minimum bearing clearance specification.

Record the indicator readings, and mark the high spots on the journal. The highest of the runout readings indicates the point of the bend. Marking the points of highest runout on each journal provides a general plane of the bend. All of the marks will usually line up. If they do not, this indicates multiple bends in the shaft.

While the crankshaft is set up in the V-blocks, move the dial indicator to measure the runout of the vibration dampener journal. Next, measure the runout of the flywheel flange. Compare these results to specifications.

Measuring Crankshaft Journals

Main and connecting rod journals are machined to very close tolerances and require thorough inspection. Use an outside micrometer to measure the diameter, taper, and out-of-round of each journal (Figure 12-39). Measure each journal at two or three locations and in two different directions (90 degrees apart) at each location (Figure 12-40). Compare the measurements with specifications.

If any of the main bearing journals are out of specifications, all main bearing journals should be ground to the next undersize. This ensures that the journals are on the same centerline. Another alternative is to build up the crankshaft journal using special welding techniques, then grinding the journal to its original size.

The lateral movement of the crankshaft is controlled by a thrust bearing and a thrust surface on the crankshaft (Figure 12-41). Inspect the thrust surface for indications of excessive contact and wear. If the thrust surface is worn, causing excessive end play, the crankshaft will have to be replaced, the surface built up, or an undercut radius machined into the journal (Figure 12-42).

SERVICE TIP:
A crack can sometimes be detected by "ringing" the counterweights with a light tap from a hammer. A dull sound indicates a crack is present.

SERVICE TIP:
Excessive torsional vibrations usually result in a crack near the number one piston connecting rod journal. Causes of excess vibration include an unbalanced crankshaft, a defective vibration damper, a wrong flywheel, a damaged or improper converter drive plate, or an improperly balanced torque converter.

SPECIAL TOOLS
V-blocks
Dial indicator

The total movement of the dial indicator (above and below zero) is the **total indicator reading (TIR)**.

SERVICE TIP:
If V-blocks are not available, it is possible to measure crankshaft warpage by installing the front and rear upper bearing inserts, then placing the crank shaft into the block. Set a dial indicator to measure the middle journal. This procedure will only be accurate if the crankshaft saddles are straight.

FIGURE 12-39 Measure the crankshaft journals to be sure the crank is within specifications and can be reused.

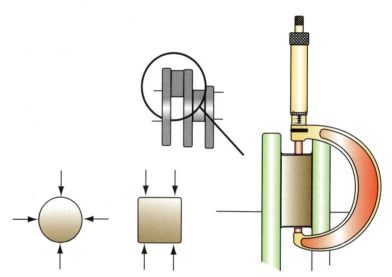

FIGURE 12-40 Measuring the journals for wear, taper, and out-of-round.

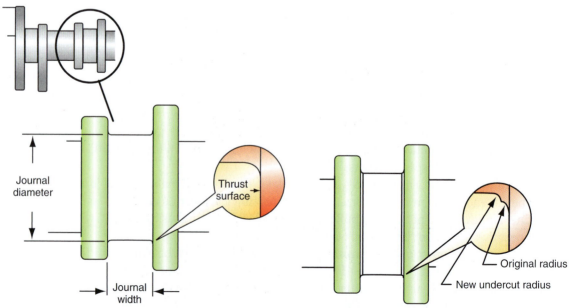

Journal diameter

Thrust surface

Journal width

FIGURE 12-41 Crankshaft thrust is controlled by a thrust bearing and a thrust surface on the crankshaft.

Original radius

New undercut radius

FIGURE 12-42 Undercut radius restores the thrust surface.

INSPECTING THE HARMONIC BALANCER AND FLYWHEEL

The harmonic balancer (vibration dampener) is inspected for signs of wear in its center bore. Also inspect the rubber mounting for indications of twisting and deterioration. Replace any harmonic balancer that has slipped.

If the flex plate or flywheel is to be reused, inspect it closely for stress or heat cracks. Flex plates generally crack around the mounting hole area. Flywheels can develop cracks as a result of overheating because of a slipping clutch. If the cracks are deep or the flywheel is excessively blue, replace the flywheel.

Flywheels with light scoring and surface check cracks can be resurfaced (Figure 12-43). In addition, warpage can be removed during the resurfacing procedure. The clutch disc must have a flat surface to ride against. If the flywheel is warped, the contact surface for the disc is reduced, causing it to chatter on engagement or slip.

Flywheels and flex plates also contain the starter ring gear. Inspect the gear for damage. If a flex plate ring gear is damaged, replace the flex plate. Most flywheel ring gears can be removed from the flywheel and a new one can be installed.

SERVICING THE CYLINDER BLOCK ASSEMBLY

The cylinder block assembly may require some machining operations to assure proper sealing and performance. Although not all machining operations listed in this section will have to be performed on all engine block assemblies, it is important for today's technician to be aware of the machining options available. This section discusses the procedures for performing some of the most popular methods of reconditioning the engine block, crankshaft, connecting rods, and pistons. If you are working in a general repair facility, you will have many of these processes performed at a machine shop.

It is not the intent of this text to replace the operating manuals for the equipment being used. The procedures performed in this manual are general procedures and are designed to familiarize you with the procedures. Always refer to the equipment manufacturer's instructions and the engine service manual for proper machining procedures.

Line Boring and Align Honing

The main bearing bores can be resized and realigned by line boring or align honing. Line boring is performed if the bearing bores are warped or are excessively worn, while align

FIGURE 12-43 This milling machine can be used to resurface cylinder heads, block decks, or flywheels.

FIGURE 12-44 Set the honing mandrel so the centering pins are resting on the center saddle.

SPECIAL TOOLS

Line boring machine
Dial bore gauge

Align honing is used to make small corrections to the bores.

Cylinder boring makes it possible to bore cylinders to accept over-size pistons.

honing is used to make small corrections to the bores. The main purposes of these operations are to be sure that all bores are straight, have the correct diameter, provide the correct bearing crush, and provide proper heat transfer.

Line boring or align honing the main bearing bores is usually the first machining operation performed on the engine block since many of the other operations require the main bearing bores to be lined and sized. Line boring and align honing resize and realign the main bearing bores by slightly changing their location in the crankcase.

The first procedure to be performed when reconditioning the main bearing bores is to remove metal from the parting surface of the bearing cap. After the proper amount of metal is removed from the main bearing caps, these caps are installed on the engine block, and the retaining bolts are tightened to the specified torque. A line hone (Figure 12-44) or a line boring machine is then used to bore the main bearing bores so they are properly aligned. The main bearing bores are usually bored to their original diameter. **Align honing** or line boring main bearing bores is normally done in an automotive machine shop, not in an automotive repair shop.

Resurfacing the Deck

Resurfacing the deck gives the cylinder head gasket a flat, machined surface to seal against. Even if the deck is not warped, a finish surface should be cut. A minimum amount of material should be removed to provide a new surface finish for proper gasket sealing.

This operation is usually performed after the main bearing bores are machined and before cylinder boring. This is because many boring bars and indexing plates are mounted to the deck surface. The exception is if the cylinders are being sleeved. In this case, the sleeves are installed first, then the deck is surfaced.

Generally, there are three methods used to recondition the deck surface of the engine block: grinding, broaching, and milling. The same milling machine shown in Figure 12-43 can be used to resurface cylinder decks. Resurfacing the block deck is usually done in an automotive machine shop, not in an automotive repair shop.

Cylinder Reconditioning

The cylinder must be the proper diameter for the piston, be free of taper, and have no out-of-round. In addition, the surface finish must allow the rings to seal. Depending on the amount of wear, cylinder wall reconditioning can be performed using one of the following methods:

- Deglazing
- **Cylinder boring**
- Honing
- Sleeving the cylinder

Before reconditioning the cylinder bores, use an abrasive wheel to cut a chamfer at the top of the cylinder. Make sure the angle of the chamfer does not cut into the area of upper ring travel.

The finished cylinder wall surface should have a crosshatch pattern. It is the proper crosshatch that assists in sealing the piston rings. The crosshatch leaves thousands of small diamond-shaped reservoirs for oil and maintains the oil film for the rings to glide on.

After the cylinders are reconditioned, they must be thoroughly cleaned. Use hot soapy water and a stiff brush to remove all of the residue. Rinse and dry the block, then lightly coat each cylinder with engine oil. Do not use a solvent tank or steam cleaner to clean the cylinders. You must get into each bore and remove all traces of grit and metal particles. If any grit or metal shavings are left in the bores, the rings will be damaged.

Deglazing. If the original crosshatch is still visible, clean the varnish using lacquer thinner. Achieving the surface finish required by today's engines is impossible using the old methods of glaze breaking. If the crosshatch is not present, hone the cylinder with a rigid hone.

Cylinder Boring. If it is determined that the cylinders are worn beyond specifications, it may be possible to bore the cylinders to accept oversize pistons. This operation is possible only if the cylinder bore diameter is not beyond the maximum allowed. If the bore is too large, the cylinder wall is thinner than required to maintain its shape and to prevent cracks. Whenever a cylinder is bored, the pistons will have to be replaced with the correct oversize. If any of the cylinders require boring, all of the cylinders should be bored to the same oversize to keep cylinder output equal.

Use the inspection notes to determine the oversize of the bores required to clean up the worst cylinder. Refer to the parts supplier's catalog and engine service manual to determine available oversize pistons. The cylinders should be bored to the smallest oversize piston that will clean the worst cylinder bore. All cylinders will be bored to this oversize.

It is a good practice to match the pistons to the bore. The pistons will have some differences in size due to manufacturing tolerances. Measure the exact size of each of the replacement pistons. Then determine the desired finished size of the cylinder and how much will be bored, leaving 0.003 to 0.005 in. (0.075 to 0.125 mm) for finishing by honing.

When setting up the boring machine, there are some special considerations. First, centering the cutting bit is vital. Center is not based on the center of the original bore. Instead, the bit must be centered to the crankshaft centerline. One method of centering the bit is to use an indexing plate attached to the deck of the block (Figure 12-45). The indexing plate

SPECIAL TOOLS
Boring bar
Dial bore gauge
Indexing plate
Torque plate

Blueprinting is a technique of building an engine using stricter tolerances than those used by most manufacturers. This results in a smoother running, longer lasting, and higher output engine.

FIGURE 12-45 An indexing plate is used to center the boring bar.

aligns the cutting bit to bore the cylinders based on the position of the head bolts. This method is used by many high-performance machine shops. Another method is to center the cutting bit using the bottom of the cylinder where no wear has occurred.

Once the boring bar is located at the centering location, turn the control knob to expand the centering fingers. The bar may have three or four fingers. The fingers contact the indexing plate or cylinder wall and are pushed out tightly. With the bar centered, clamp the machine to the engine block by inserting an anchor assembly through the cylinder adjacent to the one being bored. Next, raise the boring bar out of the cylinder.

Another consideration is that the boring machine must be square to the engine block. If the machine is not square, the cylinders will not be parallel. Also, keep the cutting bit dressed and lapped.

Install the cutting bit into the tool holder using a micrometer that has been set to the desired dimension of the cylinder. Fit the tool holder into the boring bar head and adjust it to the required setting using a special boring bar micrometer. The set screw locks the tool holder into the head.

Before cutting the bore, set the feed stop to prevent the boring bit from going past the bottom of the cylinder. Finally, set the spindle speed and feed rate. The settings used will depend upon the type of machine used, the type of material the block is constructed of, and the type of bit used. The boring machine manufacturer provides guidelines to use in determining speed and feed rate.

Finally, turn on the motor and engage the feed lever. The cutting bit will work its way down the cylinder as it cuts the bore. When the bore bar reaches the bottom of its travel, turn off the motor. Remove or relocate the cutting bit so the bore bar can be raised out of the cylinder without damaging the new cylinder wall surface. Some machines use a long cutting bit for cutting a chamfer at the top of the cylinder. If a chamfer was not already cut, do so now. Check the bottoms of the cylinders for chamfer. Some chamfer should remain. A sharp edge at the bottom of the cylinder can scrape oil off of the piston skirt.

Honing Cylinder Walls. A rigid hone can be used to correct slight taper or out-of-round conditions in the cylinder. Honing is also recommended to bring the cylinder bore to its final size after boring. This is done to remove the very rough surface the boring bar leaves (Figure 12-46).

FIGURE 12-46 Notice the torque plates bolted to the deck to prevent the honing process from distorting the cylinders.

FIGURE 12-47 The plateaus provide a bearing-type surface for the rings to glide over.

Cylinder honing can be performed using manual or automatic feed machines. Follow Photo Sequence 24 for general cylinder honing operations using a manual honing machine. When honing a cylinder to size, set the stop so the stones are stroked just above the deck. Do not allow the stone to come out of the top of the cylinder or a taper will be cut.

If a manual machine is used, stroke the honing head with a fast and steady up-and-down movement to achieve the correct crosshatch pattern. On either system, allow the stone to run free of drag after the bore has been honed to its desired size. This results in desirable plateaus that act as bearing surfaces for the rings.

After cylinders are honed, they must be thoroughly cleaned. Use hot soapy water and a stiff nonmetallic brush to remove all of the residue. Rinse and dry the block; then lightly coat each cylinder with engine oil.

Plateau Honing. Plateau honing removes the peaks left by normal honing operations (Figure 12-47). With the peaks flattened out, the rings will seal and wear in faster. You will need to vary the machine settings to match the specific bore diameter you are working with. Set the spindle speed and strokes per minute to achieve a crosshatch angle between 30 and 40 degrees. Start with a 200- or 220-grit stone. Hone the cylinder to within 0.0005 in. (0.013 mm) of the finished size. Change to a 280-grit stone, and leave the machine's spindle speed, stroke rate, and feed the same as when finished with the previous stone. Use this stone to bring the cylinder bore to its final size.

Leaving the machine at the same settings, install a plateau honing tool, soft honing tool, or 600-grit stone. Run the machine for about 45 seconds to remove the peaks. You should always consult the service information for your particular engine, as the surface finish requirements can vary. Different types of rings will also require different stone grits. Many machinists use the Automotive Engine Rebuilders Association (AERA) publications to locate up-to-date engine machining specifications that are not easily found in traditional service information sources.

> **CUSTOMER CARE:** It is in the best interest of the customer to do a little research to make sure that high-quality rings are fitted into cylinders with the correct surface finish. Opting for the cheapest and easiest option may not provide the level or length of performance that she expects.

Sleeve Installation

Many aluminum and some cast-iron blocks use replaceable liners in their cylinders (Figure 12-48). **Dry sleeves** are thin-walled liners that are pressed into the block, while wet sleeves are thicker and are held in place by supporting flanges in the block. Each type of sleeve is serviced differently. In addition, dry sleeves can be used to restore worn cylinders.

Replacing Dry Sleeves. Dry sleeves are pressed into the block with 0.001 to 0.002 in. (0.025 to 0.050 mm) interference. To remove the sleeve, it is bored to make its walls very thin. The thin shell is then pried away from the block and lifted out. The new sleeve is installed by lubricating the outer diameter with high-pressure lubricant, then pressing it or driving it into the block (Figure 12-49). The sleeve is longer than the cylinder, requiring the remainder of the sleeve to be cut off using a hacksaw or a cutoff bit. The deck is machined

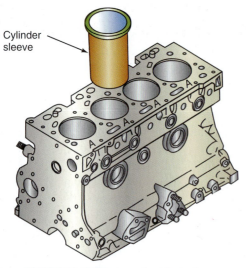

FIGURE 12-48 Some engine blocks have replaceable sleeves. Liners can also be used to correct excessive wear problems in a cylinder.

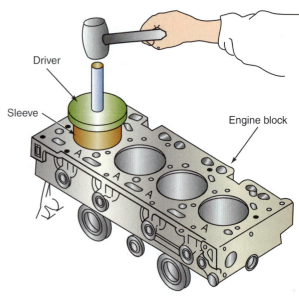

FIGURE 12-49 The sleeve is driven into the cylinder bore with an interference fit. A cut-down axle works well as a driver.

SPECIAL TOOLS

Boring bar
Dial bore gauge
Torque plate
Sleeve driver
Rigid hone

Wet sleeves are thicker sleeves constructed to support the forces of combustion; coolant jackets surround them.

Sleeving is the process of boring the cylinder to accept a sleeve. Sleeving is used to restore the cylinder bore to its original size.

after the sleeves are installed to blend the top of the sleeve with the deck. The sleeve bore diameter is smaller than a standard bore, so it must be bored and honed to original size.

Replacing Wet Sleeves. Many engines originally fitted with **wet sleeves** call for replacement of any sleeves that have any scoring, taper, or out-of-round. Wet sleeves are supported in the block only at the top and bottom of the cylinder. Engine coolant surrounds the sleeve and is kept out of the crankcase by O-rings. To remove wet sleeves, the supporting flanges in the block are bored out. A special puller may be required to remove the sleeve (Figure 12-50). The puller is fitted inside the sleeve and then expanded so it is tight against the walls. Some pullers use a slide hammer attachment to force the sleeve out.

Sleeving. Even engine blocks that are not equipped with sleeves may be repaired using a dry sleeve. This is usually done if one or two of the cylinders are worn or damaged to the extent that boring cannot be done, or when the expense of boring the cylinders is too high. Sleeving also provides a method of repairing a cylinder that is cracked. Dry sleeves are typically installed as an alternative to boring all the cylinders and installing oversize pistons when a low mileage engine has suffered a catastrophic failure in one or two cylinders. The other cylinders would need to have very little wear, or boring all the cylinders would make

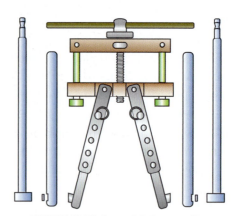

FIGURE 12-50 A special sleeve puller.

TYPICAL PROCEDURE FOR CYLINDER HONING

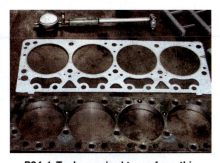

P24-1 Tools required to perform this task include a variety of hone stones, a torque plate, a new head gasket, bolts to attach the torque plate, a honing machine, a setting gauge, and a bore gauge.

P24-2 Install and torque the main bearing caps to the crankcase. This stresses the lower end of the engine block.

P24-3 Place a new head gasket on the block deck and install the torque plate. Make sure the bolts used to attach the plate go the same distance into the block as the original cylinder head bolts.

P24-4 Install the engine block into the honing machine and level it.

P24-5 Locate the setting gauge in the cylinder to be bored and center it. Once it is centered, tighten the knob, and move the graduated slide to zero.

P24-6 Select the appropriate stone to rough cut the cylinder to within 0.0005 inch of the finished size. A stone grit of 150 to 220 is usually used.

P24-7 Place one of the stones on the setting gauge and add shims to remove any slack. A stone that is properly shimmed will slide in and out of the setting gauge easily. Both stones are shimmed using the same thickness.

P24-8 Place one of the guides in the setting gauge to determine the size of the shim required. Select the second guide shim in the same manner.

P24-9 Install the stones into positions 1 and 2 of the honing head. The main guide goes into position 3, and the centering guide into position 4.

TYPICAL PROCEDURE FOR CYLINDER HONING

P24-10 With the stones fully retracted, insert the honing head into the cylinder. Adjust the cradle so the stones protrude about 3/8 inch out from the top of the cylinder.

P24-11 Select the proper stroke length, spindle speed, and stroking rate.

P24-12 Use the feed wheel to expand the stones gradually. The stones should be just snug in the cylinder.

P24-13 Turn on the motor and adjust the coolant flow over the work.

P24-14 Engage the clutch to begin the stroking of the hone head.

P24-15 Hone for about 20 to 30 seconds, then disengage the clutch and measure the bore. Once the bore diameter is to within 0.0005 inch of the desired finished diameter, change stones to provide a finer cut. The finish stone used will depend upon the type of piston rings selected.

If sleeves are to be installed, perform this operation before resurfacing the block deck.

Note that the amount to be bored is over standard, not the amount to be removed from the existing cylinder bore size.

the most sense. An example of a situation when installing a dry sleeve would be advisable is if an engine with 34,000 miles on it had a connecting rod come loose and it severely scored just that cylinder. To install one sleeve on the low-mileage engine would be much less expensive than boring all the cylinders and purchasing new oversize pistons.

Most sleeving operations are performed using dry sleeves that are pressed into the block with a 0.001 to 0.002 inch (0.025 to 0.050 mm) interference. A general rule of thumb is 0.0005 in. (0.013 mm) interference per inch (25.4 mm) of cylinder bore diameter. Dry sleeves are available in 3/32 or 1/8 in. (2.5 or 3.0 mm) wall thickness. The cylinder is sleeved by boring it to a size 0.002 inch (0.050 mm) less than the outside diameter of the sleeve. If any finer adjustment is required to the cylinder, use a hone. The sleeve may not be perfectly round. Measure it in three directions at the top and bottom, then add the six measurements together. Divide the sum of the measurements by six to get the average outside diameter (Figure 12-51).

Generally, a 3/32 in. (2.5 mm) sleeve requires the cylinder wall to be bored about 0.1875 inch (4.8 mm) over the standard size. For the 1/8 in. (3 mm) sleeve, the cylinder will need to be bored about 0.250 in. (6.5 mm) over standard. When the cylinder is bored, leave

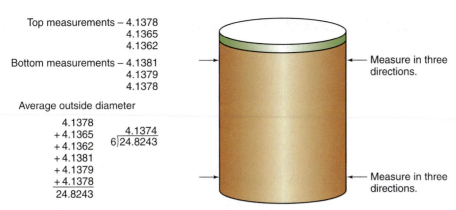

Top measurements –	4.1378
	4.1365
	4.1362
Bottom measurements –	4.1381
	4.1379
	4.1378

Average outside diameter

```
   4.1378
 + 4.1365      4.1374
 + 4.1362    6)24.8243
 + 4.1381
 + 4.1379
 + 4.1378
  24.8243
```

Measure in three directions.

Measure in three directions.

FIGURE 12-51 Determining the outside diameter of the new sleeve.

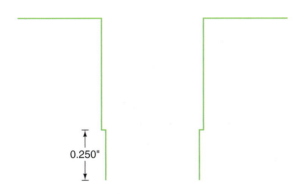

0.250"

FIGURE 12-52 A step is left at the bottom of the bore so the sleeve cannot drop into the crankcase.

a step about 0.250 inch (6.5 mm) from the bottom to prevent the sleeve from pushing out (Figure 12-52). Chamfer the top of the cylinder to assist in sleeve installation. In addition, cut a chamfer at the bottom of the sleeve.

Coat the sleeve with high pressure lubricant, then drive it into the cylinder. Cut off any excess sleeve with a hacksaw or cutoff bit. After all sleeves are installed into the block, machine the deck to blend the top of the sleeve with the deck.

Since pressing the sleeves into the block distorts the cylinders, install all sleeves before honing any of the bores. If only some of the cylinders are to be sleeved, install the sleeves before machining any of the other cylinders. The inside diameter of the sleeve is smaller than a standard bore. After all the sleeves are installed, they can be bored and honed to the desired size using the methods already discussed. Match the piston to the cylinder to achieve the best results.

Lifter Bore Repair

To complete the machining of a pushrod (OHV) engine block, the lifter bores are reconditioned. The lifters must be able to rotate in the bores. If they cannot, the bottom of the lifter will wear along with the camshaft lobe.

Use a small glaze breaker or hone to remove any burrs or glazing inside the bore. Be careful not to enlarge the bore; only clean it up. If the bore is too large, oil will leak past the lifters and accelerate wear. It is also possible to remove varnish from the lifter bore with a small brush and lacquer thinner. If the bores are excessively damaged, some manufacturers and aftermarket suppliers provide oversize lifters.

SERVICE TIP:
Thermal heating the block to about 200°F (93°C) and freezing the sleeve before pressing the sleeve will make installation easier.

Classroom Manual
Chapter 12

SPECIAL TOOLS
Small glaze breaker or hone

CAMSHAFT AND BALANCE SHAFT BEARING SERVICE

Camshaft Bearing Measurement, Removal, and Replacement

SERVICE TIP:

If the block has oil grooves around the camshaft bearing bores, install the camshaft bearings with the oil hole in the bearing at the 3 o'clock position. This position allows oil to enter the bearing easily as the camshaft rotates.

SERVICE TIP:

The camshaft bearings should require moderate force to drive them into the bearing bores. If the bearings can be installed easily, the bearing may have the wrong outside diameter.

Worn camshaft bearings cause low oil pressure. During an engine rebuild, the camshaft bearings are usually replaced. To determine the camshaft bearing clearance, measure the camshaft bearing diameter with an inside micrometer or a telescoping gauge, and measure the matching camshaft journal with a micrometer. Subtract the two readings to obtain the bearing clearance. If the bearing clearance is more than specified, the camshaft bearings must be replaced.

A special camshaft bearing removal and replacement tool is used to remove and replace the camshaft bearings. This tool consists of a guide cone (1), driving washers (2 or 3), expander bearing drivers (4–8), driver bars (9 or 10), expander jaws (11), expander sleeve (12), expander cone (13), expander shaft (14), and expander assembly (15) (Figure 12-53). The expansion plug in the rear camshaft bearing bore located in the rear of the block must be removed before camshaft bearing removal and replacement. Use the proper expander/driver, driver bar, and a hammer to remove the front and rear camshaft bearings (Figure 12-54). Some engine manufacturers recommend the use of a special puller to remove all the camshaft bearings. Always follow the instructions in the vehicle manufacturer's service manual. Use the proper driver bar, expander/driver, and expander cone to remove the center camshaft bearing (Figure 12-55). The centering cone centers the driver so the camshaft bearing is driven straight out of the bore.

Be sure the camshaft bearing bores are clean and smooth prior to installing the camshaft bearings. In some engine blocks, the camshaft bearings are different sizes, with the largest bearing at the front of the block and the bearings progressively smaller toward the rear of the block. Be sure to install the camshaft bearings in the proper bores. Use the proper driving tools and centering cone to install the center camshaft bearing. Be sure the oil hole in the bearing is aligned with the oil passage in the block on all the camshaft bearings. If these oil holes are not aligned, the camshaft bearing does not receive adequate lubrication, and the bearing will fail very quickly. When the bearing fails because of lack of lubrication, the camshaft journal will be severely scored, so the camshaft must be replaced.

Install the front and rear camshaft bearings with the proper driving tools. Some front camshaft bearings are installed past the front surface of the bearing bore to provide oil flow from this bearing bore to the timing chain and gears.

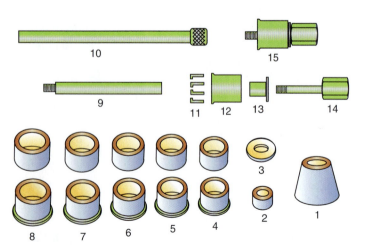

FIGURE 12-53 Camshaft bearing puller and driver with attachments.

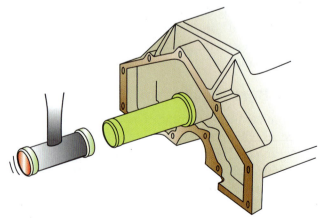

FIGURE 12-54 Removing or replacing front or rear camshaft bearings.

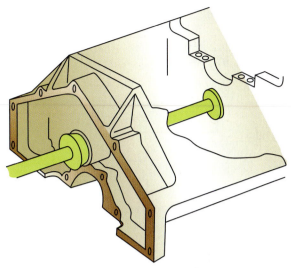

FIGURE 12-55 Removing or replacing a center camshaft bearing.

Balance Shaft Bearing Measurement, Removal, and Replacement

Use a C-clamp to mount a dial indicator above the rear of the balance shaft (Figure 12-56). Position the dial indicator stem against the balance shaft, and zero the dial indicator. Grasp the balance shaft and pull upward and push downward on this shaft while observing the shaft to rear bearing play. If this play exceeds specifications, the balance shaft bearing and/or balance shaft must be replaced. After the balance shaft is removed, measure the balance shaft journal diameter with a micrometer to determine if this journal is worn. Use the C-clamp to mount the dial indicator near the front of the balance shaft, and position the dial indicator stem against the surface of the balance shaft. Zero the dial indicator. Grasp the balance shaft and pull upward and push downward on the balance shaft while observing the shaft to block play on the dial indicator. If this play is excessive, the balance shaft requires replacement, because the balance shaft journal directly contacts the block bore. After the balance shaft is removed, measure the front block bore to determine if the block bore is worn. Since there is no load on the balance shaft, wear in the front balance shaft block bore is unlikely.

Use a C-clamp to mount the dial indicator on the front of the block so the dial indicator stem may be positioned against the retaining bolt in the front of the balance shaft (Figure 12-57). Grasp the balance shaft and move it forward and rearward while observing the end play on the dial indicator. If the end play is excessive, the balance shaft retainer may be worn.

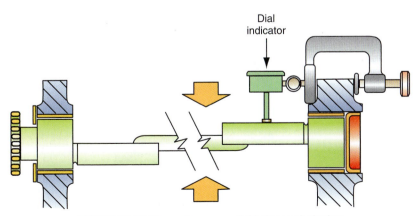

FIGURE 12-56 Measuring rear radial balance shaft play.

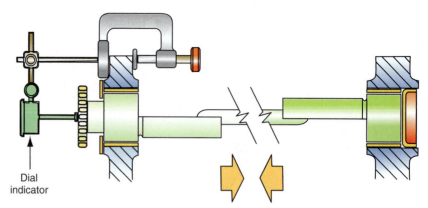

FIGURE 12-57 Measuring balance shaft end play.

Position the dial indicator so the stem is contacting one of the balance shaft gear teeth, and zero the dial indicator (Figure 12-58). Grasp the balance shaft and try to turn it back and forth while observing the gear lash on the dial indicator. If the gear lash is excessive, the balance shaft drive and driven gears must be replaced. Use a special puller to remove the balance shaft (Figure 12-59).

Use the special removal tool to remove the rear balance shaft bearing (Figure 12-60). Be sure the rear balance shaft bearing bore is clean and smooth. A special bearing installation tool is used to install the rear balance shaft bushing (Figure 12-61). Use the proper driving tool

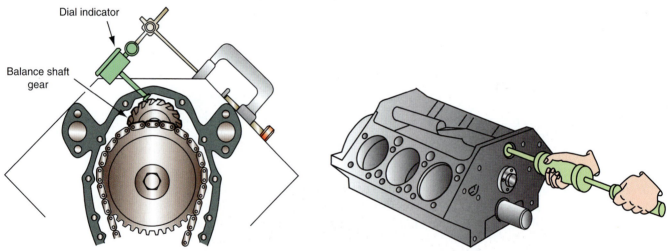

FIGURE 12-58 Measuring balance shaft gear lash.

FIGURE 12-59 Removing the balance shaft.

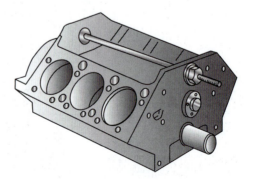

FIGURE 12-60 Removing the rear balance shaft bearing.

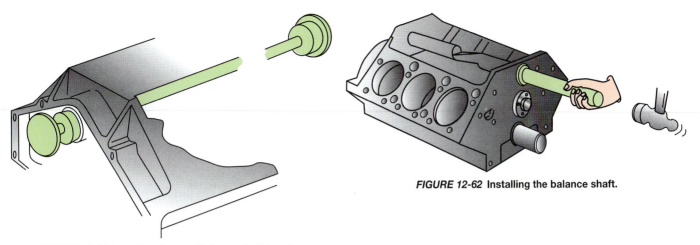

FIGURE 12-61 Installing the rear balance shaft bearing.

FIGURE 12-62 Installing the balance shaft.

to install the balance shaft (Figure 12-62). When installing the balance shaft drive and driven gears, these gears must be properly timed, or severe engine vibrations will occur.

RECONDITIONING CRANKSHAFTS

A worn crankshaft does not always require replacement, though it is often the more cost-effective option. Depending on the severity of the damage, and the material the crankshaft is constructed from, it may be reconditioned with no adverse affects. As discussed in the Classroom Manual, crankshafts are constructed from forged steel or cast iron. Forged steel crankshafts are easily repaired. Until recently, cast crankshafts were too brittle to be reconditioned, other than resizing the bearing journals. Modern materials have made cast crankshaft reconditioning possible, including straightening.

Proper interpretation of the crankshaft inspection results will provide the necessary information regarding the extent of crankshaft reconditioning required. The journal surfaces can be reconditioned to provide a smooth surface, and if necessary, they can be built up using a special welding process and then machined to restore original size. In addition, many crankshafts can be straightened; however, do not attempt repairs on a cracked crankshaft (Figure 12-63).

FIGURE 12-63 Despite this severe damage, this specially made racing crankshaft will be repaired.

When reconditioning the crankshaft, a proper order should be followed. First, if the crankshaft is warped, it must be straightened before any other machining operations are performed. Second, if the bearing journals are tapered, out-of-round, or worn, they may be ground to accept a standard undersize bearing. After the journals are ground to the desired size, they should be polished.

If the journals are not worn beyond specifications, they should be polished to provide a smooth surface.

Crankshaft Straightening

WARNING: **Always wear approved eye protection when performing this task.**

If the inspection of the crankshaft indicates excessive warpage, the shaft may be straightened. The labor costs of this operation must be balanced against the replacement cost. If the crankshaft is to be straightened, this procedure must be done before any other machining operations are performed. There are two methods for straightening the crankshaft: peening and using straightening presses. Both methods work on forged and cast shafts; however, pressing cast shafts requires extra caution. In addition, it is important to understand exactly what the operation of crankshaft straightening is. Straightening the crankshaft is done by relieving the stress tension built up in the shaft. The shaft is not bent back into alignment. Removing the stress within the crankshaft will automatically return it to its original shape. Stress is removed by striking the crankshaft in the direction of the bend.

A hammer and a rounded bronze chisel with its end dressed to fit the fillet radius of the rod journals are used for peening (Figure 12-64). Striking the journal in this location prevents damage to the bearing surface. The chisel is struck with a hard, sharp blow from the hammer.

The crankshaft straightening press applies hydraulic pressure to the bend. A dial indicator is attached to the journal to indicate the amount of overbend being applied. At this time, the chisel and hammer are used to strike the fillet radius of the bent journal in the direction of the bend. This is done to relieve stress in the shaft.

With either procedure, the crankshaft is placed into V-blocks to support the ends at the rear seal diameter and dampener mounting surface. Following is the procedure for **peening** the crankshaft.

All straightening is done at the rod journals, not the main journals. Once the TIR and location of the bend are identified and the crankshaft is properly supported, position it so the high point of runout (the marks) are facing down. Straighten the shaft starting at the journal indicating the greatest amount of TIR. Work from this journal down the shaft in both directions to remove the warpage (Figure 12-65). Locate V-blocks or clamps on the

SPECIAL TOOLS
V-blocks
Dial indicator
Bronze chisel

Peening is the process of removing stress in a metal by striking it in the direction of the bend.

Hydraulic pressure is used to assist in relieving pressure, not to straighten the shaft.

SERVICE TIP:
If the shaft has multiple bends, treat the shaft as if it were separate shafts and remove each bend individually.

FIGURE 12-64 Peening a crankshaft to remove warpage.

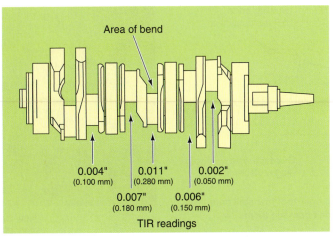

Area of bend

0.004" (0.100 mm)   0.011" (0.280 mm)   0.002" (0.050 mm)

0.007" (0.180 mm)   0.006" (0.150 mm)

TIR readings

FIGURE 12-65 **The journal with the greatest amount of TIR is the location point of the bend. Start at this location and move outward to remove the bend.**

CAUTION:
Do not attempt to straighten a crankshaft by striking it against the direction of the bend. Doing this causes the shaft to be bent back into position and weakens the shaft.

main bearing journals adjacent to the actual bend location. Frequently check the progress of the work to prevent bending the crankshaft in the opposite direction.

Journal Grinding

The importance of journal surface condition to forming the correct oil barrier between the journal and the bearing was discussed in the Classroom Manual. The manufacturer often applies a hardening treatment to the journals to protect them from wear; however, even with hardened surfaces, the bearing journals can become scored or scuffed due to improper lubrication, excessive heat, contamination, or improper installation. It may be possible to restore the journal surfaces by grinding them to a standard undersize. Grinding of the crankshaft journals is done to correct any of the following conditions:

- Out-of-round
- Taper
- Improper oil clearances
- Scratches, scoring, or nicks
- Damaged thrust surfaces

The crankshaft shown in Figure 12-36 has scoring deep enough that it should be reground or replaced.

INSTALLING OIL GALLERY PLUGS AND CORE PLUGS

After thorough cleaning, install the oil gallery plugs. Make sure the threads in the block and on the plug are clean. Apply pipe sealant to the threads and torque the plugs in to their specification. Check to be sure that each gallery hole is properly sealed.

Make sure there are no deposits or corrosion in the core plug bores. Any dirt in the bores or on the new plugs can cause leakage. Apply a hardening sealant as specified to the outside diameter of the core plug. Use a core plug tool or bearing driver to install the core plugs (Figure 12-66). The driving face should be just slightly smaller than the edges of the plug. Do not hit the core plug in the center area; this will collapse it and distort the sealing edge. Using a punch to tap the plug in will also deform the core plug and cause leaks (Figure 12-67).

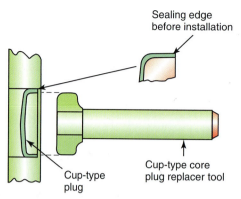

FIGURE 12-66 Use a proper tool to install core plugs so you don't damage them.

Sealing edge before installation

Cup-type plug

Cup-type core plug replacer tool

New oil gallery plug

New core plug

FIGURE 12-67 With new core plugs and oil gallery plugs installed, the block is clean and ready for reassembly.

TERMS TO KNOW

Align honing

Cylinder boring

Deck

Dry sleeves

Harmonic balancer

Impact screwdriver

Interference fit

Line boring

Magnafluxing

Out-of-roundness

Peening

Pilot bushing

Polished

Ring ridge

Sleeving

Taper

Total indicator reading

(TIR)

Wet sleeves

CASE STUDY

Inspection of a V8 engine block indicated that one cylinder was excessively scuffed and damaged. The technician researched the options available and discussed them with the customer. It was decided that the cost of a new block was not feasible, and having to bore each cylinder would require the purchase of new pistons. In this instance, it was decided the most cost-effective method of repairing the block was the installation of a sleeve. After completing all the required machining operations on the engine block, the technician was ready to turn her attention to the crankshaft. The inspection notes taken indicated one of the main bearing journals required grinding to restore its surface finish. All other main bearing journals were good. Realizing it is unusual to have only one bearing fail in this manner, the technician inspected the old bearing very closely and discovered the original bearing was undersize even though the journal was not ground. Someone had attempted to remove an engine noise by simply installing a thicker bearing to take up clearance. The extra friction caused the journal to score. In order to maintain proper crankshaft position in the block, all main bearing journals were ground to the next standard undersize, new bearings were installed, and oil clearances were checked. By taking the time to find what is in the best interest of the customer, your reputation as an honest technician will grow.

ASE-STYLE REVIEW QUESTIONS

1. *Technician A* says to reverse the tightening sequence when loosening the cylinder head.

   Technician B says to loosen the main caps starting at the front of the engine and moving toward the rear.

   Who is correct?

   A. A only   C. Both A and B

   B. B only   D. Neither A nor B

2. *Technician A* says that you can pry most harmonic balancers off with two big pry bars.

   Technician B says that you can damage the crankshaft if you don't protect the threads while using a puller.

   Who is correct?

   A. A only   C. Both A and B

   B. B only   D. Neither A nor B

3. *Technician A* says to rotate the engine to TDC number 1 before removing the timing mechanism.

   Technician B says to make a mental or written note of the location of the timing marks.

   Who is correct?

   A. A only   C. Both A and B

   B. B only   D. Neither A nor B

4. *Technician A* says the pistons should come out the top of the block.

   Technician B says to drive on the edge of the piston skirt with a punch to remove the pistons.

   Who is correct?

   A. A only   C. Both A and B

   B. B only   D. Neither A nor B

5. The crankshaft has been removed for inspection.

   Technician A says the area around the fillet is a common location for stress cracks.

   Technician B says a crack near the number one piston connecting rod journal may indicate a faulty vibration damper.

   Who is correct?

   A. A only   C. Both A and B

   B. B only   D. Neither A nor B

6. The cylinder block is ready for inspection.

   Technician A says deck warpage can be checked using a precision straightedge and feeler gauge.

   Technician B says main bearing saddle alignment can be checked with a precision straightedge and a feeler gauge.

   Who is correct?

   A. A only   C. Both A and B

   B. B only   D. Neither A nor B

7. While servicing camshaft bearings:

   A. The largest bearing is installed at the rear of the block.

   B. Worn camshaft bearings may cause a knocking noise at idle speed.

   C. A hammer and brass punch may be used to remove camshaft bearings.

   D. Worn camshaft bearings cause low oil pressure.

8. While measuring and servicing balance shaft bearings:

   A. Balance shaft end play may be measured with Plastigage.

   B. Balance shaft gear lash may be measured with a dial indicator.

   C. Balance shaft upward and downward movement may be measured with a machinist's ruler.

   D. Balance shaft bushings may be replaced with a slide hammer-type puller.

9. *Technician A* says the cylinder boring bar is centered to the crankshaft centerline.

   Technician B says an indexing plate can be used to center the boring bar.

   Who is correct?

   A. A only   C. Both A and B

   B. B only   D. Neither A nor B

10. Cylinder bore inspection is being discussed.

    Technician A says the greatest amount of wear will be toward the bottom of the ring travel area.

    Technician B says if any of the cylinders are out of specifications, all cylinders should be bored to a standard oversize.

    Who is correct?

    A. A only   C. Both A and B

    B. B only   D. Neither A nor B

ASE CHALLENGE QUESTIONS

1. If pistons are removed without removing the ring ridge, the following may result:
   A. The piston skirt may be damaged.
   B. The piston pin may be broken.
   C. The connecting rod bearings may be damaged.
   D. The piston ring lands may be broken.

2. While discussing cylinder measurement:

   Technician A says cylinder taper is the difference between the cylinder diameter at the top of the ring travel compared to the cylinder diameter at the center of the ring travel.

   Technician B says cylinder out-of-round is the difference between the axial cylinder diameter at the top of the ring travel compared to the thrust cylinder diameter at the bottom of the ring travel.

   Who is correct?
   A. A only   C. Both A and B
   B. B only   D. Neither A nor B

3. *Technician A* says that core plugs are generally replaced during a high mileage overhaul.

   Technician B says that oil galleries should be blown out with an air gun to clean them.

   Who is correct?
   A. A only   C. Both A and B
   B. B only   D. Neither A nor B

4. *Technician A* says to mark the main caps before disassembly.

   Technician B says that if main caps are installed in the wrong direction, the crankshaft could seize.

   Who is correct?
   A. A only   C. Both A and B
   B. B only   D. Neither A nor B

5. *Technician A* says that a failed harmonic balancer can cause a crankshaft to crack.

   Technician B says that a cracked crankshaft is generally welded to repair it.

   Who is correct?
   A. A only   C. Both A and B
   B. B only   D. Neither A nor B

JOB SHEET

Name _____ **Date** _____

DISASSEMBLING AN ENGINE BLOCK

Upon completion of this job sheet, you should be able to properly disassemble the bottom end of the engine.

ASE Correlation

This job sheet is related to ASE Engine Repair Test content area: Engine Block Assembly Diagnosis and Repair; tasks: Disassemble engine block; clean and prepare components for reassembly.

Inspect engine block for visible cracks, passage condition, core and gallery plug condition, and surface warpage; determine necessary action.

Tools and Materials

Hand tools
Rubber hose
Number punch or paint stick

Describe the vehicle being worked on:

Year _____ Make _____ Model _____
VIN _____ Engine type and size _____

Procedure

Task Completed

Your instructor will provide you with a specific engine. Write down the information, then perform the following tasks:

1. Locate the service information for engine disassembly and read through the instruction. ☐

2. Draw a diagram of the head bolt loosening sequence below:

3. Set the engine at TDC number 1 firing and note the markings on the camshaft if you are working on an OHC engine. Remove the cylinder heads and label all components and bolts as needed. ☐

4. Is your engine an OHV or OHC engine?

   _____

5. What sort of timing mechanism is used?

   _____

6. Remove the timing cover and observe the timing marks. Describe the marks and their positions below: _____

_____

_____

_____

7. Remove the timing chain, belt, or gears. Carefully inspect the chain, belt, or gears, the sprockets, the guides, and the tensioners. List below the condition of each of the components and whether they should be reused or replaced: _____

_____

_____

_____

_____

_____

_____

8. Feel for a ring ridge with your fingernail in each of the cylinders. Which cylinders have a noticeable ridge?

_____

☐ Use the ring ridge remover to remove the ridge before removing the pistons.

9. Are the connecting rod caps marked?

_____

If not, use a paint stick or a number punch to mark them. Then, loosen the connecting rod nuts for one cylinder, place pieces of rubber hose over the rod bolts, and drive the piston out of the top of the engine using the plastic or wooden end of the hammer on the rod. Be sure to put the rod cap back on the piston assembly. Repeat for all pistons.

10. Locate the main cap loosening sequence and draw it below:

11. Loosen the main caps in the proper sequence. Loosen each bolt one turn, then repeat
☐ the sequence to fully loosen the bolts. Set the caps aside and remove the crankshaft.

12. Visually inspect the crankshaft for scoring or cracks; describe your results:

_____

_____

_____

_____

13. Visibly inspect the block assembly for cracks, severe wear, and core and oil gallery plug condition. Describe your results: _____

_____

_____

_____

_____

14. Based on your initial inspections, what is your recommendation for further measurement and possible repair of the block assembly? _____

_____

_____

_____

Instructor's Response _____

_____

_____

Name _____ **Date** _____

Engine Block Crack Detection

Upon completion of this job sheet, you should be able to properly check a cylinder block for cracks.

ASE Correlation

This job sheet is related to ASE Engine Repair Test content area: Cylinder Head and Valvetrain Diagnosis and Repair; tasks: Inspect engine block for visible cracks, passage condition, core and gallery plug condition, and surface warpage; determine necessary action.

Tools and Materials

Service manual
Technician's tool set
Magnetic particle inspection equipment
Penetrant dye crack detection equipment

Describe the vehicle being worked on:

Year _____ Make _____ Model _____
VIN _____ Engine type and size _____

Procedure for Iron Engine Blocks

Using an assigned engine and the proper service manual, you will identify and perform the required steps to detect cracks on an iron engine block.

Task Completed

1. Check for technical service bulletins for articles related to engine crack detection. Describe what you found.

   _____

2. After cleaning the cylinder block, place the electromagnet in one area of the cylinder block gasket surface. ☐

4. Turn the electromagnet on and dust the iron powder in between the magnets. ☐

5. If a surface or near-surface crack is present, the powder will collect near the crack. Did this occur?

   _____

6. Continue to perform this testing method on the remainder of the cylinder block, paying special attention to cylinders, crankshaft saddles, core and gallery plugs, and oil/coolant passages. Describe what the results of this test were.

   _____
   _____

Procedure for Aluminum Engine Blocks

Using an assigned engine and the proper service manual, you will identify and perform the required steps to detect cracks on an iron engine block.

1. Check for technical service bulletins for articles related to engine crack detection. Describe what you found.

   _____

☐ 2. After cleaning the cylinder head using conventional methods, spray the suspected area with the supplied spray bottle of special cleaner and wipe it dry.

☐ 4. After the cleaner dries, spray the area with the penetrant and wait 5 minutes.

☐ 5. After the penetrant dries, spray the area with the developer and wait 5 minutes.

6. If a surface or near-surface crack is present, the developer will outline the crack in a different color. Use of a black light will aid in seeing any cracks. Did this occur?

   _____

6. Continue to perform this testing method on the remainder of the cylinder block, paying special attention to cylinders, crankshaft saddles, core and gallery plugs, and oil/coolant passages. Describe what the results of this test were.

   _____

   _____

Instructor's Response _____

_____

_____

Name _____ **Date** _____

MEASURING DECK WARPAGE

Upon completion of this job sheet, you should be able to properly inspect and measure the engine block deck for warpage.

ASE Correlation

This job sheet is related to the Engine Repair Test's content area: Engine Block Diagnosis and Repair; tasks: Visually inspect engine block for cracks, corrosion, passage condition, core and gallery plug holes, and surface warpage; determine necessary action.

Tools and Materials

Straightedge
Feeler gauge
Engine block

Describe the vehicle being worked on:

Year _____ Make _____ Model _____
VIN _____ Engine type and size _____

Procedure

Using the engine you have disassembled and the proper service manual, you will identify and perform the required steps to inspect and measure the engine block deck for warpage.

Task Completed

1. Specification for maximum warpage allowed. _____

2. Special tools required to perform this procedure?
   _____

3. Make sure the deck surface is clean. ☐

4. Note any damage (nicks, scratches, etc.) to the deck surface.
   _____
   _____
   _____

5. Actual warpage measurement.
   Left bank _____ Right bank _____

6. Based on your findings, what is your recommendation?

_____

_____

_____

Instructor's Response _____

_____

_____

Name _____ Date _____

Cylinder Block Thread Cleaning

Upon completion of this job sheet, you should be able to clean the block threads with a tap and die set.

ASE Correlation

This job sheet is related to ASE Engine Repair Test content area: General Engine Diagnosis; Removal and Reinstallation: Perform common fastener and thread repair, to include: remove broken bolt, restore internal and external threads, and repair internal threads with a thread insert.

Tools and Materials

Technician's tool set
Tap and die set
Thread repair kit

Describe the vehicle being worked on:

Year _____ Make _____ Model _____
VIN _____ Engine type and size _____

Procedure

_____
Task Completed

Using the engine block from the engine you have disassembled and the proper service manual, you will identify and clean the threads listed below with a tap and die set.

1. Gather the engine thread repair kit and the tap and die set. Describe the thread size and pitch for the following components.

Fastener	Thread Diameter	Thread Pitch
1. Cylinder head bolt threads in block		
2. Oil pan bolt threads in block		
3. Timing cover bolt threads in block		
4. Main bearing cap bolt threads in block		
5. Main bearing cap bolts		
6. Connecting rod bolt threads		
7. Connecting rod nuts		
8. Cylinder head bolts		
9. Rocker arm stud/shaft		

2. Clean the threads listed above by using the proper thread file, tap and/or die and an air blow gun. Make sure to blow the dirt and metal filings out before and after you clean up the threads.   ☐

3. Are any of the threads damaged beyond repair? _____

4. If yes, what repair method are you going to have to do to restore them?

_____

_____

Instructor's Response _____

_____

_____

JOB SHEET

Name _____ Date _____

INSPECTING AND MEASURING THE CYLINDER WALLS

Upon completion of this job sheet, you should be able to properly inspect and measure the cylinder walls for wear.

ASE Correlation

This job sheet is related to the Engine Repair Test's content area: Engine Block Diagnosis and Repair; tasks: Inspect and measure cylinder walls; remove cylinder wall ridges; hone and clean cylinder walls; determine need for further action.

Tools and Materials

Dial bore gauge
Engine block

Describe the vehicle being worked on:

Year _____ Make _____ Model _____
VIN _____ Engine type and size _____

Procedure

Using the block from the engine you have disassembled and the proper service manual, you will identify and perform the required steps to inspect and measure the cylinder walls for wear.

1. Do any of the cylinders have visible signs of wear or damage?   ☐ Yes   ☐ No
   If YES, which one(s)?
   _____
   _____

2. Is the cross-cut pattern still visible on all cylinder walls?   ☐ Yes   ☐ No
   If NO, on which ones is it no longer visible?
   _____
   _____

3. What is the specification for the bore? _____

4. At what location in the cylinder is the bore measurement to be taken?
   _____

5. What is the specification for the maximum amount of taper? _____

6. At what locations in the cylinder is the taper measured?
   _____

7. What is the specification for the maximum amount of out-of-round? _____
   _____

8. At what location in the cylinder is out-of-round measured?

_____

9. Measure each cylinder at the locations required, and record your results.

	Bore	Taper	Out-of-Round
Cylinder 1	_____	_____	_____
Cylinder 2	_____	_____	_____
Cylinder 3	_____	_____	_____
Cylinder 4	_____	_____	_____
Cylinder 5	_____	_____	_____
Cylinder 6	_____	_____	_____
Cylinder 7	_____	_____	_____
Cylinder 8	_____	_____	_____

☐

10. Put an * (or more than one if necessary) in the chart above to indicate any readings that are out-of-specification.

11. Have the cylinders been bored to an oversize?   ☐ Yes   ☐ No
    If YES, what size? _____

12. Based on your readings, what is your recommendation?

    _____

    _____

    _____

Instructor's Response _____

_____

_____

Name _____ Date _____

INSPECTING AND MEASURING MAIN BEARING BORE ALIGNMENT

Upon completion of this job sheet, you should be able to properly inspect and measure the main bore alignment.

ASE Correlation

This job sheet is related to the Engine Repair Test's content area: Engine Block Diagnosis and Repair; tasks: Inspect and measure main bearing bores and cap alignment and fit.

Tools and Materials

Straightedge
Feeler gauge
Engine block

Describe the vehicle being worked on:

Year _____ Make _____ Model _____
VIN _____ Engine type and size _____

Procedure

Using the block from the engine you have disassembled and the proper service manual, you will identify and perform the required steps to inspect and measure the main bores for proper alignment.

1. Do any of the main bearings show signs of wear caused by bore misalignment?
   ☐ Yes   ☐ No
   If YES, which one(s)?

   _____

   _____

2. Do any of the saddles or caps have visible signs of wear or damage?
   ☐ Yes   ☐ No
   If YES, which one(s)?

   _____

   _____

3. What is the maximum amount of bore misalignment allowed? _____

4. Describe the method used to determine bore misalignment.

   _____

   _____

5. Actual amount of bore misalignment. _____

6. Based on your findings, what is your recommendation?

_____

_____

_____

Instructor's Response _____

_____

_____

Chapter 13

SHORT BLOCK COMPONENT SERVICE AND ENGINE ASSEMBLY

BASIC TOOLS

Basic mechanics tool set
Service manual

UPON COMPLETION AND REVIEW OF THIS CHAPTER, YOU SHOULD BE ABLE TO:

- Inspect main and rod bearings and perform accurate failure analysis.
- Visually inspect and measure pistons and determine necessary repairs.
- Inspect piston pins and determine condition.
- Perform a complete inspection of the connecting rods, including big-end and small-end bores and bends and twists.
- Recondition connecting rods.
- Install press-fit and full-floating piston pins.

- Properly install piston rings, including measuring and correcting ring end gap, checking side clearance, and staggering end gaps.
- Properly install camshaft bearings and camshaft into an OHV engine.
- Correctly install the crankshaft and measure end play.
- Install main and rod bearings, and measure bearing oil clearances.
- Correctly install the piston assemblies into the engine block.
- Properly complete the engine assembly.

With the engine block disassembled, cleaned, and repaired, it is now time to inspect the internal components. The engine bearings, pistons, rings, and connecting rods must all be carefully inspected. You will reassemble the engine using new rings and bearings and any other components that have significant wear. Proper reassembly of the engine block includes several critical measurements: ring end gap and side clearance, bearing oil clearances, camshaft and crankshaft end play, and rod side clearance when applicable. Proper torque sequences and specifications are essential for professional work.

> **Classroom Manual**
> Chapter 13, page 000

INSPECTING CRANKSHAFT BEARINGS

As discussed in the Classroom Manual, the crankshaft does not rotate directly on the main or rod bearings. Instead it rides on a thin film of oil trapped between the bearing and the crankshaft. If the journals are worn or become out-of-round, tapered, or scored, the oil film will not form properly. This will result in direct contact with the bearing and eventual damage to the bearing and/or crankshaft. Soft materials are used to construct the bearings in an attempt to limit crankshaft wear. Crankshaft bearings are always softer than the metal structure they support.

When an engine is reconditioned, the main and rod bearings are replaced. However, inspection of the old bearings provides clues to the cause of an engine failure; for example, the soft material used to construct the bearings allows impurities to embed into it. Excessive metal flakes may alert the technician that there was metal-to-metal contact between moving parts within the engine.

FIGURE 13-1 Bearing sizes are stamped on the back of the insert.

Inspect the main and rod bearings for wear patterns, and record your determination in your notebook or on the work order. Note any unusual wear patterns indicating crankcase or crankshaft misalignment, lack of oil, and so forth.

Inspection of the back side of the bearing may indicate previous machining was performed to the crankshaft or main bearing bores. Most bearings are stamped to indicate standard, **oversize,** or **undersize** (Figure 13-1).

Bearing Failure Analysis

Bearing failure can result from a variety of causes. The most common are oil contamination, oil breakdown, or lack of oil. Other causes include improper installation, improper crush, and worn or bent components.

Normal bearing wear is usually indicated by a smooth, gray appearance. Any wear should be confined to the center of the bearing insert, with little wear located by the parting surface. Under normal conditions, the outer material can wear away and expose some of the inner layers.

Dirt intrusion is identified by fine scoring on the bearing surface (Figure 13-2). The most likely cause for dirt intrusion is a dirty air filter that has not been replaced during regular maintenance intervals, a loose oil dipstick, a loose oil filler cap, or overextended oil change intervals.

Metal particles cause wide grooves to be dug into the bearing surface. Small amounts of the particles are usually embedded in the bearing. Excessive metal particles indicate parts failure. If the bearing failed due to metal particles, carefully inspect the oil pump since it will usually be damaged also.

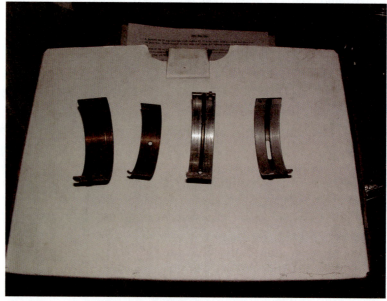

FIGURE 13-2 The bearings on the left are worn down to the copper underlayer. The one on the right shows moderate scoring.

Another common cause for bearing failure is loading. Bearings are under low load conditions when the engine is operated within its effective rpm range. This range is usually 200 to 400 rpm above maximum torque output, and 200 to 400 rpm below the maximum horsepower output. Normal loads in the effective rpm range come from combustion pressure, centrifugal force from the piston assembly, and inertia of the piston assembly. These loads are actually a continuous series of loadings.

The lower main bearing and the upper connecting rod bearing inserts carry most of the load (Figure 13-3). Excessive wear on the lower main bearings indicates the engine lugging, excessive idling, or preignition. Overloading indicated by wear on both insert halves indicates excessive engine speeds.

When the loading and friction becomes excessive, the bearing can spin in its bore. This will destroy the crankshaft and the bore (Figure 13-4). A bearing that has spun in its bore will be loose in its cap. Both the inner and outer sides of the bearing will be scored. You will be able to see scoring in the bore as well. This requires repair to the bore; do not attempt

Combustion force

FIGURE 13-3 Combution forces will cause the lower main bearing and the upper rod bearing inserts to wear more than the other bearing half inserts.

FIGURE 13-4 The bearing spun in its bore and destroyed the crankshaft and the connecting rod.

Dry starts refer to an engine increasing its speed and load right after it has been started, before the oil has been properly pressurized and circulated.

to simply install a new bearing if the old one has spun. A repeat failure is almost inevitable. This fault with the block could also make replacement an effective alternative to overhaul. Also make sure that the oil supply holes in the block and crankshaft are clear. Oil starvation is a common cause of spun bearings. Severely worn bearings will usually damage the crankshaft journals; inspect them carefully.

The specific bearings that indicate wear will also provide hints that point to the cause of the initial failure; for example, if the bearing the farthest from the oil pump has more wear than the other main bearings, this may indicate **dry starts.** Another example is if all of the lower main bearing inserts, except the front one, show wear and the front bearing shows wear on the upper half; the cause may be an excessively tight accessory drive belt.

> **CUSTOMER CARE:** Dry starts are very hard on engine components. If bearing wear indicates dry starts, inform the customer of the need to allow the engine to runlong enough to achieve oil pressure and circulation before it is loaded.

When inspecting the bearings, look closely at the wear pattern. Asymmetric wear patterns can provide information concerning a variety of engine problems. If the bearing appears to have localized smears, this may indicate contamination was between the bearing and saddle (Figure 13-5). The contamination deformed the bearing to the point where it contacted the journal during its rotation. This type of wear can also indicate improper assembly. Main bearing wear in different locations on all bearings indicates the saddles are warped (Figure 13-6). A bearing cap that is not properly aligned to the saddle is indicated by

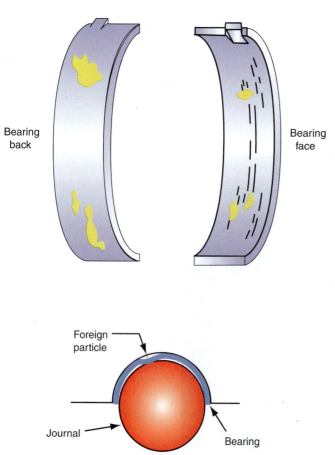

FIGURE 13-5 Foreign particles under the bearing cause localized areas of wear.

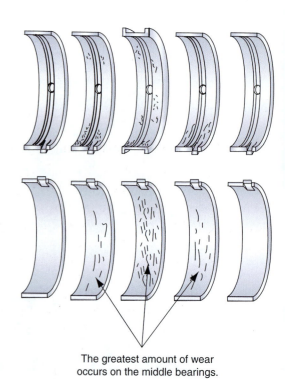

The greatest amount of wear occurs on the middle bearings.

FIGURE 13-6 Main bearing wear indicating warped main bearing saddles.

FIGURE 13-7 This thrust bearing is not badly worn, but the bearings are always replaced during an engine overhaul.

excessive wear on the edges of the bearing shell. Connecting rod bearings displaying wear patterns on opposite sides of the upper and lower bearings indicate the possibility of a bent connecting rod. A shifted bearing cap is indicated by wear patterns on the parting edges.

Lack of lubrication will cause the bearing to smear. Corrosion from acid formation in the oil will cause a pitted surface in the bearing. Corrosion is an indication of improper maintenance schedules, the use of the wrong type of oil, or a contamination of the oil by another fluid such as coolant or fuel.

It is also important to inspect the back side of the bearing for wear. If the bearing spun in the saddle, the back of the bearing will have scoring or may look polished. If this is the indication, the bearing did not have the proper clearance between the shell and journal. It is always important to check the manufacturer's TSBs for information on diagnosing abnormal bearing wear that may be common to that engine or may help you after rebuilding the engine.

The thrust bearing will also show signs of wear through normal use (Figure 13-7). An excessively worn thrust bearing can cause damage to the crankshaft and block. Check the areas on the crank and block where the thrust bearing rides. If you can see wear in these areas, repairs will be necessary. Thrust bearing failure can be caused by a defective clutch **pressure plate,** harsh shifting, or riding the clutch when the engine is coupled to a manual transmission. On an engine mated to an automatic transmission, fluid pressure in the torque converter normally pushes the crank toward the front of the engine. If the pressure is excessive, the torque converter can swell or balloon under loads and damage the thrust bearing. Check the clutch pressure plate transmission fluid pressure, or discuss driving habits with the customer if you find excessive thrust bearing wear on an otherwise sound engine.

INSPECTING THE PISTONS AND CONNECTING RODS

The piston assembly is subject to severe temperature changes, extreme pressures, and very high inertia loads. Over the life of the engine, these forces may cause the piston to deform and become unserviceable. In addition, as the cylinder bore size increases (due to wear), the piston may be allowed to rock within the cylinder and cause skirt scuffing as it contacts the wall. Many of the pistons from modern engines do not have a skirt that protrudes downward on the piston as much as those used in past engines. This lighter weight design leads to more piston rocking when the engine is at colder temperatures (often known as the engine noise "piston slap"). Although some cylinders are made of a hardened material, this may cause excessive wear in the cylinder. Before the piston is reused, it must be inspected and measured to assure its serviceability. The areas of concern are the head, the pin boss, the lands, the ring grooves, and the skirt.

If the piston has not been cleaned already, remove the carbon from the top of the piston using a scraper. Do not dig into the piston head; remove only the heavy deposits. Remove the remainder of the carbon by soaking the piston in a cold tank.

The **pressure plate** couples and uncouples the clutch disc from the flywheel on a manual transmission to transmit engine torque to the transmission.

Classroom Manual
Chapter 13, page 000

CAUTION:
Using a wire wheel to remove carbon from the top of the piston may result in the removal of some metal and rounding off of the piston head. Do not usea wire wheel to clean the skirt or ring grooves.

Piston Ring Removal and Groove Cleaning

WARNING: The rings can have sharp edges. Be very careful when removing the rings from the piston. The piston rings will most likely have a lot of carbon buildup on them, adding to the difficulty of removing them. Clean this area with a solvent as much as possible before attempting removal. Note the direction and placement of the rings as you remove them. In addition, rings are very brittle and can break easily. The edges of the break can be sharp and can easily cause injury.

SPECIAL TOOLS

Ring expander
Ring groove cleaner
Torch tip cleaner

The piston rings must be removed to inspect and clean the grooves. The carbon at the back of the groove must be removed to allow the new rings to compress as they are worked into the cylinder. The rings must be removed in a manner that will not damage the lands. Begin by cleaning the piston in a cold tank to remove any sludge or carbon that may inhibit ring removal. The rings can usually be removed by hand by spreading the ring gap and working the ring over the land. A ring expander can be used to remove piston rings if needed (Figure 13-8).

After the rings are removed, clean the piston grooves using a ring groove cleaner (Figure 13-9). Select the proper size bit for the groove, then locate the tool over the piston with the bit in the groove. Adjust the depth of the bit to the same depth as the groove. Work the tool around the piston while checking your work carefully. Be careful not to remove any metal from the groove. Use a torch tip cleaner to remove carbon and debris from the oil drain holes of the bottom groove.

CAUTION:
If a bead blaster is used to clean the piston, it must be separated from the connecting rod first to facilitate cleaning and prevent connecting rod bearing damage.

Checking Piston Pin Wear

To check for any piston pin wear, clamp the connecting rod in a vise with a shop rag around it. Pull up and push down firmly on the piston. There should be absolutely no detectable movement between the rod and the piston. If there is, disassemble the piston, pin, and connecting rod to repair the problem (Figure 13-10).

Removing the Piston Pin

Before removing the piston from the connecting rod, determine if the piston is offset on the rod. This is generally detectable if the piston head has a notch or arrow indicating the

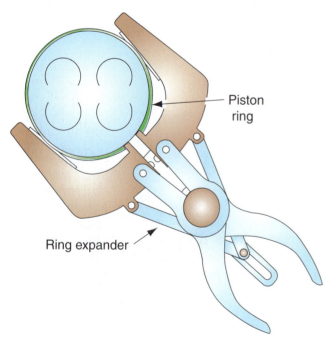

FIGURE 13-8 Use a ring expander to remove the ring from the piston.

Piston ring

Ring expander

FIGURE 13-9 Use a ring groove cleaner to remove carbon from the engine.

FIGURE 13-10 Checking for movement between the rod and piston.

direction of installation (Figure 13-11). The procedure for removing the piston from the connecting rod differs between full-floating and press-fit piston pins.

Full-Floating Pin Removal. Full-floating piston pins use snap rings or locks to hold the pin in the piston pin boss. The clearance between the piston and pin, and the pin and the small end of the rod is enough to allow the pin to be removed without the use of a press. Remove the snap rings from the pin boss and push the pin out of the piston and connecting rod (Figure 13-12). Note the direction the snap rings came out of the boss. Some are tapered or chamfered and must be reinstalled the same way to prevent the pin from coming loose. If the piston pin is too tight to allow removal by hand, use a brass drift and light hammer blows. Carefully inspect the snap ring groove, and replace any piston exhibiting wear or damage in this area.

Press-Fit Pin Removal. Most press-fit piston pins have an interference fit in the connecting rod and float in the piston. The interference fit necessitates the use of a press and special adapters to remove the pin (Figure 13-13). The special adapters prevent piston distortion and rod cocking. The lower adapter must support the piston and allow enough room for the pin to fit through it as it is pressed out. The pin driver must sit fully on the pin without contacting the piston. Once the rod is removed from the piston, store it by hanging it.

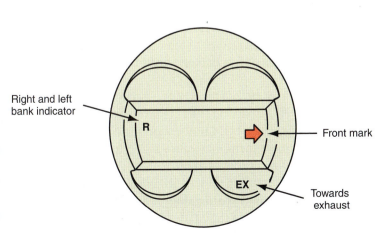
FIGURE 13-11 A notch or arrow on the piston head indicates which side faces the front of the engine.

Right and left bank indicator

R

Front mark

EX

Towards exhaust

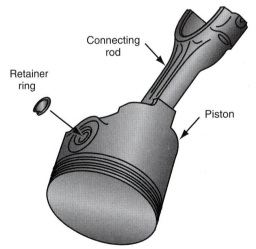
FIGURE 13-12 Full-floating piston pins can usually be pushed out after removing the snap rings.

Connecting rod

Retainer ring

Piston

Classroom Manual

Chapter 13, page 000

The **head** of the piston is the top piston surface.

A piston **land** is the area between the ring grooves.

The piston **skirt** is the area from the lowest ring to the bottom of the piston.

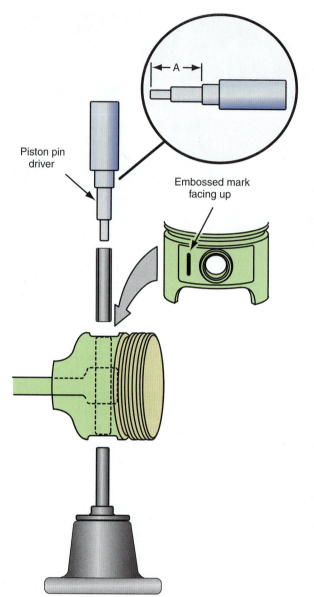

FIGURE 13-13 Special tools must be used to remove press-fit pins to prevent damage to the piston.

Piston Inspection

The **head** of the piston is subjected to the greatest amount of heat and pressure. Visually inspect the head for any damage. Light pitting or nicks on the top of the piston generally do not necessitate replacement; however, the causes of this damage must be determined and corrected.

Visually inspect the **land** and groove area for nicks and cracks. Damage in this area usually necessitates piston replacement. When inspecting the ring grooves, look for a step on the lower, inner portion of the groove (Figure 13-14). Combustion pressures force the ring down and outward (Figure 13-15). This movement may cause a step to wear in the groove and increase the ring side clearance. Worn grooves are reason to discard the piston.

Visually inspect the **skirt** for indications of scuffing (Figure 13-16). Normal skirt wear is a slightly polished symmetrical pattern. Light scoring does not necessitate piston replacement; however, if the skirt is heavily scored or scuffed, the piston must be replaced and the cause must be identified. Scuffing in a diagonal direction on the skirt indicates a bent connecting rod (Figure 13-17).

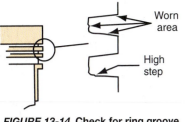

FIGURE 13-14 Check for ring groove wear.

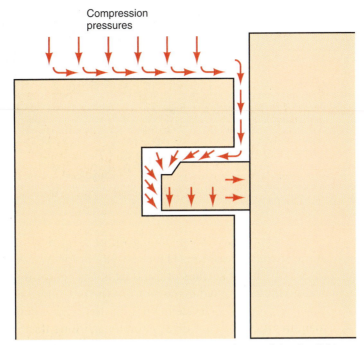

Compression pressures

FIGURE 13-15 Compression pressures work to seal the ring tighter.

FIGURE 13-16 This piston has some moderate vertical scuffing on the major thrust side.

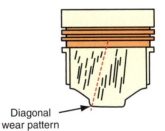

Diagonal wear pattern

FIGURE 13-17 Diagonal scuffing due to a bent connecting rod. The piston cannot travel a true path in the cylinder.

The **pin boss** area is a common area for stress cracks.

Classroom Manual
Chapter 13, page 000

SPECIAL TOOLS
Micrometer
Dial bore gauge

Finally, inspect the **pin boss** area for indications of cracks. This is a common area for stress cracks.

Measuring the Piston. If the piston passes visual inspection, follow the manufacturer's recommended procedures for measuring the piston. Measure the piston to determine clearance and to check for collapsed skirts. The clearance depends on piston materials and engine applications; thus, there are no general specifications. Piston clearance is determined by measuring the size of the piston skirt at the manufacturer's **sizing point.** This measurement is subtracted from the size of the cylinder bore. If the piston clearance is not within specifications, it may be necessary to bore the cylinder to accept an oversize piston.

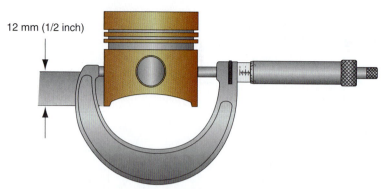

12 mm (1/2 inch)

FIGURE 13-18 Measure the piston diameter at the specified location.

Since most pistons are cam ground, it is important to measure the piston diameter at the specified location (Figure 13-18). Some manufacturers require measurements across the thrust surface of the skirt at the centerline of the piston pin. Others require measuring a specified distance from the bottom of the oil ring groove. Always refer to the appropriate service manual for the engine you are servicing. In addition, some manufacturers may require piston measurements at several locations (Figure 13-19). A piston that has smaller-than-specified diameters should be replaced. A smaller-than-specified skirt diameter may indicate a collapsed skirt. A common cause of collapsed skirts is piston pin seizure. The seized pin prevents the piston skirt from conforming properly to the bore.

Some pistons are designed with tapered skirts. The bottom of the skirt may be 0.001 to 0.0015 in. (0.03 to 0.04 mm) larger than the top of the skirt.

As the piston moves up and down in the cylinder, it is constantly changing speeds. This action can hammer hard on the piston pin and pin boss. Before reusing the piston, measure the pin boss bore for size and out-of-round. Then compare the results with specifications. Some machine shops will hone a worn bore to accept a larger piston pin.

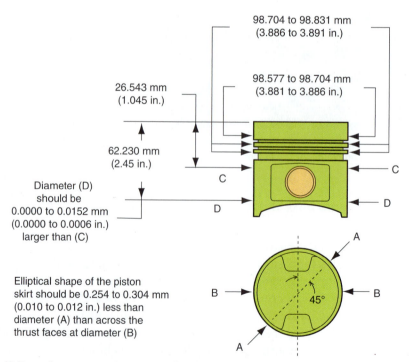

98.704 to 98.831 mm (3.886 to 3.891 in.)

98.577 to 98.704 mm (3.881 to 3.886 in.)

26.543 mm (1.045 in.)

62.230 mm (2.45 in.)

Diameter (D) should be 0.0000 to 0.0152 mm (0.0000 to 0.0006 in.) larger than (C)

Elliptical shape of the piston skirt should be 0.254 to 0.304 mm (0.010 to 0.012 in.) less than diameter (A) than across the thrust faces at diameter (B)

45°

FIGURE 13-19 It may be necessary to measure the piston at several locations to determine wear or damage.

FIGURE 13-20 Measure the piston pin boss for wear.

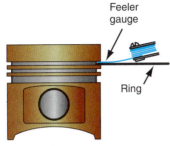

FIGURE 13-21 Measuring the ring groove clearance.

A dial bore gauge is the easiest to use for making these measurements. Measure the outside diameter of the piston pin and zero the gauge to this size. Insert the gauge into the pin bore, and note the dial reading as the gauge is rotated and worked up and down the bore (Figure 13-20).

As a final check, measure the ring groove for wear. This can be done by installing a new ring backward in the groove and using a feeler gauge to measure the clearance (Figure 13-21). Check the groove at several locations around the piston. If the side clearance is excessive, replace the piston. Excessive side clearance can result in ring breakage.

The top ring groove generally wears more than the others. Normal ring-to-groove side clearance is between 0.002 and 0.004 in. (0.05 to 0.10 mm). With a new ring located in the groove, attempt to slide in a feeler gauge the size of the maximum clearance specification. If the gauge slides in, the clearance is excessive.

Groove depth can also be checked using the new ring. Locate the ring into the ring groove in the same manner as checking clearance. Roll the ring around the entire groove while observing for binding. The ring depth should be consistent.

Piston Wear Analysis. The wear indicated on the piston can be a help in detecting the cause of the failure. Rarely does the piston cause a failure; usually something else causes the piston to fail. Studies indicate that about 41 percent of piston failures are attributable to contamination or lack of oil. The second leading cause of piston failure is scuffing. About 26 percent of piston failure is attributable to this condition. Scuffing can be caused by cylinder bores warping due to overheating, bent connecting rods, or momentary welding of the piston to the cylinder wall at TDC. In addition, scuffing on only one side of the piston may indicate excessive idling or engine **lugging.**

CAUTION: Make sure the proper rings are used when measuring clearance. While making the measurement, ensure that the piston ring is installed right side up. Usually there is a marking that indicates the top side of the ring. Do not fault the piston until proper ring size is verified. Improper rings may have excessive groove clearance.

Lugging of the engine results from improper gear selection for the load and acceleration or overloading the engine, such as towing in excess of the recommended limit.

Incorrect installation accounts for about 11 percent of all piston failures. Use the following chart to help analyze piston wear:

Condition	Cause
1. Burned piston top. Identified by a smooth hole in the top of the piston or on the outer edge of the crown (Figure 13-22).	Abnormal combustion (detonation and preignition).
2. Jagged hole in piston head.	Coolant ingestion into the combustion chamber. The rapid cooling of the head as the coolant enters causes the head to crack and create a hole. Valve contact with top of piston.
3. Nicks on piston head.	Valve contact with piston. Foreign material in the intake manifold.
4. Broken second land.	Detonation. Too thin a piston ring.
5. Skirt scuffing near pin boss.	Piston pin seized in bore.
6. Collapsed piston skirt.	Piston pin seized in bore.
7. Erosion around pin boss.	Broken piston pin lock ring causing piston to move on the pin and contact the cylinder wall.
8. Heavy scuffing of the skirt (scuffing on only one or two pistons).	Improper piston clearance. Bent connecting rod.
9. Heavy scuffing on most pistons (Figure 13-23).	Lack of lubrication. Contaminated oil. Oil breakdown. Piston overheating. Cylinder bore distortion caused by overheating or improper head bolt torques. Bent connecting rods.

FIGURE 13-22 This piston has been badly damaged by detonation.

FIGURE 13-23 This piston shows severe scuffing.

Inspecting the Piston Rings

A visual inspection is usually all that is required for the rings. Visually inspect the old rings for breakage, scoring, scuffing, and glazing. The major cause for ring wear is abrasion. When inspecting the rings for wear, look for the following conditions:

Condition	Cause
1. Top ring indicates more wear than the second ring and has vertical lines.	Dirty air entering through the air intake system.
2. Lower rings and cylinder walls show more wear than the top ring.	Abrasives in the lubrication system.
3. The bottom side of the ring indicates wear, leaving a lip on the outside edge.	Abrasives.
4. Abrasive wear (dull gray appearance with vertical lines).	Oil contamination. Lack of oil. Oil breakdown. Dirty air cleaner element. Vacuum leaks.
5. Scuffing or scoring (scratches and voids in the ring face).	Lack of oil. Oil breakdown. Too rich air-fuel mixture causing the oil film to be washed away and allowing the formation of carbon deposits. Overheating. Distorted cylinder bore. Improper cylinder wall finish. Improper piston clearance. Improper ring end-gap (indicated by polished ring ends). Detonation. High-speed operation.
6. Broken rings.	Worn piston ring grooves. Detonation. Excessive cylinder wall ridge. Improper installation.
7. Glazing.	Improper ring seating.

When new rings are installed, they should be checked for side clearance as discussed, and end-gap measurements.

Inspecting Piston Pins

As discussed earlier, the piston pin and pin boss are subjected to severe operating conditions. Compounding this is the difference in expansion rates between the steel piston pin and the aluminum piston. A steel pin expands about 0.0003 in. (0.008 mm) for every 50°F (28°C) (increase in temperature). The diameter of the bore may increase 0.0006 in. (0.02 mm) during the same temperature increase. Consequently, proper clearance is critical.

Manufacturers vary concerning recommended procedures for checking pin clearance and how much clearance is allowed. For example, Chevrolet suggests measuring the pin diameter and bore diameter; any clearance over 0.001 in. (0.025 mm) requires replacement of the pin and piston. Pontiac says the pin will fall through the bore with 0.0005 in. (0.01 mm) clearance but will not with 0.0003 in. (0.008 mm). Some import manufacturers check pin and bore

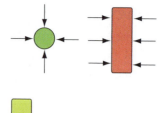

FIGURE 13-24 Measuring piston pin clearance.

wear by holding the piston and attempting to move the connecting rod up and down. Any movement felt indicates wear. Feel is not always a reliable method. If you have any doubts, measure the clearance.

Measure the pin using the recommended procedure for the engine being serviced (Figure 13-24). With the pin removed, visually inspect it for wear. Use your hands to feel for scuffing or scoring. If there is any wear, the pin must be replaced.

Seized or scored piston pins are generally caused by overheating of the piston. This can be caused by poor cooling system operation and improper combustion due to preignition or detonation. These overheating conditions cause lubrication failures of the pin.

Inspecting Connecting Rods

The connecting rod is subjected to extreme operating conditions. As discussed earlier, the speed of the piston assembly is constantly changing between TDC and BDC. These speed changes result in g-loads that literally stretch and compress the connecting rod every stroke; for example, an engine running at 6,000 rpm can exert 2,440 gs of tension at TDC and compress the rod with 1,300 gs at BDC. These forces can deform the two bores and/or bend the connecting rod. Under extreme conditions, the connecting rod can break apart.

Before a connecting rod is accepted for reuse, it must be carefully inspected and measured. Both the **big-end** and **small-end** bores should be inspected for clearance, out-of-round, and taper. Most of the wear of the big-end bore occurs in the cap and in a vertical direction. In addition, the rod should be checked for center-to-center length.

When measuring the inside diameter of the big end, assemble the cap to the rod, leaving the nuts loose. Locate the rod into a soft jaw vise or a special rod vise, with the jaws covering the parting line (Figure 13-25), and torque the cap nuts to specifications. This procedure ensures that the cap and rod are properly aligned with each other. The easiest way of measuring the bore is to use a dial bore gauge. If an inside micrometer or telescoping gauge is used, measure the diameter in two or three directions and near each end of the bore (Figure 13-26) to obtain out-of-round and taper measurements. The greatest amount of out-of-roundness will occur in the cap and in a vertical direction. Note: Some manufacturers will specify directions for measuring connecting rods for out-of-round and taper at different areas in the big end bore.

Inspect the small-end bore in the same manner as the big-end bore. Use a dial bore gauge to measure clearance, out-of-round, and taper. If the piston assembly is designed with a press-fit pin in the connecting rod, the bore must be the correct size to provide an

FIGURE 13-25 Using a rod vise to align the rod to the cap before torquing the cap bolts.

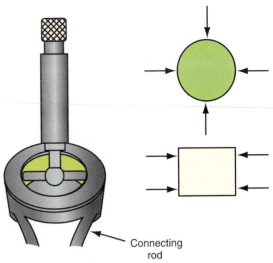

FIGURE 13-26 Measure the big-end bore for wear, taper, and out-of-round.

interference fit. In addition, any scuffing or nicks in the bore may inhibit the piston from rocking properly. Sometimes the bore can be honed to a larger size to accept an oversized piston pin.

Full-floating piston pins ride in a bronze bushing pressed in the pin bore of the connecting rod. These bushings should be replaced as a matter of standard procedure. The old bushing is pressed out and a new one is pressed in. The new bushing should be measured for clearance and corrected by honing if necessary. After honing the bushing, recheck it for taper. Taper can result if the hone is not perfectly straight with the bore. If there is more than 0.0002 in. (0.005 mm) taper, press out the bushing and install another one.

Bores should be checked for parallelism and twist (Figure 13-27). To perform this task, install the piston pin into the rod without the piston, attach the cap to the rod, and torque the bolts while the cap and rod are held in alignment. Install the rod assembly onto a special alignment fixture with the thrust side of the rod facing out (Figure 13-28). Next, locate

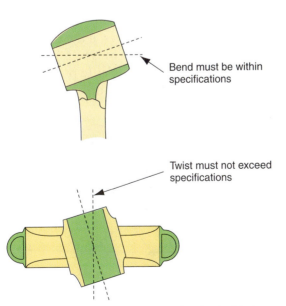

Bend must be within specifications

Twist must not exceed specifications

FIGURE 13-27 Connecting rod bend and twist must be checked before reusing the rod.

547

FIGURE 13-28 Using a rod aligner to measure twist and bend. When installing the rod to the aligner, the thrust side must be facing outward.

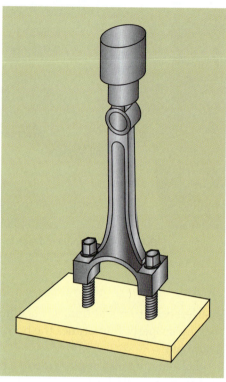

FIGURE 13-29 Pressing the cap bolts from the connecting rod.

the bend indicator fixture against the piston pin. The fixture straddles the connecting rod, and this fixture should fit squarely against the piston pin. If a feeler gauge can be inserted between the fixture and the piston pin, the rod is twisted or bent.

Some manufacturers recommend the cap bolts or nuts holding the connecting rod assembly together be replaced whenever the piston assembly is removed from the engine. Check the TSBs from the manufacturer for specific information. After completing the big-end bore measurements, remove the bolts by pressing them from the rod (Figure 13-29). Inspect the area around the bolt holes for indications of stress cracks. This area is the weakest part of the connecting rod and is subject to fractures. If there are any cracks, replace the rod. Most manufacturers recommend that connecting rods be replaced as a set.

Some connecting rods have oil jets or "spit" holes to provide cylinder wall lubrication (Figure 13-30). Clean these holes with a proper size drill bit or torch tip cleaner.

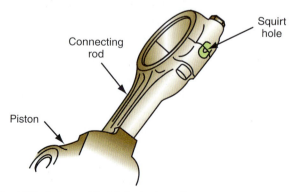

FIGURE 13-30 A "squirt" hole is used to lubricate the cylinder wall and keep the piston head cool.

The following chart lists some common connecting rod failures and their causes:

Failure	Cause
1. Big-end breakage.	Improper cap bolt torquing. Bolt breakage. Connecting rod failure. Excessive bearing clearance. Nicks in the rod. Hydrostatic lock. Detonation.
2. Rod bolt failure.	Improper torquing. Reusing rod bolts.
3. Bent or twisted rods.	Hydrostatic lock. Seized piston pin. Excessive compression ratio. Detonation.

PISTON ASSEMBLY SERVICE

Service to the piston assembly is usually confined to connecting rod reconditioning; however, the piston may require the installation and sizing of piston pin bushings. Some procedures such as piston pin bore enlarging, skirt knurling, and ring groove repairs can also be performed. These operations are not very common and are mainly used for special applications, such as heavy-duty diesel and marine engine applications.

Reconditioning Connecting Rods

If the connecting rod fails inspection, it is not always necessary to replace it. A connecting rod can often be reconditioned. The following procedures may need to be performed on a connecting rod before it is returned to service:

- Replacing rod bolts
- Straightening
- Resizing the big-end bore
- Resizing the small-end bore
- Installing bushings into the small-end bore
- Beaming

Before reconditioning the rods, inspect the thrust surfaces on the sides of the big-end bore. If there are any nicks or gouges, they should be dressed to provide proper seating in the grinder. Nicks and gouges are also stress risers; if they are not removed, the rod may fail prematurely. Connecting rod reconditioning is usually done at a machine shop. A technician should always consider the costs associated between reconditioning rods and purchasing new rods.

Cap Bolt Replacement. It is a good practice to replace the connecting rod bolts whenever the rods are removed from the engine. If the big-end bore requires reconditioning, the bolts will have to be removed to perform some of the machining operations; however, if the rod requires straightening, straighten the rod before replacing the bolts.

Most connecting rod bolts are press-fit into the connecting rod. It is best to press both bolts out at the same time instead of attempting to knock them out with a hammer. They should come out with little pressure. If excessive pressure is required, back off the press ram. Continuing to increase pressure may break or bend the connecting rod. Use a connecting rod heater to heat the rod, and try pressing the bolts out again.

After the caps and rods are machined, new bolts can be installed. When installing the bolts, it is important to protect the parting surface of the rod. A fixture can be made or purchased for installing the bolts (Figure 13-31). With the rod located over the fixture, the bolts can be seated with a punch and hammer.

WARNING: Wear approved eye protection when performing this task. Do not stand in front of the connecting rod; stand to the side.

Straightening Connecting Rods. Even though slight warpage of the connecting rod can be corrected by straightening, do not perform this operation if replacement rods are readily available. Liability and cost are factors you need to consider when deciding on this procedure.

If the inspection of the rod indicates it is bent or twisted, it may be possible to straighten it. One method of correcting bend and twist is cold bending. This can be manually performed using a special holding fixture (Figure 13-32). When performing this procedure, it is important not to nick or scratch the rod. Nicks and scratches weaken the connecting rod and may lead to possible breakage.

Begin by determining the direction of bend or twist. Most bends in the rod will be located near the small-end bore. Place the connecting rod into the straightening fixture using the correct size big-end and small-end adapters. Install the cap, and torque the cap bolts to specifications. Select the correct size bending bar. It should closely fit the rod to prevent nicks and scratches. Use the bar to bend the rod in small increments. Measure the progress on a rod alignment fixture.

Resizing the Big-End Bore. Big-end bores can be reconditioned using two methods: oversizing the bore or returning the bore to original size. The method used depends on the amount of wear and damage in the bore.

WARNING: Always wear approved eye protection when grinding the caps and rods.

If it is determined that the bore is to be returned to its original size, grind the parting surfaces of the rod and cap to make the vertical diameter of the bore about 0.002 in. (0.050 mm) smaller. When grinding the parting surface of the rod, remove only enough material to restore the surface.

The procedures for removing material from the cap and rod are the same. First, remove the cap bolts. Place the workpiece in the grinder so it is resting on the squaring

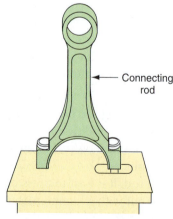

Connecting rod

FIGURE 13-31 Using an installation block to press in the new rod bolts protects the parting surface.

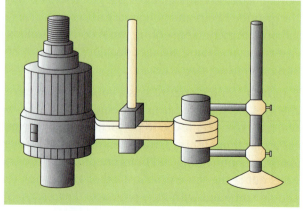

FIGURE 13-32 Cold straightening a connecting rod using a special holding fixture.

rod (Figure 13-33). Tighten the machine's vise to secure the workpiece. It is important that the workpiece be secured so the grinding wheel removes an even amount from both ends. If unequal amounts are removed, the cap and rod will not be square. Back off the feed wheel to be sure the grinding wheel will not contact the workpiece at this time. Position the workpiece over the grinding wheel, and adjust the feed wheel until the parting surface just comes into contact with the stone. Swing the workpiece out of the way. Turn on the motor and swing the workpiece over the grinding wheel. This is the zero cut. Adjust the feed wheel to remove 0.002 in. (0.050 mm), and take another pass over the stone with the workpiece. If more metal must be removed, adjust the feed wheel to the desired amount, and swing the cap over the stone. Do not remove more than 0.002 in. (0.050 mm) at a time.

Then install the connecting rod cap and torque it to the specified value. To keep the cap and rod assembly straight, use a rod vise when installing the cap. Use a dial bore gauge to determine how much material must be honed. Then select the honing mandrel that fits the bore diameter. Install the mandrel to the machine, and test the feed of the stones by turning the machine on and adjusting the feed back and forth. The stones should go in and out.

Set the spindle speed. If the bore diameter is over 2 in. (50 mm), the speed should be set to 200 rpm. Next, set the cutting pressure. On many machines, use a dial setting of 2 for rough honing and 1 to finish hone the bore. Then rotate the spindle until the mandrel shoes are facing up, and loosen the clamp that secures the shoes.

Place a connecting rod onto the connecting rod reconditioner (Figure 13-34). Depress the foot pedal and adjust the feed dial until the shoes expand and lock the rod in place. Retighten the shoe clamp screw; then back off the feed until the dial indicator reads zero. This provides a zero set to determine how much the bore is being honed.

Adjust the torque rod to support the flat surface of the rod as close to the small-end bore as possible (Figure 13-35). Adjust the coolant flow over the workpiece. Use the foot pedal to

> Throughout this procedure, the cap or rod is referred to as the workpiece.

FIGURE 13-33 Installing the rod into the grinder. Make sure it is square so equal amounts are removed from both ends.

FIGURE 13-34 This rod reconditioning machine can resize the big- or little-end bore.

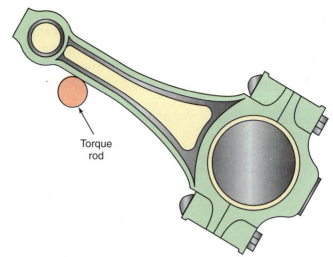

FIGURE 13-35 Position the rod against the torque rod. The torque rod should be on a flat surface as close to the small-end bore as possible.

control the machine. Stroke the rod the entire length of the stones while maintaining stone pressure by turning the feed dial. Observe the honing dial to measure the amount of material removed.

Use a slower speed to provide a smooth finish. Do not use any additional stone pressure. After the bore has been resized, measure the bore diameter.

It is recommended that the bores be rechamfered to restore the original oil throw-off properties. Remember these considerations when reconditioning the big-end bore:

- Attempt to remove most of the metal from the rod cap instead of the rod.
- Removing metal from the rod reduces the center-to-center distance of the rod and the compression ratio of the cylinder.
- The center-to-center distance between rods should not vary more than 0.010 in. (0.254 mm).

WARNING: Always wear approved eye protection when honing the connecting rod bore.

If returning the bores to their original size will result in excessive center-to-center length reduction, it may be possible to bore or hone the big-end bore to accept an oversize bearing. This can be done only if oversize bearings are available.

Reconditioning the Small-End Bore. The reconditioning method used on small-end bores depends upon the type of piston pin fit. Press-fit piston pins require about 0.001 in. (0.025 mm) interference. If the bore is worn, it will need to be resized to accept an oversize pin. This will require the pin boss in the piston to be oversize as well. Before performing this operation, be sure to weigh the cost of replacement versus the labor cost involved.

WARNING: Always wear approved eye protection when performing this operation. Do not wear jewelry, and secure any loose clothing.

Full-floating piston pins ride in a bushing pressed into the small-end bore. This bushing should be replaced as a matter of common practice whenever the engine is being rebuilt. Press out the old bushing using an adapter to prevent damage to the connecting rod. Press the new bushing into the bore, making sure any oil holes in the rod and bushing align to each other. If the bushing uses parting lines or slots to direct oil flow, install these lines to the **minor thrust side** of the rod.

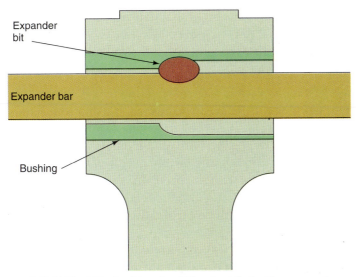

FIGURE 13-36 The bushing should be sized first with an expander bit to set the bushing into the rod properly.

Once the new bushing is installed, it is sized to the pin. The first step is to expand the bushing using an expander bit (Figure 13-36). This seats the bushing into the bore. If this process is not done, proper heat transfer will not occur. To expand the bushing, select the correct size mandrel for the bushing. Set the spindle speed to 200 rpm, and rotate the pressure-setting dial all the way off. Locate the small-end bore over the mandrel. Direct the coolant flow into the workpiece. Use the foot pedal to start the spindle and stroke the connecting rod back and forth over the mandrel. Adjust the feed dial until you feel the expander bit contact the bushing. Increase the feed dial 2 numbers while stroking the connecting rod. Remove the rod, and inspect the work. Next, use the facing cutter to remove any material from the outside of the bore.

Finally, hone the bushing to fit the pin. The clearance is usually between 0.0003 and 0.0005 in. (0.008 and 0.013 mm). To do this procedure, start by determining the amount to be honed from each bushing. Adjust the pressure setting to 1, and follow the same procedure for expanding the bushing to hone the bushing to final size.

Check piston pin fit on each bushing. If a bushing is too loose, remove it and install a new one.

Piston and Connecting Rod Reassembly

If the engine is not going to be balanced, the pistons and connecting rods can be assembled at this time. If the engine is going to be balanced, assemble the pistons and connecting rods after each component is weighed separately and matched. Remember to keep all pistons, rods, and pins identified to the cylinder bore in which they will be installed. In addition, check the service manual to determine if the pistons are offset and the proper orientation of the rod to the piston. Also, the connecting rod may have oil spit holes or jet valves that must be installed in the proper direction. The spit hole is used to squirt oil onto the cylinder walls and under the piston head to keep it cool. V-type engines typically use offset connecting rods. If these are installed backward, the engine may lock up and not rotate. Rod bearings are usually offset in the rod bore toward the middle of the crank journal. This is so the edges of the bearings will not rub on the fillets.

Before beginning the reassembling procedure, double-check the rods and pistons. They must be absolutely clean. As each piston is fitted to the rod, double-check the clearances. Rod bearing clearance can be checked by installing the rod bearings into the connecting rods and torquing the cap bolts to specifications. Measure the inside diameter of the bearing and the outside diameter of its crankshaft journal. Compute the difference between these readings, and divide it by 2 to obtain the oil clearance.

Piston–to–connecting rod installation methods depend on whether the pin is full-floating or press-fit.

Full-Floating Pins. Installing pistons with full-floating pins does not require any special tools. These types of pins can generally be pushed in using your hands. Coat the pin, piston bore, and small-end bore with light oil or assembly lube prior to installation. Fit the piston pin into one side of the piston bore until it enters into the center of the piston about 1/16 in. Next, locate the rod into the piston, making sure it is facing the proper direction. Align the small-end bore with the protruding piston pin, and push the piston pin through the connecting rod bore and into the other piston bore. If necessary, tap the pin in with light hits with a plastic hammer. Check to be sure the pin is free to rotate in both the piston bosses and connecting rod.

Finally, install the snap rings. It is very important to install the retaining clip properly. If it is installed improperly, the pin can come out and damage the cylinder bore. Handle the snap rings carefully. Do not overstress them by compressing them more than necessary. The open end of the snap ring should face toward the bottom of the piston. This places the strongest part of the snap ring in the portion of the pin boss receiving the highest stress. In addition, some snap rings are tapered or chamfered and must be reinstalled in the proper direction to prevent the pin from coming loose.

Press-Fit Piston Pins. Press-fit piston pins can be installed using a hydraulic press or by heating. Be sure to use the proper tools and equipment when pressing in the pin. The heating method expands the small-end bore of the connecting rod to allow the piston pin to be slip fit. The small end of the connecting rod is raised to a temperature of about 425°F (204°C) using a special rod oven.

Installing Piston Rings

Installing piston rings is usually an easy task; however, improper procedures can lead to early engine failure. Take the time to do the task properly and make all required measurements and checks. First, double-check to be sure the piston ring grooves are thoroughly cleaned. Next, check and correct ring end gap. Proper ring gap is critical due to expansion of the piston, rings, and cylinder bore during engine running conditions. If end gap is too wide, compression can escape; if it is too narrow, scuffing or ring breakage can result.

To check ring end gap, place the ring into the cylinder bore that it will be installed into later. Use the piston to slide the ring to the specified depth in the cylinder (Figure 13-37). The piston head will keep the ring square in the bore so accurate measurements can be made. If the cylinder has been honed or bored to correct parallelism, the ring can be installed anywhere in the cylinder. If the bore has not been reconditioned, locate the ring where specified in the service manual. Some manufacturers require gap checks to be made at the top of ring travel, while others require it at the bottom of ring travel. Use a feeler gauge to measure the width of the gap (Figure 13-38). The end of the gap is tapered, so be sure to measure at the outer edge to obtain accurate measurements. If the gap is too wide, use oversize rings. If the gap is too narrow, use a file or ring grinder to carefully remove some material. Check all rings that are to be installed into that cylinder. If the rings are not to be installed immediately onto the piston, identify them to the cylinder to maintain order.

Ring gap specifications vary according to the material the ring is constructed of and the diameter of the piston. Following is a general specification chart that may be used if manufacturer specifications are not available:

SPECIAL TOOLS
Press
Piston pin
installation set
Rod heater

Classroom Manual
Chapter 13,
page 000

SPECIAL TOOLS
Ring expander

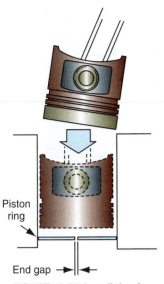

Piston ring

End gap

FIGURE 13-37 Install the ring into the bore toward the bottom of ring travel using a piston to be sure it is square.

FIGURE 13-38 Measure the ring end gap using feeler gauges.

SERVICE TIP:
A rule of thumb for ring end gaps is 0.003 to 0.004 inch for each inch of cylinder bore diameter.

SERVICE TIP:
You cannot file or grind the end of most chrome rings; the coating will chip away. Follow the instructions that are provided in the ring package.

SERVICE TIP:
It may be tempting sometimes to install oversize rings to compensate for worn cylinders instead of boring the cylinder. This is not a good idea because the taper of the worn cylinder will cause the rings to lock up as they travel down the cylinder bore.

Standard Iron Rings

Diameter	Ring Gap Clearance
2.000" to 2.999"	0.007" to 0.017"
3.000" to 3.999"	0.010" to 0.020"
4.000" to 4.999"	0.013" to 0.025"
5.000" to 5.999"	0.017" to 0.028"
6.000" to 6.999"	0.020" to 0.032"

High-Strength Iron Rings

3.000" to 3.999"	0.010" to 0.026"
4.000" to 4.999"	0.014" to 0.030"
5.000" to 5.999"	0.017" to 0.034"
6.000" to 6.999"	0.021" to 0.038"

If any of the end gaps are too narrow, the gap can be increased by a file or ring gap cutter. If a file is used, cut from the outside of the ring toward the center. Correct all rings requiring wider gaps. If the ring gap is too wide, it is possible the gap was checked with the ring in a slightly worn area of the cylinder bore; for example, if the ring were into an area of the bore that was 0.002 in. (0.05 mm) oversize, the gap would be increased by 0.006 in. (0.15 mm). As a result, a ring with a 0.018 in. (0.46 mm) gap would measure 0.024 in. (0.61 mm) clearance. Also, if the bore diameter were increased, oversize rings would have to be used. Rings are available for standard 0.020, 0.030, 0.040, and 0.060 inch oversizes. If a 0.010 inch oversize piston is installed, a standard size ring is used. Metric oversizes are 0.50, 0.75, 1.0, and 1.5 mm.

The oil control rings can generally be installed by hand (Figure 13-39). Install the expander ring first. The two ends of the expander must contact each other without any gap. Make sure the two ends do not overlap. Rotate the expander until the ends are 90 degrees from the piston pin. Next, install the upper oil ring side rail. Start the ring by placing one end between the ring groove and the expander. Hold this end in place while working the

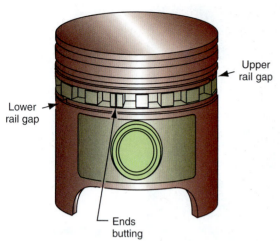

FIGURE 13-39 Installing the oil control ring assembly.

FIGURE 13-40 Use a ring expander, so you do not twist the ring or scrape the piston.

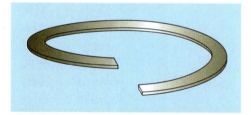

FIGURE 13-41 A damaged compression ring is the result of trying to roll the ring into the ring groove.

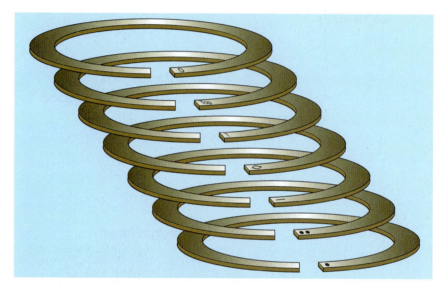

FIGURE 13-42 Most directional rings are marked in some way to identify proper installation direction.

ring around the piston and into the groove. Locate the end gap of this ring 45 degrees from the end of the expander. Now install the lower side rail in the same manner. Locate the end gap of this rail 180 degrees from the end gap of the top rail. The oil ring side rails must be free to rotate after installation.

Starting with the bottom compression ring, use a ring expander to install the rings into the piston grooves (Figure 13-40). The ring expander prevents the ring from being rolled onto the piston. Rolling a ring will result in distortion (Figure 13-41). Be careful not to overexpand the ring; doing so may cause it to crack. Be sure to note any markings indicating that the rings are directional (Figure 13-42). With the rings installed, use a feeler gauge to measure ring-to-groove clearance, and compare the measurement to specifications (Figure 13-43).

After the rings are installed, the end gaps should be staggered. This is done to prevent compression leakage. Follow the manufacturer's recommendations for proper location of ring end gaps (Figure 13-44).

FIGURE 13-43 The proper ring side clearance allows the ring to seat properly in the piston.

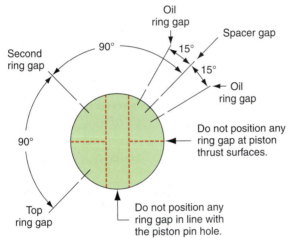

FIGURE 13-44 Position the ring end gaps to prevent compression leakage.

ENGINE BLOCK PREPARATION

Before assembling components in the engine block, use the following checklist to be sure they are properly prepared:

1. Chamfer the edge of the cylinder head bolt holes to prevent cylinder head gasket deformation during the torquing process.
2. Inspect all sealing and gasket surfaces for irregularities, nicks, scratches, and so forth.
3. Clean the threads of all bolt holes using the correct size bottoming tap.
4. Inspect all bearing surfaces for nicks, galling, or sharp edges.
5. Check all notes made during the inspection to be sure all machining operations have been performed.
6. Clean the engine block.
7. Install the core and oil gallery plugs (Figure 13-45).

If the engine uses any bushings inside of bores, such as the distributor shaft bore, use a bushing driver to replace them. It may be necessary to hone the bushing to obtain proper fit. Driving the bushing may cause it to deform slightly on the edge. Use a knife or scraper to remove this ledge.

SPECIAL TOOLS
Tap set
Core plug driver
Bushing driver
Hone kit

SERVICE TIP:
Do not install the core plug for the back of the camshaft until the camshaft is installed.

SERVICE TIP:
Some manufacturers require threaded inserts be installed on aluminum blocks before reassembly.

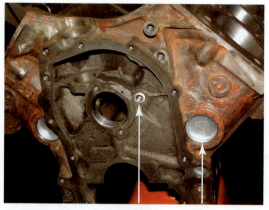

New oil gallery plug   New core plug

FIGURE 13-45 With new core plugs and oil gallery plugs installed, the block is clean and ready for reassembly.

Installing OHV Camshafts, Bearings, and Balance Shafts

Installing OHV Camshaft Bearings. The camshaft bearings should be installed before installing the crankshaft assembly. Camshaft bearings are press-fit into the engine block and are often sized differently for the different bearing journals.

Installing OHV camshaft bearings requires special tools. The most common type of tool used is a universal installation tool. This type of tool uses an expanding mandrel that fits snug and tight against the bearing. The mandrel is connected to a driver and a hammer is used to press the bearings in. Follow Photo Sequence 25 for installing OHV camshaft bearings.

Installing OHV Camshafts and Balance Shafts

When sliding the camshaft into the bearings, be careful to protect the journals and bearings. Following is a guide to typical procedures for installing the camshaft into an OHV engine:

1. Determine oil clearance between the installed camshaft bearings and the camshaft journals and compare with specifications.
2. Use a rag soaked in solvent to remove any coating on the camshaft journals. Be careful not to remove the coating on the lobes. Lubricate the journals with engine oil or assembly lube before fitting the camshaft into the block.
3. Install the camshaft, preventing any contact between the lobes and the bearings. When the camshaft is fully seated in place, spin it to assure free rotation. If rotation is impaired because of high spots on the bearings, they can be removed using a three-cornered scraper (Figure 13-46). Slight corrections to the bearings can be made with a flex hone. Make sure to clean the engine block after performing these operations. If the bearings are not the cause of the binding, the camshaft may be bent.
4. Apply sealer to the outer edge of the rear core plug, and install the plug with the convex side out (Figure 13-47). Drive the plug to the specified depth. Once the core plug is installed, use a dull punch to deform it slightly. This increases the tension applied to the outer edges to provide a positive seal. After the core plug is installed, rotate the camshaft to be sure the plug is not causing interference.
5. Install the camshaft retainer plate.
6. Measure **camshaft end play** and correct it if needed. Install a dial indicator to measure the lateral movement of the camshaft; then use a screwdriver to gently pry the camshaft back and forth. The total indicated reading (TIR) of the indicator is the

TYPICAL PROCEDURE FOR CHECKING MAIN BEARING OIL CLEARANCES

P25-1 After thoroughly cleaning the engine block and bearing bores, gather the tools you will need. This includes oil, the service manual, Plastigage™, and a torque wrench.

P25-2 Install the new bearings in the block. Avoid touching the front and back sides of the bearings and do not lubricate the bearing just yet. Using tight-fitting rubber gloves is helpful for keeping the grease from your fingers off the bearings.

P25-3 Cut off a small piece of Plastigage and install it lengthwise on the crankshaft journal.

P25-4 After placing Plastigage on the other journals, torque the crankshaft main bearing cap bolts to specification using the manufacturer's provided sequence. Avoid turning the crankshaft during this step!

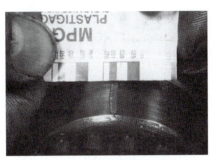

P25-5 Remove the main bearing caps, in sequence, and check the width of the Plastigage. The plastic strip will widen when it is squeezed due to the bearing cap providing pressure on it.

P25-6 After recording all of your measurements, remove the Plastigage by scraping it off with your fingernail in a clean shop rag.

P25-7 After cleaning all of the Plastigage up from the bearings and caps, thoroughly lubricate the new bearings before final assembly. Some technicians use a heavier engine assembly lube.

P25-8 Torque down the main caps one at a time using the manufacturer's provided sequence. Check to make sure that the crankshaft turns after each one is fastened down.

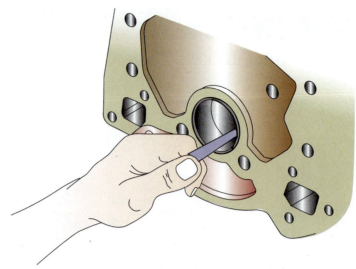

FIGURE 13-46 Use a three-corner scraper to remove high spots in the camshaft bearing.

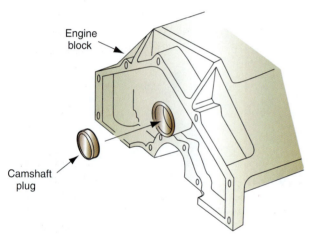

FIGURE 13-47 The camshaft plug is installed at the rear of the block.

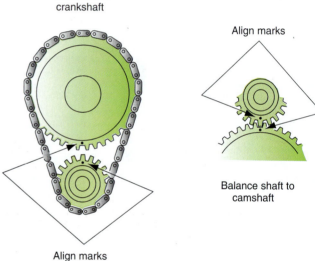

Camshaft to crankshaft

Align marks

Balance shaft to camshaft

Align marks

FIGURE 13-48 Proper balance shaft and camshaft timing.

Camshaft end play is the measure of how far the camshaft can move lengthwise in the block.

amount of end play. Some engines use shims to correct end play. If shims cannot be used, the camshaft may need to be replaced.

7. Apply special camshaft break-in high-pressure lubricant onto the lobes, being careful not to get any onto the journals.

8. Lubricate the balance shaft journals and bearings with the vehicle manufacturer's specified engine oil, and install the balance shaft.

9. Apply sealer to the outer edge of the balance shaft rear expansion plug. Position this plug with the convex side out, and drive the plug into the rear expansion shaft bearing bore to the specified depth. Use a hammer and punch to strike the expansion plug in the center until it is slightly deformed.

10. Install the camshaft and balance shaft matching gears so the timing marks are aligned.

11. Install the camshaft drive gear and timing chain so the timing marks are properly aligned (refer to Figure 13-48).

12. Tighten the balance shaft gear and camshaft gear retaining bolts to the specified torque.
13. Lubricate the camshaft gears, balance shaft gear, timing chain, and crankshaft gear with the specified engine oil.

INSTALLING THE MAIN BEARINGS AND CRANKSHAFT

It is a good practice to replace the pilot bushing or bearing installed in the end of crankshafts used with manual transmissions. The old bushing or bearing is pulled out using a special puller adapter (Figure 13-49). Remove any burrs inside the cavity, being careful not to enlarge the size of the bore. Then drive the new bushing or bearing into place (Figure 13-50). Be careful to start the pilot straight. The pilot should enter the cavity with moderate hammer blows. Do not use excessive force to drive the pilot in. Soak the bushings in oil before installing them. Pilot bearings may have an integral seal that must be installed in the correct direction. Most pilot bearings are lubricated during assembly and do not require any additional lubrication during installation.

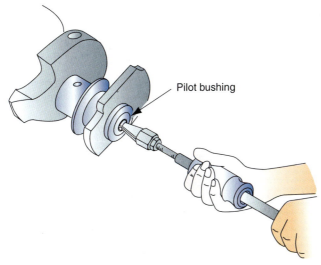

Pilot bushing

FIGURE 13-49 Using a puller to remove the pilot bushing or bearing.

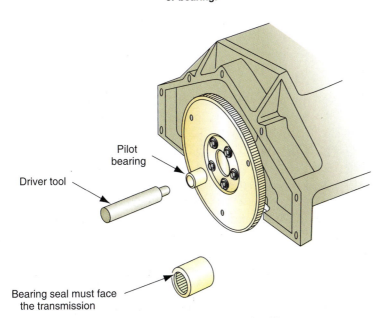

Pilot bearing

Driver tool

Bearing seal must face the transmission

FIGURE 13-50 Installing the pilot bushing.

SPECIAL TOOLS
Torque wrench
Plastigage
Dial indicator
Pilot bushing/
bearing puller
Pilot bushing/
bearing driver
Micrometer

SERVICE TIP:
If the balance shaft is not properly timed, severe engine vibrations will occur.

⚠

CAUTION:
If engine components are not properly lubricated during engine assembly, rapid component wear occurs during the first few seconds of engine operation.

Make sure the proper main bearings are selected for the engine. If no machining was required on the crankshaft journals or bearing bores, refer to the service manual for proper bearing selection. Some manufacturers use a select fit bearing selection procedure. A code that is used to select the bearings is usually stamped on the engine block and the crankshaft (Figure 13-51). A chart is then referenced for the proper bearing selection (Figure 13-52). Some manufacturers use color coding instead of numbers.

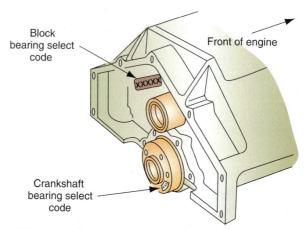

FIGURE 13-51 Engines with select bearings usually have codes stamped on the block and crankshaft to identify the proper selection of bearings.

Cylinder Block Code

	98	99	00	01	02	03	04	05	06	07	08	09	10	11	12	13	14	15	16	17	18	19	20	21
91	1	1	1	1	1	1	1	1	1	1	1	1	1	1	1	1	1	2	2	2	2	2	2	2
90	1	1	1	1	1	1	1	1	1	1	1	1	1	1	1	1	2	2	2	2	2	2	2	2
89	1	1	1	1	1	1	1	1	1	1	1	1	1	1	1	2	2	2	2	2	2	2	2	2
88	1	1	1	1	1	1	1	1	1	1	1	1	1	1	2	2	2	2	2	2	2	2	2	2
87	1	1	1	1	1	1	1	1	1	1	1	1	1	2	2	2	2	2	2	2	2	2	2	2
86	1	1	1	1	1	1	1	1	1	1	1	1	2	2	2	2	2	2	2	2	2	2	2	2
85	1	1	1	1	1	1	1	1	1	1	1	2	2	2	2	2	2	2	2	2	2	2	2	2
84	1	1	1	1	1	1	1	1	1	1	2	2	2	2	2	2	2	2	2	2	2	2	2	3
83	1	1	1	1	1	1	1	1	1	2	2	2	2	2	2	2	2	2	2	2	2	2	3	3
82	1	1	1	1	1	1	1	1	2	2	2	2	2	2	2	2	2	2	2	2	2	3	3	3
81	1	1	1	1	1	1	1	2	2	2	2	2	2	2	2	2	2	2	2	2	3	3	3	3
80	1	1	1	1	1	1	2	2	2	2	2	2	2	2	2	2	2	2	3	3	3	3	3	3
79	1	1	1	1	1	2	2	2	2	2	2	2	2	2	2	2	2	3	3	3	3	3	3	3
78	1	1	1	1	2	2	2	2	2	2	2	2	2	2	2	3	3	3	3	3	3	3	3	3
77	1	1	1	2	2	2	2	2	2	2	2	2	2	2	3	3	3	3	3	3	3	3	3	3
76	1	1	2	2	2	2	2	2	2	2	2	2	2	3	3	3	3	3	3	3	3	3	3	3
75	1	2	2	2	2	2	2	2	2	2	2	2	3	3	3	3	3	3	3	3	3	3	3	3
74	2	2	2	2	2	2	2	2	2	2	2	3	3	3	3	3	3	3	3	3	3	3	3	3
73	2	2	2	2	2	2	2	2	2	2	3	3	3	3	3	3	3	3	3	3	3	3	3	3
72	2	2	2	2	2	2	2	2	2	3	3	3	3	3	3	3	3	3	3	3	3	3	3	3
71	2	2	2	2	2	2	2	2	3	3	3	3	3	3	3	3	3	3	3	3	3	3	3	3
70	2	2	2	2	2	2	2	3	3	3	3	3	3	3	3	3	3	3	3	3	3	3	3	3
69	2	2	2	2	2	2	3	3	3	3	3	3	3	3	3	3	3	3	3	3	3	3	3	3
68	2	2	2	2	2	3	3	3	3	3	3	3	3	3	3	3	3	3	3	3	3	3	3	3

FIGURE 13-52 Bearing selection chart.

In preparation for installation of the main bearings and the crankshaft, use the engine stand to rotate the block so the crankcase is facing up. Before installing the crankshaft main bearings, clean the block saddles, caps, and bearings thoroughly. When installing the bearing inserts, hold them between the thumb and forefinger and slightly squeeze the ends together. Install the bearing insert and push in on the ends to be sure it is fully seated. When installing the crankshaft, lift it by the harmonic balancer journal and the bolts threaded into the rear flange. As the crankshaft is lowered into the bearings, keep it parallel to the bearings and lower it straight down. If it is installed properly, the crankshaft will seat squarely and will be supported by the bearings.

Main bearing oil clearance must be checked and corrected, if needed. There are three common methods of measuring oil clearance. The first is to install the bearings into the block and torque the cap bolts to the assembled torque specification (the crankshaft is not installed at this time). Measure the inside diameter of the bearings, and record the reading. It is important to measure the bearing vertically in the engine since most main bearings are thinner at the parting lines. Subtract the outside diameter size of the corresponding main bearing journals to get total clearance. Actual clearance is obtained by dividing this value by 2.

The second, and most common, method of measuring oil clearance is to use Plastigage. Although it is not as accurate as measuring oil clearance with a micrometer, Plastigage is usually accurate enough for most engine applications. Follow Photo Sequence 26 for using Plastigage to determine oil clearance.

The third method requires several measurements but is the most accurate. Begin by measuring the wall thickness of the bearing insert. Measure the insert in the center where there is no taper. Double the measured dimension so both sides of the journal are considered. Also measure the journal diameter and bore diameter. Subtract all measurements from the bore diameter to obtain total clearance (Figure 13-53). Divide total clearance by 2 to obtain actual clearance.

Reference Photo Sequence 26 and use the following list as a guide to installing the main bearings and the crankshaft (Figure 13-54):

1. Make sure the bearings are the correct size for the journals and bores.
2. Install all upper bearing inserts into the block. These inserts have oil holes and grooves that must be properly aligned with the block. Make sure to install the thrust washer into the specified saddle.
3. Install the lower bearing inserts into the caps.
4. Inspect the cap bolts for damage.
5. Measure bearing clearance.
6. If the rear main seal is a two-piece lip seal, install the pieces into the block and rear cap. Make sure the lip is facing into the crankcase.
7. Lubricate the inside surfaces of the bearing inserts with a film of engine oil.
8. Install the crankshaft, caps, and bolts. If required by the manufacturer's service instructions, apply anaerobic sealer between the rear cap and cylinder block (Figure 13-55). Place the main bearing caps over the crankshaft journals starting at the rear journal and working toward the front of the engine. Make sure each cap is installed in the correct location and in the proper direction. **Register** the main bearing caps by tightening the cap bolts by hand while rotating the crankshaft.

Registering the main bearing caps is done by tightening the cap bolts using hand pressure only while rotating the crankshaft. This squares the cap to the block before the bolts are torqued.

Housing bore	2.124"
Bearing wall thickness (0.061) x 2	-0.122"
Bearing assembled inside diameter	2.002"
Shaft diameter	-2.000"
Oil clearance	0.002"

FIGURE 13-53 Example of mathematically figuring oil clearance.

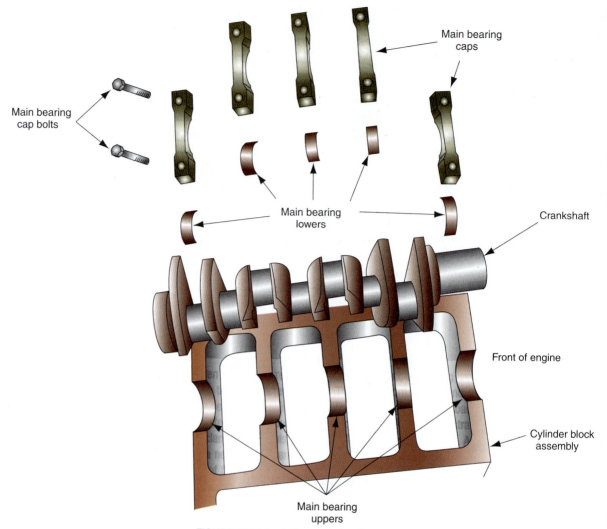

Main bearing cap bolts

Main bearing caps

Main bearing lowers

Crankshaft

Front of engine

Cylinder block assembly

Main bearing uppers

FIGURE 13-54 Installing main bearing caps.

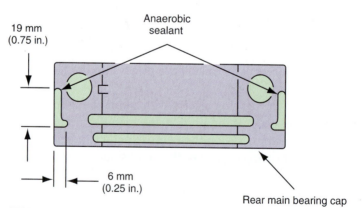

19 mm (0.75 in.)

Anaerobic sealant

6 mm (0.25 in.)

Rear main bearing cap

FIGURE 13-55 Apply sealant to the rear main bearing cap, if indicated.

9. With the main bearing cap bolts fingertight, pry the crankshaft back and forth to align the thrust bearing (Figure 13-56).

10. Engines with bedplates (one-piece main bearings) may also require the use of sealers (Figure 13-57). In some instances, installation of the bedplate must be done within 6 minutes after the sealant is applied.

11. The crankshaft should rotate freely, with no binding or restrictions. Correct any problems before continuing.

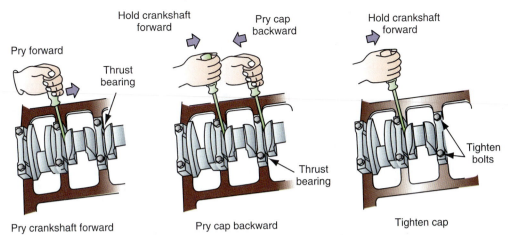

FIGURE 13-56 Aligning the thrust bearing.

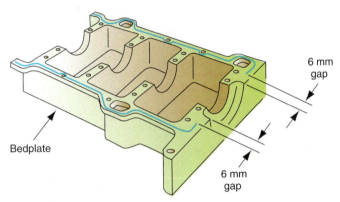

FIGURE 13-57 One-piece main bearing caps may require a special application of sealant.

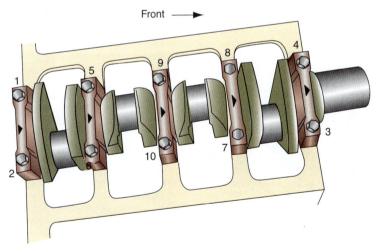

FIGURE 13-58 Typical main bearing torque sequence.

12. Torque the bolts of the first cap in sequence to the specified value (Figure 13-58). Check crankshaft rotation by rotating the crankshaft with a torque wrench. Note the amount of torque required to rotate the crankshaft during several revolutions. Read only rotating torque, not the amount of torque required to get the crankshaft to rotate.

13. Torque the next cap in the sequence and check the rotating torque of the crankshaft again.

14. Continue to torque the cap bolts in sequence, making sure to check rotating torque after each cap is tightened. Watch for large increases in the torque required to rotate

SERVICE TIP:
If all the engine needs is new main bearings, they can usually be changed with the engine still in the vehicle. When checking bearing clearance with Plastigage, use a jack to remove the load from the crankshaft so accurate measurements can be taken.

the crankshaft. If there is a large increase in the torque requirement, stop and locate the cause of the problem.

15. After all the cap bolts are torqued, check the rotating torque of the crankshaft. It should not require more than 5 lb.-ft. (6.75 N•m) to rotate the crankshaft.

16. If the engine uses a one-piece lip seal, install it (Figure 13-59). Some manufacturers require the use of a special pilot tool for installation. After the seal is located over the pilot tool, it is tapped into place using a plastic hammer. Other manufacturers may require the use of special drivers.

17. Check crankshaft end play (Figure 13-60). With the dial indicator set up to read lateral movement of the crankshaft, use a pry tool to move the crankshaft as far to the rear of the block as possible. Zero the dial gauge at this point; then force the crankshaft as far forward as it will go. The TIR must be within specifications.

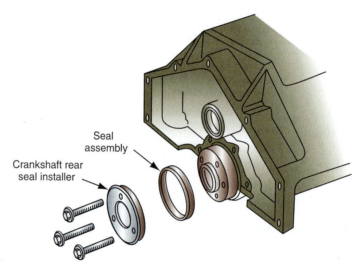

FIGURE 13-59 Installing a one-piece rear main seal.

FIGURE 13-60 Check to be sure that crankshaft end play is within specifications. Too much end play will cause knocking on acceleration; too little could cause the crank to seize.

INSTALLING THE PISTONS AND CONNECTING RODS

Begin the piston installation procedure by installing the bearing inserts into the connecting rods and caps, making sure the oil holes align and the tabs are properly located (Figure 13-61). Coat the cylinder walls with clean engine oil, and soak the piston head in a container of clean engine oil. If oil clearance is being checked using a micrometer, it is easier to measure the inside diameter of the bearing before the piston assembly is installed into the engine block. The cap bolts must be torqued to the specified value while the connecting rod and cap are held in a rod vise to maintain alignment. In addition, oil clearance can be checked using Plastigage. If bearing clearance is being measured using Plastigage, do not apply oil to the bearings. Similar to crankshaft main bearings, connecting rod bearing clearances should always be checked with Plastigage before continuing with reassembly. Do not apply any oil to these bearings during the measurement and measure one connecting rod at a time. After measuring and correcting for the bearing clearance on one rod, lubricate the bearing with engine oil or assembly lube and torque the cap down to specification. Check the rotating torque of the crankshaft assembly after each connecting rod is torqued to specification. This method will allow you to pinpoint any problems with a single connecting rod or bearing during reassembly.

It may be easier to install the connecting rod caps if the crankshaft throw is located at bottom dead center (BDC) as each piston is installed, and to rotate the engine on the stand so the bore is horizontal. Cover the connecting rod bolts with a special guide or pieces of rubber hose to protect the cylinder walls and crankshaft rod journals from damage (Figure 13-62). To prevent damage to the piston and rings, use a ring compressor to squeeze the rings together. Double-check to be sure that each piston is being installed into the correct corresponding bore and that it faces the correct direction. The pistons should go into the bore with a slight restriction. Using a hammer handle to tap the piston assembly into the cylinder will make installation easier (Figure 13-63). When installing the rod caps to the connecting rods, make sure to match the correct units, and make sure they are facing the correct direction.

When installing and torquing the cap bolts for the final assembly, coat the bolt threads with a thread locking compound. As each connecting rod is installed onto the crankshaft, tighten the cap bolts only fingertight. Check the rotating torque of the crankshaft as each piston is installed. This gives an indication of the interference between the piston rings and

SPECIAL TOOLS

Torque wrench
Plastigage
Micrometer
Rod protectors
Ring compressor
Dial indicator

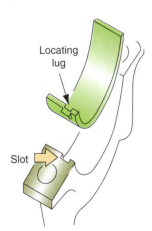

FIGURE 13-61 Install the rod bearings into the cap and rod. Be sure that the tab locks into the slot.

FIGURE 13-62 Cover the connecting rod bolts to protect cylinder walls and crankshaft rod journals from damage.

FIGURE 13-63 Make sure that the piston ring compressor's arrow is pointing in the correct direction and make sure that the piston is facing the correct direction. Use a soft-handled hammer and tap the pistons in place. If you encounter resistance, stop and see if a ring has popped out; you can easily break a ring in this procedure.

FIGURE 13-64 Be certain to torque each of the rod bolts properly; one loose bolt could ruin an engine.

FIGURE 13-65 Use a feeler gauge to measure the clearance between two rods sharing a journal.

cylinder walls. If any of the pistons causes a large increase in turning torque, repair the cause before continuing. After all the pistons are installed, check the rotating torque and record it.

Starting with the first connecting rod in sequence, torque the cap bolts (Figure 13-64). As the bolts are torqued, tighten them in small increments, alternating back and forth between the two bolts. After each piston assembly is properly torqued, rotate the crankshaft to check for binding. With all connecting rods torqued to specifications, recheck turning torque. The total increase in torque should not be more than 5 lb.-ft. (6.75 N•m) above the value recorded earlier.

After all piston assemblies are installed, use a feeler gauge to check connecting rod side clearance (Figure 13-65). Finally, flip the engine over and double check to be sure all your pistons are facing in the correct direction (Figure 13-66).

COMPLETING THE BLOCK ASSEMBLY

Most of the measurement checks have been completed, and the block is ready to receive the oil pump, cylinder heads, valvetrain, valve timing units, and manifolds.

Classroom Manual

Chapter 13, page 000

FIGURE 13-66 You can see the notches on the right of the pistons all facing forward on this engine.

Installing the Oil Pump

The oil pump can usually be installed next. If the oil pump is integral to the front cover, it will be necessary to install the valve timing components first. Always replace the front cover oil seal using an appropriate driver (Figure 13-67). Use the following list to check progress as the oil pump is installed (Figure 13-68):

SPECIAL TOOLS
Torque wrench

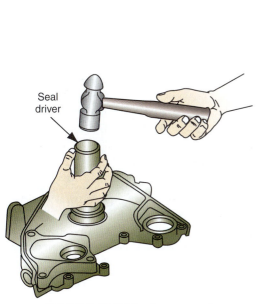

FIGURE 13-67 Driving in a new seal.

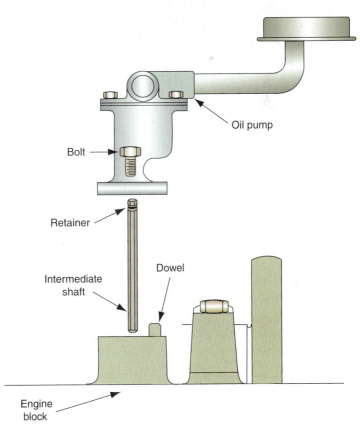

FIGURE 13-68 Oil pump installation.

TYPICAL PROCEDURE FOR INSTALLING THE MAIN BEARINGS AND CRANKSHAFT

P26-1 Thoroughly clean the engine block and bearing bores.

P26-2 Install the new bearings in the block. Avoid touching the front and back sides of the bearings and do not lubricate the back sides.

P26-3 Install the lower bearing halves into the bedplate or the main caps.

P26-4 Thoroughly lubricate the new bearings after they have been put into their bores and before laying the crankshaft in place. Some technicians use a heavier engine assembly lube.

P26-5 Lay the perfectly clean crankshaft in place.

P26-6 Use the proper sealant as specified on the crankshaft bedplate.

P26-7 Torque the bedplate (or main caps) in the proper sequence and stages to the final torque. Double check the final torque to be sure that you didn't miss any bolts. It is a good idea to spin the crankshaft after each main cap is fastened.

P26-8 This engine uses six bolt main caps; be sure to torque all the bolts properly.

P26-9 With the crankshaft bolted in, confirm that the crankshaft rotate smoothly and has end play within the specified range. Use a dial torque wrench to check rotating (spinning) torque.

1. Prime the oil pump before installing it by submerging it into a pan of clean engine oil and rotating the gears by hand. When the pump discharges a stream of oil, the pump is primed.
2. If a gasket is installed between the pump and the engine block, check to make sure no holes or passages are blocked by the gasket material. Soak the gasket in oil to soften the material and allow for good compression.
3. Install the oil pump, making sure the drive shaft is properly seated. Torque the fasteners to the specified torque value (Figure 13-69).
4. Bolt-on **pickup tubes** may use a rubber O-ring to seal them. Do not use a sealer in place of the O-ring. Lubricate the O-ring with engine oil prior to assembly, then alternately tighten the attaching bolts until the specified torque is obtained.

The **pickup tube** is used by the pump to deliver oil from the bottom of the oil pan. The bottom of the tube has a screen to filter larger contaminants.

Installing the Cylinder Head

If the engine is an OHC type, the cylinder heads must be installed before the timing components. On OHV engines, the timing components can be installed before the heads. The procedures for cylinder head installation are similar. On some OHC engines, the camshaft and valvetrain components can be installed into the head before the head is placed onto the engine block. Use the following checklist along with the appropriate service manual as the cylinder heads are installed.

1. Any bolts that enter into coolant or oil passages must have approved sealer applied to the threads.
2. Install any dowels that were originally equipped on the deck (Figure 13-70).
3. Locate any markings on the cylinder head gasket that identify the proper direction for installation (Figure 13-71).
4. Position the new head gasket over the dowels and check for proper fit.
5. Locate the crankshaft a few degrees before TDC so all pistons are below deck height.
6. Position the cylinder head over the dowels. On V-type engines, make sure the correct head is installed onto the correct bank.

SPECIAL TOOLS
Torque wrench
Alignment dowels

FIGURE 13-69 This oil pump tightening specification was given in inch-pounds. Torque to the correct value; the manufacturer has engineered the engine carefully.

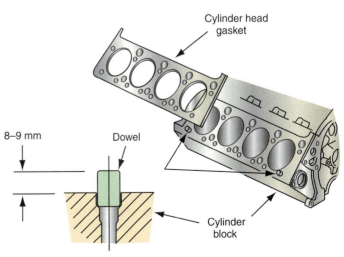

FIGURE 13-70 Alignment dowel installation.

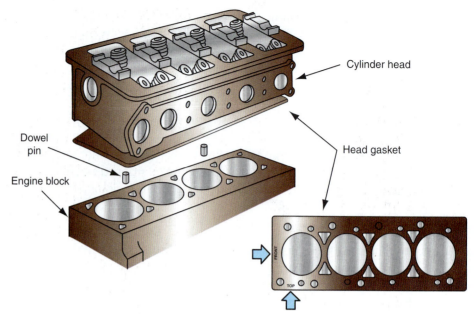

FIGURE 13-71 Most cylinder head gaskets have markings to indicate proper installation direction.

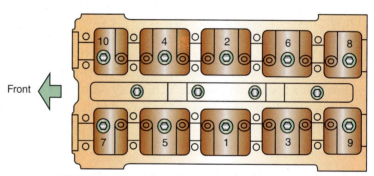

FIGURE 13-72 Typical cylinder head torque sequence.

7. Install and torque the cylinder head bolts. Most manufacturers specify multiple steps to torquing the head bolts. Follow these instructions to prevent warpage (Figure 13-72). Also follow the correct torque sequence. Torque-to-yield cylinder head bolts are used in many engines. These bolts must be tightened to the specified torque and then rotated a specific number of degrees. Torque-to-yield bolts must be replaced each time the cylinder head is removed and replaced.

VALVE TIMING AND INSTALLING VALVETRAIN COMPONENTS

On OHV engines, the valve timing components can be installed after the camshaft and crankshaft. On OHC engines, these components are installed after the cylinder head and the valvetrain components. Regardless of the type of engine, it is important to install the valve timing components properly and to index them properly. Valve timing is usually done by aligning index marks. This procedure was discussed in Chapter 11 of this manual. If a balance shaft is used, it will also have timing marks, and it must be properly indexed.

When installing the valve timing components, be sure that all idler and tensioner pulleys are free to rotate. In addition, the water pump may be driven from the timing belt, requiring installation before the timing belt.

On many OHC engines, the valvetrain components can be installed before the cylinder head. If they are already installed, pour clean engine oil over them.

OHV Valvetrain Installation

After the timing chain or belt is properly installed, the timing chain cover is installed next. Apply a small bead of sealant around the outer diameter of the timing cover seal, and install it with the lip facing toward the engine block. Lubricate the lip of the seal with engine oil. Locate the timing cover gasket onto the engine block, or place a bead of sealer onto the block. Install the timing cover on the engine block. Torque the fasteners to the specified value.

If the water pump is located outside the timing cover, it is installed next. Begin by placing the water pump gasket(s) onto the engine block, using adhesive if necessary. Install the bypass hose onto the nipple of the water pump. Place the hose clamps over the hose. Position the water pump to the engine block while installing the bypass hose to the block nipple. Torque the fasteners to the specified valve. Tighten the hose clamps.

The remaining valvetrain components are installed next. Following is a typical procedure used on most OHV engines:

1. Dip the lifters into clean engine oil.
2. Install the lifters into their bores.
3. When properly installed, the lifters should rest on the camshaft lobes.
4. Rotate each lifter to be sure it is not bound in the bore.
5. Lubricate the pushrods with engine oil or assembly lube.
6. Install the pushrods, making sure they seat properly in the lifters.
7. If oil baffles or pushrod guides are used, install them.
8. Place the rocker arms or shaft assemblies over the pushrods, making sure the pushrods seat properly in the rocker arms.
9. If the rocker arms are nonadjustable, torque the attaching bolts to the specified torque.

OHC Engine Valvetrain

Following is a typical procedure for installing valvetrain components on OHC engines:

1. Dip the lash adjusters into clean engine oil.
2. Install the lash adjusters into their bores.
3. Rotate each lash adjuster to make sure it is not bound in the bore.
4. Place the followers over the lash adjusters. Make sure the valve stem is properly seated.
5. Install the camshaft. If caps are used, torque them to proper values.
6. Install the camshaft sprockets, aligning the keyway, and torque the fastener. It may be necessary to install the rear portion of the front cover before installing the sprockets.
7. Pour engine oil over the valvetrain assembly.

Most OHC engines run the water pump from the timing belt, requiring it to be installed before the belt. Use adhesive to locate the gaskets onto the engine block. Place the water pump over the gasket, and torque the fasteners to the specified value.

The timing belt is usually installed next (Figure 13-73); however, on some engines, it may be necessary to install the flex plate or flywheel before setting valve timing. This will be necessary if the ignition timing marks are located on the flex plate or flywheel (Figure 13-74). When installing the flex plate or flywheel, align the scribe marks made for reference during disassembly.

Install the valve timing components and set valve timing using the procedures discussed in Chapter 11.

SERVICE TIP:
On some engines, the water pump can be installed onto the timing cover before the cover is installed.

CAUTION:
If the original lifters are being reused, they must be reinstalled in their original positions on the camshaft.

CAUTION:
If the original followers and lash adjusters are being reused, they must be reinstalled in their original positions on the camshaft.

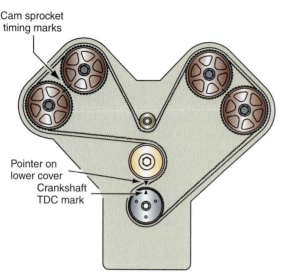

Cam sprocket
timing marks

Pointer on
lower cover

Crankshaft
TDC mark

FIGURE 13-73 Timing belt installation on an OHC engine.

TDC mark

Inspection
hole

Flywheel

Pointer
on block

FIGURE 13-74 Some engines have timing
marks on the flywheel.

CHECKING AND ADJUSTING VALVE CLEARANCE

After the valvetrain components are installed and the valves are properly timed, valve clearance should be checked (and in some instances adjusted). Refer to Chapter 10 of this manual for these procedures.

Valve clearance is easier to adjust on V-type engines when the intake manifold is still off. This is because lifters and pushrod positions can be seen. The correct hydraulic lifter adjustment is obtained by removing the lash between the lifter and the pushrod, then depressing the pushrod a specified distance into the lifter.

COMPLETING THE ENGINE ASSEMBLY

Completion of the assembly process requires the installation of such items as the oil pan, harmonic balancer, manifolds, valve covers, flywheel, fuel pump, and various accessories. Inspect the crankshaft vibration damper for wear on the sides of the pulley grooves. Replace the vibration damper if there is wear in this area. Be sure the key, and keyways, in the crankshaft and vibration damper are not worn. Check for scoring on the seal contact area on the vibration damper. Replace the vibration damper if scoring or wear is present in this area. Some vibration dampers may be machined to a smaller size, and a sleeve may be pressed over the seal contact area. If the rubber ring in the vibration damper is cracked, oil soaked, or deteriorated, replace the damper. The vibration damper should also be replaced if the outer ring has slipped on the rubber ring, because this places the timing mark in the wrong position if this mark is located on the outer ring.

Inspect the oil pan flanges for a bent condition and metal stretching around the bolt holes. Bent flanges or stretched metal around the bolt hole may be corrected with a hammer and a block of wood. Be sure that all old gasket material is cleaned from the mating surfaces on the oil pan, valve covers, water pump, and manifolds. Following is a guide for completing the engine assembly:

1. Install the oil pan and gasket, and torque the fasteners to the specified value (Figure 13-75). Use sealant only if specified by the manufacturer (Figure 13-76).
2. Coat the crankshaft vibration damper sealing surface with engine oil and apply sealer to the keyway.
3. Install the damper onto the front of the crankshaft. A special tool may be required to press the damper onto the crankshaft.

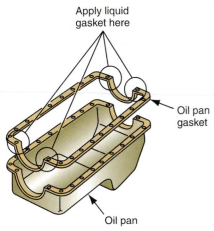

FIGURE 13-75 Oil pan and gasket.

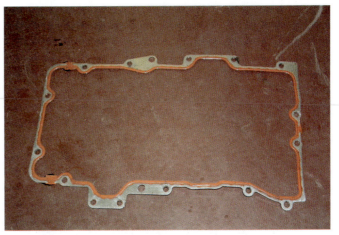

FIGURE 13-76 This oil pan gasket is installed without any additional sealant.

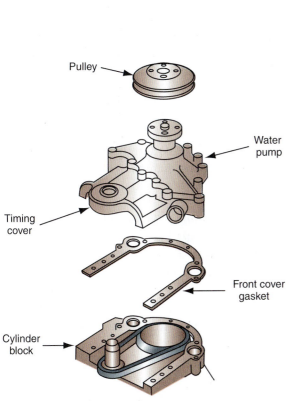

FIGURE 13-77 Water pump, pulley, and timing cover installation.

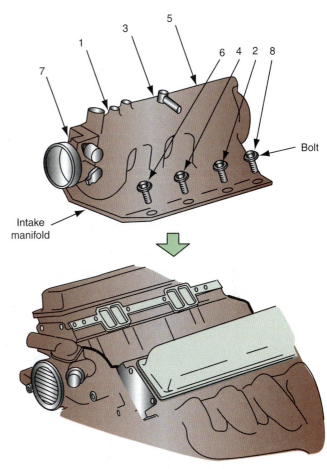

FIGURE 13-78 Intake manifold and typical torque sequence.

4. Torque the fastener to specifications.
5. Install the lower pulley to the vibration damper.
6. Install the water pump pulley (if equipped) and tighten the fasteners to specifications (Figure 13-77).

The intake manifold is installed next (Figure 13-78). It is usually easier to adjust the valves before installing the intake manifold on V-type engines. If valve clearance has not been set, set it at this time. First check for proper fit of the intake manifold gasket.

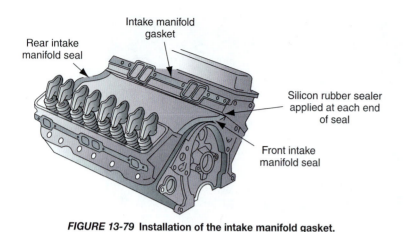

FIGURE 13-79 Installation of the intake manifold gasket.

Flange gasket

Alignment tabs

Cylinder head gasket

FIGURE 13-80 Some intake gaskets must be indexed to the head gasket.

Most gaskets are marked to indicate proper direction for installation. After testing the fit, remove the intake gasket. Apply sealer at any locations directed in the service manual. On V-type engines, the four intersection points of the cylinder heads and the engine block usually require sealer. On V-type engines, locate the front and rear end seals onto the block (Figure 13-79). In addition, the intake manifold gasket may have alignment tabs that must fit into the head gasket for proper installation (Figure 13-80).

The studs will hold the gasket in place while the manifold is installed on most in-line engines. On V-type engines, the gasket may slip as the manifold is lowered into place. To prevent this, use an adhesive to hold the gasket in place. Carefully lower the manifold into position. Then install the fasteners and torque to specifications.

WARNING: If the cylinder heads on a Vee engine have been machined, the intake manifold must be machined accordingly so. Failure to machine the intake manifold before reassembly may cause oil and coolant leaks or consumption as well as reduced performance. Refer to Chapter 9 for more information.

On fuel-injected engines, install the fuel rail and the intake plenum and gasket. Torque the fasteners to the specified value. Next, install the throttle body assembly (Figure 13-81).

Next, install the exhaust manifold and gasket. Replace all mounting hardware used with the exhaust manifold. The old fasteners may have been weakened as a result of the extreme temperatures. When installing oxygen sensors into the exhaust manifold, use an approved antiseize compound on the threads. Be sure to install all original heat shields and brackets onto the exhaust system.

Some technicians prefer to leave the valve covers off until after the engine has been preoiled so they can see if oil is reaching the upper engine and to assist in still timing of the engine. Whenever the covers are installed, the following procedures apply:

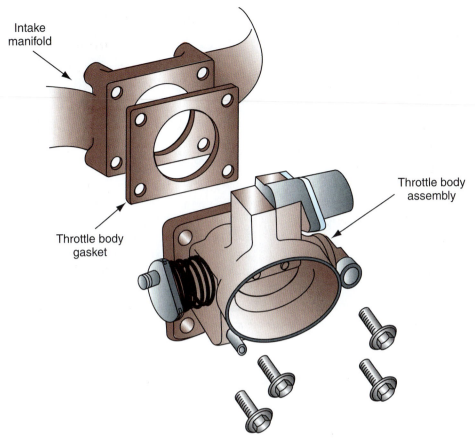

FIGURE 13-81 A typical throttle body installation.

1. Before installing a stamped steel cover, use a ball-peen hammer and a block of wood to remove deformities in and around the bolt holes.
2. If a plastic or cast aluminum cover is used, it must be replaced if it is distorted.
3. Torque the valve cover attaching fasteners using the manufacturer's specified torque (Figure 13-82). If no pattern to reference is available, use a cross pattern to prevent warpage to the cover. Overtightening the valve covers can result in oil leakage and can possibly crack the cover.
4. Always follow the manufacturer's torque specification value and sequence when tightening. Overtightening the cylinder head (valve) covers can cause leakage from an unevenly pressed gasket. It may also bend or crack the cover.

Installation of the flex plate or flywheel is next. Inspect the flywheel and crankshaft flange mating areas for metal burrs. Remove any metal burrs with a fine-toothed file. Metal burrs in this area cause excessive flywheel runout and improper clutch or transmission operation such as a grabbing clutch. Inspect the flywheel ring gear for worn teeth and cracks. Worn ring gear teeth result in improper starting motor engagement. If these teeth are worn or the ring gear is cracked, ring gear replacement is necessary. Use the alignment marks inscribed during disassembly to properly index the flywheel to the crankshaft. Many manufacturers offset the bolt hole pattern so the flywheel can be installed in one position only. With the bolts properly torqued, use a dial indicator to check flywheel runout (Figure 13-83). Check runout by holding the crankshaft as far forward or rearward as it will go so end play is not indicated. Set the dial to read zero, then turn the flywheel one complete revolution while observing the dial indicator gauge. The TIR must be within specifications. If it is not, the flywheel may need to be resurfaced or replaced; however, before condemning the flywheel, remove it from the crankshaft flange and install the dial indicator

SERVICE TIP:
Staking the engine block where the end strip gaskets are installed with a center punch will cause small bits of metal to be raised. This raised metal will catch and hold the strips in place while the intake manifold is being installed.

CAUTION:
The fasteners used to secure the intake manifold must be properly torqued to prevent vacuum leaks. Use the tightening sequence and procedure recommended in the service manual.

FIGURE 13-82 Torque the valve covers; a careless installation could cause an oil leak and an unhappy customer.

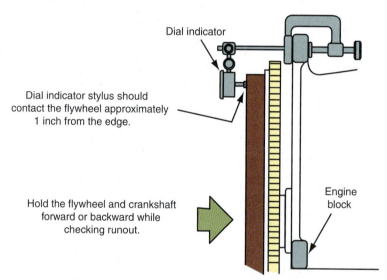

Dial indicator

Dial indicator stylus should contact the flywheel approximately 1 inch from the edge.

Hold the flywheel and crankshaft forward or backward while checking runout.

Engine block

FIGURE 13-83 Using a dial indicator to measure flywheel runout.

SPECIAL TOOLS

Dial indicator

to check flange runout. Also make sure there are no burrs on the mounting flange or the mounting surface of the flywheel.

Install all accessory items that attach to the block. These may include:

- The accessory drive belts, generator, power steering pump, air injection reaction (AIR) pump, air-conditioning compressor, and water pump
- Ignition coil and spark plugs, exhaust gas recirculation (EGR) assembly, thermostat housing, and thermostat
- All electrical sensors and switches (Figure 13-84)
- Engine mounts

If the engine uses a mechanical fuel pump operated off an eccentric on the camshaft, install it. On some designs, the arm of the fuel pump must ride on top of the eccentric, while on other designs, it must ride on the bottom of the eccentric. Some engines use a rod between the camshaft eccentric and pump arm. The rod must be held up against the camshaft as the fuel pump is installed.

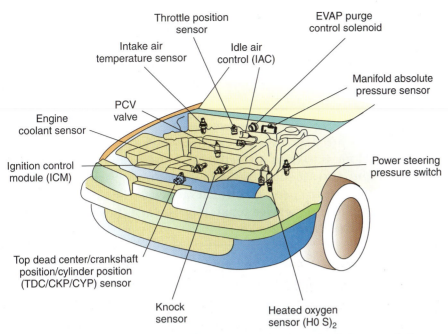

Throttle position
sensor

EVAP purge
control solenoid

Intake air
temperature sensor

Idle air
control (IAC)

Manifold absolute
pressure sensor

PCV
valve

Engine
coolant sensor

Power steering
pressure switch

Ignition control
module (ICM)

Top dead center/crankshaft
position/cylinder position
(TDC/CKP/CYP) sensor

Knock
sensor

Heated oxygen
sensor (H0 S)$_2$

FIGURE 13-84 Modern vehicles have several sensors and switches mounted on the engine assembly.

FIGURE 13-85 The technician double checked her work, repainted the old components, and is anxious to hear her freshly built engine run.

SERVICE TIP:
It may be easier on some engines to leave the spark plugs and distributor out until the engine is installed into the vehicle. Leaving the spark plugs out will make it easier to rotate the engine if needed.

CAUTION:
Overtightening the oil filter may cause seal damage and result in oil leakage.

To complete the assembly, lubricate the oil filter gasket and fill the filter with oil. Install the oil filter. Do not overtighten the filter. Turn the filter three-quarter turn after the seal makes contact.

With the engine fully assembled, double check your work before installation (Figure 13-85). Follow the installation and break-in procedures provided in Chapter 8 of the Shop Manual to ensure that your hard work is fully successful.

TERMS TO KNOW

Big end

Camshaft end play

Dry starts

Head

Land

Luagging

Minor thrust side

Oversized bearings

Pickup tubes

Pin boss

Pressure plate

Register

Sizing point

Skirt

Small end

Undersized bearings

CASE STUDY

Many hours of labor have been put into the rebuilding of the customer's engine. The technician's reputation is on the line with every job performed. Taking pride in a job well done is a big step towards building a good reputation. The technician has seen his efforts restore a worn engine into a new power plant. Now he is ready for the moment of truth. The engine has been installed, everything has been double-checked. The key is turned over and the engine roars to life. Before turning the vehicle back over to the owner, the technician takes a few extra moments to clean the vehicle and reset the radio. A few days later he calls the customer to confirm all is right and remind her of the scheduled service.

ASE-STYLE REVIEW QUESTIONS

1. The main bearings indicate wear on the lower inserts on all journals.

   Technician A says this indicates lack of oil.

   Technician B says this indicates excessive idling.

   Who is correct?

   A.   A only
   C.   Both A and B
   B.   B only
   D.   Neither A nor B

2. Removing press-fit piston pins is being discussed.

   Technician A says the pin can be carefully driven out with a punch and hammer while securing the piston in a vise.

   Technician B says removing the pin requires a press and special adaptors.

   Who is correct?

   A.   A only
   C.   Both A and B
   B.   B only
   D.   Neither A nor B

3. Piston ring end gap is being discussed.

   Technician A says an end gap that is too wide can result in scuffing.

   Technician B says that an end gap that is too narrow can allow blowby.

   Who is correct?

   A.   A only
   C.   Both A and B
   B.   B only
   D.   Neither A nor B

4. *Technician A* says that a bend in a connecting rod will be noticeable during a visual inspection.

   Technician B says that a bent connecting rod will produce diagonal scuffing on the piston.

   Who is correct?

   A.   A only
   C.   Both A and B
   B.   B only
   D.   Neither A nor B

5. The camshaft bearings are being installed in the block of an OHV engine.

   Technician A says the camshaft bearings should be installed with a lubricant between the bearing and the journal of the block.

   Technician B says that the camshaft bearings usually require a special tool to install them because they are press-fit.

   Who is correct?

   A. A only   C. Both A and B
   B. B only   D. Neither A nor B

6. *Technician A* says excessive camshaft end play may cause timing chain noise.

   Technician B says camshaft end play is checked using a feeler gauge.

   Who is correct?

   A. A only   C. Both A and B
   B. B only   D. Neither A nor B

7. Main bearing installation is being discussed.

   Technician A says many manufacturers use a select fit bearing selection procedure.

   Technician B says most manufacturers identify the use of select bearings by codes on the engine block and the crankshaft.

   Who is correct?

   A. A only   C. Both A and B
   B. B only   D. Neither A nor B

8. *Technician A* says Plastigage strips should be placed length-wise on the journal.

   Technician B says Plastigage measures total bearing clearance.

   Who is correct?

   A. A only   C. Both A and B
   B. B only   D. Neither A nor B

9. *Technician A* says most piston assemblies are nondirectional.

   Technician B says the notch on the piston head usually faces toward the front of the engine block.

   Who is correct?

   A. A only   C. Both A and B
   B. B only   D. Neither A nor B

10. If a harmonic balancer outer ring has slipped on the rubber mounting, the technician should:

    A. Twist the outer ring back to the original position.
    B. Secure the outer ring by welding it in place.
    C. Replace the harmonic balancer.
    D. Press on a new outer ring.

ASE Challenge Questions

1. If piston rings are installed without removing the ring ridge, the following may result:
    A. The piston skirt may be damaged.
    B. The piston pin may be broken.
    C. The connecting rod bearings may be damaged.
    D. The piston ring lands may be broken.

2. A bent connecting rod may cause:
    A. Uneven connecting rod bearing wear.
    B. Uneven main bearing wear.
    C. Uneven piston pin wear.
    D. Excessive cylinder wall wear.

3. While discussing piston ring service:

    Technician A says that the ring end gap should be measured with the ring just below the ring ridge.

    Technician B says to position the ring at the bottom of ring travel to measure the end gap.

    Who is correct?
    A. A only   C. Both A and B
    B. B only   D. Neither A nor B

4. A piston that is installed backwards will:
    A. Break the connecting rod.
    B. Wear out the rod bearings quickly.
    C. Cause piston slap.
    D. Break the piston.

5. A worn pilot bearing may cause a growling, rattling noise:
    A. While driving at a steady speed of 50 mph (80 kmh).
    B. In reverse with the clutch engaged.
    C. While accelerating in second gear.
    D. With the clutch disengaged.

JOB SHEET

Name _____ **Date** _____

INSPECTING THE PISTONS AND CONNECTING RODS

Upon completion of this job sheet you should be able to properly inspect pistons, connecting rods, and piston pins. You should also be able to remove a piston from a connecting rod.

ASE Correlation

This job sheet is related to ASE Engine Repair Test content area: Engine Block Assembly Diagnosis and Repair; tasks: Inspect and measure pistons and determine necessary action; Remove and replace piston pin.

This job sheet is also related to ASE General Engine Diagnosis; Removal and Reinstallation (R&R); tasks: Research applicable vehicle service information.

Tools and Materials

Micrometer

Vise

Press

Piston pin press tools

Describe the vehicle being worked on:

Year _____ Make _____ Model _____

VIN _____ Engine type and size _____

Procedure

Task Completed

Your instructor will provide you with a specific engine. Write down the information, and then perform the following tasks:

1. Line up the pistons and connecting rods in order on a clean bench. ☐

2. Visually inspect each connecting rod for damage, bend, cracks, nicks. Record results:
   Piston number 1 _____
   Piston number 2 _____
   Piston number 3 _____
   Piston number 4 _____
   Piston number 5 _____
   Piston number 6 _____
   Piston number 7 _____
   Piston number 8 _____

3. Remove the rod bearings and keep them in order for later inspection. ☐

4. Clean the big end bore; note any visible issues, and measure the diameter and for out-of-round. Record results:

Rod number 1 _____

Rod number 2 _____

Rod number 3 _____

Rod number 4 _____

Rod number 5 _____

Rod number 6 _____

Rod number 7 _____

Rod number 8 _____

☐

5. Remove the piston rings using a ring expander and be careful not to scrape the sides of the pistons. Lay to the side for later inspection. Clean the ring grooves using a ring groove cleaner.

6. Visually inspect the piston ring grooves, ring lands, piston heads, and piston skirts for excessive scuffing. Record results:

Piston number 1 _____

Piston number 2 _____

Piston number 3 _____

Piston number 4 _____

Piston number 5 _____

Piston number 6 _____

Piston number 7 _____

Piston number 8 _____

7. Locate the manufacturer's service information for where to measure piston diameter. Record that procedure: _____

_____

_____

_____

_____

_____

_____

8. Measure the piston diameter and compare it to specifications. Record results:

Piston number 1 diameter _____ Within specification? ☐ Yes ☐ No

Piston number 2 diameter _____ Within specification? ☐ Yes ☐ No

Piston number 3 diameter _____ Within specification? ☐ Yes ☐ No

Piston number 4 diameter _____ Within specification? ☐ Yes ☐ No

Piston number 5 diameter _____ Within specification? ☐ Yes ☐ No

Piston number 6 diameter _____ Within specification? ☐ Yes ☐ No

Piston number 7 diameter _____ Within specification? ☐ Yes ☐ No

Piston number 8 diameter _____ Within specification? ☐ Yes ☐ No

9. Hold one piston assembly and feel for free movement of the connecting rod. Place a shop rag around the connecting rod and place it in a vise. Push up and down on the piston to feel for even the slightest vertical movement. It should move sideways freely. Repeat for each piston and rod assembly. Note your results:

Piston number 1 _____

Piston number 2 _____

Piston number 3 _____

Piston number 4 _____

Piston number 5 _____

Piston number 6 _____

Piston number 7 _____

Piston number 8 _____

10. If any pistons felt loose on the rod and your shop has the proper press and piston pin adaptor, remove those pistons from the rod.
Does your shop have the proper equipment to remove the piston?   ☐ Yes   ☐ No

11. If your shop does not have the proper equipment, have your instructor check your work now.

12. If you have the equipment, note or mark the orientation of the piston on the rod. Set the piston and pin press adaptor on the vise. Be certain the piston and press rod are aligned properly. You must have safety glasses on. Press the pin from the piston and connecting rod. Repeat for as many piston assemblies as required. If the piston pins are full floating, simply remove the snap rings and push the pins out.
Did you successfully remove the pin(s)?   ☐ Yes   ☐ No

13. Set the piston and connecting rod aside. Visually inspect the piston pin for wear and measure its diameter. Record your results:

Pin number 1 visual condition and diameter: _____

Pin number 2 visual condition and diameter: _____

Pin number 3 visual condition and diameter: _____

Pin number 4 visual condition and diameter: _____

Pin number 5 visual condition and diameter: _____

Pin number 6 visual condition and diameter: _____

Pin number 7 visual condition and diameter: _____

Pin number 8 visual condition and diameter: _____

14. Compare the pins to specification. Which, if any, are worn? _____
_____

15. Inspect the connecting rod and piston bore for signs of wear. Measure the connecting rod bore using an inside micrometer or a dial bore gauge. Compare the bore to specifications. (On some piston assemblies, it will also be necessary to measure the bore in the piston; perform the work as specified in the service information.) Record results:

Bore number 1 diameter _____ Within specifications?   ☐ Yes   ☐ No

Bore number 2 diameter _____ Within specifications?   ☐ Yes   ☐ No

Bore number 3 diameter _____ Within specifications? ☐ Yes ☐ No

Bore number 4 diameter _____ Within specifications? ☐ Yes ☐ No

Bore number 5 diameter _____ Within specifications? ☐ Yes ☐ No

Bore number 6 diameter _____ Within specifications? ☐ Yes ☐ No

Bore number 7 diameter _____ Within specifications? ☐ Yes ☐ No

Bore number 8 diameter _____ Within specifications? ☐ Yes ☐ No

16. Describe any work necessary to repair improper piston to rod fit. Record results:

Piston number 1 _____

Piston number 2 _____

Piston number 3 _____

Piston number 4 _____

Piston number 5 _____

Piston number 6 _____

Piston number 7 _____

Piston number 8 _____

Instructor's Response _____

_____

_____

JOB SHEET

Name _____ Date _____

INSPECTING THE BEARING WEAR PATTERNS

Upon completion of this job sheet, you should be able to inspect main and connecting rod bearings and determine problems indicated by unusual wear patterns.

ASE Correlation

This job sheet is related to ASE Engine Repair Test content area: Engine Block Assembly Diagnosis and Repair; tasks: Inspect piston and bearing wear patterns that indicate connecting rod alignment and main bearing bore problems; inspect rod alignment and bearing bore condition.

     This job sheet is also related to ASE General Engine Diagnosis; Removal and Reinstallation (R&R) tasks: Research applicable vehicle service information.

Tools and Materials

Dial bore gauge
Straightedge
Feeler gauges

Describe the vehicle being worked on:

Year _____ Make _____ Model _____

VIN _____ Engine type and size _____

Procedure

Your instructor will provide you with a specific engine. Write down the information, and then perform the following tasks:

1. Lay the main bearing halves out with main number 1 bottom half below main number 1 top half. Line up the rest in order in the same fashion. Describe the visual wear on the front and back of each bearing. Record results:
   Bearing number 1 bottom _____
   Bearing number 1 top _____
   Bearing number 2 bottom _____
   Bearing number 2 top _____
   Bearing number 3 bottom _____
   Bearing number 3 top _____
   Bearing number 4 bottom _____
   Bearing number 4 top _____
   Bearing number 5 bottom _____
   Bearing number 5 top _____
   Bearing number 6 bottom _____
   Bearing number 6 top _____
   Bearing number 7 bottom _____

Bearing number 7 top _____

Bearing number 8 bottom _____

Bearing number 8 top _____

2. Is the bearing wear uneven around the bearing halves? ☐ Yes ☐ No
Does the pattern of bearing wear indicate the possibility of main bearing bore misalignment? ☐ Yes ☐ No

☐

3. If main bearing bore misalignment is a possibility, check it now. Torque the main caps on in the proper sequence to the proper specification.
Look up the specification for the maximum allowable main bore misalignment. Record results: _____
Lay a straightedge in the top and bottom of the bearing bores and try to fit a feeler gauge the size of the allowable misalignment under the straightedge in each cap.

Does the feeler gauge fit anywhere? ☐ Yes ☐ No

If the misalignment is too great, describe the procedure to fix the problem:

_____

4. Do your bearings show any concentrated wear on the sides, indicating that a bearing cap was misaligned? ☐ Yes ☐ No

5. Are your bearings smeared, indicating improper or inadequate lubrication, or dirt behind the bearing back? ☐ Yes ☐ No

6. Are any of your bearing backs scored or polished, indicating that they may have spun in their bore? ☐ Yes ☐ No

7. Do any of your connecting rod bearings indicate wear on the opposite sides of the upper and lower bearing, indicating a bent connecting rod? ☐ Yes ☐ No

8. Describe your overall assessment of the main and rod bearings in your engine and what corrective actions you will have to take before reassembling your engine.
Main bearing indications:

_____

_____

Rod bearing indications:

_____

_____

Corrective actions:

_____

_____

Instructor's Response _____

_____

_____

Name _____ Date _____

INSTALLING THE PISTON RINGS

Upon completion of this job sheet, you should be able to properly install piston rings.

ASE Correlation

This job sheet is related to ASE Engine Repair Test content area: Engine Block Assembly Diagnosis and Repair; tasks: Inspect, measure, and install piston rings.

This job sheet is also related to ASE General Engine Diagnosis; Removal and Reinstallation (R&R) tasks: Research applicable vehicle service information.

Tools and Materials

Ring expander
Feeler gauges

Describe the vehicle being worked on:

Year _____ Make _____ Model _____

VIN _____ Engine type and size _____

Procedure

Your instructor will provide you with a specific engine. Write down the information, and then perform the following tasks:

1. Locate the correct rings for your engine. Look up the correct installation of the ring gaps and draw a diagram showing it below:

2. Look up the specification for ring side clearance and for ring end gap. Record results:
Ring side clearance specification: _____
Ring end gap specification: _____

3. Place an oil ring in the cylinder where indicated by the service information or toward the bottom of ring travel and square it with a piston. Measure the end gap. Place the second ring in the cylinder and measure the end gap. Place the top ring in the cylinder and measure the end gap. Repeat for all cylinders. Record results:
Cyl. #1: oil gap _____ Second gap _____ Top gap _____
Cyl. #2: oil gap _____ Second gap _____ Top gap _____
Cyl. #3: oil gap _____ Second gap _____ Top gap _____
Cyl. #4: oil gap _____ Second gap _____ Top gap _____
Cyl. #5: oil gap _____ Second gap _____ Top gap _____
Cyl. #6: oil gap _____ Second gap _____ Top gap _____
Cyl. #7: oil gap _____ Second gap _____ Top gap _____
Cyl. #8: oil gap _____ Second gap _____ Top gap _____

4. Are any gaps too wide? ☐ Yes ☐ No   What, if anything, will you do to correct this?
_____
_____
_____

Are any gaps too narrow? ☐ Yes ☐ No   What, if anything, will you do to correct this?
_____
_____
_____

5. Lubricate piston number 1. Install the oil ring, expander, and second oil ring by hand, being careful not to twist them excessively. Use a ring expander to properly install the second ring. Is there a top marking? ☐ Yes ☐ No
Use a ring expander to properly install the top ring. Is there a top marking?
☐ Yes ☐ No
Repeat for all pistons.

6. Measure ring side clearance for the oil, second, and top rings as specified. Repeat for each piston. Record results:
Piston #1 oil clearance _____ Second clearance _____ Top clearance _____
Piston #2 oil clearance _____ Second clearance _____ Top clearance _____
Piston #3 oil clearance _____ Second clearance _____ Top clearance _____
Piston #4 oil clearance _____ Second clearance _____ Top clearance _____
Piston #5 oil clearance _____ Second clearance _____ Top clearance _____
Piston #6 oil clearance _____ Second clearance _____ Top clearance _____
Piston #7 oil clearance _____ Second clearance _____ Top clearance _____
Piston #8 oil clearance _____ Second clearance _____ Top clearance _____

☐

7. Space the gaps as instructed on each piston.

Instructor's Response _____
_____
_____

JOB SHEET

Name _____ Date _____

INSTALLING THE CAMSHAFT

Upon completion of this job sheet, you should be able to install the camshaft into the engine block (OHV engine) or the cylinder head (OHC engine) properly.

ASE Correlation

This job sheet is related to the Engine Repair Test's content area: Engine Block Diagnosis and Repair; tasks: Inspect camshaft bearings for unusual wear; remove and replace camshaft bearings; install camshaft, timing chain, and gears; check end play.

Tools and Materials

Dial indicator
Micrometer
Inside micrometer
Telescoping gauge
Engine block

Describe the vehicle being worked on:

Year _____ Make _____ Model _____
VIN _____ Engine type and size _____

Procedure

Using the engine block or cylinder head from the engine you have disassembled and the proper service manual, you will identify and perform the required steps to install the camshaft.

1. What are the specifications for oil clearance of the camshaft bearings?
   _____

2. Measure the oil clearance between the installed camshaft bearings and the camshaft journals, and record your results below.

Bore Number	Clearance
1	_____
2	_____
3	_____
4	_____
5	_____

3. Place an * (or more than one, if necessary) in the previous chart to indicate any clearances that are out-of-specification. If any bearing clearances are excessive, consult your instructor.

☐   4. Use a rag soaked in solvent to remove any coating on the camshaft journals. Be careful not to remove the coating on the lobes.

☐   5. Lubricate the journals with engine oil or assembly lube.

☐   6. Install the camshaft, preventing any contact between the lobes and the bearings.

⚠️

CAUTION:

Do not use emery cloth to remove high spots. The grit can embed into the bearing surfaces and damage the journals.

7. Spin the camshaft to assure free rotation. Does the camshaft spin freely?
☐ Yes ☐ No
If NO, inspect the camshaft bearings for high spots, and remove using a three-cornered scraper. If the bearings are not the cause of the binding, the camshaft may be bent.

8. What is the specification for the depth of the rear camshaft core plug?
_____

9. Apply sealer to the outer edge of the rear core plug and install the plug in the correct direction and to the correct depth. Which way does the plug face?
_____

☐   10. After the core plug is installed, rotate the camshaft to assure the plug is not causing interference.

☐   11. Install the camshaft retainer plate.

12. What is the specification for camshaft end play? _____

13. Measure camshaft end play. _____

14. If the camshaft end play is out-of-specification, how can it be corrected?
_____

☐   15. Apply special camshaft break-in high-pressure lubricant onto the lobes, being careful not to get any on the journals.

Instructor's Response _____
_____
_____

JOB SHEET

Name _____ **Date** _____

INSTALLING THE MAIN BEARINGS AND THE CRANKSHAFT

Upon completion of this job sheet, you should be able to install the main bearings and the crankshaft into the engine block properly.

ASE Correlation

This job sheet is related to the Engine Repair Test's content area: Engine Block Diagnosis and Repair; tasks: Install main bearings and crankshaft; check bearing clearances and end play; replace/retorque bolts according to manufacturer's procedures.

Tools and Materials

Dial indicator
Plastigage
Torque wrench
Engine block

Describe the vehicle being worked on:

Year _____ Make _____ Model _____
VIN _____ Engine type and size _____

Procedure

Task Completed

Using the engine block and crankshaft from the engine you have disassembled and the proper service manual, you will identify and perform the required steps to install the main bearings and the crankshaft.

1. Replace the pilot bushing or bearing installed in the end of crankshaft (used with manual transmissions). ☐

2. Make sure the proper main bearings are selected for the engine. Does this engine use select bearings? ☐ Yes ☐ No
   If YES, how are they identified?
   _____

3. Clean the block saddles, caps, and bearings thoroughly. ☐

4. Install all upper bearing inserts into the block. These inserts will have oil holes and grooves that must be properly aligned with the block. Make sure to install the thrust washer into the specified saddle. Which bore is the thrust bearing installed in?
   _____

5. Install the lower bearing inserts into the caps. ☐

6. Inspect the cap bolts for damage. ☐

7. What is the specification for main bearing oil clearance?

_____

□ 8. Is the specification for ACTUAL or TOTAL oil clearance?

9. What method of measuring oil clearance is recommended by the service manual?

_____

□ 10. If using Plastigage, install the crankshaft into the saddles. When installing the crankshaft, lift it by the harmonic balancer journal and by bolts threaded into the rear flange. As the crankshaft is lowered into the bearings, keep it parallel to the bearings, and lower it straight down. If installed properly, the crankshaft will seat square and be supported by the bearings.

11. Measure the oil clearance of each main bearing, and record your results below.

Bore Number	Clearance
1	_____
2	_____
3	_____
4	_____
5	_____

12. Place an * (or more than one, if necessary) in the previous chart to indicate any oil clearances that are out-of-specification. If any clearances are out-of-specification, what procedure can be used to correct them?

_____

_____

_____

□ 13. Correct any oil clearances to specifications as needed.

14. If the rear main seal is a two-piece lip seal, install the pieces into the block and rear cap. Which direction must the lip of the seal face?

_____

□ 15. Lubricate the inside surfaces of the bearing inserts with a film of engine oil.

□ 16. Install the crankshaft, caps, and bolts. Apply anaerobic sealer between the rear cap and cylinder block (if needed).

□ 17. Place the main bearing caps over the crankshaft journals, starting at the rear journal and working toward the front of the engine, making sure the correct cap is installed in the correct location and in the proper direction.

□ 18. Register the main bearing caps by tightening the cap bolts by hand while rotating the crankshaft.

□ 19. With the main bearing cap bolts fingertight, pry the crankshaft back and forth to align the thrust bearing.

20. Rotate the crankshaft. Does it rotate freely with no binding or restrictions?
□ Yes □ No
If NO, correct any problems before continuing.

21. What is the torque specification for the main bearing cap bolts?

_____

22. In the space below, draw the torque sequence for the main bearing cap bolts.

23. Torque the bolts of the first cap in sequence to the specified value. ☐

24. Check crankshaft rotation by rotating it with a torque wrench. Note the amount of torque required to rotate the crankshaft during several revolutions.

_____

25. Torque the next cap in the sequence, and check the rotating torque of the crankshaft again. ☐

26. Continue to torque the cap bolts in sequence, making sure to check rotating torque after each cap is tightened. Record the turning torque of the crankshaft for the remaining caps below.

_____

_____

_____

27. Were there any large increases in the torque required to rotate the crankshaft?
☐ Yes   ☐ No
If YES, stop and locate the cause of the problem.

28. With all cap bolts properly torqued, check the rotating torque of the crankshaft.
Specification _____   Actual _____

29. If the engine uses a one-piece lip seal, install it. ☐

30. Check crankshaft end play.
Specification _____   Actual _____

Instructor's Response _____

_____

_____

595

JOB SHEET

Name _____ Date _____

INSTALLING THE PISTONS AND CONNECTING RODS

Upon completion of this job sheet, you should be able to install the piston assemblies into the engine block properly.

ASE Correlation

This job sheet is related to the Engine Repair Test's content area: Engine Block Diagnosis and Repair; tasks: Inspect, measure, and install or replace piston rings; assemble piston and connecting rod; install piston/rod assembly; check bearing clearance and sideplay; replace/ retorque fasteners according to manufacturer's procedures.

Tools and Materials

Ring compressor
Torque wrench
Feeler gauge
Engine block

Describe the vehicle being worked on:

Year _____ Make _____ Model _____
VIN _____ Engine type and size _____

Procedure

Task Completed

Using the engine block and piston assemblies from the engine you have disassembled and the proper service manual, you will identify and perform the required steps to install the piston assemblies into the block.

1. Is there a difference between the bearing inserts that fit into the connecting rod and the inserts that fit into the cap?
   ☐ Yes   ☐ No
   If YES, describe which ones fit the rod and which ones fit the cap.

   _____
   _____
   _____

2. Install the bearing inserts into the connecting rods and caps, making sure the oil holes   ☐
   align and the tabs are properly located.

3. Coat the cylinder walls with clean engine oil, and soak the piston head in a container of   ☐
   clean engine oil. If bearing clearance is being measured using Plastigage, oil cannot be
   applied to the bearings.

4. Rotate the crankshaft so the connecting rod journal for the piston assembly being   ☐
   installed is at BDC.

☐　　5. Assure the piston assemblies are in the proper order and the correct caps are with the correct rods.

☐　　6. Cover the connecting rod bolts with a special guide or pieces of rubber hose to protect the cylinder walls and crankshaft rod journals from damage.

☐　　7. Use a ring compressor to squeeze the rings together.

☐　　8. Install the piston assembly into the correct bore. Double-check to assure the correct piston is being installed into its corresponding bore and it faces the correct direction. The pistons should go into the bore with a slight restriction. Using a hammer handle to tap the piston assembly into the cylinder will make installation easier.

☐　　9. As each connecting rod is installed onto the crankshaft, tighten the cap bolts only fingertight. Check the rotating torque of the crankshaft as each piston is installed. This gives an indication of the interference between the piston rings and cylinder walls. If any of the pistons causes a large increase in turning torque, repair the cause before continuing.

☐　　10. Starting with the first connecting rod in sequence, torque the cap bolts. As the bolts are torqued, tighten them in small increments, alternating back and forth between the two bolts.

　　　11. As each piston assembly is installed, fill out the following chart:
Rod cap bolt torque specification _____
Connecting rod bearing oil clearance specification _____
Actual oil clearance:

Piston #	Clearance
1	_____
2	_____
3	_____
4	_____
5	_____
6	_____
7	_____
8	_____

　　　12. After all the pistons are installed, check the rotating torque and record it.

_____
What is the specification? _____
If the actual turning torque is not within specifications, what must be corrected?

_____
_____
_____

☐　　13. When installing and torquing the cap bolts for the final assembly, coat the bolt threads with a thread locking compound.

14. What is the specification for connecting rod side clearance?

_____

15. Measure the side clearance for each connecting rod and record your results below.

Piston #	Clearance
1	_____
2	_____
3	_____
4	_____
5	_____
6	_____
7	_____
8	_____

16. What is the minimum specification for deck clearance? _____

17. Actual deck clearance: _____

Instructor's Response _____

_____

_____

Chapter 14

ALTERNATIVE FUEL AND ADVANCED TECHNOLOGY VEHICLE SERVICE

UPON COMPLETION AND REVIEW OF THIS CHAPTER, YOU SHOULD BE ABLE TO:

- Shut off fuel on a propane fuel vehicle to perform engine maintenance and repair.

- Shut off fuel on a CNG vehicle to perform engine maintenance and repair.

- Understand the safety precautions necessary for working on HEV vehicles.

INTRODUCTION

Most of the base engines in all of the alternative fuel and advanced technology vehicles described in the classroom manual are serviced using the same techniques we have discussed in the previous chapters of this shop manual. What differs here is that there are fuels under extremely high pressures and voltages above safe levels. You must receive proper instruction to work on the engines of these vehicles. This chapter is intended to offer some basic safety precautions for work on these vehicles. This information is NOT a substitute for the manufacturer's recommended procedures. You should always consult the service information for the vehicle you are working on before attempting engine repairs on an alternative fuel, HEV, or other advanced technology vehicle. Your basic knowledge of safety procedures such as always wearing appropriate safety glasses and clothing, using proper ventilation, and eliminating sources of ignition when working with fuel are still critical elements of safe vehicle service.

SERVICING PROPANE FUELED ENGINES

The engine of an LPG vehicle is nearly the same, if not identical to, that of a gas-powered vehicle. You can expect the engine wear to decrease with the use of propane, but still there will be occurrences of mechanical wear or failures within the engine. When these occur, you should evaluate the engine using the same tests—such as compression, noise analysis, and power balance—as you have learned for gasoline engines. If the vehicle has been converted, you can use the same specifications (such as compression, valve adjustment) as those listed for its gasoline counterpart. If ignition timing and fuel delivery is adjustable, those specifications will be modified and should be marked on an under-hood label or in an addition to the owner's manual. As mentioned in the classroom manual, propane fueled engines may last two to three times longer than the same engine run on gasoline. Spark plugs may also last significantly longer. But, when problems arise, you should still check these basics before suspecting problems related

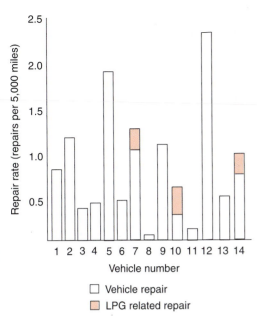

FIGURE 14-1 This chart shows that few of the required repairs to this fleet were related to the LPG system.

FIGURE 14-2 The LPG tanks hold fuel under high pressure.

to the LPG supply system (Figure 14-1). One fleet study showed that the vast majority of repairs were performed on the gas-powered vehicles rather than the propane-fueled vehicles.

When performing tests on a running propane vehicle, there should be no unusual safety precautions. If you smell propane, shut the fuel supply off immediately. Locate and repair the source of the leak before proceeding with engine diagnosis. Cooling and lubrication system testing will be the same on an LPG engine as on a gasoline engine.

Before performing mechanical repairs on a propane engine that require you to disable fuel or disconnect fuel fittings, you need to turn the propane supply off (Figure 14-2). Follow the manufacturer's procedure to disable the propane fuel system. Run the vehicle until it stalls, to remove the remaining fuel from the lines. With the propane vented from the lines and engine, you can safely work on the engine mechanical components as you would on any other vehicle.

SERVICING CNG VEHICLES

Your primary consideration when performing mechanical repairs on a CNG vehicle is safety. CNG is stored at pressures between 3,000 and 3,600 psi. You should be a qualified CNG technician before performing repairs on the CNG system. Once you learn how to disable the fuel system, you should still be able to perform the engine diagnosis and repair procedures you have already learned.

To disable the fuel system, locate the **fuel control valve** on the tank(s) and shut it off (Figure 14-3). Turn it fully clockwise to prevent fuel flow out of the cylinder. Now no fuel can flow from the tank.

Next, start the vehicle, if possible, to purge the remaining fuel from the lines. You should attempt to start the vehicle three times, and once it starts, it will run out of fuel and shut off. Now the system is free of fuel.

Finally, you should turn off the **manual shut-off valve** (Figure 14-4). It will be accessible from the exterior of the vehicle and highlighted with a label (Figure 14-5). Turn the quarter turn valve so that the handle is perpendicular to the fuel line (Figure 14-6). This stops fuel flow from this point forward to the engine.

The **fuel control valve** is a manually operated valve designed to stop fuel flow from the cylinder.

The **manual shut-off valve** stops fuel flow from the CNG cylinder(s) to the engine.

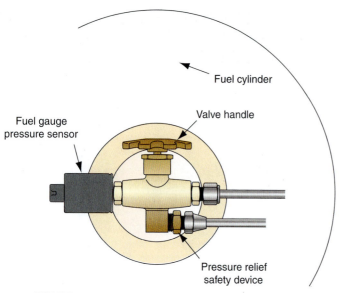

Fuel cylinder

Valve handle

Fuel gauge
pressure sensor

Pressure relief
safety device

FIGURE 14-3 Rotate the fuel control valve clockwise to
stop fuel flow.

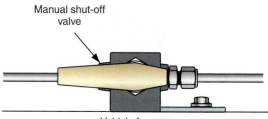

Manual shut-off
valve

Vehicle frame

FIGURE 14-4 Locate the manual shut-off valve on the
exterior of the vehicle.

MANUAL
SHUT-OFF
VALVE

CAUTION: HIGH PRESSURE

FIGURE 14-5 The manual shut-off valve
will be labeled.

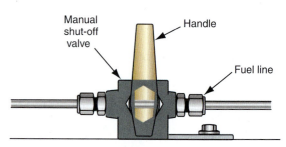

Manual
shut-off
valve

Handle

Fuel line

Vehicle frame

FIGURE 14-6 Turn the quarter turn manual shut-off
valve so it is perpendicular to the fuel line.

Even with the fuel system fully disabled, there are some precautions you should follow when working around a CNG vehicle (Figure 14-7). These include:

- Only qualified CNG technicians should service the CNG system.
- Natural gas is explosive. Do not use sources of ignition with natural gas vapors present.
- The work area should have adequate ventilation.
- Natural gas contains an odor additive. If a strong natural gas odor occurs, shut the fuel control valves off at the tank immediately. Ventilate the area to allow the gas to dissipate. Natural gas is lighter than air, so it will rise. This minimizes the risk of fire or explosion as long as there is adequate ventilation. If necessary, evacuate the area and contact emergency personnel.
- Do not undercoat or paint any fuel system line or component.
- Do not attempt to weld any fuel system component.
- Do not attempt to repair any fuel cylinder problems; they are serviced by replacement.
- Do not modify fuel system components or replace them with parts of lesser quality than those provided by the manufacturer.
- When a vehicle is disabled for a period of time, turn off the fuel control valves at the tank and the manual shut-off valve.
- Do not attempt to fill a damaged cylinder.
- Whenever a CNG vehicle has been in an accident, inspect the CNG cylinders as specified by the manufacturer. This is a law in the United States and in Canada.

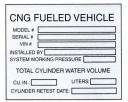

CNG FUELED VEHICLE

MODEL # [   ]
SERIAL # [   ]
VIN # [   ]
INSTALLED BY [   ]
SYSTEM WORKING PRESSURE [   ]

TOTAL CYLINDER WATER VOLUME

CU. IN. [   ]   LITERS: [   ]
CYLINDER RETEST DATE: [   ]

WARNING: DO NOT WORK ON OPEN GAS LINES OR FITTINGS WITHOUT MAKING CERTAIN ALL TANKS ARE COMPLETELY SHUT OFF.

HIGH PRESSURE NO WELDING OR CUTTING

ALL CNG COMPONENTS MUST BE SERVICED BY AUTHORIZED MECHANICS ONLY.

MANUAL SHUT-OFF VALVE

CAUTION: HIGH PRESSURE

FIGURE 14-7 Typical CNG vehicle warning labels.

FIGURE 14-8 Most CNG refilling stations are protected and need special training to operate.

- CNG cylinders must be removed and inspected every 3 years. Repairs are not allowed. A label noting the date of inspection and expiration must be affixed to the tank and sealed with epoxy.
- CNG cylinders must be replaced every 15 years.

> **CUSTOMER CARE:** When working on a customer's CNG vehicle, visually inspect the fuel system components for wear or damage. Locate the cylinder inspection label and inform the customer when the next inspection is due.

With the fuel system properly disabled and the work area ventilated, you may begin work on the CNG engine. Follow the manufacturer's procedures carefully to properly remove fuel system components for access to components such as water pumps, cylinder heads, or even for engine removal. The oil requirement may also be different for a CNG engine. Your study of this text and your work in the shop will qualify you to work on a CNG engine, just not the CNG fuel system. Remember to check manufacturer's specifications, as dedicated CNG engines have been modified from gasoline engines. The compression values for the Civic CNG with a 12.5:1 compression ratio will be higher than those for the gasoline version (Figure 14-8).

SERVICING HYBRID ELECTRIC VEHICLES

The service of HEVs is currently limited to technicians who have received factory training from the manufacturer. Manufacturers require that a dealership that sells an HEV model have at least one technician who has participated in factory training through an interactive training program and classroom and hands-on training. This text cannot train you to work on any current HEV. As a technician, you should be aware of the safety precautions of these vehicles. If you perform towing, you must be fully aware of the safety precautions before approaching a damaged HEV. These vehicles come with extended warranties, so it is likely that you will only see them in your service bay in the next few years if you are working at a dealership that sells them. Your education may well place you at the front of the line for training on these advanced technology vehicles.

SPECIAL TOOLS
Lineman's gloves

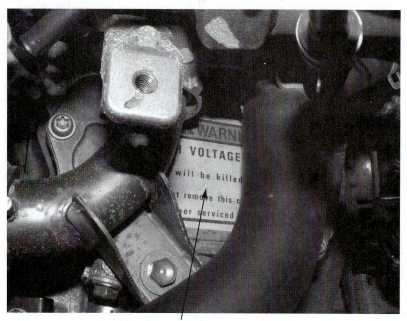

High voltage warning

FIGURE 14-9 Use caution! Look carefully at this warning label; it says "you will be killed" if you do not use appropriate safety precautions.

Safety, as with CNG vehicles, must be your first consideration when approaching an HEV (Figure 14-9). The voltages on some HEVs may reach 500 volts, and regularly store 360 volts. Voltage varies with make, but over 30 volts is potentially fatal to humans, so we are dealing with a potentially lethal mechanism. All high voltage cables are identifiable by a bright orange coating. Never touch these wires while the electrical system is live. All high voltage components are labeled as such (Figure 14-10). The battery packs are well

High voltage cables

FIGURE 14-10 The brightly colored big cables hold high voltage; components that use high voltage have warning labels.

insulated and protected in an accident. Nickel metal hydride (NiMH) batteries, as used in many of today's HEVs, are sealed in packs designed not to leak even in the event of a crash. If they do leak, the electrolyte can be neutralized with a dilute boric acid solution or vinegar.

Understanding proper operation of the vehicle you are working on is essential. Though you should have received factory training, much of this information can be found in the owner's manual. The vehicle may be "ready" and able to move with a tap of the accelerator, even when neither the electric motor nor the gasoline engine is running. The owner's manual will provide you with special operating characteristics of the vehicle and basic safety warnings.

> **CUSTOMER CARE:** It is not uncommon for a customer with a new HEV to come in with service questions. Most consumers do not take the time to read the owner's manual or view the operational CD provided. When you are working on a make of vehicle, be as familiar with it as possible, so you can competently answer questions a customer might have about his new vehicle.

SERVICING THE TOYOTA PRIUS

First, familiarize yourself with the layout of the vehicle components (Figure 14-11). The figure identifies the location of the following components:

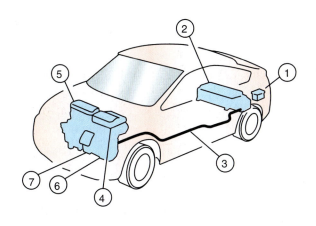

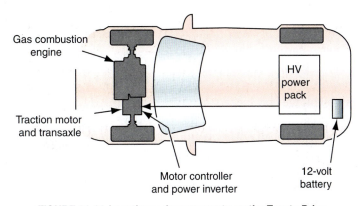

FIGURE 14-11 Locations of components on the Toyota Prius.

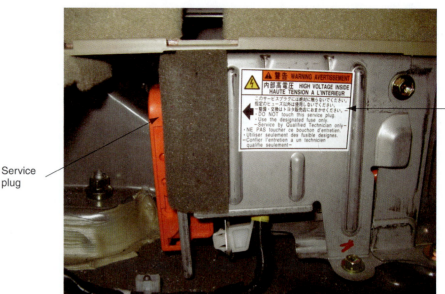

Service
plug

High voltage
and service
warning

FIGURE 14-12 The high voltage service plug is located in the trunk. Use lineman's gloves to switch power off.

Component	Location	Description
1. 12-volt battery	LR of trunk	Typical lead-acid battery used to operate accessories
2. HV battery pack	Rear seat back support, in trunk	274 NiMH battery pack
3. HV power cables	Under-carriage and engine compartment	Thick, orange-colored cables carrying HV DC and AC current
4. Inverter	Engine compartment	Converts DC to AC for the electric motor and AC to DC to recharge the battery pack
5. Gasoline engine	Engine compartment	Powers vehicle and powers generator to recharge battery
6. Electric motor	Engine compartment	AC PM motor in transaxle to power vehicle
7. Electric generator	Engine compartment	AC generator in transaxle used to recharge battery pack

Only a certified technician should disable the high voltage system by disconnecting the negative terminal of the 12-volt auxiliary battery. This is located in the left rear of the trunk. Disconnecting the auxiliary battery shuts down the high voltage circuit. For additional protection, wait 5 minutes, and then remove the service plug located through an access port in the trunk (Figure 14-12). Wear insulated rubber **lineman's gloves** when removing the service plug. Always carefully check the gloves for pin holes. Blow into them like a balloon and see that they don't leak any air. Even a very small hole could allow a deadly spark to penetrate to you. After removing the service plug, wait at least 5 minutes to allow full discharge of the high voltage condensers inside the inverter.

Once the high voltage system has fully discharged, you may diagnose and repair the Prius gas engine as you would any other engine. This is a newly designed engine; be sure to use proper procedures and service specifications.

SERVICING THE HONDA CIVIC HYBRID

The Civic hybrid runs current at 144 volts through brightly colored orange cables. Its Integrated Motor Assist (IMA) System uses a 1.3-liter gasoline engine and a DC brushless

Lineman's gloves
are highly insulated rubber gloves that protect the technician from lethal high voltages.

motor (Figure 14-13). You must shut off the high voltage electrical circuits and isolate the IMA system before servicing its components. Do not touch the high voltage cables without lineman's gloves (Figure 14-14). To isolate the high voltage circuit, turn the high voltage battery module switch to OFF. Secure the switch in the off position with the locking cover before servicing the IMA system. Wait at least 5 minutes after turning off the battery module switch, and then disconnect the negative cable from the 12-volt battery. Test the voltage between the high-voltage battery cable terminals with a voltmeter. Do not disconnect them until the voltage is below 30 volts.

Again, once you have safely disabled the high voltage system, you can use your skills and the service information to service the engine and its components. The 1.3L in-line 4-cylinder engine uses a coil-on-plug ignition (Figure 14-15). Note that Honda specifies 0W-20 motor oil in this engine (Figure 14-16)! Follow the manufacturer's service information carefully on these new engines.

FIGURE 14-13 Honda's IMA system "engine" compartment.

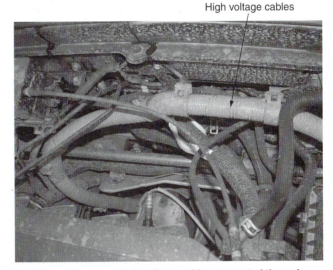

FIGURE 14-14 The high voltage cables are routed through the engine compartment, but you can't miss them with their bright orange color. Disable the high voltage system before touching them.

FIGURE 14-15 Underneath the HEV covers lies a typical four-cylinder engine that you should feel competent to work on.

FIGURE 14-16 This Honda uses 0W-20 motor oil. Follow the manufacturer's service information carefully to help you provide excellent work on any vehicle.

HEV Warranties

In order to reassure the public about the new technology and encourage consumer confidence, most HEV manufacturers offer extensive warranties on the battery packs and the hybrid components. You should always consult the manufacturer's service information before recommending expensive repairs to consumers of these vehicles, even as the vehicles age and you begin to see them in independent repair facilities. A few examples are:

Ford Escape	8 years/80,000 miles on battery pack*
Honda HEV	8 years/80,000 miles on battery pack*
Toyota Prius	8 years/100,000 miles on battery pack* and on the hybrid power plant, including the engine
	*Battery packs must be warranted for 10 years/150,000 miles in "green states" (CA, MA, VT, NY, ME)

TERMS TO KNOW

Fuel control valve

Lineman's gloves

Manual shut-off valve

CASE STUDY

A friend who owns a Honda Civic hybrid called the other day concerned because "the light that looks like an engine" was on. Although the vehicle was still under warranty, I took my scan tool and hooked it up to the powertrain control module diagnostic link connector. The diagnostic trouble code indicated a gross evaporative emissions leak. We checked the gas cap and found it loose; tightening it three clicks solved the problem. A simple explanation of the system and reassurance that it would also have happened on a conventional vehicle relieved my friend's fears that his advanced technology vehicle was going to give him hi-tech trouble.

ASE-STYLE REVIEW QUESTIONS

1. *Technician A* says that propane is not flammable unless it is under pressure.
   Technician B says that the fuel system should be disabled before performing mechanical repairs on the engine.
   Who is correct?
   A. A only
   B. B only
   C. Both A and B
   D. Neither A nor B

2. *Technician A* says that most LPG engines have higher compression pressures than gas engines.
   Technician B says that engines run on LPG may last twice as long as their gas counterparts.
   Who is correct?
   A. A only
   B. B only
   C. Both A and B
   D. Neither A nor B

3. When working on an LPG engine performance concern:
   Technician A says to check the LPG system components first.
   Technician B says more repairs are required on the base engine than on the LPG system.
   Who is correct?
   A. A only
   B. B only
   C. Both A and B
   D. Neither A nor B

4. *Technician A* says that CNG is not explosive because it is lighter than air.
   Technician B says that CNG is odorless.
   Who is correct?
   A. A only
   B. B only
   C. Both A and B
   D. Neither A nor B

5. *Technician A* says to turn the fuel control valves clockwise to shut fuel off.

   Technician B says that you should purge the fuel from the lines before disconnecting CNG components.

   Who is correct?

   A. A only
   B. B only
   C. Both A and B
   D. Neither A nor B

6. Each of the following is a service caution when working on a CNG vehicle EXCEPT:

   A. Use ventilation equipment when starting the vehicle in the shop.
   B. Do not modify CNG fuel system components.
   C. Run the vehicle with the fuel shut off to purge the fuel from the lines.
   D. Use only epoxy when repairing a CNG cylinder.

7. *Technician A* says that a CNG cylinder must be inspected every 5 years.

   Technician B says that a CNG cylinder must be removed to be inspected.

   Who is correct?

   A. A only
   B. B only
   C. Both A and B
   D. Neither A nor B

8. A minimum voltage of _____ can be lethal to humans.

   A. 24 volts
   B. 30 volts
   C. 42 volts
   D. 120 volts

9. *Technician A* says that disconnecting the 12-volt auxiliary battery on a Toyota Prius isolates the high voltage circuit.

   Technician B says that there is a service plug to isolate the high voltage system located in the trunk of a Prius.

   Who is correct?

   A. A only
   B. B only
   C. Both A and B
   D. Neither A nor B

10. *Technician A* says that voltages on an HEV may reach 500 volts.

    Technician B says that NiMH battery electrolyte can be neutralized with baking soda.

    Who is correct?

    A. A only
    B. B only
    C. Both A and B
    D. Neither A nor B

ASE CHALLENGE QUESTIONS

1. *Technician A* says that an LPG vehicle runs with fuel pressures over 200 psi.

   Technician B says that an LPG vehicle does not need spark plugs replaced at regular intervals.

   Who is correct?

   A. A only
   B. B only
   C. Both A and B
   D. Neither A nor B

2. *Technician A* says that if the fuel pressure regulator fails on a CNG vehicle, the engine may not start.

   Technician B says that compression on a CNG vehicle will be lower than on a gas engine.

   Who is correct?

   A. A only
   B. B only
   C. Both A and B
   D. Neither A nor B

3. *Technician A* says that gasoline is more explosive in the shop environment than CNG.

   Technician B says that the manual shut-off valve turns fuel off at the CNG cylinder.

   Who is correct?

   A. A only
   B. B only
   C. Both A and B
   D. Neither A nor B

4. *Technician A* says that if an electric motor won't run, one of the first things to check would be the fuse.

   Technician B says you can pull the fuse out and look at it to see if it is blown.

   Who is correct?

   A. A only
   B. B only
   C. Both A and B
   D. Neither A nor B

5. An HEV comes in on a tow truck. The HEV's owner said it just stopped moving, even though the battery indicator and fuel levels were both above half.

   Technician A says to check fuel pressure.

   Technician B says to check battery voltage.

   Who is correct?

   A. A only
   B. B only
   C. Both A and B
   D. Neither A nor B

JOB SHEET

Name _____ Date _____

SERVICING AN LPG OR CNG VEHICLE

Upon completion of this job sheet you should be able to properly disable the fuel system of an alternative fuel vehicle and perform a compression test.

Tools and Materials

Hand tools
Safety glasses
Special tools as required
Compression tester

Describe the vehicle being worked on:

Year _____ Make _____ Model _____
VIN _____ Engine type and size _____

Procedure

Task Completed

Your instructor will provide you with a specific vehicle. Write down the information, and then perform the following tasks:

1. Is the vehicle you are working on a dedicated or retrofitted alternative fuel vehicle?

_____

2. Locate the service information for disabling the fuel system. Describe the procedure below: _____

_____

_____

_____

_____

_____

3. Disable the fuel system according to the manufacturer's recommended procedure and have your instructor check your work.

4. Locate the procedure and specifications for performing a compression test on the vehicle you are working on. What is the compression specification?

_____

5. Disable the ignition system, block open the throttle, and remove the spark plugs. ☐

6. Perform a compression test as instructed by the service information, and record your results below:

Cylinder number 1 _____
Cylinder number 2 _____
Cylinder number 3 _____

Cylinder number 4 _____

Cylinder number 5 _____

Cylinder number 6 _____

Cylinder number 7 _____

Cylinder number 8 _____

7. Are any of the cylinders below specification?   ☐ Yes   ☐ No
   If there are cylinders below specification, perform a wet compression test or leak-down test as directed by your teacher.

☐

8. Properly reinstall the spark plugs, close the throttle, and enable the ignition system.

9. Locate the instructions in the service information for enabling the fuel system, and describe the procedure below: _____

   _____

   _____

   _____

☐

10. Enable the fuel system properly, and have the instructor check your work. Task completed properly?

☐

11. Attach exhaust equipment; set the parking brake; start the vehicle, and ensure that it is running properly. Clean up all tools and equipment.

Instructor's Response _____

_____

_____

Name _____ **Date** _____

RESEARCHING AN ADVANCED TECHNOLOGY VEHICLE

Upon completion of this job sheet, you should be able to research and locate research relevant automotive information and understand the basic operation of an advanced technology vehicle.

Tools and Materials

Internet or library access

Procedure

Choose a production hybrid electric vehicle that is currently being sold in North America for your research.

Describe the vehicle being researched:

Year _____ Make _____ Model _____

1. Provide a basic description of the power plant: _____
   _____
   _____
   _____
   _____

2. What type(s) of power sources are used? _____
   _____
   _____

3. What are the engine classifications? Displacement _____
   Cylinders _____ Configuration _____ Fuel used _____
   Torque _____ HP _____
   Estimated fuel economy: city _____ highway _____
   Emissions rating _____ (ULEV, SULEV, AT-PZEV)

4. What type of electric motor(s) is used? _____
   What is its torque rating? _____

5. Describe the basic power flow of the electric motor(s) and engine from a stop sign up through acceleration to cruising at highway speeds. _____
   _____
   _____
   _____
   _____
   _____

6. What is the base price of this vehicle? _____

7. If you were considering buying this vehicle, what advantages and disadvantages would affect your decision? _____

_____

_____

_____

_____

_____

_____

_____

_____

Instructor's Response _____

_____

_____

JOB SHEET

Name _____ **Date** _____

DISABLING THE HIGH VOLTAGE BATTERY PACK ON A HEV

Upon completion of this job sheet, you should be able to safely and properly disable the high voltage battery on a hybrid electric vehicle.

Tools and Materials

Lineman's rubber gloves (high voltage protective gloves)
Hybrid electric vehicle

Describe the vehicle being worked on:

Year _____ Make _____ Model _____
VIN _____ Engine type and size _____

Procedure

Task Completed

Your instructor will provide you with a specific vehicle. Write down the information, and then perform the following tasks:

1. Set the parking brake on the vehicle. ☐

2. Move the shift lever to park, turn the ignition switch to the OFF position, and place the key on a bench outside of the vehicle. ☐

3. Disconnect the negative terminal from the 12-volt auxiliary battery. ☐

4. On some models, you need to open the trunk and remove some panels to gain access to the disabling switch. Describe where the switch is located on the vehicle you are working on.

_____

_____

_____

5. While wearing the special gloves, turn the high voltage switch to the OFF position. ☐

6. Using the service manual as a guide, describe how long it takes to depower the high voltage battery pack so that it is safe.

_____

_____

Instructor's Response _____

_____

_____

JOB SHEET

Name _____ Date _____

PERFORMING A COMPRESSION TEST ON A HEV

Upon completion of this job sheet, you should be able to properly perform a compression test on a hybrid electric vehicle.

Tools and Materials

Compression tester
Hybrid electric vehicle
Manufacturer's scan tool
Oil can
Spark plug socket
Spark tester

Describe the vehicle being worked on:

Year _____ Make _____ Model _____
VIN _____ Engine type and size _____

Procedure

Your instructor will provide you with a specific vehicle. Write down the information, and then perform the following tasks:

Task Completed

1. Check the oil level and correct as needed. ☐

2. Locate the compression specifications for the engine you are working on, and record them below: _____
   _____

3. Start the engine, and allow it to idle until the engine is moderately warm. Allow it to cool slightly before removing the spark plugs.
   (NOTE: You may have to drive a HEV at highway speed in order to get the engine to turn on.)

4. Remove the spark plugs, and note their condition.
   Describe your results: _____
   _____
   _____
   _____

5. Connect the manufacturer's scan tool or handheld tester to the vehicle. ☐

6. Install the compression tester adaptor into the number 1 spark plug hole, and crank the engine over five times using the scan tool or handheld tester.
   (NOTE: Refer to the operator's manual or the manufacturer's service manual for specific directions on operation.)
   Record your results: _____
   Repeat for each cylinder, and describe your results.
   Cylinder number 2 _____
   Cylinder number 3 _____
   Cylinder number 4 _____
   Cylinder number 5 _____
   Cylinder number 6 _____
   Cylinder number 7 _____
   Cylinder number 8 _____

7. Analyze the results of the compression test, and note any recommended repairs or tests.
   _____
   _____
   _____
   _____
   _____
   _____

8. Identify the cylinder(s) with the lowest compression. Squirt 2 tablespoons of oil into the weak cylinder, and perform another compression test on that cylinder.
   Note the new results, and describe what problem this indicates: _____
   _____
   _____
   _____

9. Based on your tests and results, what are your recommendations for further testing and repairs?
   _____
   _____
   _____

Instructor's Response _____
_____
_____

1. Two technicians are discussing the markings on the heads of bolts (cap screws).

   Technician A says the higher the number of lines, the higher the strength of the bolt.

   Technician B says the higher the number on metric bolts, the higher the grade.

   Which technician is correct?

   A. A only   C.   Both A and B

   B. B only   D.   Neither A nor B

2. A metric bolt size of M8 means that _____.

   A. The bolt is 8 mm long.

   B. The bolt is 8 mm in diameter.

   C. The pitch (the distance between the crest of the threads) is 8 mm.

   D. The bolt is 8 cm long.

3. On a metric bolt sized M8 × 1.5, the 1.5 means that _____.

   A. The bolt is 1.5 mm in diameter.

   B. The bolt is 1.5 cm long.

   C. The bolt has 1.5 mm between the crest of the threads.

   D. The bolt has a strength grade of 1.5.

4. Prevailing torque (lock) nuts should be replaced rather than reused after removal.

   A. True

   B. False

5. If the bore of an engine is increased without any other changes except for the change to proper-size replacement pistons, the displacement will _____ and the compression rate will _____.

   A. Increase; increase

   B. Increase; decrease

   C. Decrease; increase

   D. Decrease; decrease

6. A battery is being tested.

   Technician A says that the surface charge should be removed before the battery is load tested.

   Technician B says that the battery should be loaded to two times the CCA rating of the battery for 15 seconds.

   Which technician is correct?

   A. A only   C.   Both A and B

   B. B only   D.   Neither A nor B

7. A starter motor cranks the engine too slowly to start.

   Technician A says that the cause could be a weak or defective battery.

   Technician B says that the cause could be loose or corroded battery cable connections.

   Which technician is correct?

   A. A only   C.   Both A and B

   B. B only   D.   Neither A nor B

8. The charging system voltage is found to be lower than specified by the vehicle manufacturer.

   Technician A says that a loose or defective drive belt could be the cause.

   Technician B says that a defective alternator could be the cause.

   Which technician is correct?

   A. A only   C.   Both A and B

   B. B only   D.   Neither A nor B

9. An engine uses an excessive amount of oil.

   Technician A say that clogged oil drain-back holes in the cylinder head could be the cause.

   Technician B says that worn piston rings could be the cause.

   Which technician is correct?

   A. A only   C.   Both A and B

   B. B only   D.   Neither A nor B

10. Battery voltage during cranking is below specifications.

    Technician A says that a defect in the engine may be the cause.

    Technician B says that the starter motor may be defective.

    Which technician is correct?

    A. A only   C.   Both A and B

    B. B only   D.   Neither A nor B

11. Two technicians are discussing torquing cylinder head bolts.

    Technician A says that many engine manufacturers recommend replacing the head bolts after use.

    Technician B says that many manufacturers recommend tightening the head bolts to a specific torque, then turning the bolts an additional number of degrees.

    Which technician is correct?
    A. A only
    B. B only
    C. Both A and B
    D. Neither A nor B

12. Engine ping (spark knock or detonation) can be caused by _____.
    A. Advanced ignition timing
    B. Retarded ignition timing

13. Two technicians are discussing the diagnosis of a lack-of-power problem.

    Technician A says that a jumped timing chain could be the cause.

    Technician B says that retarded ignition timing could be the cause.

    Which technician is correct?
    A. A only
    B. B only
    C. Both A and B
    D. Neither A nor B

14. An engine equipped with a turbocharger is burning oil (blue exhaust smoke) all the time.

    Technician A says that a defective wastegate could be the cause.

    Technician B says that a plugged PCV system could be the cause.

    Which technician is correct?
    A. A only
    B. B only
    C. Both A and B
    D. Neither A nor B

15. An engine idles roughly and stalls occasionally.

    Technician A says that using fuel with too high an RVP level could be the cause.

    Technician B says that using winter-blend gasoline during warm weather could be the cause.

    Which technician is correct?
    A. A only
    B. B only
    C. Both A and B
    D. Neither A nor B

16. *Technician A* says that coolant flows through the engine passages and does not flow through the radiator until the thermostat opens.

    Technician B says that the temperature rating of the thermostat indicates the temperature of the coolant when the thermostat is opened fully.

    Which technician is correct?
    A. A only
    B. B only
    C. Both A and B
    D. Neither A nor B

17. *Technician A* says the higher the concentration of antifreeze, the better.

    Technician B says that a 50/50 mix of antifreeze and water is the ratio recommended by most vehicle manufacturers.

    Which technician is correct?
    A. A only
    B. B only
    C. Both A and B
    D. Neither A nor B

18. *Technician A* says that the radiator pressure cap is designed to raise the boiling point of the coolant.

    Technician B says that the radiator pressure cap helps prevent cavitation and, therefore, improves the efficiency of the water pump.

    Which technician is correct?
    A. A only
    B. B only
    C. Both A and B
    D. Neither A nor B

19. *Technician A* says that used antifreeze coolant is often considered to be hazardous waste.

    Technician B says that metals absorbed by the coolant when it is used in an engine are what makes antifreeze harmful.

    Which technician is correct?
    A. A only
    B. B only
    C. Both A and B
    D. Neither A nor B

20. A water pump has been replaced three times in three months.

    Technician A says that the drive belt(s) may be installed too tightly.

    Technician B says that a cooling fan may be bent or out of balance.

    Which technician is correct?
    A. A only
    B. B only
    C. Both A and B
    D. Neither A nor B

21. The "hot" light on the dash is being discussed by two technicians.

    Technician A says that the light comes on if the cooling system temperature is too high for safe operation of the engine.

    Technician B says that the light comes on whenever there is a decrease (drop) in cooling system pressure.

    Which technician is correct?
    A. A only
    B. B only
    C. Both A and B
    D. Neither A nor B

22. An engine noise is being diagnosed.

    Technician A says that a double knock is likely to be due to a worn rod bearing.

    Technician B says that a knock only when the engine is cold is usually due to a worn piston pin.

    Which technician is correct?

    A. A only   C. Both A and B

    B. B only   D. Neither A nor B

23. A compression test gave the following results:
    Cylinder number 1 = 155, cylinder number 2 = 140, cylinder number 3 = 110, cylinder number 4 = 105.

    Technician A says that a defective (burned) valve is the most likely cause.

    Technician B says that a leaking head gasket could be the cause.

    Which technician is correct?

    A. A only   C. Both A and B

    B. B only   D. Neither A nor B

24. Two technicians are discussing a compression test.

    Technician A says that the engine should be turned over with the pressure gauge installed for "two puffs."

    Technician B says that the maximum difference between the highest-reading cylinder and the lowest-reading cylinder should less than 20 percent.

    Which technician is correct?

    A. A only   C. Both A and B

    B. B only   D. Neither A nor B

25. *Technician A* says that oil should be squirted into all of the cylinders before taking a compression test.

    Technician B says that if the compression greatly increases when some oil is squirted into the cylinders, it indicates defective or worn piston rings.

    Which technician is correct?

    A. A only   C. Both A and B

    B. B only   D. Neither A nor B

26. During a cylinder leakage (lead-down) test, air is noticed coming out of the oil fill opening.

    Technician A says that the oil filter may be clogged.

    Technician B says that the piston rings may be worn or defective.

    Which technician is correct?

    A. A only   C. Both A and B

    B. B only   D. Neither A nor B

27. A cylinder leakage (leak-down) test indicates 30 percent leakage, and air is heard coming out of the air inlet.

    Technician A says that this indicates a slightly worn engine.

    Technician B says that one or more intake valves are defective.

    Which technician is correct?

    A. A only   C. Both A and B

    B. B only   D. Neither A nor B

28. Two technicians are discussing a cylinder power balance test.

    Technician A says the more the engine rpm drops, the weaker the cylinder.

    Technician B says that all cylinder rpm drops should be within 50 rpm of each other.

    Which technician is correct?

    A. A only   C. Both A and B

    B. B only   D. Neither A nor B

29. *Technician A* says that cranking vacuum should be the same as idle vacuum.

    Technician B says that a sticking valve is indicated by a floating valve gauge needle reading.

    Which technician is correct?

    A. A only   C. Both A and B

    B. B only   D. Neither A nor B

30. *Technician A* says that black exhaust smoke is an indication of a too-rich air-fuel mixture.

    Technician B says that white smoke (steam) is an indication of coolant being burned in the engine.

    Which technician is correct?

    A. A only   C. Both A and B

    B. B only   D. Neither A nor B

31. Excessive exhaust system back pressure has been measured.

    Technician A says that the catalytic converter may be clogged.

    Technician B says that the muffler may be clogged.

    Which technician is correct?

    A. A only   C. Both A and B

    B. B only   D. Neither A nor B

32. A head gasket failure is being diagnosed.

    Technician A says that an exhaust analyzer can be used to check for HC when the tester probe is held above the radiator coolant.

    Technician B says that a chemical block tester color in the presence of combustion gases.

    Which technician is correct?

    A. A only   C. Both A and B

    B. B only   D. Neither A nor B

33. *Technician A* says that pistons should be removed from the crankshaft side of the cylinder when disassembling an engine to prevent possible piston or cylinder damage.

    Technician B says that the rod assembly should be marked before disassembly.

    Which technician is correct?

    A. A only
    C. Both A and B
    B. B only
    D. Neither A nor B

34. Before the valve is removed from the cylinder head, _____.

    A. The valve spring should be compressed and locks removed
    B. The valve tip edges should be filed if the tip is mushroomed
    C. The ridge should be removed
    D. Both a and b

35. A steel wire brush should be used to clean the gasket surface of an aluminum cylinder head.

    A. True
    B. False

36. *Technician A* says that all valvetrain parts that are to be reused should be kept together.

    Technician B says that valve springs should be tested.

    Which technician tension is correct?

    A. A only
    C. Both A and B
    B. B only
    D. Neither A nor B

37. A cast-iron cylinder head is checked for warpage using a straightedge and a feeler (thickness) gauge. The amount of warpage on a V-8 cylinder head is 0.002 in. (0.05 mm).

    Technician A says that the cylinder head should be resurfaced.

    Technician B says that the cylinder head should be replaced.

    Which technician is correct?

    A. A only
    C. Both A and B
    B. B only
    D. Neither A nor B

38. *Technician A* says that a dial indicator (gauge) is often used to measure valve guide wear by measuring the amount by which the valve head is able to move in the guide.

    Technician B says that a ball gauge can be used to measure the valve guide.

    Which technician is correct?

    A. A only
    C. Both A and B
    B. B only
    D. Neither A nor B

39. *Technician A* says that a worn valve guide can be reamed and a valve with an oversize stem can be used.

    Technician B says that a worn valve guide can be replaced with a bronze insert to restore the cylinder head to useful service.

    Which technician is correct?

    A. A only
    C. Both A and B
    B. B only
    D. Neither A nor B

40. *Technician A* says that worn integral guides can be repaired by knurling.

    Technician B says that worn insert guides can be replaced.

    Which technician is correct?

    A. A only
    C. Both A and B
    B. B only
    D. Neither A nor B

41. Typical valve-to-valve guide clearance should be _____.

    A. 0.001 to 0.003 in. (0.025 to 0.076 mm)
    B. 0.010 to 0.030 in. (0.25 to 0.76 mm)
    C. 0.035 to 0.060 in. (0.89 to 1.52 mm)
    D. 0.100 to 0.350 in. (2.54 to 8.90 mm)

42. Before a valve spring is reused, it should be checked for _____.

    A. Squareness
    B. Free height
    C. Tension
    D. All of the above

43. Many manufacturers recommend that valves be ground with an interference angle. This angle is the difference between the _____.

    A. Valve margin and valve face angles
    B. Valve face and valve seat angles
    C. Valve guide and valve face angles
    D. Valve head and margin angles

44. A cylinder is 0.008 in. out of round.

    Technician A says that the block should be bored and oversize pistons installed.

    Technician B says that oversize piston rings should be used.

    Which technician is correct?

    A. A only
    C. Both A and B
    B. B only
    D. Neither A nor B

45. After the engine block has been machined, the block should be cleaned with _____.
   A. A stiff brush and soap and water
   B. A clean rag and engine oil
   C. WD-40
   D. Spray solvent washer

46. Two technicians are discussing ring end gap.

   Technician A says that the ring should be checked in the same cylinder in which it is to be installed.

   Technician B says that the ends of some piston rings can be filed if the clearance is too small.

   Which technician is correct?
   A. A only   C. Both A and B
   B. B only   D. Neither A nor B

47. *Technician A* says that connecting rod caps should be marked when the connecting rod is disassembled and then replaced in the exact same location and direction on the rod.

   Technician B says that each piston should be fitted to each individual cylinder for best results.

   Which technician is correct?
   A. A only   C. Both A and B
   B. B only   D. Neither A nor B

48. A bearing shell is being installed in a connecting rod. The end of the bearing is slightly above the parting line.

   Technician A says that this is normal.

   Technician B says that the bearing is too big.

   Which technician is correct?
   A. A only   C. Both A and B
   B. B only   D. Neither A nor B

49. *Technician A* says that you should turn the fuel control valve clockwise until it stops before servicing a CNG system.

   Technician B says you must also turn off the manual shut-off valve.

   Who is correct?
   A. A only   C. Both A and B
   B. B only   D. Neither A nor B

50. *Technician A* says that hybrid electric vehicles operate with voltage levels that can be deadly.

   Technician B says that highly insulated lineman's gloves must be worn when disabling the high voltage system of a Toyota Prius.

   Who is correct?
   A. A only   C. Both A and B
   B. B only   D. Neither A nor B

Ajax Lifting Equipment,
Roseville, MI

Bear Automotive Service Equipment,
Milwaukee, WI

Blackhawk Automotive Inc.,
Waukesha, WI

Dana Corporation,
Ann Arbor, MI

Federal Mogul Corporation,
St. Louis, MO

Fluke Corporation,
Everett, WA

KD Tools Danaher Tool Group,
Lancaster, PA

K-Line Industries Inc.,
Holland, MI

Kwik-Way Manufacturing Company,
Marion, IA

Lisle Corporation,
Clarinda, IA

Mac Tools,
Washington Courthouse, OH

Magnaflux Corporation,
Chicago, IL

Mitchell International,
San Diego, CA

OTC Division, SPX Corporation,
Owatonna, MN

Sealed Power Corporation,
Muskegon, MI

Sioux Tools Inc.,
Sioux City, IA

Snap-on Tools Inc.,
Kenosha, WI

Sunnen Products Company,
St. Louis, MO

Winona-Van Norman Machine Company,
Winona, WI

Acura
See Honda.

Audi
Erwin.audi.de
800-544-8021

Bentley
www.Bentleytechinfo.com

BMW
www.bmwtechinfo.com

Chrysler/Dodge/Eagle/Jeep/ Plymouth
www.techauthority.com
800-890-4038

Ferrari
www.ferrariusa.com

Ford/Lincoln/Mercury
www.motorcraft.com

General Motors
Buick/Cadillac/Chevrolet/Geo/GMC/
Hummer/Oldsmobile/Pontiac/Saturn
www.gmtechinfo.com
800-825-5886
800-272-8876 (Saturn)

Honda/Acura
www.serviceexpress.honda.com

Hyundai
www.hmaservice.com

Infiniti
See Nissan.

Isuzu
www.isuzutechinfo.com

Jaguar
www.jaguartechinfo.com

Kia
www.kiatechinfo.com
866-542-8665
Lamborghini
781-788-0600

Land Rover
www.landrovertechinfo.com

Lexus
See Toyota.

Maserati
www.maseratiusa.com

Mazda
www.mazdatechinfo.com
800-824-9655

Mercedes-Benz
www.startekinfo.com

Mini
www.minitechinfo.com

Mitsubishi
www.mitsubishitechinfo.com

Nissan/Infiniti
www.nissantechinfo.com
www.infinititechinfo.com
440-572-0725

Porsche
www.techinfo.porsche.com

Saab
www.saabtechinfo.com

Scion
See Toyota.

Subaru
www.techinfo.subaru.com
866-428-2278

Suzuki
www.suzukitechinfo.com

Toyota/Scion/Lexus
www.techinfo.toyota.com
www.techinfo.scion.com
www.techinfo.lexus.com
800-622-2033

Volkswagen
Erwin.Volkswagen.de
800-822-8987

Volvo
www.volvotechinfo.com
800-258-6586

to convert these	to these,	multiply by:
TEMPERATURE		
Degrees Celsius	Degrees Fahrenheit	1.8 then + 32
Degrees Fahrenheit	Degrees Celsius	0.556 after − 32
LENGTH		
Millimeters	Inches	0.03937
Inches	Millimeters	25.4
Meters	Feet	3.28084
Feet	Meters	0.3048
Kilometers	Miles	0.62137
Miles	Kilometers	1.60935
AREA		
Square Centimeters	Square Inches	0.155
Square Inches	Square Centimeters	6.45159
VOLUME		
Cubic Centimeters	Cubic Inches	0.06103
Cubic Inches	Cubic Centimeters	16.38703
Cubic Centimeters	Liters	0.001
Liters	Cubic Centimeters	1000
Liters	Cubic Inches	61.025
Cubic Inches	Liters	0.01639
Liters	Quarts	1.05672
Quarts	Liters	0.94633
Liters	Pints	2.11344

to convert these	to these,	multiply by:
Pints	Liters	0.47317
Liters	Ounces	33.81497
Ounces	Liters	0.02957
Milliliters	Ounces	0.3381497
Ounces	Milliliters	29.57
WEIGHT		
Grams	Ounces	0.03527
Ounces	Grams	28.34953
Kilograms	Pounds	2.20462
Pounds	Kilograms	0.45359
WORK		
Centimeter-Kilograms	Inch-Pounds	0.8676
Inch-Pounds	Centimeter-Kilograms	1.15262
Meter Kilograms	Foot-Pounds	7.23301
Foot-Pounds	Newton-Meters	1.3558
PRESSURE		
Kilograms/Square Centimeter	Pounds/Square Inch	14.22334
Pounds/Square Inch	Kilograms/Square Centimeter	0.07031
Bar	Pounds/Square Inch	14.504
Pounds/Square Inch	Bar	0.0689
Pounds/Square Inch	Kilopascals	6.895
Kilopascals	Pounds/Square Inch	0.145

APPENDIX E

Engine Specifications Chart

General Engine Specifications

Manufacture of Engine			
Model Year			
Engine Block Casting Number(s)			
Cylinder Head Casting Number(s)			
Cylinder Block Casting Material			
Cylinder Head Casting Material			
Combustion Chamber Design			
# of Cylinders		Cyl. Arrangement	
Bore		Firing Order	
Stroke		Valve Arrangement	
Rated Horsepower		Rated Torque	
Compression Pressure		Oil Capacity	
Compression Ratio		Oil Pressure	
Cooling Capacity		Oil Type	

Engine Torque Specifications

Crankshaft Main Bearing Bolts		Rocker Arm Shaft/Stud	
Connecting Rod Bolts/Nuts		Vibration Damper	
Oil Pump Housing		Flywheel/Flexplate	
Water Pump		Flywheel	
Timing Cover		Clutch Pressure Plate	
Thermostat Housing		Camshaft Caps	
Spark Plugs		Timing Tensioner	
Cylinder Head Bolts		Camshaft Sprocket	
Intake Manifold		Oil Pan	
Exhaust Manifold		Oil Pump Bolts/Nuts	

Engine Dimensions

Camshaft			
Bearing Journal Diameter			
Base Circle Runout			
Cam Lobe lift			
Valve Lift			
Intake Valve Timing	Opens @		Closes @
Exhaust Valve Timing	Opens @		Closes @
Valve Overlap			
Duration	Intake		Exhaust

Crankshaft

End Play	
Main Bearing Journal Diameter	
Main Bearing Journal Width	
Main Bearing Clearance	
Connecting Rod Journal Diameter	
Connecting Rod Journal Width	
Maximum Out-of-Round (All Journals)	
Maximum Taper (All Journals)	

Cylinder Block

Deck Height	
Deck Clearance	
Cylinder Bore Diameter (Standard)	
Maximum Taper	
Maximum Out-of-Round	
Maximum Cylinder Block Flatness	

Connecting Rods

Total Length (Center-to-Center)	
Piston Pin Bore Diameter	
Connecting Rod Bore	
Bearing Clearance (Standard)	
Side Clearance	

Pistons

Piston-to-Bore Clearance	
Piston Ring Gap Clearance	
Piston Ring Side Clearance	
Piston Ring Groove Height	
Piston Ring Groove Diameter	
Piston Pin Diameter	
Piston-to-Pin Clearance	

Cylinder Head

Combustion Chamber Volume	
Valve Arrangement	
Valve Guide Inside Diameter	
Valve Stem-to-Guide Clearance	
Intake Valve Seat Angle	
Exhaust Valve Seat Angle	
Intake Valve Seat Width	

Exhaust Valve Seat Width	
Valve Seat Runout	
Cylinder Head Flatness	
Cylinder Head Thickness	

Rocker Arms & Push Rods	
Rocker Arm Ratio	
Push Rod Length	
Push Rod Diameter	
Lifter/Tappet Diameter	

Valves		
Intake Valve Length		
Exhaust Valve Length		
Intake Valve Stem Diameter		
Exhaust Valve Stem Diameter		
Intake Valve Head Diameter		
Exhaust Valve Head Diameter		
Intake Valve Face Diameter		
Exhaust Valve Face Diameter		
Intake Valve Face Angle		
Exhaust Valve Face Angle		
Valve Interference Angle		
Valve Margin		
Valve Lash	Cold	Hot

Valve Springs	
Intake Valve Spring Free Length	
Exhaust Valve Spring Free Length	
Intake Spring Tension (force@length)	
Exhaust Spring Tension (force@length)	
Installed Height	

Oil System		
Oil Pressure	*@ Idle RPM*	*@ 2000 RPM*
Oil Pump Gear-to-Body Clearance		
Oil Pump Gear End Clearance		

GLOSSARY
GLOSARIO

Note: **Terms are highlighted in color,** followed by **Spanish translation in bold**.

Abrasive cleaning Abrasive cleaning is a method of cleaning parts involving the use of high pressure sand blast.

Limpieza abrasiva La limpieza abrasiva es un método para limpiar partes con el uso del chorro de arena a alta presión.

Airflow restriction indicator The airflow restriction indicator displays the amount of air filter element restriction.

Indicador de restricción del flujo del aire El indicador de restricción del flujo del aire muestra la cantidad de restricción del elemento del filtro de aire.

Align honing Machining process of the main bearing journals used to remove in excess of 0.050 in., but this will result in changing the location of the crankshaft in the block. Align honing restores original bore size by removing metal from the entire circumference of the bore.

Rectificado en serie Un proceso de rectificar en máquina los muñones del cigüeñal para quitar más de 0.050 de una pulgada, pero ésto resulta en que cambia de lugar el cigüeñal en el monoblock. El rectificado en serie restaura el tamaño original del taladro removiendo el metal de toda su circunferencia.

Aligning bars Tools used to determine the proper alignment of the crankshaft saddle bores.

Barras para alinear Las herramientas que sirven para determinar el alineamiento correcto de los taladros del asiento del cigüeñal.

Aluminum hydroxide Corrosion products that are carried to the radiator and deposited when they cool off. They appear as dark gray when wet and white when dry.

Hidróxido de aluminio Los productos de corrosión que se llevan al radiador y se depositan al enfriarse. Son de color gris obscuro mojados y se blanquean al secarse.

Backlash The clearance between two parts.

Culateo/juego La holgura entre dos partes.

Base circle The base circle is the diameter across the sides of the camshaft lobe with the lobe facing upwards.

Círculo de la base El círculo de la base es el diámetro a través de los lados del compartimiento del árbol de levas con el compartimiento que va hacia arriba.

Battery terminal test A test that checks for poor electrical connections between the battery cables and terminals. Use a voltmeter to measure voltage drop across the cables and terminals.

Prueba de los terminales de la batería Una prueba que averigua las conexiones eléctricas malas entre los cables y los terminales de la batería. Usa un voltímetro para medir la caída del voltaje entre los cables y los terminales.

Bearing Soft metallic shells used to reduce friction created by rotational forces.

Cojinete Una pieza hueca de metal blanda que sirve para reducir la fricción creada por las fuerzas giratorias.

Bell housing The bell housing is the portion of the transmission that bolts to the back of the engine.

Caja de la campana La caja de la campana es la porción de la transmisión que se fija en la parte posterior del motor.

Belt surfacer A belt surfacer is used to plane cylinder head surfaces.

Cepilladora de la banda La cepilladora de la banda se usa para aplanar las superficies de la cabeza del cilindro.

Big end The end of the connecting rod that attaches to the crankshaft.

Extremo grande Refiere a la extremidad de la biela que se conecta al cigüeñal.

Blowby The unburned fuel and combustion products that leak past the piston rings and enter the crankcase.

Soplado El combustible no consumido y los productos de la combustión que escapen por los anillos de pistón y entran al cárter.

Boost pressure Boost pressure is the amount of pressure supplied by the turbocharger or supercharger to the intake manifold.

Presión impulsadora La presión impulsadora es la cantidad de presión que provee el turbocargador o el supercargador al colector de entrada.

Boring The process of enlarging a hole.

Escariar El proceso de agrandar un agujero.

Bottom-end knock A bottom-end knock is a loud, deep knocking noise heard from the lower end of the engine. It is usually an indication of serious engine problems, such as worn main or rod bearings or crankshaft, or low oil pressure.

Golpe de fondo Un golpe de fondo es un ruido fuerte que se escucha en el fondo del motor. Es usualmente una indicación de problemas serios con el motor, tales como cojinetes principales o de la varilla o del cigüeñal que están gastados, o por baja presión del aceite.

Broach To finish the inside surface of a bore by forcing a multiple-edge cutting tool through it.

Fresar con barrena Acabar una superficie interior de un taladro al atravesarla con una herramienta que tiene múltiples hojas cortantes.

Caliper Measuring tool capable of taking readings of inside, outside, depth, and step measurements in 0.001-inch increments.

Calibre Una herramienta de medir capaz de tomar las medidas interiores, exteriores, de profundidad y de paso en incrementos de 0.001 de una pulgada.

Camshaft The shaft containing lobes to operate the engine valves.

Arbol de levas Un eje que tiene lóbulos que operan las válvulas del motor.

Camshaft end play Camshaft endplay is the measure of how far the camshaft can move lengthwise in the block.

Juego longitudinal del árbol de levas El juego longitudinal del árbol de levas es la medida de hasta donde se puede mover el árbol de levas longitudinalmente en el bloque.

Camshaft position actuator A camshaft position actuator is a hydraulically actuated device that changes the camshaft timing in relation to the crankshaft.

Biela de accionamiento de la posición del árbol de levas La biela de accionamiento de la posición del árbol de levas es un dispositivo ac-tuado hidráulicamente que cambia la puesta a punto del árbol de levas en relación con el cigüeñal.

Capacity test A test that checks the battery's ability to perform when loaded.

Prueba de capacidad Un prueba que averigua la habilidad de la bateria a funcionar bajo carga.

Carbon monoxide An odorless, colorless, and toxic gas that is produced as a result of incomplete combustion.

Monóxido de carbono Un gas sin olor, sin color y tóxico que se produce como resultado de una combustión incompleta.

Catalytic converter A catalytic converter reduces tailpipe emissions of carbon monoxide, unburned hydrocarbons, and oxides of nitrogen.

Catalizador Un catalizador reduce las emisiones del tubo de escape de monóxido de carbono, hidrocarburos no quemados y óxidos de nitrógeno.

Chemical block tester A chemical block tester uses a chemical that reacts to combustion gases. When air is passed across the solution from on top of the radiator, no combustion gases should be present. If they are, a problem with the head or head gasket is indicated.

Probador de bloque químico Un probador químico en tubo usa un químico que reacciona ante los gases de combustión. Cuando se pasa aire a través de la solución por encima del radiador, no debe estar presente ningún gas de combustión. Si hay, se está indicando que existe un problema con la cabeza o con la cabeza de la junta.

Chemical cleaning The process of using chemical action to remove soil contaminants from the engine components.

Limpieza química El proceso de usar la acción química para des-prender los contaminantes sucios de los componentes del motor.

Compression The reduction in volume of a gas.

Compresión La reducción del volumen de un gas.

Compression gauge A compression gauge is used to check cylinder compression.

Calibre de compresión Un calibre de compresión se usa para revisar la compresión del cilindro.

Compression testing A diagnostic test to determine the engine cylinder's ability to seal and to maintain pressure.

Prueba de compresión Una prueba diagnóstica que determina la habilidad del cilindro del motor de sellar y mantener la presión.

Core plugs Metal plugs screwed or pressed into the block or cylinder head at locations where drilling was required or sand cores were removed during casting.

Tapones del núcleo Los tapones metálicos fileteados o prensados en el monoblock o en la cabeza de los cilindros ubicados en donde se había requerido taladrar o quitar los machos de arena durante la fundición.

Crankshaft end play The measure of how far the crankshaft can move lengthwise in the block.

Juego en el extremo del cigüeñal La medida de cuánto puede moverse el cigüeñal a lo largo del monoblock.

Current draw test A test that measures the amount of current that the starter draws when actuated. It determines the electrical and mechanical condition of the starting system.

Prueba de carga al corriente Una prueba para medir la cantidad del corriente que consuma el motor de arranque al ser puesto en acción. Determina la condición eléctrica y mecánica del sistema de arranque.

Cylinder boring Cylinder boring makes it possible to bore cylinders to accept oversize pistons.

Perforación del cilindro La perforación del cilindro hace posible perforar los cilindros para que acepten pistones de gran tamaño.

Cylinder bore dial gauge An instrument used to measure the cylinder bore for wear, taper, and out-of-round.

Calibre carátula del taladro del cilindro Un instrumento de medida que sirve para averiguar si el taladro del cilindro está gastado, cónico, o ovulado.

Cylinder leakage test A test that determines the condition of the piston ring, intake or exhaust valve, and head gasket. It uses a controlled amount of air pressure to determine the amount of leakage.

Prueba de fugas en el cilindro Una prueba para determinar la condición del anillo de pistón, la válvula de entrada o escape, y la junta de la cabeza. Se emplea una cantidad controlada de presión de aire para determinar la cantidad de la fuga.

Deck The top of the engine block where the cylinder head is attached.

Cubierta La parte superior del monoblock en donde se conecta la cabeza del cilindro.

Detonation A defect that occurs if the air/fuel mixture in the cylinder is burned too fast.

Detonación Un defecto que ocurre si la mezcla aire/combustible en el cilindro se quema demasiado rápido.

Diagnostic link connector (DLC) The diagnostic link connector (DLC) is a connector, usually located under the dash, that provides access to diagnostic information.

Conectador de bieleta de diagnóstico El conectador de bieleta de diagnóstico es un conectador que generalmente se localiza debajo del tablero de instrumentos, que proporciona acceso a la información de diagnóstico.

Diagnostic trouble code (DTC) A diagnostic trouble code (DTC) is a coded output from the PCM to identify system faults.

Código de diagnóstico de problemas Un código de diagnóstico de problemas es una salida en código desde el MCM para identificar las fallas del sistema.

Dial calipers Vernier style calipers with a dial gauge installed to make reading of the vernier scale easier and faster.

Calibres de carátula Los calibres tipo vernier equipados con una carátula para facilitar y acelerar la lectura de un escala vernier.

Dial indicator A measuring instrument consisting of a dial face with a needle. The dial is usually calibrated in 0.001-inch increments. A spring loaded plunger or toggle lever transfers movement to the dial needle.

Indicador de carátula Un instrumento de medida que consiste de una carátula con una aguja. La carátula suele ser calibrada en incrementos de 0.001 de una pulgada. Un pistón tubular cargado de resorte o por una palanca acodada transfere el movimiento a la aguja de la carátula.

Die A tool used to repair or cut new external threads.

Matriz Una herramienta para reparar o cortar las roscas exteriores.

Digital pyrometer A digital pyrometer is an electronic device that measures heat.

Pirómetro digital Un pirómetro digital es un dispositivo electrónico que mide el calor.

Dry sleeves Replacement cylinder sleeves that do not come into contact with engine coolant. They are surrounded by the cylinder bore.

Camisas secas Las camisas de repuesto del cilindro que no se ponen en contacto con el fluido refrigerante del motor. Se ajusten en el taladro del cilindro.

Dry starts Dry starts refer to increasing the engine speed right after engine starting, before oil has filled the system.

Arranque en seco Los arranques en seco se refieren a aumentar la velocidad del motor justo después de encenderlo, antes de que el aceite haya llenado el sistema.

Duration The length of time, expressed in degrees of crankshaft rotation, the valve is open.

Duración La cantidad del tiempo, representado por los grados de rotación del cigüeñal, que esta abierta la válvula.

Engine hoist A special lifting tool designed to remove the engine through the hood opening. Most are portable or fold up for easy storage. The lifting of the boom is performed by a special long reach hydraulic jack.

Grúa Una herramienta especial de izar diseñada para remover el motor por la apertura del cofre. Suelen ser portátiles o se doblan fácilmente para guardarse. El brazo de la grúa se levanta con un gato hidráulico de larga extensión.

Engine stand A special holding fixture that attaches to the back of the engine, supporting it at a comfortable working height. In addition, most stands allow the engine to be rotated for easier disassembly and assembly.

Bancada para motor Un accesorio de apoyo especial que se conecta a la parte trasera del motor. Permite que se soporta el motor en una altura ideal para trabajar. Además, la mayoría de las bancadas permiten la rotación del motor para facilitar el desmontaje y montaje.

Environmental Protection Agency (EPA) The Environmental Protection Agency (EPA) is a department of the federal government dedicated to reducing pollution and improving environmental quality.

Secretaría de Protección del Ambiente La Agencia de Protección del Ambiente es una agencia del gobierno federal que se dedica a reducir la contaminación y a mejorar la calidad del ambiente.

Exhaust gas analyzer A four-gas exhaust analyzer enables the technician to look at the effects of the combustion process. It will measure hydrocarbons, carbon monoxides, carbon dioxide, and oxygen levels in the exhaust. A five-gas analyzer also measures oxides of nitrogen.

Analizador del gas de emisión Un analizador de emisión de 4 gases permite al técnico observar los efectos del proceso de la combustión. Medirá los hidrocarburos, los monóxidos de carbono, el dióxido de carbono y los niveles de oxígeno en las emisiones. Un analizador de 5 gases también mide los óxidos de nitrógeno.

Exhaust gas recirculation (EGR) valve The exhaust gas recirculation (EGR) valve is an emission control device that cools the combustion chamber to reduce emissions of oxides of nitrogen.

Válvula de recirculación del gas de emisión La válvula de recirculación del gas de emisión es un dispositivo de control de emisiones que enfría la cámara de combustión para reducir las emisiones de los óxidos de nitrógeno.

Exhaust valve An engine part that controls the expulsion of spent gases and emissions out of the cylinder.

Válvula de escape Una parte del motor que controla la expulsión de los gases consumidos y los emisiones del cilindro.

Extractor An extractor is a special tool that is used to remove a broken bolt from a bolthole.

Extractor Un extractor es una herramienta especial que se usa para quitar un tornillo de un hoyo de tornillo.

False guides Devices that are similar to inserts.

Guías falsas Los dispositivos muy parecidos a las piezas insertas.

Feeler gauge A measuring tool consisting of a series of metal strips cut to precise thicknesses.

Galga calibrada Una serie de hojas metálicas cortadas a los espesores precisos.

Flat rate pay Flat rate pay means that you are paid for each hour of billable work as determined by the manufacturer, an aftermarket labor-estimating guide, or your shop.

Tarifa bloque La tarifa bloque significa que se pagará por cada hora de trabajo realizado como lo determina el fabricante, una guía de estimación de trabajo por horas extras, o por el taller.

Flex fans Fans designed to widen their pitch at slow engine speeds when more air flow is required through the radiator fins. At higher engine speeds, the pitch is decreased to reduce the horsepower required to turn the fan.

Ventiladores ajustables Un diseño de los ventiladores que amplían su paso en las velocidades más bajas que requieren más flujo del aire por los devanados del radiador. En las velocidades más altas el paso se disminuye reduciendo los requerimientos del par de dar las vueltas al ventilador.

Floor jack A portable hydraulic tool used to raise and lower a vehicle.

Gato Una herramienta portátil hidráulica que sirve para levantar y bajar un vehículo.

Front pipe The front pipe connects the exhaust manifolds to the catalytic converter or muffler.

Tubo frontal El tubo frontal conecta los colectores de emisión con el catalizador o mofle.

Fuel control valve The fuel control valve on a CNG vehicle is a manually operated valve designed to stop fuel flow from the cylinder.

Válvula de control del combustible La válvula de control del combustible en un vehículo GNC es una válvula que se maneja manualmente diseñada para parar el flujo del combustible que está en el cilindro.

Fuel pressure gauge A fuel pressure gauge is used to measure the pressure in the fuel system.

Calibre de la presión del combustible Un calibre de la presión del combustible se usa para medir la presión en el sistema de combustible.

Fuel pressure regulator A fuel pressure regulator is typically a mechanical device that maintains the correct fuel pressure in the fuel rail the injectors sit in.

Regulador de la presión del combustible Un regulador de la presión del combustible es típicamente un dispositivo mecánico que mantiene la presión correcta del combustible en el riel de combustible en donde se asientan los inyectores.

Gallery plugs Metal plugs used to cap the drilled oil passages in the cylinder head or engine block. They can be threaded or pressed into the hole.

Tapones de la canalización de aceite Los tapones metálicos que sirven para estancar los pasajes taladrados del aceite en la cabeza del cilindro o en el monoblock. Pueden ser fileteados o prensados en el agujero.

Harmonic balancer (vibration dampener) A component attached to the front of the crankshaft used to reduce the torsional or twisting vibration that occurs along the length of the crankshaft.

Equilibrador armónico (amortiguador de vibraciones) Un componente conectado a la parte delantera de un cigüeñal que sirve para reducir las vibraciones torsionales que ocurren por la longitud del cigüeñal.

Head The head of the piston is the top piston surface.

Cabeza La cabeza de un pistón es la superficie superior del pistón.

Heel The camshaft heel is roughly 180° from the lobe high point. This is where the valve will be closed.

Talón El talón del árbol de levas está a aproximadamente 180 grados del punto más alto del compartimiento. Allí es donde se cerrará la válvula.

Hydraulic press A hydraulic press is a tool used to remove and install components with a tight press fit.

Prensa hidráulica La prensa hidráulica es una herramienta que se usa para quitar e instalar componentes con un cierre apretado.

Hydrostatic lock The result of attempting to compress a liquid in the cylinder. Since liquid is not compressible, the piston is not able to travel in the cylinder.

Bloqueo hidrostático El resultado de un intento de comprimir un líquido en el cilindro. Como no se puede comprimir un líquido, el pistón no puede moverse en el cilindro.

Impact screwdriver An impact screwdriver twists the screw as you hit the end of the driver with a hammer. It works very well to remove tightly fastened, rusty, or recessed screws.

Destornillador de golpe Un destornillador de impacto le da vuelta al tornillo mientras se golpea la orilla del destornillador con un martillo. Trabaja muy bien para quitar tornillos muy apretados, oxidados o en retranqueo.

Intake manifold Component that delivers the air or air/fuel mixture to each engine cylinder.

Múltiple de admisión Un componente que entrega el aire o la mezcla del aire/combustible a cada cilindro del motor.

Interference angle Valve design in which the seat angle is 1° greater than the valve face angle to provide a more positive seal.

Angulo de interferencia Un diseño de la válvula en el cual el ángulo del asiento de la válvula es un grado además del ángulo de la cara de la válvula para proveer un sello más positivo.

Interference fit Interference fit components are slightly smaller than the shaft they fit onto or the bore they fit into. This makes a puller or a press necessary for disassembly and reassembly.

Ajuste duro Los componentes de ajuste duro son un poco más pequeños que el eje en el que caben o el orificio en el que están. Esto hace necesario el uso de un arrancador o de una prensa para desensamblarlo y reensamblarlo.

Jack stands (safety stands) Support devices used to hold the vehicle off the floor after it has been raised by the floor jack.

Torres (soportes de seguridad) Los dispositivos de soporte que se emplean para mantener el vehículo levantado del piso despues de que haya sido levantado del piso por un gato hidráulico.

Jet valves Valves used by some manufacturers to direct a stream of oil to the underside of the piston head. Jet valves are common on turbocharged engines to keep the piston cool to prevent detonation and piston damage.

Válvulas de chorro Las válvulas empleados por algunos fabricantes para dirigir un chorro de aceite a la parte inferior de la cabeza del pistón. Las válvulas de chorro suelen usarse en los motores turbocargados para mantener frío al pistón y para prevenir la detonación y los daños al pistón.

Knock sensor A knock sensor is bolted into the engine and senses engine vibrations that resemble knock. It relays this information to the PCM so it can adjust ignition timing to try to reduce knocking.

Sensor de golpe Un sensor de golpe está atornillado al motor y percibe las vibraciones del motor que parezcan golpes. Le envía esta información al MCM para que ajuste el encendido de arranque y trate de reducir el golpeteo.

Knurl A special bit that rolls a thread into the guide and causes the metal to rise.

Moleta Una broca especial que enreda un hilo en la guía y causa que se levanta el metal.

Knurling A machining process that decreases the size of a bore by forcing a bit that swells the metal much like a tap does when it cuts threads.

Moletear Un proceso de rebajar a máquina el tamaño de un taladro forzando una broca que hincha al metal de una manera muy parecida un macho cortando las roscas.

kPa kPa stands for kilopascals.

kPa kPa significa kilo pascales.

Land A piston land is the area between the ring grooves.

Apoyo, contacto El apoyo o contacto del pistón es el área entre las rodadas de los anillos.

Leakdown The relative movement of the lifter's plunger in respect to the lifter body.

Tiempo de fuga El movimiento relativo del émbolo de levantaválvulas con relación al cuerpo del levantaválvulas.

Lifters Mechanical (solid) or hydraulic connections between the camshaft and the valves. Lifters follow the contour of the camshaft lobes to lift the valve off its seat.

Levantaválvulas Las conexiones mecánicas (sólidas) o hidráulicas entre el árbol de levas y las válvulas. Los levantaválvulas siguen el contorno de los lóbulos del árbol de levas para levantar la válvula de su asiento.

Line boring A machining process of the main bearing journals that restores the original bore size by cutting metal using cutting bits.

Taladrar en serie Un proceso de rectificación a máquina de los muñones del cigüeñal que restaura el tamaño original del taladro cortando el metal con brocas para cortar.

Lineman's gloves Lineman's gloves are highly insulated rubber gloves that protect the technician from lethal high voltages.

Guantes de operador de línea Los guantes de operador de línea son guantes de hule altamente aislantes que protegen al técnico de un alto voltaje letal.

Liners Thin tubes placed between two parts.

Manguitos Los tubos delgados puestos entre dos partes.

Lobelift Lobe lift is the distance that a cam lobe moves the contacting valve train component as the cam lobe rotates.

Realce del lóbulo El realce del lóbulo es la distancia que un lóbulo del eje mueve el componente del tren de la válvula de contacto mientras gira el lóbulo del eje.

Lugging Lugging of the engine results from too high a gear selection for the load.

Arrastre El arrastre de un motor resulta de la selección de una velocidad muy alta para la carga.

Machinist's rule A multiple scale ruler used to measure distances or components that do not require precise measurement.

Regla de acero Una regla con graduaciones múltiples para medir las distancias o los componentes que no requieren una medida precisa.

Magnafluxing Magnafluxing is a crack detection method that can be used on any ferrous metal component. Magnets are placed at either end of the component and liquid with metal filings is poured over the component. If there is a crack, the filings line up across the crack and are highlighted under a black light.

Aplicación de la magna flujo La aplicación de la magna flujo es un método para detectar grietas que puede usarse en cualquier componente metálico ferroso. Los imanes se colocan en cualquier terminal del componente y un líquido con partículas metálicas se vacía sobre el componente. Si hay una grieta, las partículas se forman a lo largo de la grieta y se realzan bajo una luz negra.

Malfunction indicator light (MIL) The malfunction indicator light (MIL) is an amber light on the dash used to alert the driver about an emissions-related powertrain problem. It is on all 1996 and newer passenger cars and light duty trucks sold in the United States and Canada.

Luz indicadora de falla La luz indicadora de falla es una luz color ámbar en el tablero de instrumentos y se usa para alertar al conductor sobre problemas con una emisión relacionada con el motor. Está presente en los automóviles de pasajeros y en las camionetas de carga li-gera desde los modelos de 1996 en adelante que se venden en EU. y Canadá.

Manual shut-off valve The manual shut-off valve stops fuel flow from the CNG cylinder(s) to the engine.

Válvula de apagado manual La válvula de apagado manual para el flujo de combustible de los cilindros GNC al motor.

Material Safety Data Sheet (MSDS) A sheet that contains detailed information concerning hazardous materials. The MSDS must be maintained by the employer.

Hoja de Dato de Seguridad de los Materiales (MSDS) Una hoja que contiene la información detallada referente a los materiales peligrosos. El patrón debe conservar el MSDS.

Meter One meter is a little more than three inches longer than a yard.

Metro Un metro es un poco más de 3 pulgadas más largo que una yarda.

Microinch One millionth of an inch. The microinch is the standard measurement for surface finish in the American customary system.

Micropulgada La millonésima parte de una pulgada. Una micropulgada es una unidad común para el acabado de la superficie en el sistema de medida americana.

Micrometer One millionth of a meter. The micrometer is the standard measurement for surface finish in the metric system.

Micrómetro Una millonésima parte de un metro. La medida común para el acabado de la superficie en el sistema métrica.

Micrometer Precision measuring instrument designed to measure outside, inside, or depth measurements.

Micrómetro Un instrumento de medidas precisas diseñada para tomar medidas exteriores, interiores o de profundidad.

Minor thrust side The side opposite the side of the rod that is stressed during the power stroke.

Lado de empuje menor El lado opuesto del lado cuyo biela recibe el esfuerzo durante la carrera de potencia.

Misfire A misfire is when a combustion event is incomplete or does not occur at all.

Fallo de encendido El fallo de encendido es cuando un evento de combustión está incompleto o no sucede del todo.

Muffler The muffler is a device, mounted to the exhaust system behind the catalytic converter, that reduces engine noise.

Mofle El mofle es un dispositivo montado en el sistema de emisiones detrás del catalizador, que reduce el ruido del motor.

National Institute of Automotive Service Excellence (ASE) The National Institute of Automotive Service Excellence (ASE) offers nationally recognized certifications to automobile and heavy duty truck technicians through a combination of bi-annual testing and proof of professional experience.

Instituto Nacional para la Excelencia del Servicio Automovilístico El Instituto Nacional para la Excelencia del Servicio Automovilístico ofrece certificaciones reconocidas a nivel nacional a los técnicos de automóviles y camionetas de carga pesada mediante una combinación de exámenes bianuales y prueba de experiencia profesional.

Noid light A noid light can be used to determine if the fuel injector is receiving its proper voltage pulse from the computer.

Luz Noid Una luz Noid puede usarse para determinar si el inyector de combustible está recibiendo un pulso de voltaje apropiado de la computadora.

Nose The nose is the area just before and just after the peak of the camshaft lobe.

Boca La boca es el área justo antes y después del tope del lóbulo del árbol de levas.

Occupational Safety and Health Administration (OSHA) The Occupational Safety and Health Administration (OSHA) is a part of the U.S. government tasked with protecting the health and safety of workers.

Administración de la Salud y Seguridad Ocupacional La Administración de la Salud y Seguridad Ocupacional es una agencia del go-bierno de EU. que se encarga de proteger la salud y la seguridad de los trabajadores.

Oil pressure test Test to determine the condition of the bearings and other internal engine components.

Prueba de presión de aceite Una prueba que se efectua para determinar la condición de los cojinetes u otros componentes interiores del motor.

Oil pump A rotor or gear type positive displacement pump used to take oil from the sump and deliver it to the oil galleries. The oil galleries direct the oil to the high wear areas of the engine.

Bomba de aceite Un rotor o una bomba de tipo deplazamiento positivo de engranajes que sirve para tomar el aceite de un suministro y entregarlo a las canalizaciones de aceite. Las canalizaciones de aceite dirigen el aceite a las áreas del motor que imponen gastos severos.

Open circuit voltage test Test to determine the battery's state of charge. It is used when a hydrometer is not available or cannot be used.

Prueba de voltaje de circuito abierto Una prueba para determinar el estado de carga de la bateria. Se usa cuando no es disponible o no se puede usar un hidrómetro.

Open pressure Spring tension when the valve spring is compressed and the valve is fully open.

Presión abierta La tensión de un resorte de la válvula al estar comprimido con la válvula completamente abierta.

Out-of-round The condition when measurements of a diameter differ at different locations. The term out-of-round applies to inside or outside diameters.

Ovulado Una condición en la cual las medidas de un diámetro varían an lugares distinctos. El término ovulado aplica a los diámetros inte-riores o exteriores.

Out-of-round gauges Instruments used to measure the concentricity of connecting rod bores.

Calibradores de ovulado Los instrumentos para medir la concentricidad de los taladros de las bielas.

Overlap Overlap is the number of degrees, as measured in degrees of crankshaft rotation, when both the intake and exhaust valves are open near TDC on the exhaust stroke.

Solapado El solapado es el número de grados, como se mide en grados de la rotación del cigüeñal, cuando ambas válvulas de entrada y de sa-lida están abiertas cerca del PMS en el golpe de salida.

Oversize bearings Bearings that are thicker than standard to increase the outside diameter of the bearing to fit an oversize bearing bore. The inside diameter is the same as standard bearings.

Cojinete de medidas superiores Los cojinetes que son de un espesor más grueso de lo normal para aumentar el diámetro exterior del cojinete para quedarse en un taladro que rebasa la medida. El diámetro interior es lo mismo del cojinete normal.

Oxygen sensor An oxygen sensor sits in the exhaust stream and measures the amount of oxygen present in the exhaust. One use of this sensor is to help the computer to determine whether the engine is running rich or lean and adjust accordingly.

Sensor de oxígeno El sensor de oxígeno se asienta en la corriente de salida y mide la cantidad de oxígeno presente en la salida. Un uso de este sensor es el ayudar a la computadora a determinar si el motor está trabajando en alta o en baja combustibilidad y a hacer los ajustes correctos.

Particulates Particulates are small carbon particles present in diesel exhaust and considered to be pollutants.

Micropartículas Las micropartículas son pequeñas partículas de carbono presentes en las emisiones diesel y se les considera contaminantes.

Peening The process of removing stress in a metal by striking it.

Martillar El proceso de quitar la fatiga de un metal golpeandolo.

Pickup tube A tube used by the oil pump to deliver oil from the bottom of the oil pan. The bottom of the tube has a screen to filter larger contaminants.

Tubo de captación Un tubo empleado por la bomba de aceite para entregar el aceite del fondo del colector de aceite. La parte inferior del tubo tiene una rejilla para filtrar los contaminantes más grandes.

Pilot bushings The pilot bushings or pilot bearings are pressed into the centerline of the crankshaft at the back, or the flywheel. They are used to support the front of the transmission's input shaft.

Casquillos piloto Los casquillos piloto o cojinetes piloto están apretados hacia adentro de la línea central del cigüeñal al fondo, o del volante. Se usan para apoyar el frente del eje de entrada de la transmisión.

Pin boss A bore machined into the piston that accepts the piston pin to attach the piston to the connecting rod.

Mamelón Un taladro tallado a máquina en el pistón que acomoda la espiga del pistón para conectarlo a la biela.

Piston slap A sound that results from the piston hitting the side of the cylinder wall.

Golpeteo del pistón Un ruido resultando del pistón golpeando contra el muro del cilindro.

Plastigage A string-like plastic that is available in different diameters used to measure the clearance between two components. The diameter of the plastigage is exact, thus any crush of the gage material will provide an accurate measurement of oil clearance.

Plastigage Un plástico en forma de hilo disponible en diámetros distinctos que se emplea en medir la holgura entre dos componentes. El diámetro del calibre plástico es preciso, asi cualquier aplastamiento del material del calibre provee una medida precisa de la hogura del aceite.

Pneumatic tools Tools powered by compressed air.

Herramientas neumáticas La herramientas que derivan su poder del aire bajo presión.

Polishing The process of removing light roughness from the journals by using a fine emery cloth. Polishing can be used to remove minor scoring of journals that do not require grinding.

Bruñido El proceso de quitar la aspereza ligera de los muñones usando una tela abrasiva muy fina. El bruñido puede emplearse en quitar las rayas superficiales de los muñones que no requieren rectificación.

Positive crankcase ventilation (PCV) system An emission control system that routes blowby gases and unburned oil/fuel vapors to the intake manifold to be added to the combustion process.

Sistema (PCV) de ventilación positiva de la caja del cigüeñal Un sistema de emisión que lleva los gases soplados y los vapores del aceite/combustible no quemados al múltiple de entrada para que se pueden añadir al proceso de combustión.

Power balance test Test used to determine if all cylinders are producing the same amount of power output. In an ideal situation, all cylinders would produce the exact same amount of power.

Prueba del equilibrio de fuerza Una prueba para determinar si todos los cilindros producen la misma cantidad de potencia de salida. En una situación ideal, todos los cilindros producirían exactamente la misma cantidad de poder.

Power tools Tools that use other forces than that generated from the body. They can use compressed air, electricity, or hydraulic pressure to generate and multiply force.

Herramientas de motor La herramientas que usan otras fuerzas que las producidas por el cuerpo. Pueden usar el aire bajo presión, la electricidad, o la presión hidráulica para engendrar y multiplicar la fuerza.

Powertrain control module (PCM) The powertrain control module (PCM) is the vehicle's computer that controls engine performance and related functions such as fuel and spark control.

Módulo de control del motor (MCM) El modulo de control del motor (MCM) es la computadora del vehículo que controla el funcionamiento del motor y las funciones relacionadas tales como el control del combustible y de la chispa.

Preignition Defect that is the result of spark occurring too soon.

Autoencendido Un defecto que resulta de una chispa que ocurre demasiado temprano.

Prelubrication Prelubrication entails lubricating the engine or turbocharger bearings or other components before starting the engine.

Pre-lubricación La pre-lubricación implica la lubricación del motor o de los cojinetes del turbosobrealimentador u otros componentes antes de encender el motor.

Pressure plate The pressure plate couples and uncouples the clutch disc from the flywheel on a manual transmission to transmit engine torque to the transmission.

Placa de presión La placa de presión acopla y desacopla los discos del embrague del volante en una transmisión manual para transmitir par motor a la transmisión.

Profilometer A tool capable of electrically sensing the distances between peaks to determine the finish of a cut.

Perfilómetro Una herramienta capaz de detectar electrónicamente las distancias entre dos puntos altos para determinar cuando terminar un corte.

Prussian blue Prussian blue is thick ink used to show the area of seat contact on the valve face.

Azul de Prusia El azul de Prusia es una tinta fuerte que se usa para mostrar el contacto de asiento en la cara de una válvula.

Purge solenoid The purge solenoid is part of the evaporative emissions control system and is used to purge fuel vapors from the fuel tank into the intake manifold to be burned.

Solenoide de purga El solenoide de purga es parte del sistema de control de emisiones evaporativas y se usa para purgar los vapores del combustible desde el tanque de la gasolina hacia el colector de admisión para quemarse.

Pushrod A connecting link between the lifter and rocker arm. Engines designed with the camshaft located in the block use pushrods to transfer motion from the lifters to the rocker arms.

Varilla de presión Una conexión entre la levantaválvulas y el balancín empuja válvulas. Los motores diseñados con el árbol de levas en el bloque usan las varillas de presión para transferir el movimiento de la levantaválvulas a los balancines.

Quick disconnect fittings Quick disconnect fittings are fuel line fittings that are removable with a special tool rather than threaded fittings.

Montaje de desconectado rápido Los montajes de desconectado rápido son montajes de la línea del combustible que se pueden quitar con una herramienta especial en vez de montajes enroscados.

Radiator A component consisting of a series of tubes and fins that transfer the heat from the coolant to the air.

Radiador Un componente que consiste de una serie de tubos y aletas que transferen el calor del fluido refrigerante al aire.

Registering Registering the main bearing caps is done by tightening the cap bolts using hand pressure only while rotating the crankshaft. This squares the cap to the block before the bolts are torqued.

Ajuste El ajuste de las tapas de los cojinetes principales se hace al apretar las tapas de los tornillos usando sólo la presión de la mano mientras gira el cigüeñal. Esto escuadra la tapa al bloque antes de que se tuerzan los tornillos.

Relief valve Valve used to prevent excessive oil pressure. Since the oil pump is positive displacement, pressures could increase to a hazardous level at higher engine speeds. The relief valve opens to return oil to the sump and drop pressure in the system.

Válvula de rebose Una válvula que previene una presión excesiva de aceite. Como la bomba de aceite es de desplazamiento positivo, las presiones podrían aumentar a un nivel peligroso en las velocidades más altas. La válvula abre para regresar el aceite al resumidero y bajar la presión del sistema.

Ridge reamer A cutting tool used to remove the ridge at the top of the cylinder.

Escariador de reborde Una herramienta de cortar que sirve para quitar el reborde en la parte superior del cilindro.

Right-To-Know Laws Right-To-Know Laws inform workers regarding exposure to hazardous materials.

Leyes de derecho de información Las leyes de derecho de información informan a los trabajadores sobre el riesgo de exponerse a los materiales peligrosos.

Ring ridge A ring ridge develops at the top of the cylinder above where the rings scrape and wear down. Carbon also builds up in this area to create a smaller diameter region of the cylinder bore.

Canal del anillo El canal del anillo se desarrolla en la parte superior del cilindro encima donde los anillos se tallan y se gastan. El carbón también se junta en esta área para crear una región de un diámetro más pequeño de la perforación del cilindro.

Rocker arm Pivots that transfer the motion of the pushrods or followers to the valve stem.

Balancín Un punto pivote que transfiere el movimiento de las levantaválvulas o de los seguidores al vástago de la válvula.

Rod aligner A rod aligner measures connecting rod twist and bend.

Ajustador de la biela Un ajustador de la biela mide la torcedura y doblez de la biela.

Rod honer A rod honer machines the large connected rod bore.

Afilador de la biela Un afilador de la biela tornea la perforación grande de la biela.

Safety goggles Safety devices that provide eye protection from all sides. Goggles fit against the face and forehead to seal off the eyes from outside elements.

Gafas de seguridad Proporciona la protección a los ojos de todos lados siendo que quedan apretados contra la cara y el frente para formar un sello para los ojos contra los elementos exteriores.

Scale The distance of the marks from each other on a measuring tool.

Escala La distancia entre las marcas en una herramienta de medir.

Scan tool A scan tool communicates with the powertrain control module to display diagnostic trouble codes and engine data.

Herramienta de análisis Una herramienta de análisis se comunica con el MCM (modulo de control del motor) para mostrar códigos de problemas de diagnóstico y datos del motor.

Seat concentricity Seat runout or concentricity is a measure of how circular the valve seat is in relation to the valve guide.

Concentricidad de alojamiento El alcance de alojamiento o concentricidad es la medida de cuán circular es el alojamiento de la válvula con relación a la guía de la válvula.

Seat pressure Term that indicates spring tension with the spring at installed height and the valve closed.

Presión del asiento Un término que indica la tensión del resorte en su altura de instalación con la válvula cerrada.

Self-tapping insert A self-tapping insert provides a new thread when installed in an opening with damaged threads.

Inserto rosca cortante Un inserto rosca cortante proporciona una nueva rosca cuando se instala en una abertura con enroscado dañado.

Sizing point The location the manufacturer designates for measuring the diameter of the piston to determine clearance.

Punto de calibración El lugar indicado por el fabricante para efectuar las medidas del diámetro del pistón para determinar la holgura.

Skirt The piston skirt is the area from the lowest ring to the bottom of the piston.

Bisel El bisel del pistón es el área desde el anillo inferior hasta el fondo del pistón.

Small-hole gauge An instrument used to measure holes or bores that are smaller than a telescoping gauge can measure.

Calibre de taladros chicos Un instrumento que se emplea para medir los agujeros o taladros que son demasiado pequeños para medirse con un calibrador telescópico.

Spring free length The height the spring stands when not loaded.

Longitud libre del resorte La altura del resorte cuando no tiene carga.

Spring shims Components used to correct installed height of the valve spring. Spring shims are used to correct for machining tolerance.

Chapas de relleno Los componentes que se emplean para ajustar la altura de instalación del resorte de la válvula. El sobreespesor del maquinado se puede ajustar por medio de estas chapas.

Spring squareness Refers to how true to vertical the entire spring is.

Escuadrado del resorte Refiere a si la posición del resorte entero esta en línea recta al vertical.

Starter quick test The starter quick test will isolate the problem area: i.e., if the starter motor, solenoid, or control circuit is at fault.

Prueba rápida de encendido La prueba rápida de encendido aislará el área problemática; esto es, si tiene falla el motor de encendido, el solenoide o el circuito de control.

Steam cleaners A pressure washer that uses a soap solution, heated under pressure to a temperature higher than its normal boiling point. The superheated solution boils once it leaves the nozzle as it shoots against the object being cleaned.

Limpiadoras de vapor Una limpiadora a presión que utiliza una solución de jabón, calentado bajo presión a una temperatura más elevada de su punto de ebullición. La solución sobrecalentada hierve al salir de la boquilla projectada hacia el objeto para limpiar.

Straight time pay Straight time pay means that you are paid for every hour you are physically at work, regardless of your productivity.

Pago por el número de horas acostumbrado por un período de trabajo El pago por el número de horas acostumbrado por un período de trabajo significa que se paga por cada hora que se está físicamente en el trabajo, sin reparar en la productividad.

Sulfuric acid Sulfuric acid is a corrosive acid used in automotive batteries.

Ácido sulfúrico El ácido sulfúrico es un ácido corrosivo que se usa en las baterías de automóviles.

Supercharger A supercharger is a belt-driven device that forces more air into the cylinders.

Sobrealimentador Un sobrealimentador es un dispositivo manejado por una banda que fuerza más aire hacia los cilindros.

Surface grinder A surface grinder may be used to machine cylinder surfaces.

Afiladora de superficies Una afiladora de superficies puede usarse para tornear las superficies del cilindro.

Tailpipe The tailpipe conducts exhaust gases from the muffler to the rear of the vehicle.

Tubo de escape El tubo de escape conduce los gases de emisión del mofle hacia la parte posterior del vehículo.

Tap A tool used to repair or cut new internal threads.

Macho Una herramienta que sirve para reparar o cortar las roscas interiores nuevas.

Taper Taper is the difference in diameter at different locations in a bore or on a journal.

Conicidad La conicidad es la diferencia en diámetro en diferentes lugares en un orificio o en una muñequilla.

Telescoping gauge A precision tool used in conjunction with outside micrometers to measure the inside diameter of a hole. Telescoping gauges are sometimes called snap gauges.

Calibrador telescópico Una herramienta de precisión que se emplea junta con los micrómetros exteriores para medir los diámetros interiores de un agujero. Tambien se llaman calibradores de brocha.

Thermal cleaning Thermal cleaning is a method of cleaning that bakes off or oxidizes dirt.

Limpieza térmica La limpieza térmica es un método de limpieza que rompe u oxida la mugre.

Thread insert (heli-coil) A device that allows for major thread repairs while keeping the same size fastener.

Inserto prerescado (heli-coil) Un dispositivo que permite las reparaciones completas de las roscas manteniendo los retenes del mismo tamaño.

Throating Machining process using a 60-degree stone to narrow the contact surface.

Rectificación Un proceso de rebajar a máquina empleando una muela de 60 grados para disminuir la superficie de contacto.

Thrust bearing A double-flanged bearing used to prevent the crankshaft from sliding back and forth.

Cojinete de empuje Un cojinete con doble brida que sirve para prevenir que el cigüeñal desliza de un lado a otro.

Topping Refers to the use of the 30-degree stone to lower the contact surface on the valve face.

Despunte Refiere al use de una muela de 30 grados para rebajar al superficie de contacto en la cara de la válvula.

Torque plates Metal block about 2 in. thick that are bolted to the cylinder block at the cylinder head mating surface to prevent twisting during honing and boring operations.

Chapas de torsión Los bloques de metal midiendo unas dos pulgadas de grueso empernados al monoblock en la superficie de contacto de la cabeza del cilindro que previenen que se retuerce durante las operaciones de esmerilado y taladreo.

Torque wrench Wrench that measures the amount of twisting force applied to a fastener.

Llave de torsión Una llave que sirve para medir la cantidad de la fuerza de torsión que se aplica a una fijación.

Torque-to-yield A bolt that has been torqued to its yield point can be rotated an additional amount without any increase in clamping force. When a set of torque-to-yield fasteners is used, the torque is actually set to a point above the yield point of the bolt. This assures that the set of fasteners will have an even clamping force.

De torsión a rendimiento Un tornillo que se ha torcido hasta su punto de rendimiento puede girarse un poco más sin aumentar su fuerza de apriete. Cuando se usa un juego de sujetadores de torsión a rendimiento, la torsión se asienta a un punto encima del punto de rendimiento del tornillo. Esto asegura que la fijación de los sujetadores tendrá una fuerza de sujeción más pareja.

Total indicator reading (TIR) The total movement of the dial indicator (above and below zero) is the total indicator reading (TIR).

Lectura total del indicador El movimiento total del indicador graduado (encima y por debajo de cero) es una lectura total del indicador.

Turbocharger A turbocharger is a device that is driven by exhaust pressure and that forces air into the cylinders.

Turboalimentador El turboalimentador es un dispositivo al que lo impulsa la presión de emisión y que fuerza aire hacia los cilindros.

Undersize bearing Bearing with the same outside diameter as standard bearings but constructed of thicker bearing material in order to fit an undersize crankshaft journal.

Cojinete de dimensión inferior Un cojinete que tiene el diámetro exterior del mismo tamaño que los cojinetes normales pero que se ha fabricado de una material más gruesa para que puede quedar en un muñón de cigüeñal más pequeño.

United States Customary (USC) United States Customary (USC) measuring system is commonly referred to as the English system.

Acostumbradas en EU. El sistema de medidas acostumbradas en EU comúnmente se refiere como el sistema inglés.

Valve guide A part of the cylinder head that supports and guides the valve stem.

Guía de válvula Una parte de la cabeza del cilindro que apoya y guía al vástago de la válvula.

Valve guide bore gauge Instrument that provides a quick measurement of the valve guide. It can also be used to measure taper and out-of-round.

Verificador del calibrador para la guía de válvula Un instrumento que provee una medida rápida de la guía de la válvula. Tambien puede servir para medir lo cónico y lo ovulado.

Valve seat runout gauge Instrument that provides a quick measurement of the valve seat concentricity.

Calibrador de la excenticidad del asiento de la válvula Un instrumento que provee una medida rápida de la concentricidad del asiento de la válvula.

Valve spring tension tester Device used to measure the open and closed valve spring pressures.

Probador de tensión del resorte de válvula Un dispositivo que se emplea en medir las presiones de los resortes de las válvulas en la posición abierta o cerrada.

Valve stem height Valve stem height is the height of the valve tip as measured from the base of the head to the top of the valve tip.

Altura del vástago de la válvula La altura del vástago de la válvula es la altura del punto de la válvula como se mide desde la base de la cabeza hasta la parte superior de la punta de la válvula.

Valvetrain clatter Valvetrain clatter is a light tapping noise from the top of the engine, associated with wear of valve-related components.

Matraqueo del tren de la válvula El matraqueo del tren de la válvula es un ruido ligero de golpeteo de la parte superior del motor, asociado con el desgaste de los componentes relacionados con la válvula.

Vernier calipers Calipers that use a vernier scale which allows for measurement to a precision of ten to twenty-five times as fine as the base scale.

Pie de rey Los calibres que emplean una escala vernier permitiendo una medida precisa de diez a veinticinco veces más finas que la escala fundamental.

Warranty repair A warranty repair means that the manufacturer reimburses the shop for the parts and labor to make a repair.

Reparación de garantía Una reparación de garantía significa que el fabricante le reembolsa al taller las partes y el trabajo para realizar una reparación.

Waste gate A waste gate is a valve that opens to bypass some exhaust around the turbine wheel to limit turbocharger boost pressure.

Válvula de expulsión La válvula de expulsión es una válvula que se abre para desviar una poca emisión alrededor del volante de la turbina para limitar la presión de empuje del turboalimentador.

Wet sleeves Replacement cylinder sleeves that are surrounded by engine coolant.

Camisas húmedas Las camisas de repuesta del cilindro que se rodean por el fluido refrigerante del motor.

Workplace Hazardous Materials Information Systems (WHMIS) Workplace Hazardous Materials Information Systems (WHMIS) list information about hazardous materials.

Sistemas informativos de materiales peligrosos en el centro de trabajo Los sistemas informativos de materiales peligrosos en el centro de trabajo tienen una lista con información sobre materiales peligrosos.

INDEX

Note: page references with *f* notation refer to figure on that page

A

Abrasive cleaning, 14
Advanced technology vehicles, researching (job sheet), 587–588
Aerobic sealers, 207–208
Air cleaner. *See also* Air filter inspecting and servicing (job sheet), 247
Air filter, 219*f*. *See also* Air cleaner steps for service or replacement of, 219–221, 219*f*
Airflow restriction indicator, 219, 220*f*
Align boring equipment, 88
Align honing, 88, 501–502
Aligning bars, 73
Alignment, 87, 88
  acceptable, 496
  camshaft bore, 73, 354
  fixture, rod, 546
  inspecting and measuring (job sheet), 531–532
  main bearing bore, 495–496, 496*f*, 501–502
  procedure for checking, 502
  problems, preventing, 88
Allen wrenches, 199
Aluminum hydroxide deposits, 146
Anaerobic sealers, 207–208
Antifreeze
  checking condition of, with digital voltmeter, 145
  content in coolant, testing, 141
ASE. *See* National Institute of Automotive Service Excellence (ASE)
Automotive Engine Rebuilders Association (AERA), 505

B

Backlash, 452
  checking for, 485, 486*f*
Balance shafts
  bearings, measurement, removal, and replacement, 511–513
  installing OHV, 558–561
Base circle, camshaft, 401
Basic testing, 31–36
  engine component review (job sheet), 39–40
Batteries
  inspection of, 122–123
  need for caution when charging, 18
  testing, 122, 124–126. *See also Specific tests*
Battery capacitance test, 124–125
Battery case test, 125
Battery terminal voltage drop test, 125
Bearings, 134. *See also* Camshaft bearings
  balance shaft, 511–513
  failure analysis, 534–537
  foreign particles under, 536*f*
  inspecting crankshaft, 533–534
  inspecting wear patterns (job sheet), 587–588

  installing, 570
  oversized, 498
  pilot, 561, 561*f*
  prelubricating turbocharger, 241–242
  removers and installers for camshaft, 84. *See also* Camshaft bearings
  selection chart, 562*f*
  undersized, 534, 534*f*
Bell housing, 283
Bellmouthing, 355*f*, 356
Belt surfacers, 83
Belt tension gauge, 55, 56*f*
Big-end bores, 550
  resizing, 549, 550, 551*f*
Black smoke, 271
Blowby, 154
  out-of-round, 494
Blue smoke, 235, 266–267, 425, 425*f*
Bolts
  flywheel, 321
  torque-to-yield, 75
Boost pressure
  low turbocharger, effects of, 238
  testing, 238
  (job sheet), 257–258
  typical procedure for, 237*f*
Boring, 85
Bottom-end knock, 283, 285
Broach, 83
Bushings, 547, 557

C

Calipers, 62–64
Camshaft
  base circle, 401
  bore alignment, 354
  heel, 401, 402*f*
  with high duration, 401, 401*f*
  inspecting, 401–403
  (job sheet), 437–438
  installing
  (job sheet), 591–592
  OHV, 558–560
  -in-the-block engines, timing mechanism replacement in, 464–467
  nose, 401
  removing, 483
  on DOHC engine with VTEC, 398–399
  and replacing on OHC engine, 399–401, 400*f*
  and replacing on OHV engine, 398–399
Camshaft bearings
  measurement, removal, and replacement of, 510, 510*f*
  removers and installers, 84

Camshaft end play
    measuring, 559–560
        and bearing clearance, on DOHC engine with VTEC, 399–401
Camshaft position actuators, 469
Capacity test, 124–125
Carbide seat cutters, 368–369
Carbon monoxide (CO), 17
Carrying, safety precautions, 4–5
Case studies, 19, 36, 101, 163, 208, 245, 285, 329, 372, 430,
    471, 516, 580, 608
Casting plugs, 151–153
Catalytic converter, 230
    testing, 233–234
Challenges to technician
    identifying repair area, 31–32
    lack of communication (customer), 31
    service history, 33
Chassis ear machine, 33, 33f
Chemical block tester, 269
Chemical cleaning, 13–14
Cherry-picker, 12
Clothing, safety precautions, 3–4
Combustion
    gases, leaks of, 269
    improper, 272–273
Combustion chamber
    coolant in, 269f
    signs of potentially serious trouble in, 262
Compensation, 93–94
Compressed air
    guidelines for working with, 6
    tools that use, 8
Compressed natural gas (CNG) vehicles, 601–603
    servicing (job sheet), 611–612
    warning labels, 603f
Compression
    of air-fuel mixture, 50
    leaks, 50, 51, 353
    low cylinder, 236
Compression gauges, 50–51
    types of, 50–51
Compression testers, 50–51, 269. See also Compression tests
Compression tests, 273–274. See also Compression testers
    analyzing results of, 276
    cranking, 274
    performing (job sheet), 297–298
    photo sequence, 275
    running, 277–278
    wet, 276–277
Computers, 98f
Connecting rods, 567
    cap bolt replacement, 549–550
    failures and causes, 549
    inspecting, 546–549
        (job sheet), 583–586
    installing, 567–568
        (job sheet), 583–586

reassembly, 553–554
reconditioning, 549–553
    equipment for, 89–90
straightening, 550
Connectors, labeling, during engine removal, 306f
Coolant
    checking antifreeze content in, 141–142
    in combustion chamber, 276
    leaks, 156–157, 158f
        diagnosing (job sheet), 177–178
    level, checking, 141
    premixed with water, 145
Coolant hydrometer, 53–54
Cooling system, 140–150
    combustion gases leaking into, 269
    components, 141
        testing of, 141–146
    draining, 143–145
    exhaust gases in, 269
    filling, 145
    flushing, 145–146
    important parts of, 151–153
    inspection, 141–143
    maintenance, 141–146
    photo sequence, 158–159
    pressure tester, 53
    temperature warning system, 161–162
Core plugs, 152–153, 152f, 337
    installation of, 515
    removal of, 490, 490f
Corrosion, 318f
Cranking compression test, 274
    photo sequence depicting, 275
    steps for performing, 275
Crankshaft
    balancers, 88–89
    bearings
        failure analysis, 534–537
        inspecting, 533–534
    grinder, 88
    inspecting, 498–500
    installing, 558–566
        (job sheet), 593–595
        typical procedure for, 563–565
    journals, measuring, 499, 500f
    peening, 514
    polishers, 88
    pulley, removal of, 316, 453
    reconditioning, 513–515
        equipment for, 88–89
    removal, 316, 453
    straightening, 514–515
    thrust, 499, 500f
    warpage
        checking, 499
        preventing, 482

Crankshaft end play, 283–284
Crate engines, 96
Current draw, 133
Current draw test, 132–133
Cylinder
  boring, 84
  finding malfunctioning, 259
  hone, 87f
  with low cranking compression, 277, 279
  out-of-roundness, 494
  oversized, 493
  reconditioning, 502–505
  wear, 493. See also Cylinder walls
Cylinder block assembly
  cylinder reconditioning, 502–505
  line boring and align honing, 501–502
  resurfacing deck, 502
  servicing, 501–509
  sleeve installation, 505–509
Cylinder block inspection, 492–498
  checking for deck warpage, 492, 492f
  checking main bearing bore alignment, 495–496
  engine block bore measurements, 495–498
  inspecting and measuring cylinder wall wear, 493–495
Cylinder bore dial gauges, 71–72
Cylinder boring, 502, 503
Cylinder head
  assembly in OHC engine, 416–417
  disassembling, 342–344
    (job sheet), 379–380
    typical procedure for, 345–346
  equipped with inserts, 357
  inspection, 353–358
  installing, 423–424, 570–571
    checklist for, 571–572
  procedure for installing valves into, 413–414
  reassembly, 413–417
    installed spring height, 413
    installing valve stem seals, 413–415
    (job sheet), 433–434
  reconditioning equipment, 77–84
  removal, 338, 482–483
    (job sheet), 379–380
    from OHC engine, 340–342
    from OHV engine, 338–342
    procedure for, 482–483
  straightening aluminum, 359
  valve removal, 344–349
Cylinder head gasket, installation of, 204
Cylinder leakage test, 51, 279–281. See also Cylinder leakage tester
  performing
    (job sheet), 301–302
    steps for, 279
  photo sequence depicting, 281

Cylinder leakage tester, 51. See also Cylinder leakage test
Cylinder walls
  boring, 503–504
  deglazing, 503
  inspecting and measuring
    (job sheet), 529–530
    wear on, 493–495
  measuring wear on (job sheet), 111
  reconditioning methods for, 503–504
  sleeving, 506–509

D

Deck
  resurfacing, 502
  warpage
    checking for, 492–493
    measuring (job sheet), 425–426
Detonation, 263
  common causes of, 272
Diagnostic link connector (DLC), 48
Diagnostic tree, 102f
Diagnostic trouble code (DTC), 49, 236
Dial calipers, 62f, 64
  reading (job sheet), 109
Dial indicators, 60, 70
  using, 70
Dies, 200
Diesel fuel, storage of, 18
Digital pyrometer, 223
Digital voltmeter, 145
DOHC engine with variable timing electronic control (VTEC)
  inspecting and servicing rocker arms and shafts on, 409
  measuring camshaft end play and camshaft bearing clearance on, 399–401
  removing camshafts and rocker arms from, 398–399
  service of timing mechanism on (job sheet), 479–480
Dry sleeves, 505
  replacing, 505–506
Dry starts, 536
Duration, 401, 402f

E

Ear protection, 4
Education, 91–92
Emergency telephone numbers, 18
Engine analyzers, 59
  power balance testing using, 263–264
Engine assembly. See also Engine block, assembly
  completing, 574–579
    guide for, 574–575
  sensors and switches mounted on, 578, 579f
Engine block. See also Engine block disassembly
  assembly, completing, 568–571
  bore measurements, 496–498
  components, cleaning, 481–482
  flushing, 145–146
  installing accessory items that attach to, 578

Engine block. *See also* Engine block disassembly (*continued*)
  preparation, 557
    checklist for, 557
      reconditioning equipment, 84–88
Engine block disassembly, 487–491, 491*f*
  camshaft removal, 487
  core and oil gallery plugs, 490–491
  crankshaft removal, 489
  (job sheet), 519–521
  piston removal, 488–489
  ring ridge removal, 488
Engine crane, 12
Engine installation, 303–307
  front-wheel-drive vehicles, 308–309
  (job sheet), 336–336
  rear-wheel-drive vehicles, 310–312
Engine mounts, removal of, 310*f*
Engine removal, 308–309
  front-wheel-drive vehicles, 308–309
    typical procedure for, 308–309
  (job sheet), 331–332, 333–334
  preparing for, 303–308
    steps to complete in process of, 304–307
  rear-wheel-drive vehicles, 310
  swap, and installation, 303–336
  typical procedure for, 311–312
Engine(s)
  analyzer, 58–60, 264
  camshaft-in-the-block, 464–467
  crate, 317
  diagnostic tools, 48–58
  disassembly, 481
    preparation for, 481–482
  installation of, 322–325
    remanufactured, 315–321
  lugging, 543
  materials, identifying (job sheet), 217
  measuring tools, 60–74. *See also Specific tools*
  mounting, on stand, 312–314, 313*f*
  noise diagnosis, 282–285. *See also* Noises, engine
  OHC. *See* OHC engine
  OHV. *See* OHV engine
  operating systems, diagnosing and servicing, 121–195
  performance concerns, diagnosing, 259–302
  preoiling, procedure for, 140
  propane fueled, 600–601
  rebuilding tools and skills, 45–120
  removal. *See* Engine removal
  repair, skills required for, 96–98
  safely lifting, 12
  sensors, switches, and solenoids, 318*f*
  smoking, 266–271
  starting, at initial start-up and break-in, 326–328
  that use mechanical followers, 421–422
  with VVT systems, 469–471

Engine stand, 312–314
  securing engine on, 481–482
Engine start-up and break-in, 325–328
  final checks, 327
  five-hundred-mile service, 327–328
  initial, 325
  priming lubrication system, 325–326
  road test, 327
  starting engine, 326–327
English system. *See* United States Customary (USC) measuring system
Environmental Protection Agency (EPA), 16
  guidelines for hazardous waste, 16–17
  toxic pollutants regulated by, 58
Equipment
  align boring, 84
  block resurfacing, 84
  cleaning, safety, 12–16
  connecting rod and piston reconditioning, 89–90
  crankshaft reconditioning, 88–89
  cylinder head
    reconditioning, 77–84
    resurfacing, 83–84
  engine block reconditioning, 84–88
  grinding, guidelines for, 365–367
  special reconditioning tools and, 77–94
  valve guide renewing, 79–83
Exhaust gas analyzers, 58–60, 135
  four-gas, 160
Exhaust gas recirculation (EGR) valve, 317–318
Exhaust system
  components, replacing, 230
  functions of, 218
  gaskets and seals, replacing, 231–233
  inspection, 230
    and test (job sheet), 251–252
  manifold and pipe servicing, 230–231
  service, 229–234
Extractor, 197, 198*f*
  steps for using, 197
Eye protection, 2

F

False guide, 362, 362*f*
Fan(s)
  inspection of, 147–150
  shroud, 149, 149*f*, 151
Fasteners. *See also Specific Types of Fasteners*
  loosening rusted (job sheet), 213–214
  methods for removing, 154
Feeler gauges, 60
  used as "go, no-go" gauges, 354
  using, 61
    (job sheet), 107–108
Fire extinguishers, 18
  guide to selection of, 19*f*

Flat rate pay, 93
Flex fans, 148
Flex plates, 501
Floor jacks, 9*f*
Flywheel, 316
  bolts, 316
  holder, 316*f*
  inspecting, 501
Front pipe, 231*f*
Front-wheel-drive vehicles
  engine removal, 308–310
    typical procedure for, 308–310
  installation of engine, 322
    undercarriage installation procedures, 322–323
    upper engine procedures, 323
Fuel
  air-, mixture, compression of, 50
  compressed natural gas, 601
  diesel, 18
  under extremely high pressure, 602*f*, 603*f*
  -fouled spark plugs, 262
  gasoline, 18
  natural gas, 602
  propane, 600–601
Fuel control valve, 601
Fuel pressure gauges, 57
Fuel pressure regulator, 271
Full-floating piston pins, 539, 554
  installing, 554
Fusible link, 102*f*, 131

G
Gallery plugs, 337
  removal of, 490–491, 490*f*
Gases
  combustion, 269
  escape of, during charging of battery, 18
  exhaust, 230
Gaskets
  form-in-place, 207
  installation of, 204–206
  oil pan, 205
  procedure for removing and replacing intake manifold, 224–225
  replacing exhaust system, 230
  steps for preparing to replace, 204
    cylinder head, 204
Gasoline, storage of, 18
Gauges
  belt tension, 55, 56*f*
  cylinder bore dial, 71–72
  feeler, 60, 353*f*, 354*f*
  fuel pressure, 56–57
  oil pressure, 55
  out-of-round, 72–73
  small-hole, 70

telescoping, 69
vacuum, 59, 219
valve guide bore, 71
valve seat runout, 71
General Motors, 133
Grinding
  equipment guidelines, 365–367
  journal, 515
Grinding stones, 80

H
Hair, safety precautions, 4
Hand tools
  safety steps for, 6
  special, 74–77
Harmonic balancers, 317*f*
  inspecting, 501
  removal of, 316, 316*f*, 484–486, 484*f*
Hazardous materials, 15. *See also* Hazardous waste
  proper handling of (job sheet), 25–26
Hazardous waste, 16–17. *See also* Hazardous materials
Head, piston, 538–539, 538*f*
Head resurfacing equipment, 83–84
Heel, camshaft, 401, 402*f*
Hoists, 8, 9*f*. *See also* Lifts
  typical procedure for lifting vehicle on, 10
Honda Civic hybrid, 606–607
Honing, 87–88, 87*f*
  cylinder walls, 504–505, 504*f*
    typical procedure for, 507–508
  plateau, 505
Hoses
  broken or disconnected, 221
  labeling of, during engine removal, 305, 306*f*
Hotline services, 60
Hybrid electric vehicles (HEVs), 603–605, 604*f*
  Honda Civic hybrid, 606–607
  Toyota Prius, 605–606
  warranties, 608
Hydraulic lifters, 419, 419*f*
Hydraulic presses, 7
Hydrostatic lock, 127

I
Idler pulleys, 149, 150*f*
Ignition systems
  coil-on-plug, 265
Impact screwdriver, 487
Infrared testers, 58
Insert guides, 362
Insert valve seats, 363–364
Intake manifold, 324
  gasket, procedure for removing and replacing, 228–229
  reasons for replacing, 224–225
  removal procedure, 224–225
  service, 224–226
  vacuum, 52, 223

Intake system
  air cleaner, 218–221
  diagnosis and service, 218–229
  functions of, 218
  intake manifold, 224–229
  vacuum system, 221
Integral guides, 362, 362f
Integral valve seats, 364–365
Interference angle, 365
Interference fit, 484
Internet, 98
Initial inspection and service writing, 31–36

J

Jack stands, 11
Jet valves, 139
Jewelry, safety precautions, 4
Journals, 402, 402f
  grinding, 515
  inspecting, 498
  measuring crankshaft, 499–500
  polishing, 498

K

Keepers, installing, 415
Knock sensor (KS), 272
Knurling, 80
  typical procedure for, 360–361
kPa (metric kilopascals), 50

L

Land, piston, 540
Lash adjusters, inspecting (job sheet), 441–442
Leak-down, 404
  test, performing, 404–405
Leaks
  combustion gas, into cooling system, 268–269, 269f
  compression, 51, 52, 353, 557f
  coolant, 142, 142f, 156–157
  detection of, using cranking compression test, 274f
  detectors for, 52–53
  diagnosis of, 153–157
    photo sequence depicting, 158–159
  in exhaust system, likely spot to find, 231
  intake manifold gasket, 224–225
  oil, 153–157
  use of smoke leak detector to find, 223
  vacuum, 52, 222
    methods for testing, 224
Lifters
  disassembling, 404–405
  hydraulic
    adjusting adjustable, 418, 419–420
    internal components of, 405f
  inspecting, 403–405
    (job sheet), 441–442
  mechanical or solid, adjusting, 418
  roller, 402

Lifting
  engine, 12
  safety precautions, 4–5
  vehicle
    (job sheet), 27–28
    typical procedure for, 10
Lifts. See also Hoists
  engine, safety for, 12
  safety measures for, 8–10
Line boring, 489, 496, 498
Lineman's gloves, 606
Liners, 362
Liquid petroleum gas (LPG) vehicles, 600
  servicing (job sheet), 611–612
Liquids, volatile, 18
Load test, 123–124, 124f
Lobe lift, 402
Lubrication system
  important parts of, 151
  maintenance, 139–140
  priming, for start-up and break-in, 325–326
  testing and service, 134–136
Lugging, 543

M

Machine shop services, 358–364
Machinist's rule, 60, 62
Magnafluxing, 492
Main bearing bores
  checking alignment of, 495–496
    typical procedure for, 497
  inspecting and measuring alignment of (job sheet), 531–532
  resizing and realigning, 501–502
Main bearings
  installing, 561–566
    guide list for, 563–566
    (job sheet), 593–595
  wear on, 536f
Major repair indication
  blue smoke, 34
Malfunction indicator light (MIL), 49–50, 233
Manual shut-off valve, 602f, 603f
Material Safety Data Sheets (MSDS), 15
Measurement
  of length, 47
  of mass, 47
  units of, 46–47
  valve, 372
  of volume, 47–48
Mechanical followers, adjusting, 421–422
Meter (m), 46
Metric Conversion Act (1975), 46
Metric system, 46–47
  common equivalents between USC system and, 46
  common prefixes in, 47
Microinch, 84

Micrometers, 64–66, 67
  reading, 66–67, 67*f*
    typical procedure for, 68–69
Misfire, 49, 58
  corrosion as cause of, 318*f*
Muffler, 231

N

National Institute of Automotive Service Excellence (ASE), 94
  blue seal of excellence, 18
  certification program, 94–95
Natural gas, 602
Noid lights, 58
Noises, engine, 282–283
  bottom-end knock, 283, 284
  crankshaft end play, 283–284
  diagnosing (job sheet), 289–290
  piston pin knock, 284
  piston slap, 284
  timing chain clacking or rattling, 285
  use of stethoscope to detect, 55, 56
  valvetrain clatter, 284
Nose, camshaft, 401
Nut splitter, 196, 197

O

Occupational Safety and Health Administration (OSHA), 15
OHC engine
  with balance shafts, timing chain replacement on, 458–462
  bearing bore alignment, 359
  camshaft bearing wear in, 354
  cylinder head assembly, 416–417
  cylinder head removal from, 338
  removing and replacing camshaft on, 397–398
  replacing timing belt on, 462–464
  timing chain replacement on, 467
  valve timing components, installing, 572–573
  valvetrain installation, 573
OHV engine
  camshaft removal from, 397–398
  cylinder head removal from, 338
  inspecting and servicing valvetrain components on, 397–399
  installing camshafts and balance shafts on, 558–561
  removing and replacing camshaft on, 397–398
  valve timing components, installing, 572–573
  valvetrain installation, 573
  valvetrain timing (job sheet), 479–480
Oil
  change intervals, improper, 342–343
  consumption due to worn valve seals, 425
  coolant mixed with, 268
  jet valves, 139
  leaks, 153–157, 153*f*
    diagnosis of (job sheet), 177–178
    from valve cover gaskets, 205
  pressure, causes of low, 510

Oil gallery plugs, 337, 359
  installation of, 515–516
Oil pan gaskets, 575*f*
  installation of, 205
  replacing (job sheet), 215–216
Oil pressure gauge, 55
Oil pressure test, 134. *See also* Oil pressure testing
Oil pressure testing, 134, 163. *See also* Oil pressure test
    (job sheet), 175–176
Oil pump, 139
  crank-driven, 485*f*
  installing, 569–571
  service, 136–139
  typical procedure for disassembly and inspection of rotor-type, 137–138
On-board diagnostic (OBD) system, 233, 271
On-car service, 425–429
  valve adjustment, 426–427
  valve seal replacement, 425–426
Open circuit voltage test, 125
Open pressure, 411
O-rings, 145, 151*f*, 319*f*, 346, 571
  installing, 414–415
Out-of-round gauges, 72–73
Out-of-roundness, 494–495
Outside micrometers, 352
  using (job sheet), 113
Overboost, 218
Overhead camshaft engine. *See* OHC engine
Overlap, 401
Oversized bearings, 534
Oxygen sensor, 271

P

Particulates, 18
Peening, 514, 514*f*
Personal safety, 2–4
  warnings, 4
Pickup tube, 136, 139, 571
Pilot bearings, 561
Pilot bushings, 498
Pin boss, 541
Pin drift, 89
Pin hole machines, 89, 90
Piston pin knock, 284
Piston pins, 89
  checking wear on, 539*f*
  full-floating, 539*f*
  inspecting, 545–546
  press-fit, 539, 554
  removing, 539
Piston rings
  inspecting, 545
  installing, 554–556
    (job sheet), 589–590
  removal and groove cleaning, 538–539

Piston(s)
  assembly service, 549–556
  inspecting, 553–554
    (job sheet), 583–586
  installing, 554–556, 567–568
    (job sheet), 589–590
    typical procedure for, 567–568
  measuring, 541–543
  oversizes, common, 493
  reassembly, 553–554
  reconditioning equipment, 89–90
  removal of, 488–489
  wear analysis, 543
    chart, 544
Piston skirt, 540
Piston slap, 284
Plastigage, 73–74, 73*f*
Plateau honing, 505
Plugs, 151–153, 337, 343
Pneumatic tools, 8
Polishing, 498
Pollutants, 18, 58
Poppet valves, 349
Positive crankcase ventilation system, 154
Power balance testing, 259
  analyzing results of, 265
  analyzing spark plugs and performing (job sheet), 291–293
  manual, 264–265
  uses of, 264
  using engine analyzer or scan tool, 264
Power tools, 6–7
Powertrain control module (PCM), 48
Preignition, 272
  common causes of, 272
Prelubrication, 241–242
  of engine after oil change, 241
Press-fit piston pins, 539, 554
  installing, 554
Pressure plate, 537
Professionalism, 92–93
Professional technicians
  ASE certification, 94–95
  compensation, 93–94
  education, 91–92
  professionalism, 92–93
  skills required for engine repair, 96–98
  working as, 90–98
Profilometers, 84
Propane, 600, 601
Prussian blue, 371, 496
psi (pounds per square inch), 50
Purge solenoid, 318
Pushrods, inspecting, 405
  (job sheet), 443–444

Q

Quick disconnect fittings, 226, 226*f*

R

Radiator, 141
  flushing, 145–146, 145*f*
  removal, 150–151
Rear-wheel-drive vehicles
  camshaft removal with engine in vehicle, 397
  engine removal, 308
  installation of engine, 322
Registering, 563
Relief valve, 136
Remanufactured engines, installing, 315–321
Repair order (RO), 35
  customer's concern, 36
  finding solution, 36
  legal document, 35
  3Cs, 36
  as vital information, 36
Resource Conservation and Recovery Act (RCRA), 13
Restricted exhaust, 234
Ridge reamers, 84–85
Right-to-Know Laws, 15
Ring compressors, 77
Ring end-gap grinders, 89–90
Ring expanders, 76
Ring groove cleaners, 77
Ring ridge, 488
Rings. *See also* Piston rings
  damaged compression, 556
  installing, typical procedure for, 554–557
  marking of directional, 556
  tools for, 89
Road test, for initial start-up and break-in, 327
Rocker arm geometry, 405
  correcting, 408
Rocker arms. *See also* Rocker arm geometry
  adjustable, adjusting valves using, 422–423
  inspecting and servicing, 405–406
    shaft-mounted, 407
    and shafts, DOHC engine with VTEC, 409
    stud-mounted, 407–408, 407*f*
  removing, on DOHC engine with VTEC, 398–399
  replacing studs of, 406–408, 406*f*
Rod aligners, 90
Rod cap grinders, 90
Rod heaters, 90
Rod honers, 90, 90*f*
Roller lifters, 404
Room-temperature vulcanizing (RTV) sealant, 207–208
Rule 66 (California) for clean air, 14
Running compression test, 277–278
  analyzing results of, 278–279
  performing (job sheet), 299–300
  steps for performing, 277–278

S

Safety
  cleaning equipment, 12–14
  compressed air equipment, 8
  concerns regarding cylinder head valve removal, 357–358, 357f
  engine lift, 12
  guidelines for toxic chemicals and solvents, 15–16
  hand tool, 6
  jack and jack stand, 11–12
  lift, 8–10
  lifting and carrying, 4–5
  mounting engine on stand, 312–313
    photo sequence, 312
  performing repairs on CNG vehicles, 604–605
  personal, 1–4
  power tool, 6–7
  proper installation of engine to stand, 204
  servicing HEV vehicles, 604, 604f
  shop practices and, 1–19
  survey, shop (job sheet), 23–24
    vehicle operation, 17–18
    work area, 18–19
Safety glasses, 2, 2f
Safety stands, 11
Salvage (flood) title, 33, 34f
Scales, 62
  on machinist's rule, typical, 62
Scan tools, 48–49
  power balance testing using, 264
Sealants
  room-temperature vulcanizing, 207–208
  using, 207–208
Seals
  installation of, 205, 207f
  replacing exhaust system, 230
  sign of worn turbocharger, 235
  valve stem, 343. See also O-rings
Seat concentricity, 357–358
Seat cutters, 77
  carbide, 368–369
  stone, 369–370
Seat grinders, 80–81
Seat inserters, 81–82
Seat pressure, 412
Seat runout, 357
Self-tapping inserts, 201
Service information, 98–101
  finding (job sheet), 117
  inspection information, 99f
  layout, 100f
  manufacturers' websites for, 98
  referring to proper, 139
  specification tables provided in, 101f
  table of contents for major component area of, 103f
  use of computers to retrieve, 98f

Service repair order
  definition, 35
  job sheet, 43–44
Service writer, 35
  communication with customers, 35
  responsibility of, 35
Shaft-mounted rocker arms, 407
Shafts, inspecting and servicing, 405–406
Shims, 133, 134
  adjusting mechanical followers with, 421–422
  spring, 410
Shroud, fan, 121, 149, 149f, 151
Sizing point, 541
Sleeving, 502, 508–509
Small-end bores, 546
  reconditioning, 552
Small-hole gauges, 70
Smoke
  black, 271
  blue, 235, 266–267, 425
  engine, 266–271
  exhaust, 235
  white, 268–271
Solenoids, 317, 318f
Solvents, safety guidelines for, 15
Spark plugs, 260, 261
  analyzing, and performing power balance test
    (job sheet), 291–293
  reading, 261–263
  removal and installation of, 260–261
  worn, 262f
Spring free length, 410
Spring shims, 410
Springs, installing, 413
Spring squareness, 410
Sprocket inspection, 456–457
Starter. See also Starter motor; Starting system
  shimming, 134
  testing (job sheet), 169–170
Starter motor. See also Starter; Starting system
  installation, 133
  R & R, 135–136
  removal, 133
Starter quick test, 131–132
Starting system. See also Starter; Starter motor
  problems, diagnostic chart used to determine,
    128f–130f
  testing, 131–132
  tests and service, 127
  troubleshooting, 127
Steam cleaner, 304
Stethoscopes, 55, 56f, 279
Stone seat cutters, 369
Straight time pay, 93
Stud-mounted rocker arms, 407–408
Studs, 348f

Sulfuric acid, 18
Superchargers, 242
 diagnosis of, 242
 installation of, steps for, 243–244
 intercooler inlet and outlet tubes, 243f
 malfunction of, 218
 mounting bolt location, 244
 removal of, steps for, 243
Surface grinders, 84f

T

Tailpipe, 230
Taper, 493
 permissible amount of, 494
Tappets, 404
Taps, 199, 200
Technician
 challenges to
  identifying repair area, 31–32
  lack of communication (customer), 31
  service history, 33
 role's of
  test driving, 33
Telescoping gauges, 69
Tensioner inspection, 457
Test driving (by technician)
 and basic inspection (job sheet), 41–42
 requirement norms, 33
Thermal cleaning, 19
Thermostat, testing, 146–147
Thread chasers, 200
Thread inserts, 199, 201
 types of, 201
Thread repair, 196–204
 (job sheet), 213–214
 photo sequence depicting, 202–203
Throating, 371, 374f
Thrust bearing, 283–284
Timing belt
 inspection, 455
  (job sheet), 473–474
 removal of, 486f
 replacement, 462–463
  on OHC engines, 465–466
 snapped, 452f
 water pump driven by, 151f
Timing chain
 guide inspection, 457–458
 inspection, 454–455
  (job sheet), 473–474
 noise, 285
 replacement
  on camshaft-in-the-block engines, 464–467
  engines with OHC and balance shafts, 467–469
  on engines with VVT systems, 469–470
  on OHC engines, 458–462

Timing gear
 inspection, 456
  replacement on camshaft-in-the-block engines, 464–467
Timing mechanism
 disassembly, 458
 jumped or broken
  diagnosing, 453
  symptoms of, 452–453
 replacement, 458
 reuse versus replacement decisions, 458
 service, 451–471
  sprocket inspection, 456–457
  tensioner inspection, 457
  timing belt inspection, 455
  timing chain guide inspection, 457–458
  timing chain inspection, 454–455
  timing gear inspection, 456
 symptoms of worn, 452
Tools. *See also* Hand tools
 engine diagnostic, 48–60
 engine measuring, 60–74
 and equipment, special reconditioning, 77–89
 exhaust system service, 229–230
 pneumatic, 8
 power, 6
 quick disconnect fitting, 226f, 229
 tied rod end removal, 309f
Top dead center (TDC), 51
Topping, 371, 372
Torque plates, 87
Torque-to-yield fasteners, 75
Torque wrenches, 74
 using, 75–76
Torquing, 76, 139
Total indicator reading (TIR), 499, 499f
Toxic chemicals, safety guidelines for, 15–16
Toyota Prius, 605–606
Turbochargers
 diagnosis of, 235–236
  guide for, 236
 failures of, common, 236
 inspection, 238
  component, 240–241
  typical procedure for, and testing boost pressure, 239
 installation and prelubrication, 241–242
 malfunction of, 218
 removal procedure for, 238–240
 service, guide for, 235
 system inspection (job sheet), 253–255
 troubleshooting guide, 237f

U

Underboost, 218
Undersized bearings, 534
United States Customary (USC) measuring system, 46–47
 common equivalents between metric system and, 46

V

Vacuum gauges, 51–52, 222
Vacuum pumps, 52
Vacuum system
  diagnosis and troubleshooting, 221
  leaks, 52, 223–224, 223*f*
    diagnosing (job sheet), 249–250
  problems, driveability symptoms related to, 221
  tests, 222–224
  typical devices and controls, 222*f*
Valve adjustment, 418–421
  intervals, 418
  symptoms of improper, 418–419
  typical procedure for, 426
Valve clearance, 421–422
  checking and adjusting, 574
    (job sheet), 433–434
Valve cover gaskets, replacement of, 205
  photo sequence depicting, 206–207
Valve faces, 337
Valve grinding equipment, 82–83
Valve guide
  bellmouthing of, 356
  bore gauges, 71
  inspection, 355–357
  measuring clearance, methods used, 355–357
  renewing equipment, 77, 79
  repair, 360
  replacement, 362
  worn, 425
Valve inspection, 349, 350–352. *See also* Valves
  inspecting valve head, 350–351
  valve guide inspection, 355–357
  valve seat inspection, 357–358
  valve stem inspection, 351–352
Valves. *See also* Valve inspection
  adjusting, using adjustable rocker arms, 422–423
  fitting, 370–372
  installing, 415–416
  measurement of, 372
  poppet, 349
  reconditioning, 365–367
    (job sheet), 387–389
  refinishing, 365–367
  removal, from cylinder head, 344–345
Valve seals
  replacing, 427–428
    on the vehicle, 425–426
  worn, 425*f*
Valve seat runout gauges, 71
Valve seat(s)
  cutters, 81
  fitting, 370–372
  grinders, 80–81
  grinding
    (job sheet), 391–392
    typical procedure for, 373–374

inserters, 81–82
  inspection, 357–358
  reconditioned, 357
  refinishing, 367–368
    using carbide seat cutters, 368–369
    using stone seat cutters, 369–370
  replacement, 363–364
    insert, 363–364
    integral, 364–365
    (job sheet), 391–392
Valve spring compressor, 78
Valve springs, 409–412
  inspecting, 409–412
  open pressure, 411*f*, 412
  replacement, 429
  seat pressure, 412
Valve spring tension testers, 78–79, 412*f*
Valve stem
  grinding equipment, 82–83
  height, 372
  oversized, 361
Valve stem seals, installing
  O-ring, 416
  positive stem, 414
  umbrella-type, 414
Valve timing components, installing, 572–573
Valvetrain
  clatter, 284, 418
  components
    installing, 572–573
    unequal forces created by, 3
  inspecting and servicing, 403–405
    inspecting camshaft, 400–402
    measuring camshaft end play and bearing clearance, DOHC engine with VTEC, 399–401
    OHV engine, components on, 403–405
    removing and replacing camshaft on OHC engine, 398
    removing and replacing camshaft on OHV engine, 397–398
  installation
    on OHC engine, 573
    on OHV engine, 573
  service, 397–430
  timing
    on OHV engine (job sheet), 475
    on SOHC engine (job sheet), 477–478
  with too much lash, 418
Vehicle(s). *See also* Vehicle service
  advanced technology, researching (job sheet), 613–614
  CNG, 601–603
    servicing (job sheet), 611–612
  lifting, 10–11
    (job sheet), 29–30
  LPG, 600
    servicing (job sheet), 611–612
  operation safety, 17–18

Vehicle service
alternative fuel and advanced technology, 600–608
CNG vehicles, 601–603
hybrid electric vehicles (HEVs), 603–605
Honda Civic hybrid, 606–607
Toyota Prius, 605–606
LPG vehicles, 600
Ventilation, need for proper, 17, 600
Vernier calipers, 63–64
Vernier scale, 63, 63*f*
Vibrations
diagnosing (job sheet), 289–290
reduction of torsional, 482
Volatile liquids, 18
VVT systems, 469–470

W

Warning systems
cooling system temperature, 161–162
diagnosis of, 136
engine oil pressure, 161
Warranty repair, 91
on HEV vehicles, 603
Wastegate, 235
Waste, hazardous, 16–17
Water pump
removal of, 150–151, 486*f*
replacing (job sheet), 187–188
Wet compression tests, 276–277
Wet sleeves, replacing, 506
White smoke, 268–269
Workplace Hazardous Materials Information Systems (WHMIS), 15
Wrenches
Allen, 199
torque, 74–75
Wrist pins, 89. *See also* Piston pins
Writing
service repair order, 35